Heat and Mass Transfer

L.P. Yarin · A. Mosyak · G. Hetsroni

Fluid Flow, Heat Transfer and Boiling in Micro-Channels

Springer

L.P. Yarin
Technion-Israel Institute of Technology
Dept. Mechanical Engineering
Technion City
32000, Haifa, Israel
meromro@techunix.technion.ac.il

G. Hetsroni
Technion-Israel Institute of Technology
Dept. Mechanical Engineering
Technion City
32000, Haifa, Israel
hetsroni@tx.technion.ac.il

A. Mosyak
Technion-Israel Institute of Technology
Dept. Mechanical Engineering
Technion City
32000, Haifa, Israel

ISBN 978-3-540-78754-9 e-ISBN 978-3-540-78755-6

DOI 10.1007/978-3-540-78755-6

Heat and Mass Transfer ISSN 1860-4846

Library of Congress Control Number: 2008936040

Cover design: deblik, Berlin
Production: le-tex publishing services oHG, Leipzig

Printed on acid-free paper

9 8 7 6 5 4 3 2 1

springer.com

Preface

The subject of the book is fluid dynamics and heat transfer in micro-channels. This problem is important for understanding the complex phenomena associated with single- and two-phase flows in heated micro-channels.

The challenge posed by high heat fluxes in electronic chips makes thermal management a key factor in the development of these systems. Cooling of microelectronic components by new cooling technologies, as well as improvement of the existing ones, is becoming a necessity as the power dissipation levels of integrated circuits increases and their sizes decrease. Miniature heat sinks with liquid flows in silicon wafers could significantly improve the performance and reliability of semiconductor devices. The improvements are made by increasing the effective thermal conductivity, by reducing the temperature gradient across the wafer, by reducing the maximum wafer temperature, and also by reducing the number and intensity of localized hot spots.

A possible way to enhance heat transfer in systems with high power density is to change the phase in the micro-channels embedded in the device. This has motivated a number of theoretical and experimental investigations covering various aspects of heat transfer in micro-channel heat sinks with phase change.

The flow and heat transfer in heated micro-channels are accompanied by a number of thermohydrodynamic processes, such as liquid heating and vaporization, boiling, formation of two-phase mixtures with a very complicated inner structure, etc., which affect significantly the hydrodynamic and thermal characteristics of the cooling systems.

The multiplicity of phenomena characteristic of flow in heated micro-channels determined the content of the book. We consider a number of fundamental problems related to drag and heat transfer in flow of a pure liquid and a two-phase mixture in micro-channels, coolant boiling in restricted space, bubble dynamics, etc. Also considered are capillary flows with distinct interfaces developing under interaction of inertia, pressure, gravity, viscous and capillary forces.

In this book we use our own results on a number of problems related to flow and heat transfer in micro-channels, as well as those of numerous theoretical and experimental investigations published in current literature. The presented materials are easily comprehensible for engineers and applied scientists with some graduate level familiarity with fluid mechanics and theory of heat transfer.

The book consists of two parts: hydrodynamics and heat transfer of single- and two-phase media in micro-channels (Chaps. 1–7), capillary flow with distinct interfaces (Chaps. 8–11).

The book is not meant to be an undergraduate text, but can be used in graduate level courses.

Acknowledgements

While writing the book we felt support and encouragement from our wives Nelly Sakharov-Yarin, Lidia Kharchenko-Mosyak and Ruthie Hetsroni who gave us the inspiration to complete the book successfully. Unfortunately, Nelly and Lidia passed away and cannot see a printed version of the book. We dedicate it to their memory; it is also dedicated to Ruthie.

We would like to express our gratitude to our colleagues Professor J. Zakin, Dr. E. Pogrebnyak, Dr. R. Rozenblit, Dr. Z. Segal, Dr. I. Tiselj, Dr. G. Ziskind, as well as D. D. Klein (M.Sc.), Y. Mishan (M.Sc.) and R. Zimmerman (M.Sc.) for their participation in the investigations of a number of problems considered in this book.

We are especially grateful to Dr. M. Fichman and Dr. G. Ziskind for many valuable discussions and comments made after reading the manuscript.

Special thanks are directed to Mr. E. Goldberg for correction of the text, as well as to Dr. E. Pogrebnyak for her help in preparation of the manuscript.

During the work on the book some of us were recipients of grants from the Committee of the Council of Higher Education, the Israel Academy of Sciences and Humanities. L.P. Yarin and A. Mosyak were also partially supported by the Center of Absorption (State of Israel) and the Israel Council for Higher Education.

Contents

Chapter 1
Introduction

1.1 General Overview

The rapid advances made during the past decade in the field of production and use of high power micro-devices led to widespread interest in the problems of micro-fluid mechanics and the need for both comprehensive and detailed treatment of the fundamental aspects of these phenomena.

Despite the fact that experimental and theoretical studies of flow in small pipes began already in the first part of the nineteenth century, systematic treatment of the vast class of problems associated with flow and heat transfer in micro-channels started only in the middle of the twentieth century. It was then that the true significance of such investigations was realized for different applications in micro-system technology, in particular, micro-scaled cooling systems of electronic devices which generate high power. Accordingly, experimental and theoretical investigations were aimed at detailed study of the flow of incompressible and compressible fluids in regular and irregular micro-channels under adiabatic conditions, conditions corresponding to intensive heat transfer with the environment, and phase change. At the same time specific problems associated with roughness, energy dissipation, heat losses, etc., were considered. As a result, important features of flow and heat transfer in micro-channels were revealed, simple models of the processes were developed, and empirical and semi-empirical correlations for drag and heat transfer coefficient were suggested. Comparison of systematic experimental data with predictions of the conventional theory based on the Navier–Stokes equation reveals the actual sources of disparity between them. The recent developments in micro-scale heat transfer and fluid flow have been discussed by e.g. Zhang et al. (2004), Celata (2004), Kakac et al.(2005), Kandlikar et al. (2005), Zhang (2007).

In spite of the progress described above, certain fundamental problems in flow and heat transfer are still unclear. This leads to difficulties in understanding the essence of micro-thermohydrodynamic phenomena.

L.P. Yarin, *Fluid Flow, Heat Transfer and Boiling in Micro-Channels*

1.2 Scope and Contents of Part I

The first part of this book deals with the characteristics of flow and heat transfer in the channels and comparison between conventional size and micro-channels, which are important to understanding micro-processes in cooling systems of electronic devices with high power density and many other applications in engineering and technology.

It contains six chapters related to the overall characteristics of the cooling systems: single-phase and gas–liquid flow, heat transfer and boiling in channels of different geometries.

Chapter 2 presents general schemes of these systems, as well as the characteristics of the micro-channels used.

In Chap. 3 the problems of single-phase flow are considered. Detailed data on flows of incompressible fluid and gas in smooth and rough micro-channels are presented. The chapter focuses on the transition from laminar to turbulent flow, and the thermal effects that cause oscillatory regimes.

Chapter 4 is devoted to single-phase heat transfer. Data on heat transfer in circular micro-tubes and in rectangular, trapezoidal and triangular ducts are presented. Attention is drawn to the effect of energy dissipation, axial conduction and wall roughness on the thermal characteristics of flow. Specific problems connected with electro-osmotic heat transfer in micro-channels, three-dimensional heat transfer in micro-channel heat sinks and optimization of micro-heat exchangers are also discussed.

The results of experimental and theoretical investigations related to study of drag and heat transfer in two-phase gas–liquid flow are presented in Chap. 5.

The concepts of boiling in micro-channels and comparison to conventional size channels are considered in Chap. 6. The mechanism of the onset of nucleate boiling is treated. Specific problems such as explosive boiling in parallel micro-channels, drag reduction and heat transfer in surfactant solutions are also considered.

Chapter 7 deals with the practical problems. It contains the results of the general hydrodynamical and thermal characteristics corresponding to laminar flows in micro-channels of different geometry. The overall correlations for drag and heat transfer coefficients in micro-channels at single- and two-phase flows, as well as data on physical properties of selected working fluids are presented. The correlation for boiling heat transfer is also considered.

1.3 Scope and Contents of Part II

The second part treats specific problems typical of capillary flow with a distinct interface. It contains four chapters in which steady and unsteady capillary flow are treated.

The quasi-one-dimensional model of two-phase flow in a heated capillary slot, driven by liquid vaporization from the interface, is described in Chap. 8. It takes into account the principal characteristics of the phenomenon, namely, the effect of inertia, pressure and friction forces and capillary pressure due to the curvature of the interface, as well as the thermal and dynamical interactions of the liquid and vapor phases.

Chapter 9 is devoted to regimes of capillary flow with a distinct interface. The effect of certain dimensionless parameters on the velocity, temperature and pressure within the liquid and vapor domains are considered. The parameters corresponding to the steady flow regimes, as well as the domains of flow instability are defined.

Chapter 10 deals with laminar flow in heated capillaries where the meniscus position and the liquid velocity at the inlet are unknown in advance. The approach to calculate the general parameters of such flow is considered in detail. A brief discussion of the effect of operating parameters on the rate of vaporization, the position of the meniscus, and the regimes of flow, is also presented.

The onset of flow instability in a heated capillary with vaporizing meniscus is considered in Chap 11. The behavior of a vapor/liquid system undergoing small perturbations is analyzed by linear approximation, in the frame work of a one-dimensional model of capillary flow with a distinct interface. The effect of the physical properties of both phases, the wall heat flux and the capillary sizes on the flow stability is studied. A scenario of a possible process at small and moderate Peclet number is considered. The boundaries of stability separating the domains of stable and unstable flow are outlined and the values of the geometrical and operating parameters corresponding to the transition are estimated.

Authors

Dr. Yarin is a Visiting Professor at the Faculty of Mechanical Engineering at the Technion–Israel Institute of Technology. He received his M.S. degree from the Polytechnic Institute of Kharkov in 1952, his Candidate of Technical Sciences (Ph.D.) degree from the Institute of Energetics, Acad. of Kazakhstan, in 1962, and his Doctor of Technical Sciences degree from the Institute of High Temperatures, Acad. Sci. USSR, in 1970. He is the author of about 200 research works (including five monographs) in the fields of combustion theory, heat and mass transfer, two-phase flows, turbulent flows, energetics, aircraft and rocket engines, experimental methods in gas dynamics and heat transfer, thermoanemometry, high temperature combustion reactors and micro-fluid mechanics. His research activities focus on detailed analysis of aerodynamics and thermal regimes of combustion in gas torches; gas, liquid fuels and coal combustion, combustion wave propagation in porous and bubbly media; aerodynamics of furnaces and combustion chambers of jet and rocket engines; gasodynamics of jet flows; hydrodynamics of stratified flows, magnetohydrodynamics; turbulent two-phase flows; the theory of chemical reactors; micro-fluid mechanics, in particular heat and mass transfer in micro-channels. He headed the Chair of Engineering Thermal Physics at the Ukta Industrial Institute, teaching undergraduate and graduate courses in hydrodynamics, heat and mass transfer and thermodynamics. He presents an undergraduate, and graduate course on "Principles of Combustion of Two-Phase Media" at the Technion–Israel Institute of Technology.

Dr. Mosyak is a Research Fellow, Senior A of the Faculty of Mechanical Engineering, Technion, Haifa, Israel. He received his M.S. degree from Polytechnic Institute of Odessa in 1960, his Candidate of Technical Sciences (Ph.D.) degree from Polytechnic Institute of Odessa in 1972. He is the author of about 100 research works in the fields of turbulent flows, two-phase flows, heat and mass transfer, microfluid mechanics and thermodynamics. He was Associate Professor of Energy Engineering, Kishinev Polytechnic Institute, USSR where he taught undergraduate and graduate courses in hydrodynamics, heat and mass transfer, and thermodynamics.

Dr. Hetsroni is an Emeritus Danciger Professor of Engineering, Faculty of Mechanical Engineering at the Technion–Israel Institute of Technology (IIT). He received his B.Sc. Cum Laude, Technion–IIT in 1957, and received his Ph.D. from Michigan State University in 1963. He was with the Atomic Power Division of Westinghouse for a few years before joining the Faculty at the Technion in 1965; since 1974 he has been the Danciger Professor of Engineering. In the USA he held positions also with the Electric Power Research Institute. At the Technion he has served as Dean of Mechanical Engineering, and as the Head of the Neaman Institute for Advanced Studies in Science and Technology. Dr. Hetsroni was also the Head of the National Council for Research and Development of Israel. He has also been a Visiting Professor at Carnegie Mellon University, Stanford University, the University of California-Santa Barbara, the University of Minnesota and the University of New South Wales. He was the Vice President for Region XIII of the ASME International and was a Governor of the ASME. He is the Founding Editor of the International Journal of Multiphase Flow. He is the author of about 250 research works (including five monographs) in the fields of multiphase flow and heat transfer, experimental and computational fluid mechanics and heat transfer, turbulent flows, thermal-hydraulic design of nuclear reactors, turbulent boundary layers, boiling and steam generators. The research activities of Professor Hetsroni have focused on detailed analysis of aerodynamics of two-phase turbulent jets, particle turbulent interactions, heat transfer in two-phase turbulent boundary layer, coherent structure of turbulent flows, direct numerical simulations of turbulent flows, boiling of surfactant solution, heat and mass transfer in micro-channels, as well as particle image velocimetry (PIV) and hot-foil infrared imaging (HFIRI) measurements in a flume. At present he is the Head of the Multiphase Flow Laboratory of Technion.

Part I
Flow and Heat Transfer

Chapter 2
Cooling Systems of Electronic Devices

We attempt here to describe the fundamental equations of fluid mechanics and heat transfer. The main emphasis, however, is on understanding the physical principles and on application of the theory to realistic problems. The state of the art in high-heat flux management schemes, pressure and temperature measurement, pressure drop and heat transfer in single-phase and two-phase micro-channels, design and fabrication of micro-channel heat sinks are discussed.

2.1 High-Heat Flux Management Schemes

With the trend towards increasing levels of integration in high-density, very large-scale integral circuits and heat sink technologies, higher level of performance are required to meet the elevated power dissipation in electronic and optical devices. Thermal design for cooling of microprocessor packages has become increasingly challenging in its thermal and fabrication aspects. Figure 2.1 shows the International Technology Roadmap for Semiconductors (ITRS) (Prasher et al. 2005). The upper line corresponds to high-performance semiconductors.

It can be seen that thermal design power (TDP) rises linearly up to about 2009–2010 and is expected to remain approximately constant afterwards. However, these data do not indicate whether new cooling technologies are needed for future packages. Due to die shrinkage and to other complexities of electronic and optical design, the heat flux will increase drastically, leading to highly non-uniform heat generation that will in turn cause localized hotspots. Breakthroughs in many semiconductor technologies are becoming increasingly dependent upon the ability to safely dissipate enormous amounts of heat from very small areas. Frequently, advanced electronic, optical, nuclear equipment and high-frequency microwave systems require cooling of some devices at heat fluxes on the order of $10^3\ \mathrm{W/cm^2}$ (Hetsroni et al. 2006a). Fusion reactors, for example, contain components that require continuous

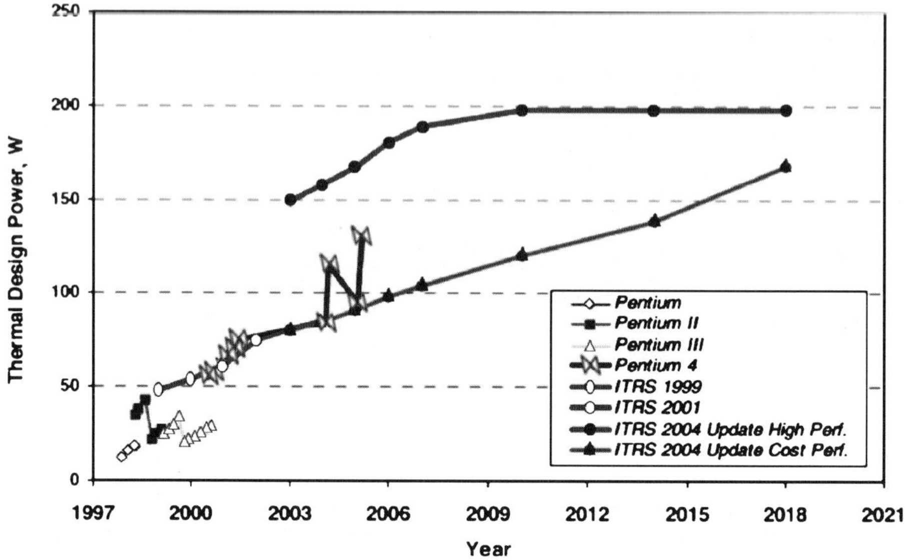

Fig. 2.1 International Technology Roadmap for Semiconductors (ITRS). Reprinted from Prasher et al. (2005) with permission

cooling on the order of $10^4\,\mathrm{W/m^2}$ (Boyd 1985). Cooling schemes exploiting recent research developments in high-heat flux thermal management were discussed and compared as to their potential heat dissipation reliability and packaging aspects by Lasance (Philips Research Laboratories) and Simons (IBM Corporation) (Lasance and Simons 2005). Some of these results are presented below.

Conduction and heat spreading

In all cooled appliances, the heat from the device's heat sources must first arrive via thermal conduction at the surfaces exposed to the cooling fluid before it can be transferred to the coolant. For example, as shown in Fig. 2.2, it must be conducted from the chip through the lid to the heat sink before it can be discharged to the ambient air. As can be seen, thermal interface materials (TIMs) may be used to facilitate this process. In many cases a heat spreader in the form of a flat plate with high thermal conductivity may be placed between the chip and the lid.

Heat spreading is a very effective means to alleviate the need for sophisticated high-heat flux cooling options. Of course, the benefit of reducing the heat flux density by increasing the area should outweigh the "penalty" of the additional layer in the path of the heat stream. Figure 2.3 shows heat spreading results for $q = 150\,\mathrm{W/cm^2}$ as a function of thermal conductivity, thickness and heat transfer coefficient. For example, using an $8 \times 8\,\mathrm{cm^2}$ heat spreader of some advanced composite with $k = 800\,\mathrm{W/m\,K}$ and thickness of $\delta = 4\,\mathrm{mm}$ results in a temperature rise of about 40 K at a heat transfer coefficient $h = 2{,}500\,\mathrm{W/m^2\,K}$.

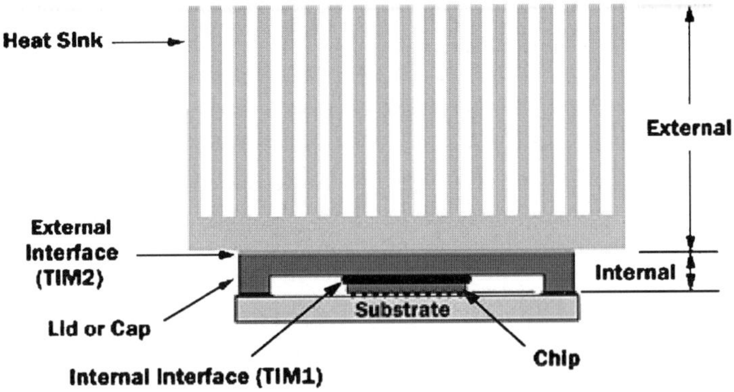

Fig. 2.2 Chip package with thermal conduction path to heat sink via TIMs. Reprinted from Lasance and Simons (2005) with permission

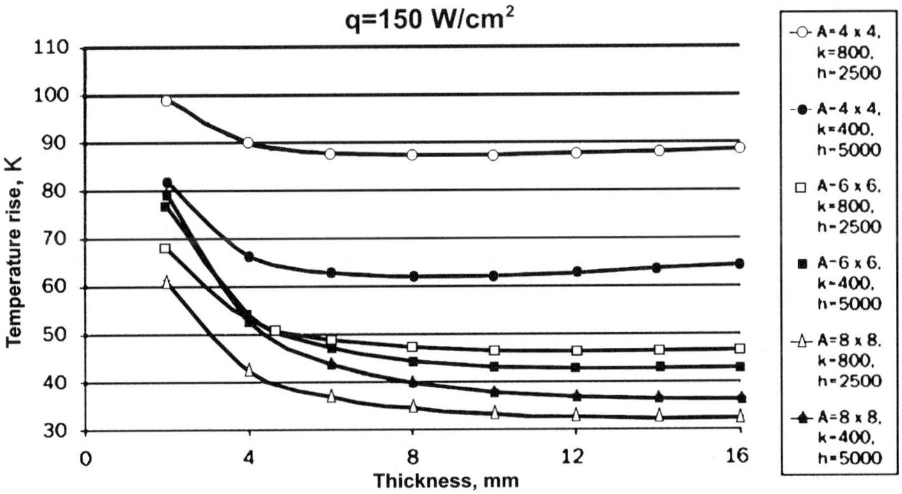

Fig. 2.3 Effect of thickness on heat spreading for different heat source areas, material thermal conductivities, and heat transfer coefficients (A in cm^2, k in W/mK, h in W/m^2K). Reprinted from Lasance and Simons (2005) with permission

Air cooling

It is generally acknowledged that traditional air-cooling techniques are about to reach their limit for high-power appliances. With standard fans a maximum heat transfer coefficient of about $150\,W/m^2\,K$ can be reached with an acceptable noise level, which is about $1\,W/cm^2$ for a $\Delta T = 60\,K$ temperature difference. Using "macro-jet" impingement, theoretically we may reach $900\,W/m^2\,K$, but with unacceptable noise levels. Non-standard fan/specialized heat sink combinations for CPU cooling are expected to have a maximum of about $q = 50\,W/cm^2$. Recently, some new initiatives have extended the useful range of air-cooling, such as piezo fans, "synthetic" jet cooling and "nanolightning."

Piezo fans

Piezoelectric fans are small, low-power, relatively low-noise, solid-state devices that provide viable thermal management solutions for a variety of portable electronic appliances, including laptop computers and cellular phones. In these fans piezoceramic patches are bonded onto thin, low-frequency flexible blades driven at resonance frequency, thereby creating an air stream directed at the electronics components. Thereby, up to 100% improvement over natural convective heat transfer can be achieved (Acikalin et al. 2004).

Synthetic jet cooling

An approach using periodic micro-jets called "synthetic jets" is still in the initial stages of study. Due to the pulsating nature of the flow, the jets create stronger entrainment than conventional steady jets with the same Reynolds number, and more vigorous mixing between the wall boundary layers and the rest of the flow. A test set-up is shown in Fig. 2.4. The jet entrains cool air from the environment, impinges on the top hot surface and returns the heated air to the environment. Radial current counter flow is created in the gap between the plates, with hot air dispersed along the top and the ambient air entrained along the bottom surface. The idea was further advanced by the development of flow actuators using micro-electromechanical systems (MEMS) technology (Beratlis and Smith 2003).

Nanolightning

An interesting new approach to increasing the heat transfer coefficient, known as "nanolightning," was also investigated. It is based on "micro-scale ion-driven airflow" using very strong electric fields created by nanotubes. As shown in Fig. 2.5, the ionized air molecules are moved by another electric field, thereby inducing secondary airflow (Peterson et al. 2003). Cooling at a heat flux level of $q = 40\,\mathrm{W/cm^2}$ has been reported.

Liquid cooling

Liquid cooling in electronics is generally divided into two main categories, indirect and direct, according to the type of contact between the coolant and the cooled components. The following sections discuss the two categories, represented by heat pipes and cold plates, and immersion cooling and jet impingement, respectively.

Heat pipes

Heat pipes are an enhanced means of transporting heat (in certain circumstances much better than copper) from a source to a heat sink where it can be transmit-

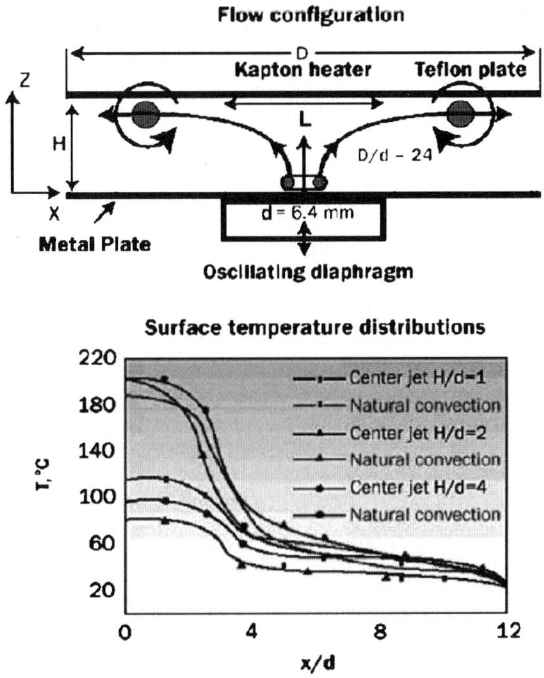

Global heat transfer coefficient

H/d	1	2	4
Jet, h, W/m²K	63	79	105
Natural convection, h, W/m²K	4	5	19

- Effective jet cooling for H/d>0.5 (D/d=24)
- Jet at 80 Hz, Re_d - 3300, L/d=15, d=6.4mm

Fig. 2.4 Flow dynamics of normal jet impingement with an oscillating diaphragm. Reprinted from Lasance and Simons (2005) with permission

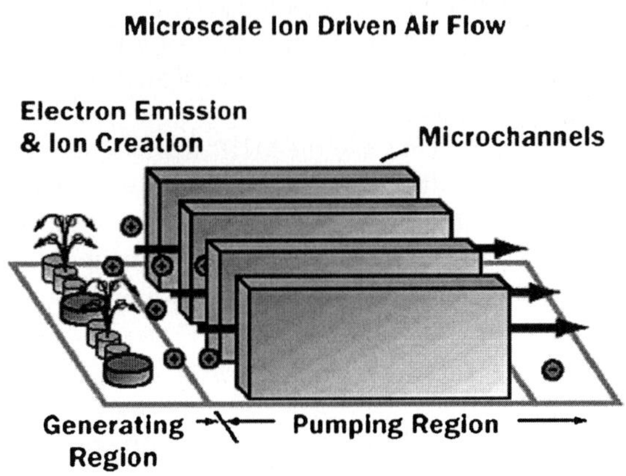

Fig. 2.5 Nanolightning sketch. Reprinted from Lasance and Simons (2005) with permission

ted to the cooling medium by natural or forced convection. They are sealed and vacuum-pumped vessels partially filled with liquid. Their internal walls are lined with a porous medium (the wick) that acts as a passive capillary pump. When heat is applied to one side of the pipe the liquid begins to vaporize. The pressure gradient causes the vapor to flow towards the cooler regions, where it condenses and is transported back by the wick structure, thereby closing the loop. Their performance scales from a heat flux of $10\,\mathrm{W/cm^2}$ to over $300\,\mathrm{W/cm^2}$. Loop heat pipes (LHP) have attracted increased attention. Their advantage over their conventional counterparts is that the vapor and liquid paths are separated, permitting much better performance of the liquid return loop. For example, Kim and Golliher (2002) showed the ability to accommodate a heat flux of $625\,\mathrm{W/cm^2}$.

Silicon micro-heat pipes (MHP) are passive systems used to transfer high heat fluxes and to increase the effective thermal conductivity of a Si substrate. Their working principle is based on two-phase heat transfer. A MHP consists of a non-circular closed channel, a few hundred micrometers wide and a few centimeters long. Similar to the preceding case, the liquid vaporizes at one end of the channel, where heat flux is applied (evaporator). The vapor, driven by the pressure gradient, flows to the other end of the channel, which is cooled (condenser). There it condenses and gives up its latent heat of vaporization. The capillary pressure difference between the evaporator and the condenser causes the liquid to flow back to the evaporator. Thereby, heat is transferred from the evaporator to the condenser by a continuous cycle of working fluid with a low thermal gradient. To increase the heat load, MHPs are implemented in arrays of several tens.

The role of a MHP is to reduce the maximum temperature of the wafer and decrease the temperature gradient across it, thereby increasing the effective thermal conductivity. Such a device was used in the study conducted by Le Berre et al. (2006). Its size was $20 \times 20\,\mathrm{mm^2}$ and consisted of a series of 27 parallel triangular channels, 500 μm wide and 350 μm deep (hydraulic diameter about 257 μm), micromachined into a silicon wafer using KOH anisotropic etching. The array volume was about $50\,\mathrm{mm^3}$, which represents a void fraction of 11%. Such a geometry with acute-angled corners is necessary for a proper heat pipe operation: the corners favor reduction of the meniscus curvature radius, which in turn increases the capillary pumping pressure.

The performances of the MHP array were evaluated for different methanol filling charges under different experimental conditions. The results indicated an increase in the effective thermal conductivity to about 200 W/m K under optimum conditions, equivalent to a 67% increase over an empty array. Performances are favored by reducing the input heat flux or increasing the cooling temperature.

Cold plates

Liquid-cooled cold plates perform a function analogous to that of air-cooled heat sinks. Unlike heat pipes, they may be considered active devices in that the liquid is usually forced through them by a pump.

Both direct and indirect liquid cooling methods can be further categorized as single-phase and two-phase methods. For example, jet impingement cooling may be a two-phase direct method, in contrast to the two-phase indirect technology of heat pipes and thermosiphons. Two-phase methods are preferable because of the high-heat transfer coefficients, although the systems tend to be more complex. In the mid-1980s, IBM employed an indirect liquid cooling technology using water for mainframes and supercomputers. This technology became the norm for high-performance computers, in which the large cold plates can be sufficiently separated from the electronics and thermally connected with the heat conduction devices. Two-phase cooling was not considered for notebook computers at that time, since most microprocessors were sufficiently cooled with metallic heat sinks and fans.

Immersion cooling

Direct liquid immersion cooling also provides a higher heat transfer coefficient, depending on the specific coolant and mode of convective heat transfer. Since there is no wall separating the heat source from the coolant, the heat can be removed directly from the chip. In these circumstances, consideration must be given to the coolant's chemical compatibility with the chip and other packaging materials exposed to the liquid. At present, with the continuous increase of heat flux at the chip level, microchannel cold plates are receiving renewed attention. The relative magnitude of the heat fluxes accommodated by liquid cooling depends on convective heat transfer processes classified as natural convection, forced convection, or boiling modes. The natural convection mode is derived from the mixing and fluid motion induced by density differences of the coolant. Its heat transfer rates are low and can be exceeded by forced convection mode in which the flow velocity of the liquid over the heated surface is increased with the aid of a pump. Boiling is a convective heat transfer process based on the phase change of the working fluid with vapor bubbles forming at the heated surface, and commonly has the form of pool boiling or flow boiling. The boiling curve (Fig. 2.6, Simons 1996) for a typical fluorocarbon coolant shows the

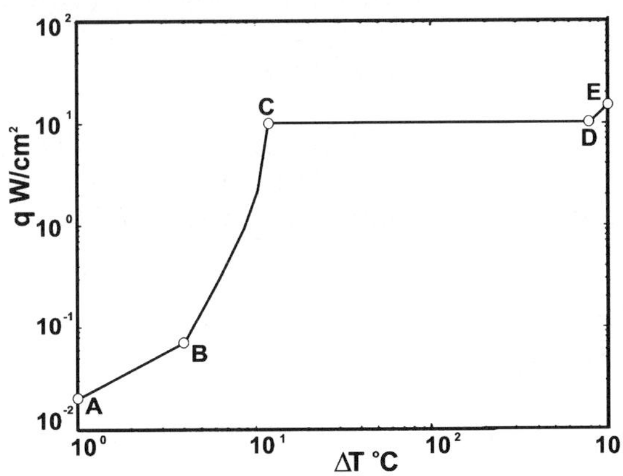

Fig. 2.6 Typical heat transfer regimes for immersion cooling with a fluorocarbon. Reprinted from Simons (1996) with permission

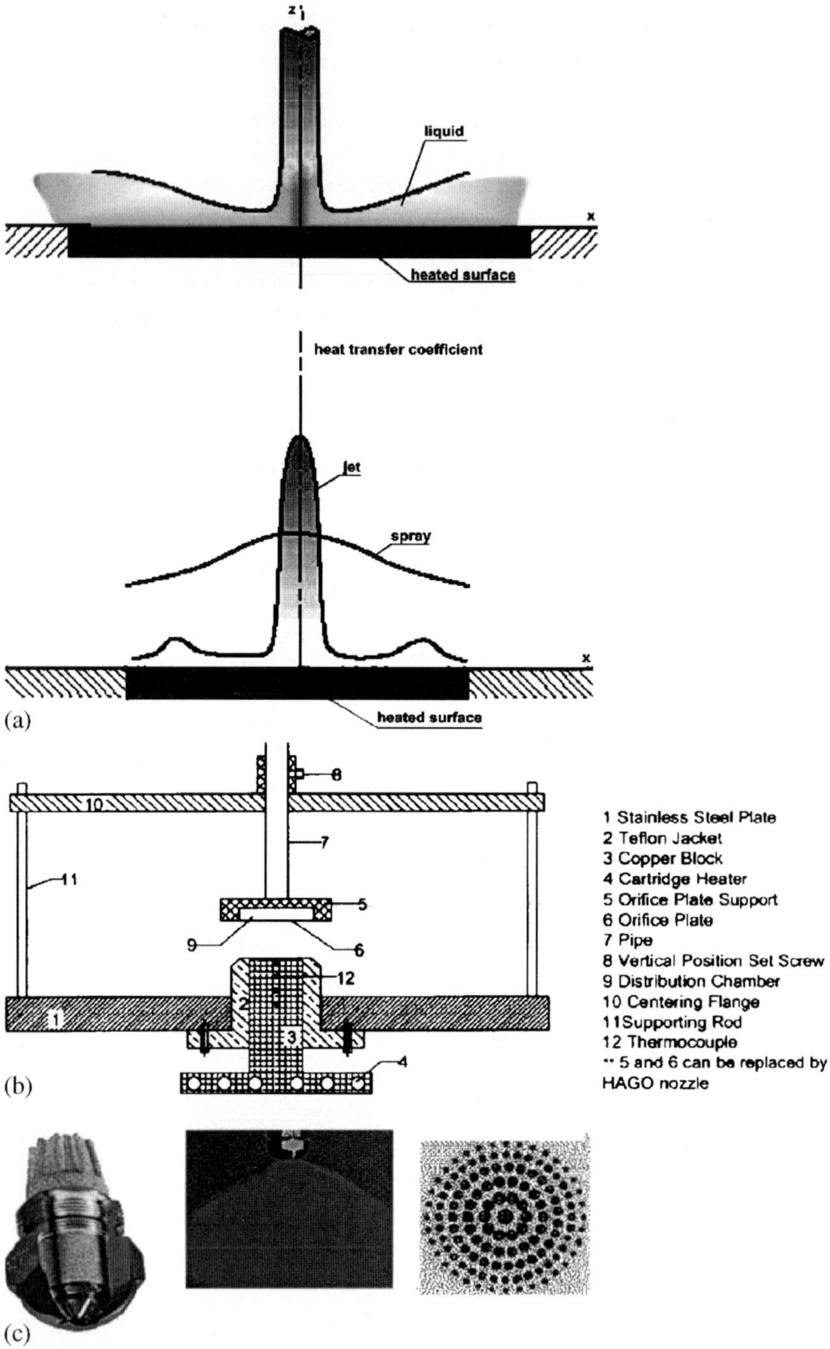

Fig. 2.7 (**a**) Spray and jet impingement cooling. Reprinted from Lasance and Simons (2005) with permission. (**b**) Details of the test section. (**c**) HAGO nozzle and spray details. Parts (**b–f**) reprinted from Fabbri et al. (2005) with permission

magnitude of heat fluxes as a function of excess temperature. At low chip powers, natural convection (A–B) initiates the heat transfer process until sufficient excess temperature is available to promote bubble growth on the surface, at which point boiling begins. As the power increases, more nucleation sites are activated and bubbles are detached at higher frequency, intensifying fluid circulation near the chip. This stretch between B and C is termed the nucleate boiling regime, where the intensified fluid circulation accommodates higher heat fluxes with minimal increase in the surface temperature. Line C–E represents nucleate boiling regime. Stretch D–E represents the transition to film boiling in which the heat transfer is realized by conduction through a vapor film. It is, however, very poor and may result in electronic failure due to the high temperatures. The most desirable mode for electronics cooling is the nucleate boiling regime. Immersion cooling is a well-established method, backed by over thirty years of university and industrial research. With natural convection two-phase flow, the critical heat flux using FC-72 is in the range of 5 to $20\,\mathrm{W/cm^2}$. However, much higher heat fluxes up to $100\,\mathrm{W/cm^2}$ can be accommodated through surface enhancement of the heat source.

Spray cooling and liquid jet impingement

Spray cooling and jet impingement (as shown in Fig. 2.7) are often considered as competing options for electronics cooling. In general, spraying requires lower flow rates but a higher nozzle pressure drop. In recent years spray cooling has received increasing attention as a means of supporting higher heat fluxes. In it the liquid is disintegrated into fine droplets that impinge individually on the heated wall. The cooling effect is achieved through a combination of thermal conduction through the liquid in contact with the surface, and vaporization at the liquid–vapor interface. The droplet impingement both enhances the spatial uniformity of heat removal and delays liquid separation at the wall during vigorous boiling.

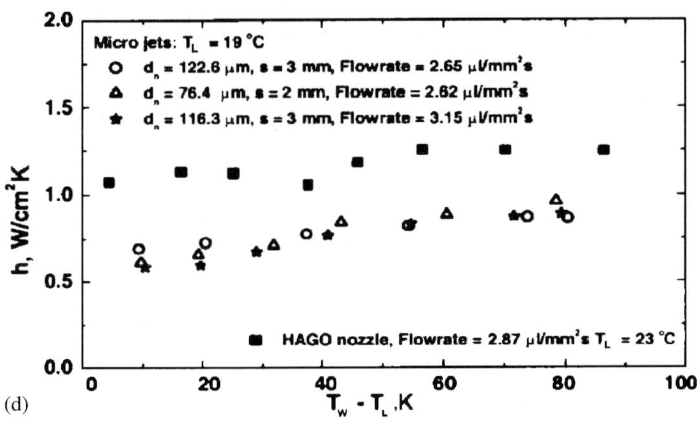

(d)

Fig. 2.7 (**d**) Comparison between spray and micro-jet performance for two flow rates 50.56 ml/min [2.87 μl/mm^2 s]. Parts (**b–f**) reprinted from Fabbri et al. (2005) with permission

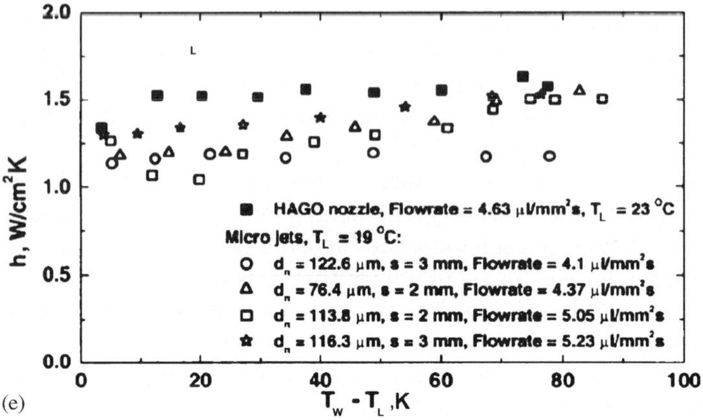

(e)

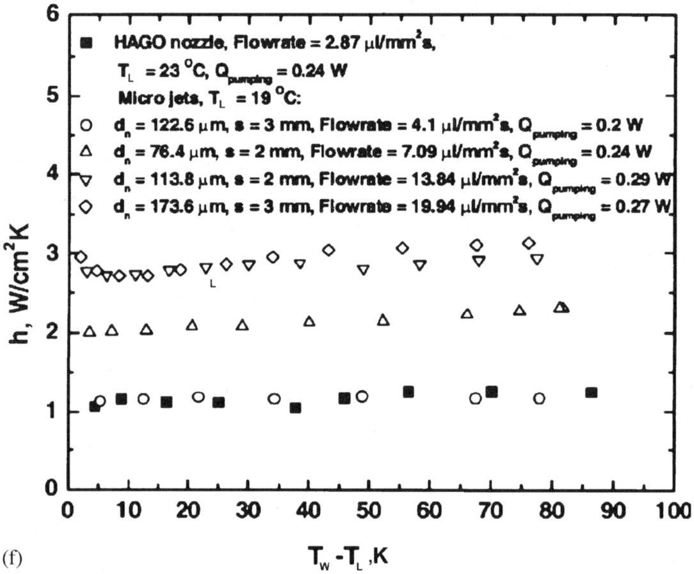

(f)

Fig. 2.7 (**e**) Comparison between spray and micro-jet performance for two flow rates 81.56 ml/min [4.63 μl/mm^{2}s]. (**f**) Comparison between spray and micro-jet performance for the same pumping power. Parts (**b–f**) reprinted from Fabbri et al. (2005) with permission

Single-phase heat transfer rates using droplet sprays and arrays of micro-jets have been compared by Fabbri et al. (2005). It was found that at a flow rate of 2.87 μl/mm^2 s spraying provides a higher heat transfer rate than any jet configuration, while at higher flow rate of 4.63 μl/mm^2 s jet arrays can perform as well as sprays.

Micro-jet arrays are usually associated with lower energy consumption rates than sprays generated by the special (HAGO) nozzle for the same flow rate. The liquid was pushed through a 0.5 mm stainless steel orifice plate to form the jets. The holes in the plate were laser drilled and were arranged in a circular pattern giving a radial and circumferential pitch of 1 mm for 397 jets, 2 mm for 127 jets, and 3 for 61 jets. The results presented here are confined to this type of nozzle, although designs may exist involving lower pressure losses at the considered flow rates. For equal pumping power and $T_W - T_L = 76$ K, they can remove fluxes as high as 240 W/cm^2, while sprays can only handle 93 W/cm^2.

The pressure drop in the HAGO nozzle quickly reaches impractical values. There is always a combination of jet diameter and jet spacing that yields the same heat transfer coefficient as the spray, but at a much lower energy cost.

Liquid micro-jet arrays have been successfully put to use. The module has proved capable of dissipating 129 W, with a heat flux of 300 W/cm^2 at a surface temperature of 80 °C, a considerable achievement at the present state of the art. Reduction of the system pressure made for lower boiling inception temperatures, thus allowing for higher heat removal rates at lower surface temperatures.

Liquid metal cooling

High-electrical conducting fluids such as liquid metals offer a unique solution to current and future cooling needs of high-power density heat sources. The two principal advantages of single-phase cooling systems based on liquid metals lie in their superior thermophysical properties and in the feasibility of moving them efficiently with silent, non-moving pumps. Closed loops based on liquid metals and the requisite pumps make possible gravity-independent high-performance cooling systems. Analytical and experimental work has been presented, using miniature pumps operating at a greater than 8 kPa maximum pressure rise, and showing heat transfer coefficients on the order of 10 W/cm^2 K (Miner and Ghoshal 2004).

An example of a liquid cooling loop is shown in Fig. 2.8.

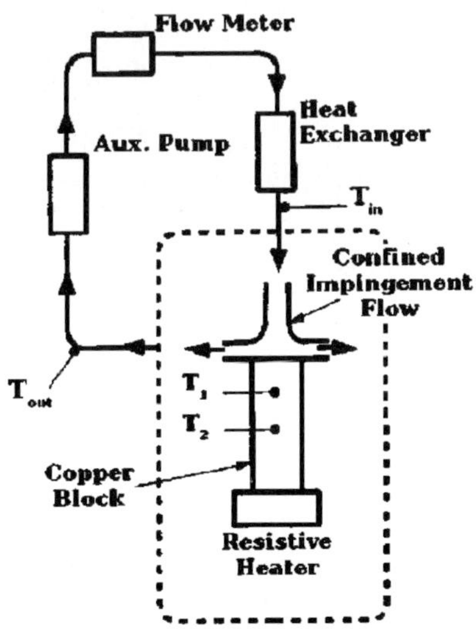

Fig. 2.8 Schematic representation of an experimental set-up for a liquid metal impingement/stagnation flow. Reprinted from Miner and Ghoshal (2004) with permission

Sintered porous inserts

The heat transfer and pressure drop in a rectangular channel with sintered porous inserts, made of stainless steels of different porosity, were investigated. The experimental set-up is shown in Fig. 2.9. Heat fluxes up to $6\,\mathrm{MW/m^2}$ were removed by using samples with a porosity of 32% and an average pore diameter of 20 μm. Under these experimental conditions, the temperature difference between the wall and the bulk water did not exceed $\Delta T = 55$ K at a pressure drop of $\Delta P = 4.5$ bars (Hetsroni et al. 2006a).

Concept of micro-channel heat sink

For flow at a given rate, the only way to significantly increase the heat transfer coefficient is to reduce the channel size, whose optimum can be calculated assuming a practical limit on the available pressure. Recourse to multiple channels, instead of continuous coolant flow over the entire back substrate surface, enables one to multiply the substrate area by a factor φ, representing the total surface area of the channel walls which are in contact with fluid. Single-row micro-channels etched dir-

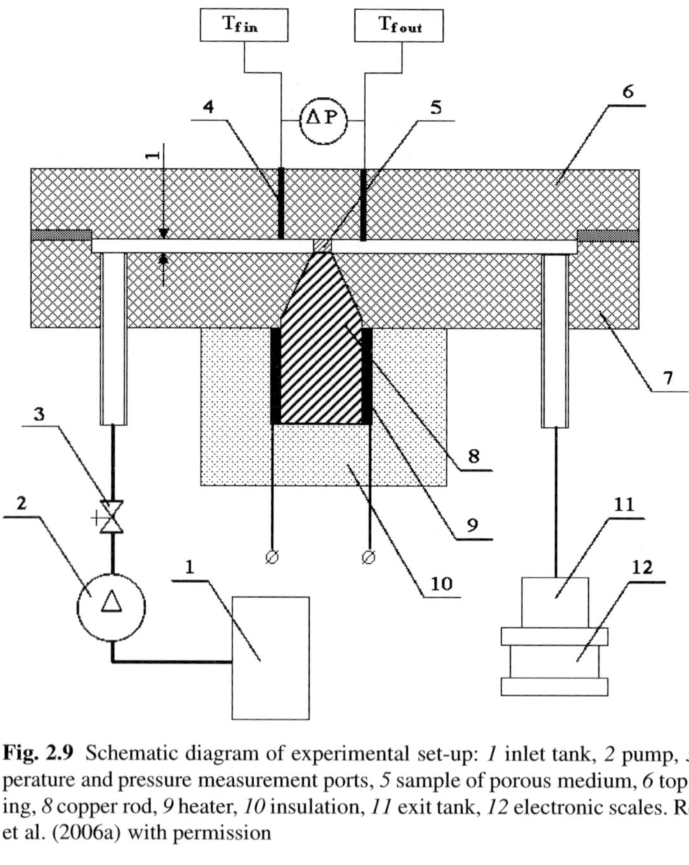

Fig. 2.9 Schematic diagram of experimental set-up: *1* inlet tank, *2* pump, *3* control valve, *4* temperature and pressure measurement ports, *5* sample of porous medium, *6* top of test section, *7* housing, *8* copper rod, *9* heater, *10* insulation, *11* exit tank, *12* electronic scales. Reprinted from Hetsroni et al. (2006a) with permission

ectly into the backs of silicon wafers were first shown to be effective by Tuckerman and Pease (1981) in which a maximum of $q = 790\,\mathrm{W/cm^2}$ was removed with a rise in water temperature $\Delta T = 71$ K at water pressure $\Delta P = 2$ bar (Fig. 2.10). The heat sink is made of deep rectangular channels of width w_c and depth H, separated by walls of thickness w_w. A cover plate is bonded onto the back, confining the coolant to the channels. The front surface of the substrate contains a planar heat source (the circuits).

The performance of a heat sink may be measured by its thermal resistance $R = \Delta T/q$, where ΔT is the temperature rise above that of the input coolant and q is the heat flux. As electronic and optical devices typically operate at a maximum $\Delta T_{max} = 50-100$ K above room temperature, their maximum power is determined by thermal resistance. In general, R is the sum of R_{cond} associated with conduction from the circuits through the substrate, package, and heat sink interface, R_{conv} with convection from the heat sink to the coolant, and R_{heat} with heating of the fluid as it absorbs the energy passing through the heat exchanger. To reduce R_{cond}, research into micro-scale heat exchangers has focused on heat sinks fabricated from a highly thermally conductive solid, such as copper or silicon, with rows of small channels fabricated into the surfaces. High solid conductivity is particularly important in multiple row structures, as the amount of heat by any given row can be large. A highly conductive medium increases heat conduction into subsequent layers where it can be transferred to the fluid. R_{heat} can be reduced by using a coolant of high volumetric heat capacity ρc_p. With these two components accounted for with relative ease, the convective thermal resistance R_{conv} becomes the dominant consideration in high-performance heat sinks. We focused on some aspects of single-phase and two-phase flow and heat transfer in small size channels.

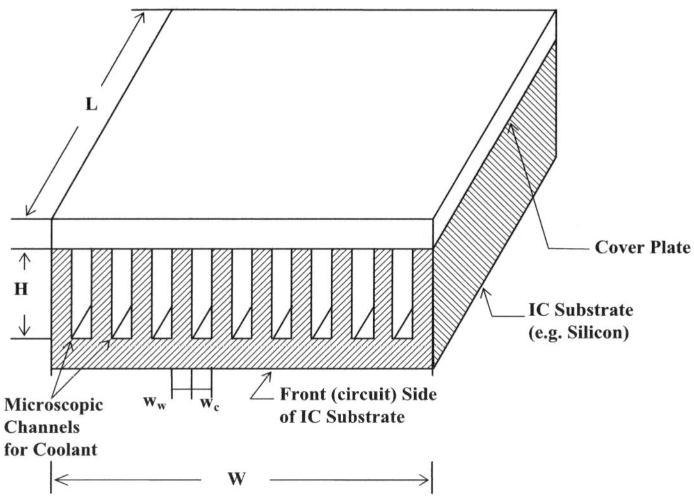

Fig. 2.10 High-performance micro-channel heat sink. Reprinted from Tuckerman and Pease (1981) with permission

Channel classification

The problems of micro-hydrodynamics were considered in different contexts: (1) drag at laminar, transient and turbulent single-phase flows; (2) heat transfer of liquids and gas flows; and (3) two-phase flows in adiabatic and heated micro-channels. As indicated by Kandlikar and Grande (2002), and by Hetsroni et al. (2005a, c), no fundamental change occurs in single-phase flow in the absence of rarefaction effects, which for gases are described by the Knudsen number, Kn, and are significant in the range $\mathrm{Kn} = 0.01-0.1$:

$$\mathrm{Kn} = \frac{\bar{\lambda}}{d_\mathrm{h}} \tag{2.1}$$

where $\bar{\lambda}$ is the mean free path for the gas:

$$\bar{\lambda} = \frac{\mu\sqrt{\pi}}{\rho\sqrt{2RT}} \tag{2.2}$$

with gas constant R in J/kg K, dynamic viscosity μ in kg/m s, density ρ in kg/m^3, temperature T in K.

The book by Karniadakis and Beskon (2002) addresses gas and liquid micro-flows with focus on the former, in which most of the deviations from macro-scales occur. The authors also treat liquid-specific phenomena and molecular dynamics.

Table 2.1 reproduced from the book by Karniadakis and Beskon, gives the mean free path for different gases at 300 K.

The classification in Table 2.2, based on the hydraulic diameter of the channel, was suggested by Kandlikar and Grande (2002).

The definition of mini-channels and micro-channels has not been clearly and strictly established in the literature although many related studies have been done. For example, for compact heat exchangers, Mehendale et al. (1999) gave a relatively

Table 2.1 Mean free path for gases at atmospheric pressure

Gas	T K	R J/kg K	ρ kg/m^3	μ kg/m s	$\bar{\lambda}$ μm
Air	300	287.0	1.1614	1.85×10^{-5}	0.068
Helium	300	2077	0.1625	1.99×10^{-5}	0.194
Hydrogen	300	4124	0.08078	8.96×10^{-6}	0.125
Nitrogen	300	296.8	1.1233	1.78×10^{-5}	0.066

Table 2.2 Channel classification

Conventional channels	$d_\mathrm{h} > 3$ mm
Mini-channels	200 μm $\leq d_\mathrm{h} \leq$ 3 mm
Micro-channels	10 μm $\leq d_\mathrm{h} \leq$ 200 μm
Transitional channels	0.1 μm $\leq d_\mathrm{h} \leq$ 10 μm
Molecular nanochannels	$d_\mathrm{h} \leq 0.1$ μm

loose definition of mini-channels, in terms of hydraulic diameter: $d_h = 1-6$ mm. In this book we will consider the channels with hydraulic diameters ranging roughly from 5 to 500 µm as micro-channels and the channels with hydraulic diameters $d_h > 500$ µm as conventional size channels. Traditional correlations may not be suitable to predict flow regimes, pressure drop and heat transfer in micro-channels and applicability of existing correlations for conventional size channels to micro-channels should be carefully examined.

Classification on the basis of the Knudsen number, as per Karniadakis and Beskon (2002), is given in Table 2.3.

The micro-electromechanical systems (MEMS) operate in a wide range of regimes covering continuum, slip and transition flows. Further miniaturization of the MEMS device components and appliances in the emerging field of nanoelectromechanical systems (NEMS) would result in high Knudsen numbers, making it necessary to study mass, momentum, and energy transport over the entire Knudsen range.

The overall performance of micro-channel heat sinks

Micro-channel heat sinks are devices that provide liquid flow through parallel channels having a hydraulic diameter of around 5–500 µm. Figure 2.11 shows the range of heat transfer coefficients attainable with different fluids and cooling schemes (Mudawar 2001). Air is the most readily available, and remains the most widely used coolant for most applications. However, its poor thermal transport properties limit its use to low-heat flux devices. Better results are obtained with fluorochemical liquids, and the most demanding cooling situations are typically managed with water.

Mudawar (2001) reviewed high-heat flux thermal management schemes, including ultra high-heat fluxes in the range of 1,000–100,000 W/cm^2. Garimella and Sobhan (2003) reviewed research on fluid dynamics and heat transfer in micro-channels up to 2000. Recent overviews were also provided by Morini (2004), Mohapatra and Loikitis (2005), Hetsroni et al. (2005a, 2006c), Thome (2006), and Cheng and Wu (2006).

A micro-channel heat sink can be classified as single-phase or two-phase according to the state of the coolant inside it. For single-phase fluid flow in smooth

Table 2.3 Classification on the basis of the Knudsen number

Range of Knudsen number	Type of flow
Kn = 0.001–0.1	Continuum flow: no rarefaction effects
Kn = 0.01–0.1	Slip flow: rarefaction effects that can be modeled with a modified continuum theory with wall slip taken into consideration
Kn = 0.1–10	Transition flow: between slip flow and free molecular flow, treated statistically, e.g., by the Boltzmann equation
Kn > 10	Free molecular flow: motion of individual molecules, that must be modeled and then treated statistically

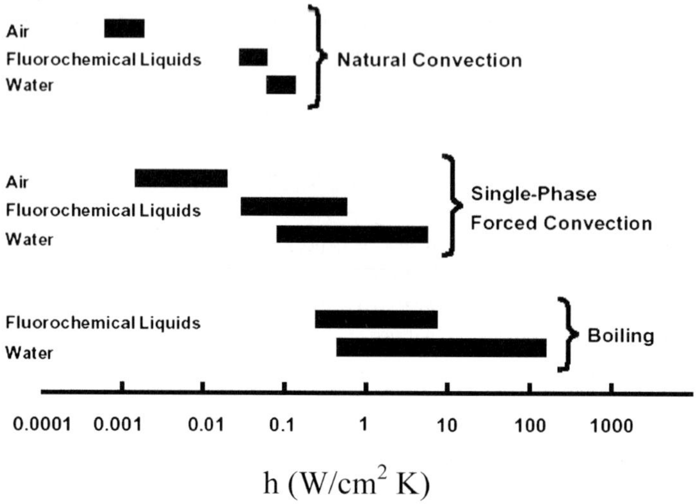

Fig. 2.11 Heat transfer coefficient for different coolants. Reprinted from Mudawar (2001) with permission

channels with a hydraulic diameter d_h from 15 to 4,000 μm, in the Reynolds number range $Re < Re_{crit}$, the Poiseuille number Po is independent of Re. For single-phase gas flow in channels with a hydraulic diameter from 100 to 4,000 μm, in the range $Re < Re_{crit}$, with Knudsen number $0.001 \leq Kn \leq 0.38$, and Mach number $0.07 \leq Ma \leq 0.84$, the experimental friction factor agrees quite well with the theoretical prediction for fully developed laminar flow. The behavior of single-phase flow in micro-channels shows no differences from macro-scale flow (Hetsroni et al. 2005a).

By contrast, two-phase flow in micro-channels of rectangular, circular, triangular, or trapezoidal cross-section showed significant differences when compared to macro-scale flow. Lee et al. (2004) studied the bubble growth dynamics in single micro-channels as small as 41.3 μm. They concluded that a conventional bubble departure model for convective boiling in conventional tubes larger than 6 mm cannot predict the growth pattern, and that the bubbles always nucleate from the corners. Chung and Kawaji (2004) investigated liquid and gas adiabatic two-phase flow in micro-capillaries of circular and square cross-sections and found that the transition of flow patterns occurred at lower surface velocities in the circular capillary than in the square one. Convective boiling in transparent single micro-channels with similar hydraulic diameters but different cross-sections was studied by Yen et al. (2006). Two types of glass micro-channels were tested: circular of $d = 210$ μm, and square of $d_h = 214$ μm. In the latter, the corners acted as active nucleation cavities, both the number of nucleation bubbles and the local heat transfer coefficient increased with decreasing vapor quality. The performance of a micro-channel heat sink is enhanced by increasing the heat transfer area and coefficient, the first being achieved by increasing the number of micro-channels and the second by reducing the hydraulic diameter, which makes for a dramatic increase in pressure drop. Calame et al. (2007)

carried out experiments on removing high heat flux from GaN-on-Sic semiconductors dies using micro-channels coolers. A wide variety of micro-channel materials and configurations were investigated. Silicon micro-channel coolers exhibited good performance at power densities of 1,000–1,200 W/cm^2. Polycrystalline chemical vapor deposited (CVD) Sic micro-channel coolers were found to be promising for higher power densities of 3,000–4,000 W/cm^2. The performance was good as a copper micro-channel cooler, but presumably without the stress problems associated with differential thermal expansion between the semiconductor and cooper.

Lee and Vafai (1999) compared jet impingement and micro-channel cooling for high-heat flux appliances. One of their conclusions is that micro-channel cooling is more effective for areas smaller than $7 \times 7\,cm^2$. Kandlikar and Upadhye (2005) showed enhanced micro-channel cooling by using off-set strip fins and a split-flow arrangement. Colgan et al. (2005) published a practical implementation of a silicon micro-channel cooler (shown in Fig. 2.12) for high-power chips. They argued that given the high cost of high-performance processor chips it is impractical to form the micro-channels directly on the chip. Instead, a separate micro-channel cold plate is bonded to the back of the chip, a design requiring very low interface thermal resistance. If the micro-cooler is based on silicon, rigid bonding dictates use of silver-filled epoxies or solder. Power densities in excess of 400 W/cm^2 are reported, for a flow of 1.2 l/min at 30 kPa.

Micro-channel heat transfer can be pushed even further by recourse to boiling. In addition to offering higher heat transfer coefficients, boiling convection in micro-channels is promising because it requires less pumping power than its single-phase liquid counterpart to achieve a given heat sink thermal resistance. For the same heat flux the pressure drops by a factor of 20. A review on boiling and evaporation in small-diameter channels was published by Bergles et al. (2003).

The vapor–liquid exchange process that is largely responsible for the effectiveness of phase-change cooling, requires uninterrupted liquid flow on the device surface. Higher heat fluxes are dissipated by a higher output of vapor bubbles per unit surface area. Unfortunately, bubble crowding may lead to significant vapor coalescence, eventually interfering with the liquid access to the device surface. Once the vapor–liquid exchange process is interrupted, the power dissipated in the device it-

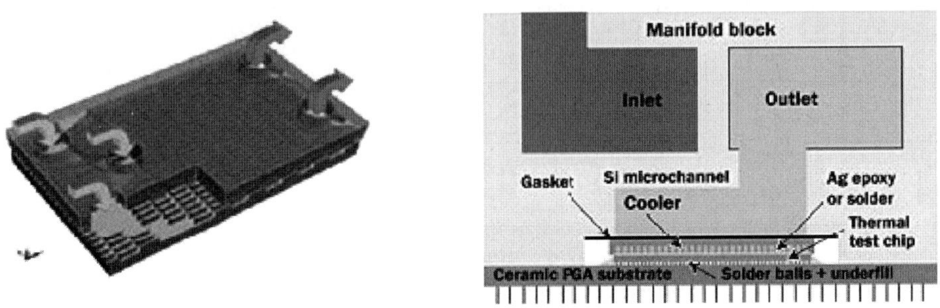

Fig. 2.12 Pictures from IBM paper showing high-performance liquid cooling technology using micro-channels. Reprinted from Colgan et al. (2005) with permission

Table 2.4 Comparison of pressure drops between PG 50% and water for the same thermal resistance

Liquid	Flow rate (ml/min)	Pressure drop (kPa)
Water	200	80
PG 50%	220	900

self will no longer be removed and the device temperature begins to escalate out of control. This is the highest point in the boiling curve, termed the critical heat flux (CHF), constituting the upper design limit for any phase-change cooling system.

The main problem in single-phase micro-channel cooling are freezing thresholds required in the electronics industry, which may be as low as 40 °C, in which case water is ruled out as cooling medium. Table 2.4 shows a comparison between the performance of the widely used propylene glycol antifreeze (1:1 PG and water) versus pure water for the same thermal performance. The micro-channel size for this study is $50 \times 300\,\mu\text{m}$ (Prasher et al. 2005). It is seen that the pressure drop for conventional antifreeze is very large due to its low thermal conductivity and high viscosity. As a result, strong forces act on the pump bearings. Therefore, for single-phase micro-channel cooling, other antifreeze coolants are needed, with high-thermal conductivity and low viscosity.

Another major problem in this context is that the coolant also has to be used as lubricant for pump bearings, as the pump has to be hermetically sealed. This creates a situation of conflicting requirements: high viscosity from lubrication viewpoint, low viscosity from pressure drop viewpoint. Figure 2.13 shows the thermal performance of the package-based micro-channel cold plate as a function of the pressure drop (Prasher et al. 2005). It can be seen that reducing the thermal resistance of the micro-channels will result in a large pressure drop. In turn, this large pressure

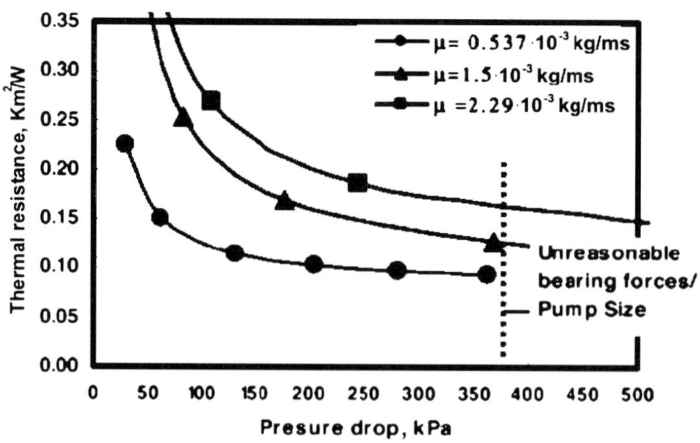

Fig. 2.13 Thermal resistance vs. pressure drop for fluids with different viscosity. Reprinted from Prasher et al. (2005) with permission

Table 2.5 Saturation thermophysical properties of some liquid coolants at 1 bar

Fluid	Saturation temperature T_{sat} [°C]	Liquid density ρ_L [kg/m^3]	Liquid specific heat $c_{p,L}$ [J/kg K]	Vapor density ρ_G [kg/m^3]	Latent heat of vaporization h_{LG} [kJ/kg]	Surface tension $\sigma \times 10^3$ [N/m]
FC-72	56.6	1600.1	1102.0	13.43	94.8	8.35
FC-87	3.0	1595.0	1060.0	13.65	87.93	14.53
PF-5052	50.0	1643.2	936.3	11.98	104.7	13.00
Water	100.0	957.9	4217.0	0.60	2256.7	58.91

drop across the device will generate significantly large forces on the bearings, thus increasing the wear and possibly reducing the life time of the pumps. In addition, the low physical size of the pump shaft may impose significant additional challenges on the bearing design.

Phase-change cooling systems make do with smaller sizes without necessarily imposing a larger pumping power requirement compared with single-phase systems.

The saturation thermophysical properties of some liquid coolants are presented in Table 2.5 (Mudawar 2001).

Poor flow distributions may result in localized dry hotspots which, absent control of the temperature fluctuations, may cause rapid overheating. Temperature and pressure fluctuations, and poor flow distribution, are the main problems that accompany the use of two-phase micro-channels.

2.2 Pressure and Temperature Measurements

Pressure measurement

Until recently pressures and temperatures were not measured directly inside the micro-channels because of size limitations. To obtain the channel inlet and exit pressures, measurements were taken in a plenum or supply line prior to entering the channel. Special coefficients were sometimes assumed to account for losses at the ends and in any piping between the channel plenums and the pressure transducers. In attempts to obviate the need for such assumptions experiments were conducted on integration of pressure sensors with a micro-channel, allowing the static pressure inside it to be measured at multiple locations. These early experiments involved either surface micro-machined channels with channel heights on the order of 1–2 μm (Shih et al. 1996; Li et al. 2000), or of conventionally machined channels, which are typically larger than $d_h = 250$ mm (Pfund et al. 2000). Due to the difficulty of getting integrated sensors to operate properly and the limited range of channel dimensions tested, the experiments provided little additional information about micro-channel flows at $Kn < 0.01$. As a further step Kohl et al. (2005), using micro-fabrication technologies, restored the integration of tap lines and pressure sensing membranes into a system consisting of three silicon chips (Fig. 2.14). The micro-channel was fabricated by etching silicon wafers in KOH, producing a rectangular cross-section.

The lower chip contains the micro-channel test section with inlet and exit plenums, eight static pressure tap lines intersecting the micro-channel at equally spaced intervals, and one tap line per plenum. The tap line to the micro-channel

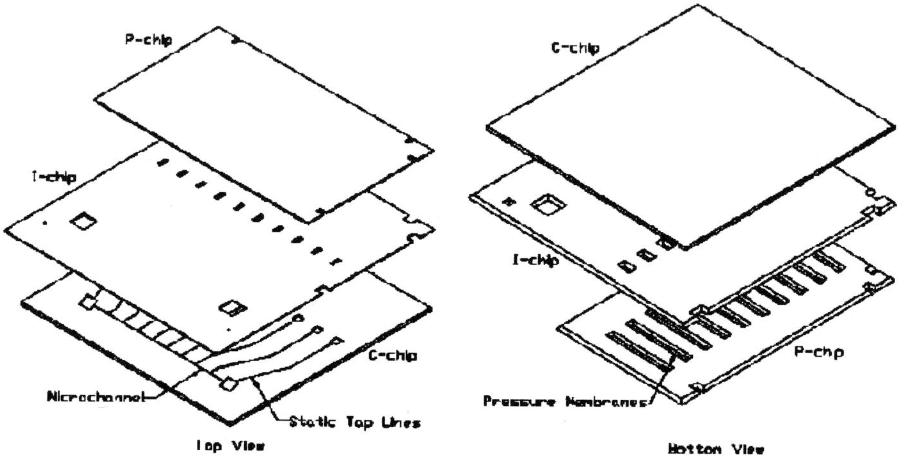

Fig. 2.14 Micro-fabricated test section components, *top* and *bottom* views. Reprinted from Kohl et al. (2005) with permission

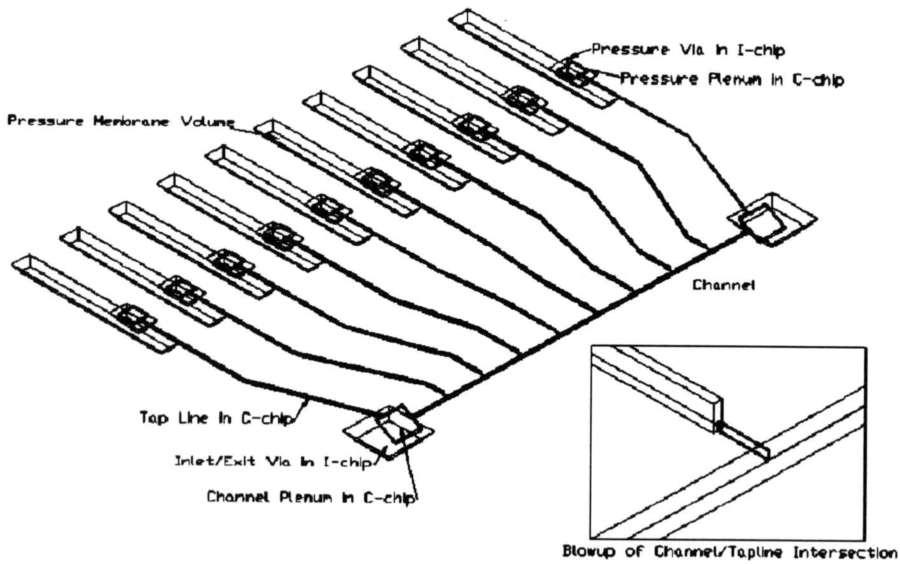

Fig. 2.15 Fluid-filled volume of the micro-channel system showing the connections between the micro-channel, static tap lines, and the fluid-filled volume of the pressure sensors. Reprinted from Kohl et al. (2005) with permission

intersection is etched by a deep silicon RIE process to a width less than 7 μm and depths on the order of 10 μm. The middle chip is used to seal the channel and tap lines and provides ports for introducing and removing fluid from the plenums, as well as for connecting the tap lines to the pressure membrane chip, which is located at the top. The pressure membrane chip contains 10 rectangular membranes for sensing pressure from the tap lines. The membranes are KOH etched out of silicon wafers and are approximately 0.564 mm wide, 10 mm long, and 50 μm thick. Figure 2.15 is a schematic of the fluid-filled volume of the micro-channel and system including the tap lines and pressure sensor volumes.

The investigation shows agreement between the standard laminar incompressible flow predictions and the measured results for water. Based on these observations the predictions based on the analytical results of Shah and London (1978) can be used to predict the pressure drop for water in channels with d_h as small as 24.9 μm. This investigation shows also that it is insufficient to assume that the friction factor for laminar compressible flow can be determined by means of the well-known analytical predictions for its incompressible counterpart. In fact, the experimental and numerical results both show that the friction factor increases for compressible flows as Re is increased for a given channel with air.

Temperature measurements

Reliable micro-scale measurement and control of the temperature are required in developing thermal micro-devices. Available measurement techniques can be largely classified into contact and non-contact groups. While the resistance thermometer, thermocouples, thermodiodes, and thermotransistors measure temperature at specific points in contact with them, infrared thermography, thermochromic liquid crystals (TLC), and temperature-sensitive fluorescent dyes cover the whole temperature field (Yoo 2006).

Resistance thermometry

Resistance thermometry, based on the variation of the resistance with temperature, is one of the most traditional techniques used for temperature measurement in the micro-scale. It is stable and applicable to a wide range of temperatures, but subject to inaccuracies due to self-heating since it involves use of electric current. The resistance temperature detector (RTD) and thermistor are the most frequently used forms. Polysilicon microthermistors have been used to study the heat transfer characteristics in micro-channels (Jiang et al. 1999a,b, 2000) and applied to the wall shear stress measurement (Lin et al. 2000). Resistance thermometry is also applied in thermocapillary pumping systems (Sammarco and Burns 1999), micro-machined chips (Yoon et al. 2002), and transient temperature measurement in thermal bubble formation (Tsai and Lin 2002). It requires carefully controlled fabrication including sensor materials of high purity, precise control of the dosage, and calibration of the sensors. In the latter, one should bear in mind that all components can be temperature-dependent.

Thermocouples

A thermocouple is one of the most common temperature sensors – inexpensive, reliable, interchangeable, and covering a wide range of temperature. For reduced size and improved spatial resolution, the micro-machined thermocouple attached to a cantilever-based probe tip such as in atomic force microscopy (AFM) has been developed. This technique, which is called scanning thermal microscopy (STM), generates a thermal map simultaneously with a topographical map by scanning the thermoelectric voltage on the surface. The spatial resolution of the technique has been reduced down to 24 nm (Luo et al. 1997). Significant improvements on the STM technique and its applications have been reviewed in detail by Majumdar (1999). Varesi and Majumdar (1998) reported a new technique called scanning joule expansion microscopy (SJEM) that could simultaneously image surface topography and material expansion due to joule heating.

Thermochromic liquid crystal

In the thermochromic liquid crystal (TLC) the dominant reflected wavelength is temperature-dependent and it has been employed for full-field mapping of temperature fields for over three decades. Although it is non-intrusive and cost effective, there are some problems in applying it to micro-scale measurements, because of size (typically tens of micrometers) and time response (from a few milliseconds to several hundred milliseconds depending on the material and the form). Examples of application are micro-fabricated systems (Chaudhari et al. 1998; Liu et al. 2002) and electronic components (Azar et al. 1991).

The liquid crystal thermographs method has been used for measuring micro-tube surface temperature with uncertainties of lower than ± 0.4 K by Lin and Yang (2007). The average outside diameter micro-tubes was 250 μm and 1,260 μm, respectively. The surface was coated with thermochromic liquid crystal (TLC). The diameters of encapsulated TLC were ranging from 5 to 15 μm. The TLC was painted on the tested tubes surface with thickness of approximately 30 μm.

Laser-induced fluorescence

Fluorescence is the capacity of certain molecules to absorb energy at a particular wavelength and to reemit it at a longer wavelength, in the range of visible light. Examples of relevant studies are Kim and Kihm (2001), and Ross et al. (2001).

Infrared thermography

Infrared (IR) thermography is one of the most advanced non-destructive (NDT) methods based on the fact that all bodies whose absolute temperature is above zero emit electromagnetic radiation over a wide spectrum of wavelengths depending on the temperature. Recently, several researchers have applied it to micro-scale temperature measurement. Hetsroni et al. (2001a) constructed a thermal micro-system and studied the effect of the geometric configuration on the flow and heat transfer. Using a flow visualization system, they observed that the heat transfer was enhanced under conditions of flow boiling. Subsequently, more detailed investigations regarding boiling in capillary tubes have been conducted (Hetsroni et al. 2003a,b). IR thermography was also used to confirm formation of thermally isolated integral micro-hotplates, which is the key element in the development of an array of micro-pilasters and conductivity-type gas sensors (Furjes et al. 2002). In the case of liquids, a problem is that the temperature is measured on the basis of the radiation emitted from an object, which means that it is necessary to know the accurate value of the emissivity, which in liquids is sensitive to the wavelength in the infrared range.

Mishan et al. (2007) developed an infrared technique to measure the temperature of the fluid and wall in a micro-channel, using a transparent cover made of sapphire glass with a wavelength range of 0.15 to 7 μm, thus permitting visual observation. When radiant energy strikes a surface, a part of the radiation is reflected, a part is absorbed and a part is transmitted. The transmission of the sapphire window with a thickness of 10 mm is shown in Fig. 2.16. From this figure one can conclude that the transmission is constant for most of the wavelength range, and for $\lambda = 5\,\mu\text{m}$ and $T = 273\,\text{K}$, it equals about 85%, a suitable value for the IR technique. The measurements were preceded by calibration, with the temperature of the water flowing into the manifolds (range 30 to 90 °C) measured by a thermocouple and IR camera with ± 1 K accuracy.

The IR technique also yielded temperature distributions (Fig. 2.17) in the symmetry plane at $\text{Re} = 30$ and $q = 19 \times 10^4\,\text{W/m}^2$. The wall temperature decreases by axial conduction through the solid walls in the last part of the micro-channel ($x/L > 0.75$) since this part is not heated. Neither the wall nor the fluid bulk temperature distribution can be approximated as linear.

The temperature distribution depends on the material and design of the module, the flow rate, and the heat flux. Figure 2.18a shows the IR image (top view) of the central part of the test module obtained at $\text{Re} = 100$ and $q = 25 \times 10^4\,\text{W/m}^2$, and Fig. 2.18b shows its counterpart in the spanwise direction. The field of view is

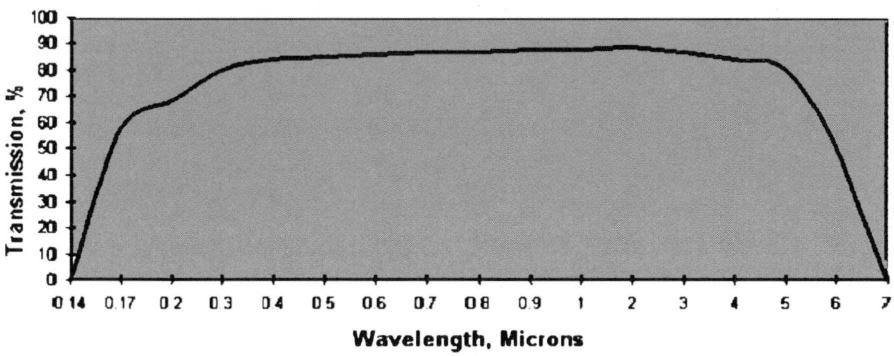

Fig. 2.16 Transmission vs. wavelength at 10 mm thick sapphire glass. Reprinted from Mishan et al. (2007) with permission

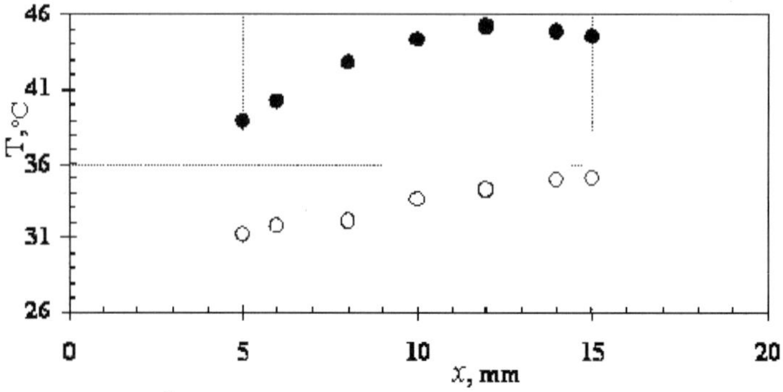

Fig. 2.17 Temperature distribution in the streamwise direction. The *solid circles* (•) represent the bottom temperature, and the *empty circles* (○) represent the fluid temperature. Reprinted from Mishan et al. (2007) with permission

3.6 mm and 3.6 mm in the streamwise and spanwise direction, respectively, the gray strips refer to the temperature on the channel walls (about 31.2 °C), the black strips to that of the water (about 29.9 °C).

Uncertainty in experimental measurements

For each result presented, the experimental data should include the following information:

1. The precision limit P. The $\pm P$ interval about a nominal result (single or average) is the region, with 95% confidence, within which the mean of many such results would fall, if the experiment were repeated under the same conditions using the same equipment. Thus, the precision limit is an estimate of the lack of repeatability caused by random errors and unsteadiness.
2. The bias limit B. The bias limit is an estimate of the constant error. It is assigned with the understanding that the true value of the bias error, if known, would be less than $|B|$ with 95% confidence.
3. The uncertainty U. The $\pm U$ interval about the nominal result is the band within which true value of the result lies with 95% confidence. It is calculated from

$$U = [B^2 + P^2]^{1/2} \tag{2.3}$$

Example

Consider an experiment in which an air-cooled device is being tested and it is desired to determine the rate of heat transfer Q to the cooling air (ASME 2000). This can be accomplished by measuring the mass flow rate m, and the inlet and outlet air temperatures, T_{in} and T_{out}, and computing:

$$Q = mc_p(T_{out} - T_{in}) \tag{2.4}$$

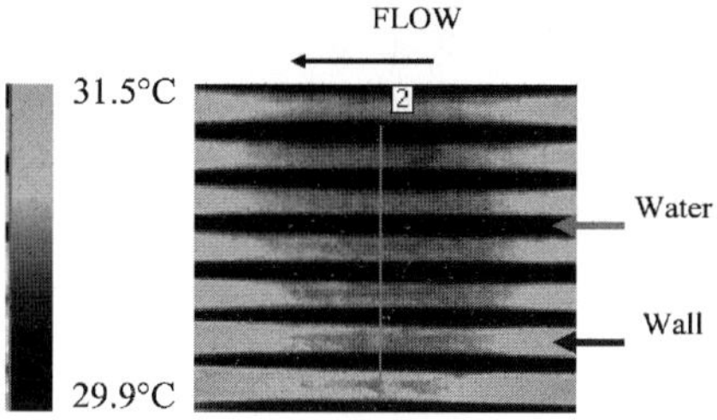

(a)

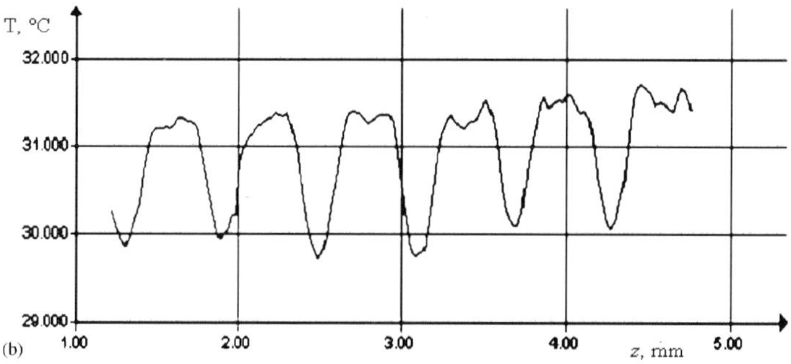

(b)

Fig. 2.18a,b Top view of the temperature field, Re = 100, $q = 25 \times 10^4\ \mathrm{W/m^2}$. (**a**) IR image. (**b**) Temperature variation in the spanwise direction. Reprinted from Mishan et al. (2007) with permission

where Q is the heat rate, c_p is the constant-pressure specific heat of air, and m is the mass flow rate in kg/s.

The 95% confidence uncertainty U_Q in the experimental result Q, is given by the following combination of the precision (random) contribution P_Q, and the bias contribution B_Q:

$$U_Q = \sqrt{P_Q^2 + B_Q^2} \tag{2.5}$$

These two contributions can be evaluated separately in terms of the sensitivity coefficients of the result to the measured quantities by using the propagation equation of Kline and McClintock (1953):

$$P_Q^2 = \left(\frac{\partial Q}{\partial m}\right)^2 P_m^2 + \left(\frac{\partial Q}{\partial c_p}\right)^2 P_{c_p}^2 + \left(\frac{\partial Q}{\partial T_{\mathrm{out}}}\right)^2 P_{T_{\mathrm{out}}}^2 + \left(\frac{\partial Q}{\partial T_{\mathrm{in}}}\right)^2 P_{T_{\mathrm{in}}}^2 \tag{2.6}$$

and

$$\begin{aligned} B_Q^2 = & \left(\frac{\partial Q}{\partial m}\right)^2 B_m^2 + \left(\frac{\partial Q}{\partial c_p}\right)^2 B_{c_p}^2 + \left(\frac{\partial Q}{\partial T_{\mathrm{out}}}\right)^2 B_{T_{\mathrm{out}}}^2 \\ & + \left(\frac{\partial Q}{\partial T_{\mathrm{in}}}\right)^2 B_{T_{\mathrm{in}}}^2 + 2\left(\frac{\partial Q}{\partial T_{\mathrm{out}}}\right)\left(\frac{\partial Q}{\partial T_{\mathrm{in}}}\right) B'_{T_{\mathrm{out}}} B'_{T_{\mathrm{in}}} \end{aligned} \tag{2.7}$$

where $B'_{T_{\mathrm{out}}}$ and $B'_{T_{\mathrm{in}}}$ are the combination of $B_{T_{\mathrm{out}}}$ and $B_{T_{\mathrm{in}}}$, which arise from identical error sources (as in the case of the thermocouples calibrated using the same standards, equipment, and procedures), and are therefore presumed to be perfectly correlated.

Using Eq. (2.4) to evaluate the derivatives, denoting $\Delta T = T_{\mathrm{out}} - T_{\mathrm{in}}$ and rearranging, one obtains

$$\left(\frac{P_Q}{Q}\right)^2 = \left(\frac{P_m}{m}\right)^2 + \left(\frac{P_{c_\mathrm{p}}}{c_\mathrm{p}}\right)^2 + \left(\frac{P_{T_{\mathrm{out}}}}{\Delta T}\right)^2 + \left(\frac{P_{T_{\mathrm{in}}}}{\Delta T}\right)^2 \tag{2.8}$$

and

$$\left(\frac{B_Q}{Q}\right)^2 = \left(\frac{B_m}{m}\right)^2 + \left(\frac{B_{c_\mathrm{p}}}{c_\mathrm{p}}\right)^2 + \left(\frac{B_{T_{\mathrm{out}}}}{\Delta T}\right)^2 + \left(\frac{B_{T_{\mathrm{in}}}}{\Delta T}\right)^2 + 2\left(\frac{B'_{T_{\mathrm{out}}}}{\Delta T}\right)\left(\frac{B'_{T_{\mathrm{in}}}}{\Delta T}\right). \tag{2.9}$$

These derivatives can be evaluated numerically, using a data reduction program, or analytically.

In Eq. (2.4) the temperatures T_{out} and T_{in}, denote the bulk mean air temperatures at the outlet and inlet cross-sections, respectively. Their representation by point measurements introduces a bias error equal to the difference between the latter and the corresponding bulk means. In evaluating it, allowance should be made for the residual uncertainty involved in the bias errors from the probe calibration, etc.

Consider now a situation in which the bias limits in the temperature measurements are uncorrelated and are estimated as 0.5 °C, and the bias limit on the specific heat value is 0.5%. The estimated bias error of the mass flow meter system is specified as "0.25% of reading from 10 to 90% of full scale." According to the manufacturer, this is a fixed error estimate (it cannot be reduced by taking the average of multiple readings and is, thus, a true bias error), and B_m is taken as 0.0025 times the value of m. For $\Delta T = 20\,^\circ\mathrm{C}$, Eq. (2.9) gives:

$$\frac{B_Q}{Q} = \sqrt{(0.0025)^2 + (0.005)^2 + \left(\frac{0.5\,^\circ\mathrm{C}}{20\,^\circ\mathrm{C}}\right)^2 + \left(\frac{0.5\,^\circ\mathrm{C}}{20\,^\circ\mathrm{C}}\right)^2} = 0.036(=3.6\%). \tag{2.10}$$

Obviously, the bias limits on the temperature measurements are dominant in this specific case. When they are totally correlated, the last term in Eq. (2.9) cancels out the third and fourth terms, and B_Q/Q equals 0.0056 instead if 0.036.

If the random errors and process unsteadiness are such that the precision limit for Q, P_Q, calculated from Eq. (2.7), is 2.7%, the overall uncertainty in the determination of Q, U_Q, is:

$$\frac{U_Q}{Q} = \sqrt{\left(\frac{B_Q}{Q}\right)^2 + \left(\frac{P_Q}{Q}\right)^2} = \sqrt{(0.036)^2 + (0.027)^2} = 0.045(=4.5\%). \tag{2.11}$$

2.3 Pressure Drop and Heat Transfer in a Single-Phase Flow

Many correlations have been proposed in literature for the friction factor and heat transfer, based on experimental investigations on liquid and gas flow in micro-channels. Garimella and Sobhan (2003) presented a comprehensive review of these investigations conducted over the past decade.

Incompressible flow: friction correlations

According to Schlichting and Gersten's (2000) equation, the friction factor is:

$$\lambda = 2\Delta P \frac{d_{\mathrm{h}}}{L} \frac{1}{\rho U^2} \tag{2.12}$$

where ΔP is the pressure drop across the channel, d_{h} is the channel hydraulic diameter, L is the channel length, ρ is the fluid density, and U is the bulk fluid velocity.

Laminar flow

For developed laminar flow in smooth channels of $d_{\mathrm{h}} > 1$ mm, the product $\lambda\mathrm{Re} =$ const. Its value depends on the geometry of the channel. For a circular pipe $\lambda\mathrm{Re} = 64$, where $\mathrm{Re} = Ud_{\mathrm{h}}/\nu$ is the Reynolds number, and ν is the kinematic viscosity.

The correlations of Wu and Little (1984), Choi et al. (1991), and Yu et al. (1995), Maynes and Webb (2002), Turner et al. (2002), Celata et al. (2004), Hetsroni et al. (2005a) predict constant values of $\lambda\mathrm{Re}$ for micro-channels. Predictions by Peng et al. (1994a) for water flow in rectangular micro-channels in the range of d_{h} from 133 to 343 μm show an altogether different trend: in all cases, $\lambda\mathrm{Re}$ decreases with an increase of the Reynolds number. While the slopes of the curves for all test cases presented by Peng et al. (1994a) are identical, the magnitude of $\lambda\mathrm{Re}$ is highest for the largest micro-channels and lowest for the smallest. Several experimental results of $\lambda\mathrm{Re}$ in the laminar regime for smooth micro-channels are shown in Fig. 2.19. The unusual results obtained by Peng et al. (1994a) are attributable, at least in part, to uncertainties of the experimental data.

Transition and turbulent flows

Transition from laminar to turbulent flow occurs when the friction factor exceeds the low $\lambda\mathrm{Re}$ range. In Fig. 2.20a the results obtained for a tube of diameter 705 μm by Maynes and Webb (2002) are compared against the value accepted for laminar flow $\lambda = 64/\mathrm{Re}$. Based on the above data, one can conclude that the transition occurs at $Re > 2{,}100$.

The friction correlations in the turbulent regime are compared against conventional correlations in Fig. 2.20b. Predictions for nitrogen flow by Choi et al. (1991) agree very well with conventional results; the Wu and Little (1984) correlation is similar to those two in its trend, but the predicted values are much higher in mag-

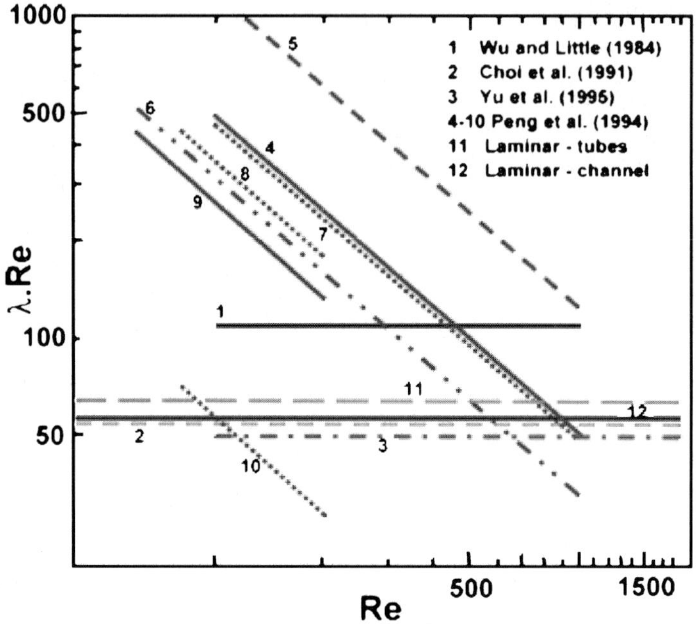

Fig. 2.19 Laminar flow and friction correlations. Reprinted from Sobhan and Garimella (2001) with permission

nitude. The correlations of Peng et al. (1994a) for water flow again exhibit a very different trend: in all cases, λRe decreases with increase of the Reynolds number in contradiction to conventional correlations. The detailed data for single-phase flow in micro-channels (experimental conditions, effect of roughness, energy dissipation) will be discussed later.

Heat transfer. Experimental data for laminar and turbulent flow regimes

The experimental studies have shown departure from the conventional theory for heat transfer. Choi et al. (1991) found that the measured Nusselt number in laminar flow exhibits a Reynolds number dependence, in contrast to the conventional prediction for fully developed laminar flow, in which the Nusselt number is constant. The heat transfer at forced convection in channels with cross-section 0.6×0.7 mm was investigated by Peng and Wang (1998), who observed an unusual trend in this dependence of the Nusselt number. Weisberg et al. (1992) and Bowers and Mudawar (1994) also noted that the behavior of fluid flow and heat transfer in micro-channels without phase change is substantially different from that which typically occurs in conventionally sized channels. Wang and Peng (1994) reported single-phase heat transfer coefficients in six rectangular channels having $0.31 < d_{\mathrm{h}} < 0.75$ mm for water and methanol. The Nusselt numbers were only 35% of those predicted by the Dittus–Boelter equation. By contrast, Webb and Zhang

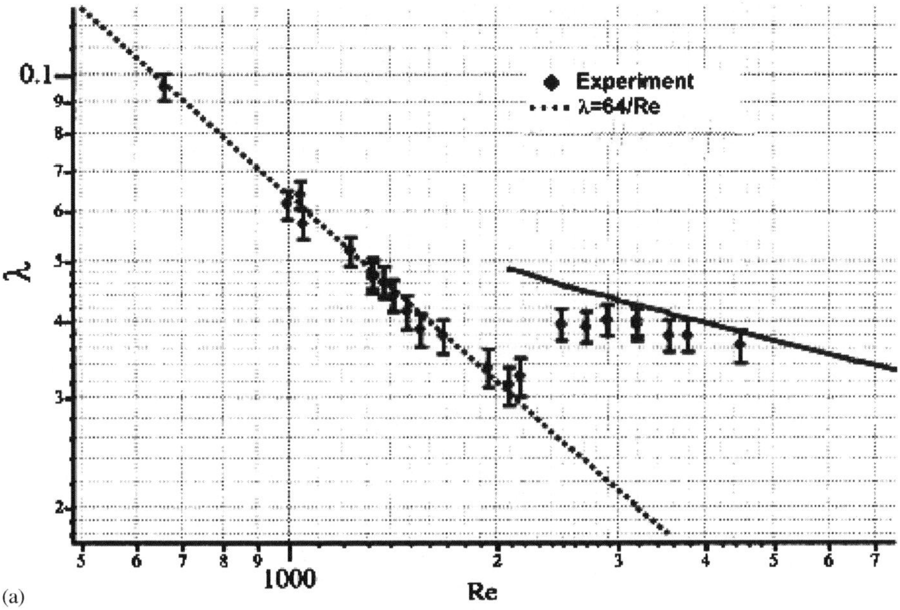

(a)

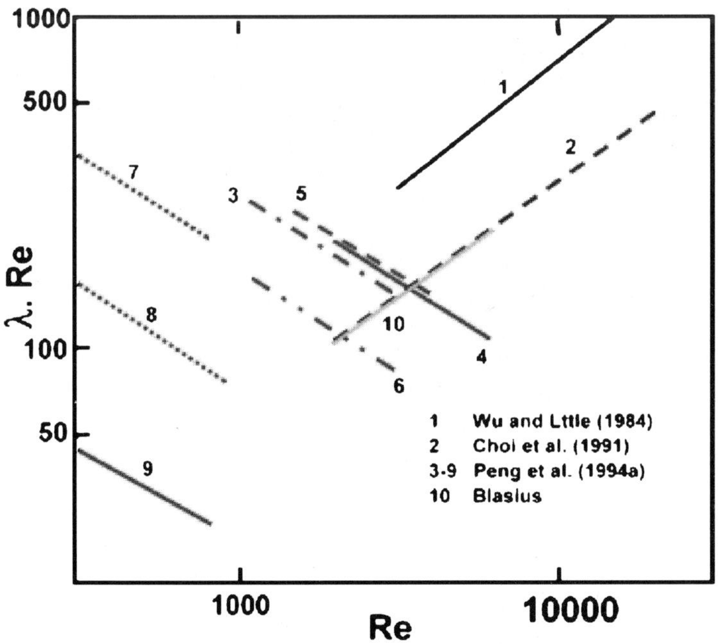

(b)

Fig. 2.20 (**a**) Dependence of the friction factor on Reynolds number for tube of diameter 705 μm. Reprinted from Maynes and Webb (2002) with permission. (**b**) Turbulent flow: friction correlations. Reprinted from Sobhan and Garimella (2001) with permission

(1998) results were adequately predicted by the commonly accepted correlations for single-phase flow in multiple tubes having hydraulic diameters between 0.96 and 2.1 mm. Wu and Little (1984) measured the flow and heat transfer characteristics for the flow of nitrogen gas in heat exchangers. The Nusselt numbers for laminar flow (Re < 600) were lower than those predicted by the correlations. Peng and Peterson (1995) showed a strong effect of the geometric configuration (aspect ratio and ratio of the hydraulic diameter to the center-to-center spacing of the micro-channels) on the heat transfer and flow characteristics in single-phase laminar flow.

The heat transfer correlations are considered separately in the laminar and turbulent regimes in Figs. 2.21 and 2.22, respectively. The dependence of the Nusselt number on the Reynolds number is stronger in all the micro-channel predictions compared to conventional results, as indicated by the steeper slopes of the former; Choi et al. (1991) predict the strongest variation of Nusselt number with Re. The predictions for all cases by Peng et al. (1996) also fall below those for a conventional channel.

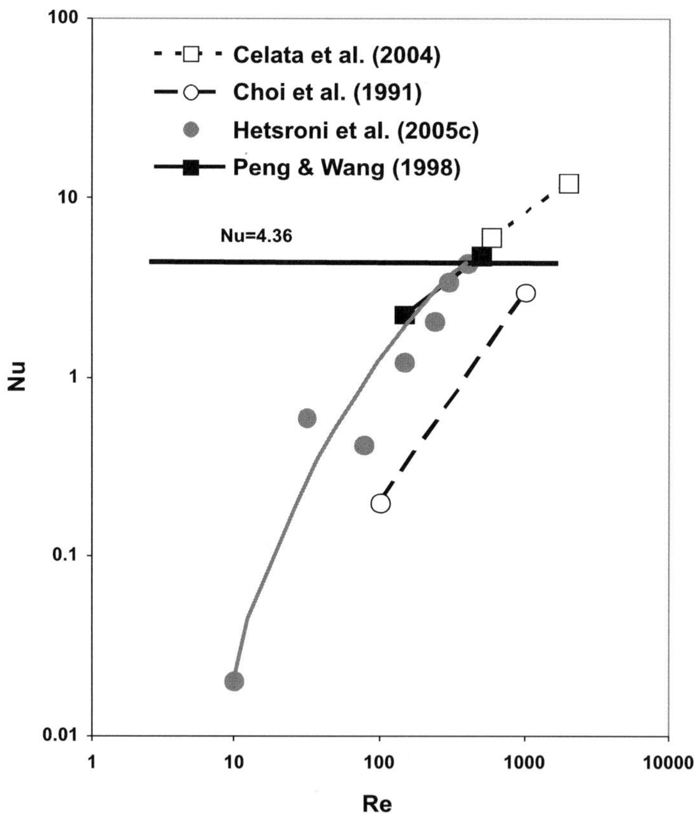

Fig. 2.21 Laminar flow: heat transfer

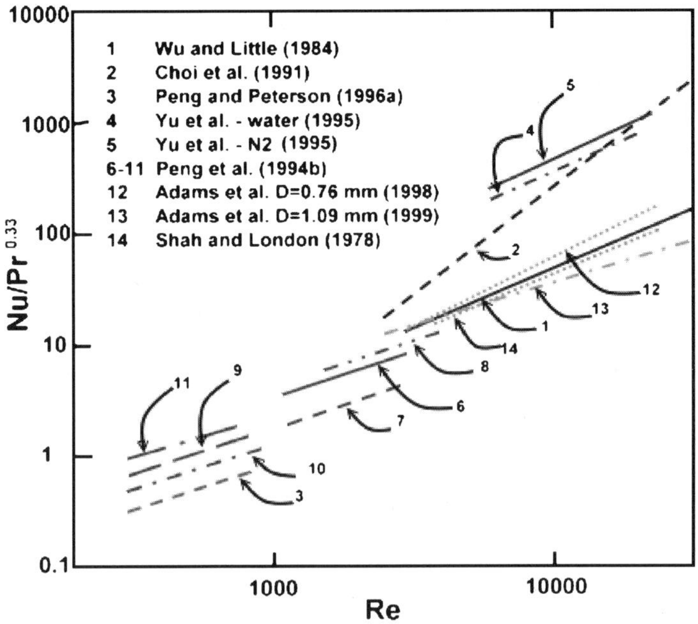

Fig. 2.22 Turbulent flow: heat transfer. Reprinted from Sobhan and Garimella (2001) with permission

Experimental and numerical study of the pressure drop and heat transfer in a single-phase micro-channel heat sink by Qu and Mudawar (2002a,b) demonstrated that the conventional Navier–Stokes and energy equations can adequately predict the fluid flow and heat transfer characteristics.

New questions have arisen in micro-scale flow and heat transfer. The review by Gad-el-Hak (1999) focused on the physical aspect of the breakdown of the Navier–Stokes equations. Mehendale et al. (1999) concluded that since the heat transfer coefficients were based on the inlet and/or outlet fluid temperatures, rather than on the bulk temperatures in almost all studies, comparison of conventional correlations is problematic. Palm (2001) also suggested several possible explanations for the deviations of micro-scale single-phase heat transfer from convectional theory, including surface roughness and entrance effects.

Effect of axial heat conduction in the channel wall

In general, the axial heat conduction in the channel wall, for conventional size channels, can be neglected because the wall is usually very thin compared to the diameter. Shah and London (1978) found that the Nusselt number for developed laminar flow in a circular tube fell between 4.36 and 3.66, corresponding to values for constant heat flux and constant temperature boundary conditions, respectively.

However, for flow in micro-channels, the wall thickness can be of the same order of channel diameter and will affect the heat transfer significantly. For example, Choi et al. (1991) reported that the average Nusselt numbers in micro-channels were much lower than for standard channels and increased with the Reynolds number.

Hetsroni et al. (2004a) obtained similar results for heat transfer in a circular stainless steel tube of inner diameter 1.07 mm and outer diameter 1.50 mm. In the turbulent regime, the predictions of all investigators with the exception of Peng et al. (1994b) and Peng and Peterson (1996) fell below the conventional channel values. In particular, those of Adams et al. (1998, 1999) and Wu and Little (1984) lay in one group. The predictions of Choi et al. (1991) and Yu et al. (1995) were also somewhat comparable, forming a different group. It should be noted that these groups differ neither in fluid type (both groups cases refer to nitrogen and water) nor in micro-channel dimensions. The rectangular micro-channels of different dimensions considered in Peng et al. (1994b) exhibited a wide range of the predicted Nusselt numbers. In all these cases, as well as in that of Peng and Peterson (1996), the predicted values lay below those of the Dittus–Boelter equation presented by Shah and London (1978).

One drawback of a single-phase micro-channel heat exchanger is a relatively high temperature rise longitudinally compared to traditional heat sinks. Furthermore, the fluid flow rate might not be evenly distributed between the parallel micro-channels. Steep temperature gradients produce thermal stresses in the elements and packages due to the differences in the coefficient of thermal expansion, thus impairing the device reliability. Heat transfer in gas flow and thermal effects will be discussed in Chaps. 3 and 4.

Special features of gas flow

In designing micro-scale devices, it is necessary to establish the physical laws governing gas flow in small conduits. Among other things, such flow may differ from its macroscopic counterpart in that relatively high, subsonic Mach numbers may be maintained concurrently with low Reynolds numbers, the surface area to the volume ratio is huge (on the order $10^6\ m^2/m^3$), the fabrication process may lead to relatively heavy surface roughness, and non-continuum effects may occur at pressures above one atmosphere. In a recent review of single-phase flow and heat transfer data at the micro-scale, Guo and Li (2003) discussed the influence of gas rarefaction, and the combined influence of channel size and surface roughness. Their primary conclusion was that the small characteristic length of the micro-channels favors wide variation of the flow properties – pressure, density, and velocity – even in the absence of rarefaction. In viscosity-dominated flows, their compressibility has to be taken into account. It was recognized that there can be significant differences between the inlet and exit Mach numbers and that flow compressibility is likely to influence the friction factor and Nusselt number. Surface effects like friction-induced compressibility, surface roughness, viscous forces and axial heat conduction were attributed to the large surface area to volume ratio of micro-channels, compared to conventional conduits. Finally, it was acknowledged that between experimental results and theoretical values there could be a result of entranced effects and measurement error rather than specific micro-scale phenomena.

Turner et al. (2004) studied the independent variables: relative surface roughness, Knudsen number and Mach number and their influence on the friction factor. The micro-channels were etched into silicon wafers, capped with glass, with hydraulic diameters between 5 and 96 μm. Their surface roughness was $0.002 < k_s <$ 0.06 μm for the smooth channels, and $0.33 < k_s < 1.6$ μm for the glass-capped ones. The surface roughness of the glass micro-channels was measured to be in the range $0.0014 < k_s < 0.003$ μm.

Continuum friction factor and entrance length

The average friction factor for the helium test cases is plotted against the Reynolds number in Fig. 2.23, with designation of the sample ($N = 319\ldots$), the hydraulic diameter d_h, the relative roughness k_s/H, and the Knudsen and Mach number ranges, indicated for each case. For the largest micro-channel, the experimental friction factor agrees with the theoretical value. As d_h decreases, so does the friction factor. This is most noticeable in the case $N = 0.72$ in which Kn = 0.01 for the entire length of the channel.

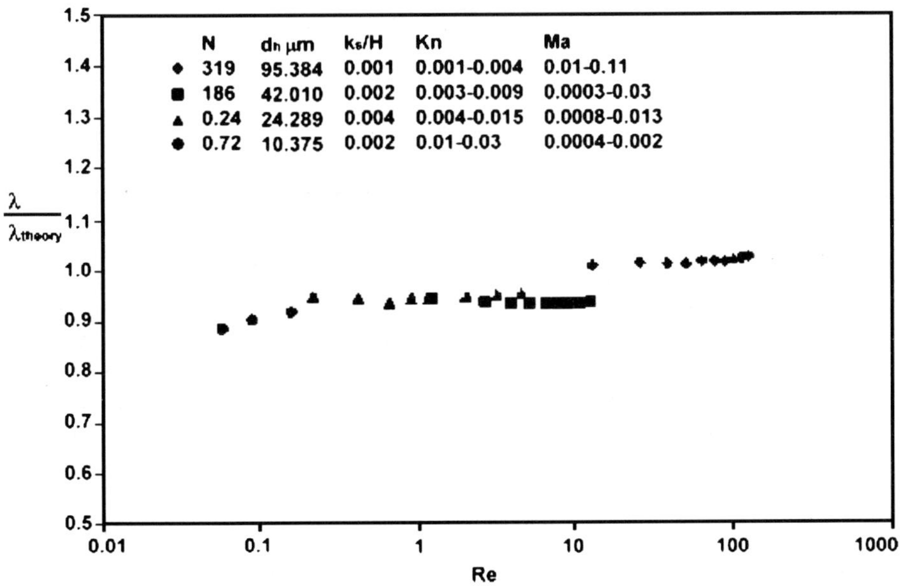

Fig. 2.23 Average friction factor for helium flow plotted against Reynolds number. Reprinted from Turner et al. (2004) with permission

Rarefaction

The influence of rarefaction was isolated from that of compressibility and surface roughness by conducting low-Re, low-Ma flow tests through smooth micro-channels. For a given micro-channel, Kn was varied by testing gases with different mean free paths ($\overline{\lambda}_{He} \approx 3\overline{\lambda}_{air}$) and by reducing the pressure to below atmospheric conditions. The Kn range was also expanded by testing micro-channels of different length. Additionally, since Kn increases along the channel length (as the pressure decreases), the local pressure measurements were used to determine the friction factor for four different values of average Kn along the channel. The experimental λ was normalized versus the incompressible λ_{theory} and is plotted versus Kn in Fig. 2.24, which shows close agreement between λ and λ_{theory} for Kn < 0.01 (continuum flow regime). As Kn was increased above 0.01, λ decreased significantly up to 50% of the continuum value at Kn $= 0.15$.

Compressibility

Experiments were conducted with air through micro-channel $N = 319$ ($d_h = 95.384\,\mu m$) to determine the effect of compressibility on the friction factor. The relative surface roughness was low ($k_s/H = 0.001$) and Kn < 0.001, thus the experiments were effectively isolated from the influence of surface roughness and rarefaction. The local friction factor is plotted versus Ma in Fig. 2.25 for air. The experimental λ increases about 8% above the theoretical λ as Ma increases to 0.35.

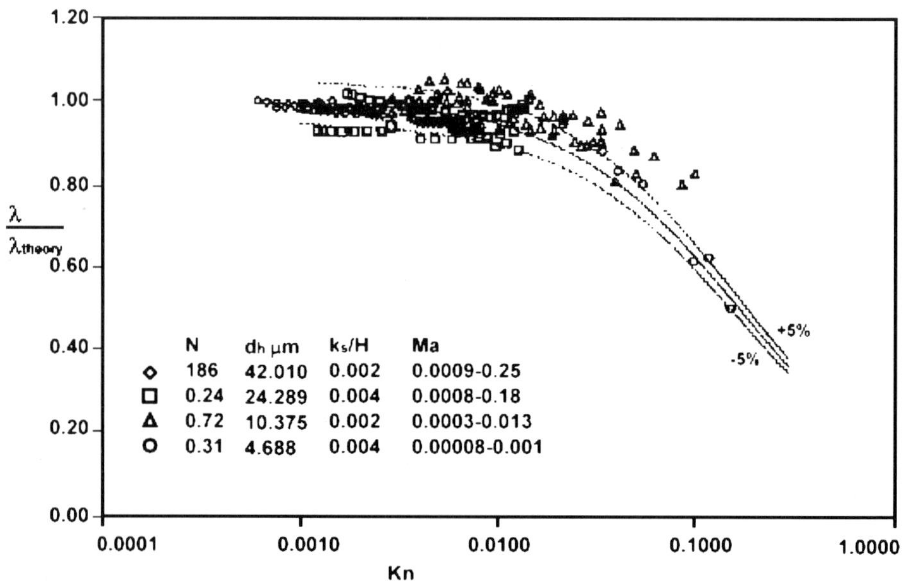

Fig. 2.24 Influence of rarefaction on the local friction factor. Reprinted from Turner et al. (2004) with permission

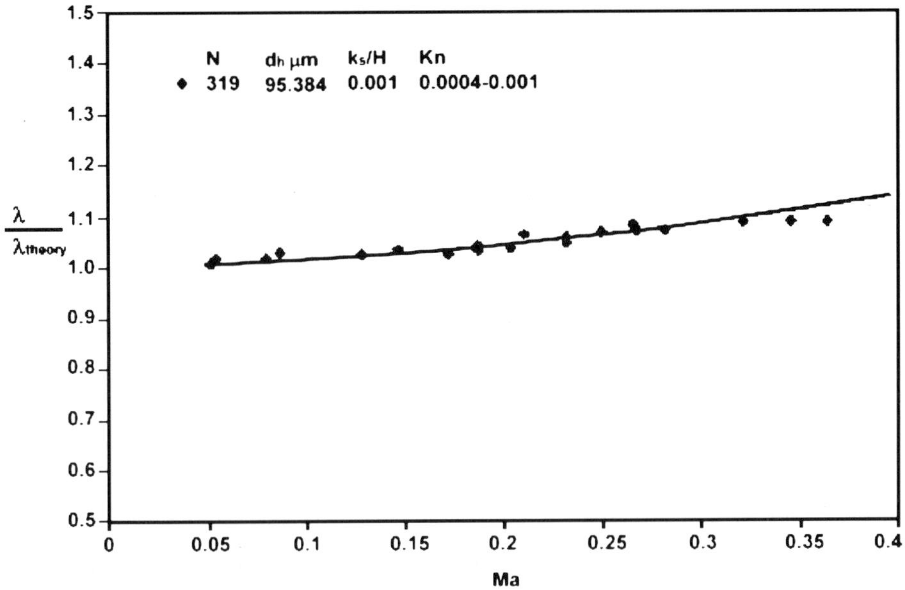

Fig. 2.25 Influence of compressibility on local friction factor for air. Reprinted from Turner et al. (2004) with permission

Celata et al. (2005) studied the gaseous flow of helium through micro-tubes of diameter between 30 and 254 μm and high L/d ratio (between 300 and 2400) at Reynolds numbers up to 500. They concluded that the quantitative behavior is satisfactory predicted by incompressible theory, and that the friction losses at the inlet and outlet give only a minor contribution to the total pressure drop. Morini et al. (2006) investigated the characteristics of gaseous flow of nitrogen in commercial stainless steel micro-tubes of $d = 762$, 558 and 127 μm and $L/d = 800\text{–}3{,}800$ at $\text{Re} = 300\text{–}4{,}000$. It was observed that in laminar regime the Poiseuille low correctly predicts the value of the pressure drop. Laminar to turbulent transition takes place at $\text{Re} = 1{,}800\text{–}1{,}900$.

Surface roughness

The average λ is plotted versus Re for $H = 10\,\mu\text{m}$ channels in Fig. 2.26. In this figure, the difference in λ between the smooth and rough channels is generally less than 3%. To investigate the effect of surface roughness on rarefied gas flow, the smooth and rough channels of $H = 5$ and 10 μm were tested with air and helium at below atmospheric pressure. The test resulted in a Knudsen number range of $0.006 < \text{Kn} < 0.11$, and friction factor λ was normalized by λ_{theory} (Kn = 0) and is plotted versus Kn in Fig. 2.27. In the slip flow regime ($0.01 < \text{Kn} < 0.1$) there is no clear difference in λ for the smooth and roughness cases. Overall, the influence of the relative surface roughness on the friction factor appears to be quite weak

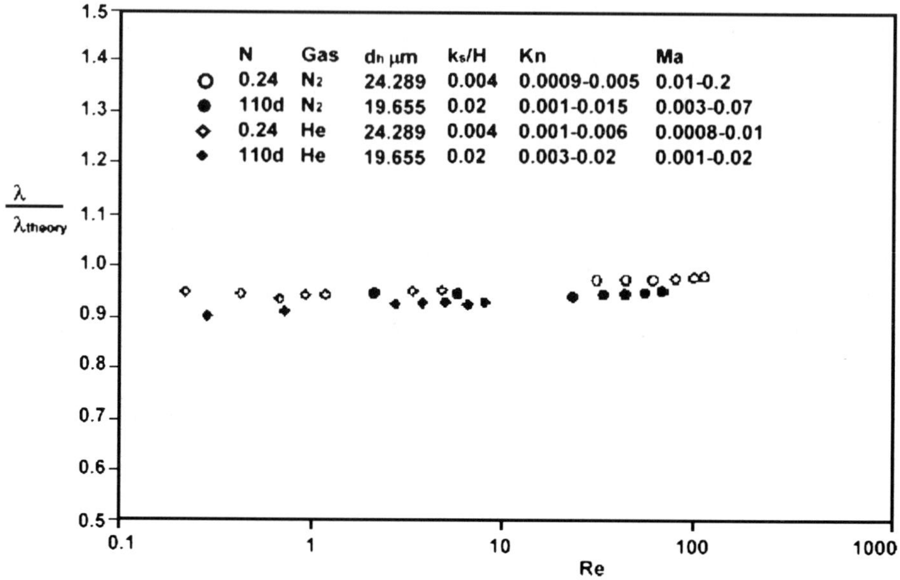

Fig. 2.26 Influence of surface roughness on friction factor. Reprinted from Turner et al. (2004) with permission

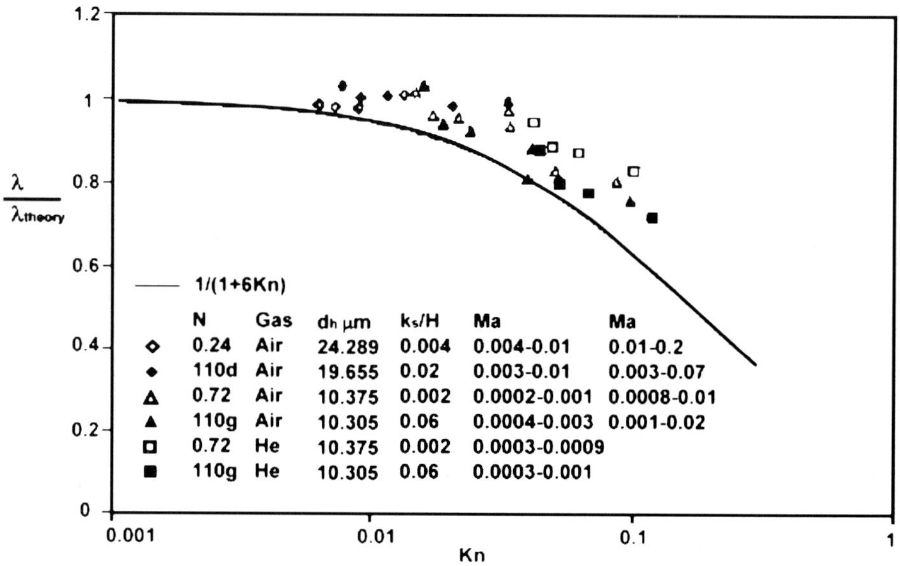

Fig. 2.27 Influence of surface roughness on rarefied flow. Reprinted from Turner et al. (2004) with permission

(within 2 to 6%). The experiments showed that λ for the smooth case is within 3% of that for the rough case, which is also within the experimental uncertainty. Considering that the rough channels have a higher uncertainty associated with the friction factor, there is no statistical difference between the two cases. The friction factor is strongly dependent on the measurement of channel height. Consequently, as its standard deviation increases (as is expected and desired for rough surfaces), so does the uncertainty of the friction factor. To accommodate for this inherent uncertainty, great care is to be taken to ensure uniform channel depth and surface roughness along both the channel width and length.

The above results show close agreement between the experimental and theoretical friction factor (solid line) in the limiting case of the continuum flow regime. The Knudsen number was varied to determine the influence of rarefaction on the friction factor with k_s/H and Ma kept low. The data shows that for $\mathrm{Kn} < 0.01$, the measured friction factor is accurately predicted by the incompressible value. As Kn increased above 0.01, the friction factor was seen to decrease (up to a 50% λ as Kn approached 0.15). The experimental friction factor showed agreement within 5% with the first-order slip velocity model.

The influence of compressibility was assessed by varying the Mach number in the range $0 < \mathrm{Ma} < 0.38$, while Kn and k_s/H were kept low. Friction factor data were reported only with $\mathrm{Ma} < 1$ at the exit, to ensure the flow rate was controlled by viscous forces alone. A mild increase in the friction factor (8%) was observed as Ma approached 0.38. This effect was verified independently by numerical analysis for the same conditions as in the experiment. The range of relative surface roughness tested was $0.001 < k_s/H < 0.06$, yet there was no significant influence on the friction factor for laminar gas flow.

2.4 Steam–Fluid Flow

To achieve higher heat dissipation rates for micro-electronic technologies, the fundamentals of two-phase heat transfer in micro-channels are being studied ever more extensively. Two-phase flow maps and heat transfer prediction methods exist for vaporization in macro-channels, but these are not applicable in micro-channels due to the small-scale phenomena involved, which have to be studied separately. A state-of-the-art overview of boiling and two-phase flows in micro-channels was published by Thome (2006). The most important aspects to control are the frequencies, lengths and velocities of the bubbles and the coalescence processes, which control the flow pattern transitions. To better predict the heat transfer coefficients in micro-cooling elements and heat spreaders for electronics cooling, for example, it is desirable to develop a map for predicting the flow regimes. Due to the predominance of surface tension over the gravity forces, the tube orientation has a negligible influence on the flow pattern. For this reason stratified flow does not exist in horizontal micro-channels (Triplett et al. 1999).

Single micro-channel

Thome et al. (2004) and Dupont et al. (2004) proposed the first mechanistic analysis for vaporization in a micro-channel, with a three-zone flow boiling model describing the transient variation in the local heat transfer coefficient during sequential and cyclic passage of (1) a liquid slug, (2) a vaporizing elongated bubble, and (3) a vapor slug created by film dryout. Figure 2.28 shows this model, in which δ_0 is the initial liquid film thickness, δ_{min} the minimum liquid film thickness, L_L the length of the liquid slug, L_G the length of the vapor bubble, L_p the length of the pair (liquid slug/bubble), L_{dry} the length of the dryout zone, L_{film} the length of the liquid film, q the heat flux and r, d the radius and diameter of the tube. The new model illustrates the strong dependence of heat transfer on the bubble frequency, the lengths of the system elements, and the liquid film thickness, and is so far only applicable to the slug flow regime.

Flow patterns

Serizawa et al. (2002) studied the flow patterns in steam–water flow. Figure 2.29 shows their observations in a 0.05 mm glass channel. Here a new pattern was identified, namely, liquid ring flow.

The micro-scale flow patterns described by Revellin et al. (2006) were categorized as follows:

1. Bubbly flow. Here the bubbles were shorter than the tube diameter and the vapor phase was distributed as discrete bubbles in a continuous liquid phase (Fig. 2.30a).
2. Bubbly/slug flow. Bubbles both longer and shorter than the channel diameter (Fig. 2.30b). The bubble frequencies increase rapidly with the heat flux in the evaporator, reached a peak 900 Hz and then decreased due to coalescence.
3. Slug flow. Vapor bubbles longer than the channel diameter, which is slightly smaller than that of the tube. The bubbles were separated from the inner channel

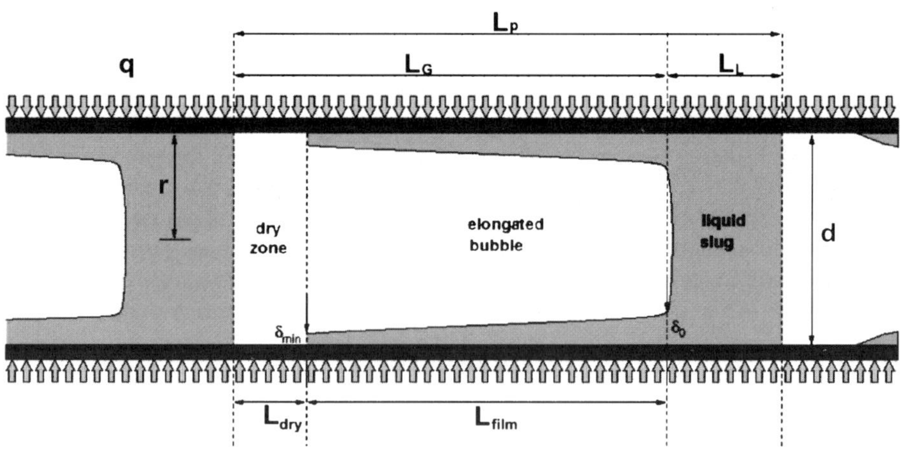

Fig. 2.28 Schematic of three-zone vaporization model. Reprinted from Thome et al. (2004) with permission

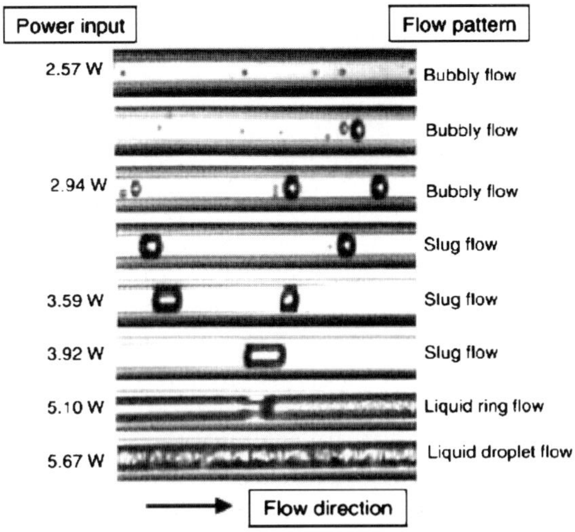

Fig. 2.29 Flow regimes in 50 μm channel for steam–water flows. Reprinted from Serizawa et al. (2002) with permission

wall by a thin film of liquid and from one another by liquid slugs (Fig. 2.30c). The bubble frequency decreased with increasing vapor quality due to coalescence but at a slower rate than in the case of small bubbles.

4. Slug/semi-annular flow. Here both slug and semi-annular flows were present. The vapor velocity increased with the heat flux and the rear of elongated bubbles began to break up (Fig. 2.30d). Coalescence was no longer clean and created a churn-like zone where the liquid slug had been.
5. Semi-annular flow. Liquid slugs were non-existent (Fig. 2.30e). A liquid film formed at the tube wall with a nearly continuous central vapor core, truncated periodically by churning liquid–vapor zones, which disappear gradually.
6. Annular flow (wavy and smooth). A liquid film flowed on the tube wall with a continuous central vapor core without churning zones (Fig. 2.30f,g).

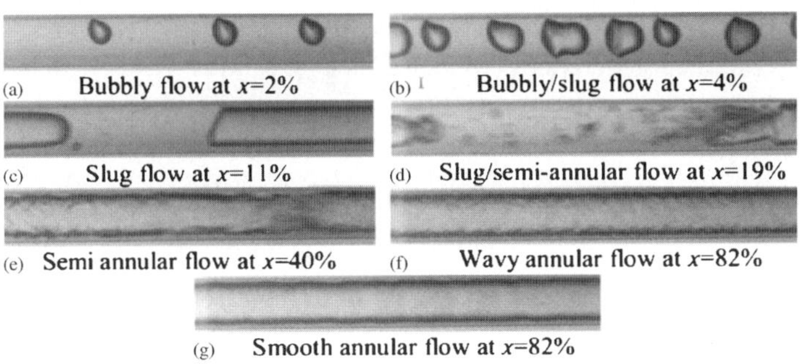

Fig. 2.30a–g Flow patterns and transitions for $d = 0.5$ mm, $L = 70$ mm, $G = 500\ \mathrm{kg/m^2s}$, $T_s = 30\ °\mathrm{C}$ and $\Delta T_{sub} = 3$ K. Reprinted from Revellin et al. (2006) with permission

Flow pattern maps

Two-phase flow pattern maps, observed by Revellin et al. (2006), are presented in Fig. 2.31 in mass flux versus vapor quality, and superficial liquid velocity versus superficial vapor velocity formats calculated from the test results as follows:

$$U_{\mathrm{LS}} = \frac{(1-\chi)G}{\rho_{\mathrm{L}}} \tag{2.13}$$

$$U_{\mathrm{GS}} = \frac{\chi G}{\rho_{\mathrm{G}}} \tag{2.14}$$

where U_{LS} and U_{GS} are the liquid and vapor superficial velocities, respectively, and χ is the vapor quality.

Notably, the higher the mass flux, the earlier annular flow is reached. Bubbly flow is more or less non-existent for mass fluxes exceeding 1,000 $\mathrm{kg/m^2\,s}$. The most important observation about the flow patterns is that their transitions are controlled primarily by the rate of coalescence, which is not recognized as a contributing factor by any of the micro-scale or macro-scale flow pattern maps.

Coalescing flow map for vaporizing flows in micro-channels

In order to create phenomenological heat transfer and frictional pressure drop models as opposed to wholly empirical methods, the two-phase flow characteristics of each type of flow need to be discernible. For example, the Thome et al. (2004) three-zone flow boiling model (Fig. 2.28) is currently applicable to slug (elongated bubble) flow and includes the effect of bubble frequency, but only assuming that no coalescence occurs. For further development of such models, it is clear that coalescence must be dealt with directly to account for the variation of the heat transfer coefficient due to the decrease in bubble frequency and increase in bubble length as the coalescence proceeds along the channel. For distinct, high-frequency elongated bubbles, the heat transfer coefficient is controlled primarily by transient conduction through the liquid film surrounding them as against convection through the liquid films in the annular flow regime. To extend the three-zone heat transfer model in the elongated bubble regime and then generalize it to include the heat transfer mechanisms controlling bubbly and annular flows, a composite coalescing flow and diabatic flow type of map is required that reflects both the bubble coalescence phenomena and the critical heat flux limitation. Furthermore, as can be observed in Fig. 2.31, conventional flow pattern maps do not provide the necessary information for development of such models here, since coalescence changes the flow pattern along the channel length and the critical heat flux limits the feasible combinations of mass velocity and vapor quality.

An example of a diabatic coalescing bubble map for vaporizing flow in microchannels is shown Fig. 2.32 (Thome et al. 2006) where Bo is the boiling number, $\mathrm{Bo} = q/Gh_{\mathrm{LG}}$, G is the mass velocity of liquid plus vapor ($\mathrm{kg/m^2\,s}$), h_{LG} is the latent heat of vaporization (J/kg), q_{crit} is the critical heat flux ($\mathrm{W/m^2}$), $\mathrm{Re_L}$ is the

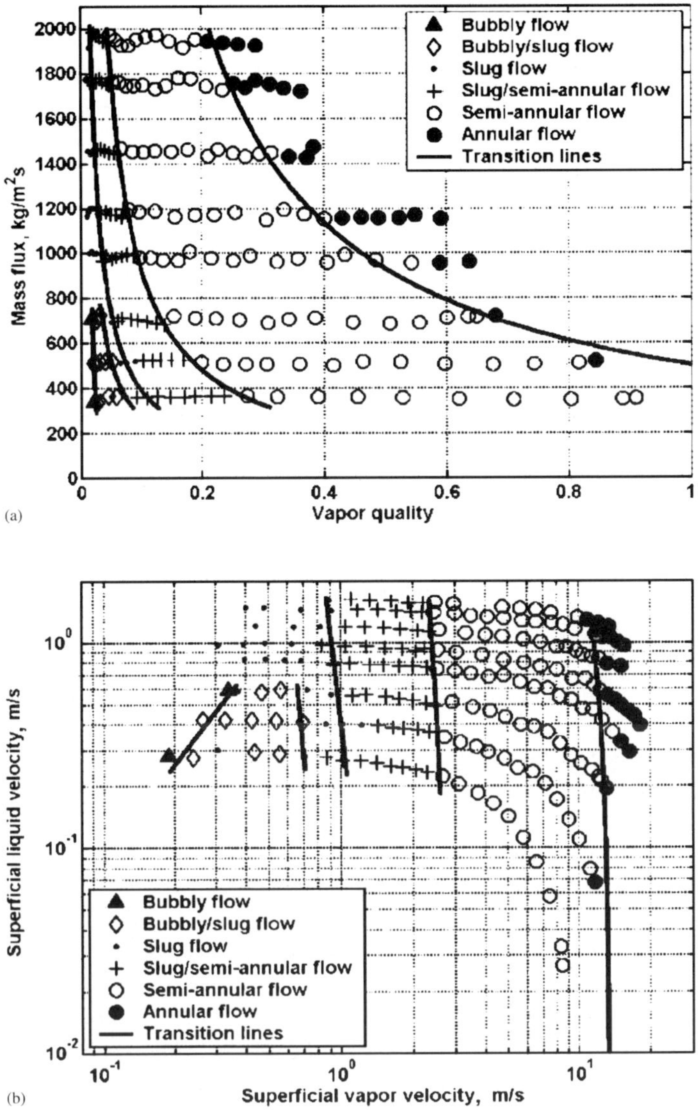

Fig. 2.31a,b Flow pattern maps with experimental transition lines for R-134a, $d = 0.5$ mm, $L = 70$ mm, $T_s = 30\,°C$ and $\Delta T_{sub} = 3$ K. Reprinted from Revellin et al. (2006) with permission

liquid Reynolds number, $Re_L = Gd/\mu_L$, We_G is the vapor Weber number, $We_G = G^2 d/(\sigma\rho_G)$, and We_L is the liquid Weber number, $We_L = G^2 d/(\sigma\rho_L)$.

The first flow pattern zone corresponds to the isolated bubble (IB) regime where the bubble generation rate is much higher than the coalescence rate. It includes both bubbly flow and/or slug flows and is present up to the onset of coalescence process domination. The second zone is the coalescing bubble (CB) regime, which is

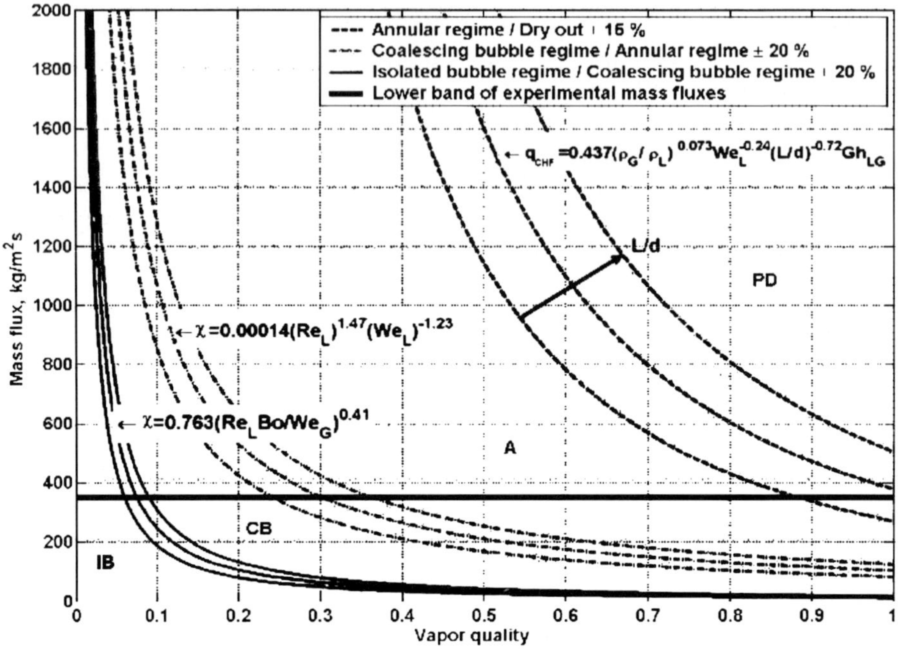

Fig. 2.32 Diabatic flow pattern map for vaporizing flow in uniformly heated micro-channel, R-134a, $d = 0.5$ mm, $L = 70$ mm, $T_s = 30\,°\mathrm{C}$, $q = 50\ \mathrm{kW/m^2}$ without subcooling at inlet. Flow patterns: isolated bubble regime (IB), coalescing bubble regime (CB), annular (completely coalesced) regime (A), post-dryout regime (PD). Reprinted from Thome et al. (2006) with permission

present up to the end of the coalescence process. This zone is present when the bubble coalescence rate is higher than the bubble generation rate. The third zone is the annular (A) zone and is limited by the fourth zone of this diabatic map, the post-dryout (PD) zone, whose transition is given by the critical vapor quality signaling the onset of critical heat flux. The transition lines in Fig. 2.32 are predicted by the methods discussed below (notably also showing the accuracy limits of the new transition predictions that are absent in prior conventional flow maps). The lower end of the transition lines below the horizontal black line represents extrapolation below the lowest mass velocities tested, where two-phase flow instabilities occur. Thus the present map does not specifically capture the threshold of two-phase instability at low mass velocities, but this feature remains as a possibility in future tests. Such a diabatic/coalescing flow pattern map with instability limits would provide a complete picture of the physical process for the thermal/hydraulic design of micro-channel evaporators.

Before describing the transition prediction methods, it is instructive to describe how a diabatic map is used. One chooses a desired mass flux and sets the heat flux to be dissipated (assumed uniform along and around the circular channel) up to the desired local length from the inlet to find the corresponding local vapor quality (from an energy balance) and thus the location of this process condition on the map. The inlet is for a saturated or slightly subcooled liquid ($\chi \approx 0$) and the exit conditions are applied to find the location (at the extreme right), and the resulting horizontal line joining these points represents the flow regimes that will be encountered from inlet to outlet. Significantly, the location of the outlet to the left or right of the dryout transition curve (indicating the vapor quality at which the critical heat flux is reached) shows whether the intended operating conditions are feasible or not. Saturation conditions and the physical properties at the inlet pressure of the micro-evaporator are used for calculating the map.

Parallel micro-channels

The behavior of the flow pattern in a parallel micro-channel is different from that in a single micro-channel. It was shown by Hetsroni et al. (2003b) that at the same value of heat flux, different flow regimes may be observed in different micro-channels, depending on the time interval. Moreover, at the same time interval different flow regimes may exist in each of the component micro-channels. In Fig. 2.33 two-phase steam–water flow in the central part of such a parallel system $d_{\mathrm{h}} = 100\,\mu\mathrm{m}$ is shown as the top view observed through a transparent cover. The

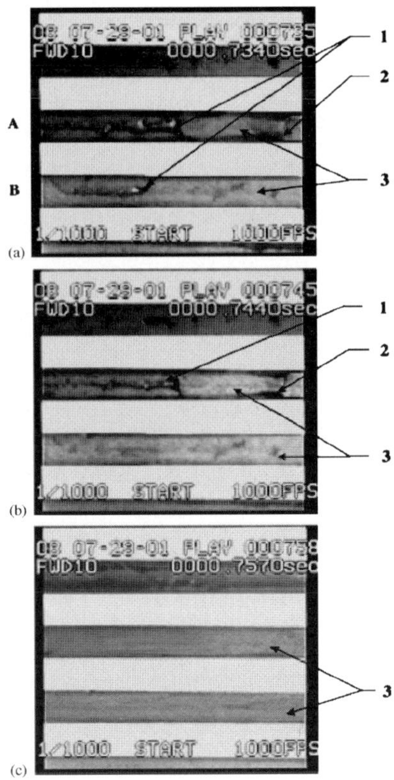

Fig. 2.33a–c Boiling in the central part of micro-channels. $U_{\mathrm{LS}} = 0.14$ m/s, $q = 220$ kW/m^2. *1* Clusters of liquid droplets at the bottom of the channel. *2* Clusters of the liquid droplets on the side-wall. *3* Steam. Reprinted from Hetsroni et al. (2003b) with permission

field of view is 2.4 mm in the streamwise direction and 2.0 mm in the spanwise direction, the flow moves from left to right. In these images three micro-channels are shown, marked by uniform gray color or by gray color with light and dark regions. In channels A and B, clusters of liquid droplets were observed (Fig. 2.33a). One of the findings of the study is the extreme shortness of the period during which the "vapor with clusters of droplets" regime was observed. The clusters may appear at the corner of the bottom as on the side-walls (Fig. 2.33a, channel A). After 0.01 s, liquid drops were not observed on the bottom of channel B (Fig. 2.33b). Figure 2.33c shows no presence of liquid droplets in both channels.

The temperature distribution on the heated wall depends on the material and design of the module, the inflow rate and the heat flux. For given values of flow rate and heat flux, the infrared (IR) image of the heated module side was clearly observed. Typical IR images of the heated side of the module, which contained 26 micro-channels of $d_h = 100\,\mu m$, are shown in Figs. 2.34 and 2.35. The area of the heater is marked as a square, the flow moved from right to left. We restricted the thermal image analysis to the marked square area of $10 \times 10\,mm^2$. The temperature distribution on the heated wall and the histogram of the thermal field at $q = 190\,kW/m^2$ and $U_{LS} = 0.14\,m/s$ are shown in Fig. 2.34a,b, respectively. In this case the mean temperature (averaged over the surface of the heater) was $T_{mean} = 93.3\,°C$ and boiling was observed in part of the parallel micro-channels only. The temperature measurements indicate that this "partial" boiling took place at $T_{mean} < T_S$ (T_S being the satu-

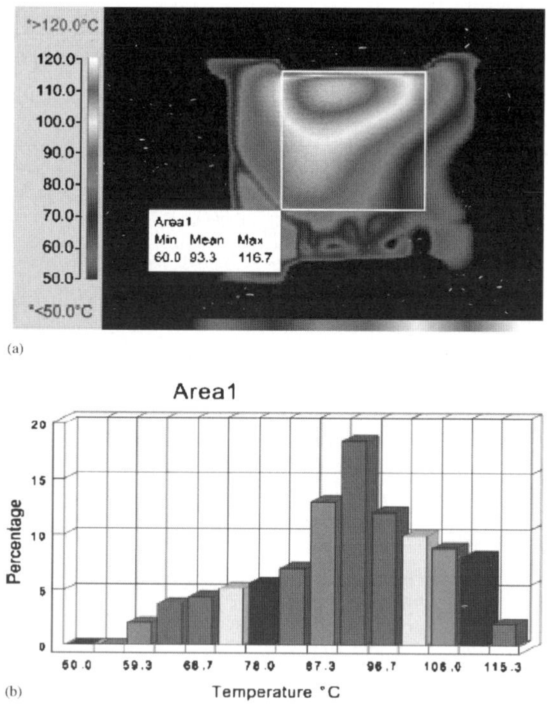

Fig. 2.34a,b Thermal field on the heated wall. $U_{LS} = 0.14\,m/s$, $q = 190\,kW/m^2$. (**a**) Infrared image. (**b**) Histogram of temperature distribution. Reprinted from Hetsroni et al. (2003b) with permission

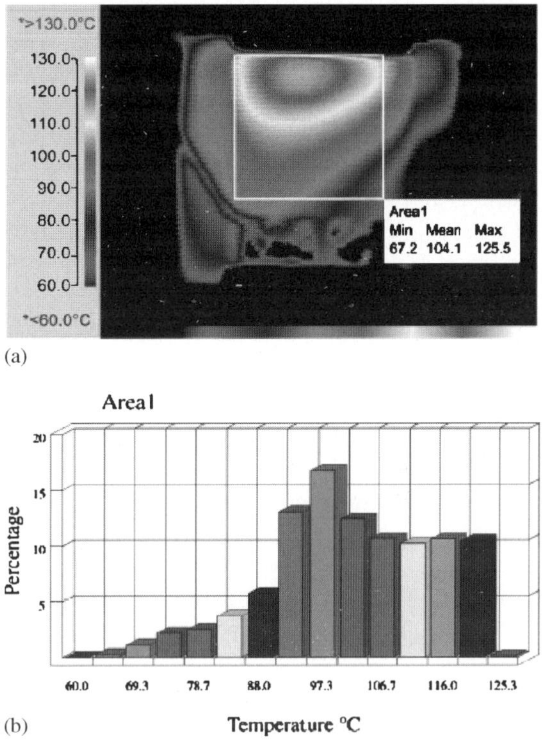

Fig. 2.35a,b Thermal field on the heated wall. $U_{LS} = 0.14\,\mathrm{m/s}$, $q = 220\,\mathrm{kW/m^2}$. (**a**) Infrared image. (**b**) Histogram of temperature distribution. Reprinted from Hetsroni et al. (2003b) with permission

ration temperature), whereas at $T_{mean} > T_S$ boiling in all parallel micro-channels was observed. The temperature distribution on the heated wall and the histogram of the thermal field at $T_{mean} = 104.1\,°\mathrm{C}$, $q = 220\,\mathrm{kW/m^2}$ and $U_{LS} = 0.14\,\mathrm{m/s}$ are shown in Fig. 2.35a,b, respectively. From Figs. 2.34 and 2.35 one can conclude that the variation of $(T_{max} - T_S)$ is more than 20 K, where T_{max} is the maximum temperature on the heated surface. It is clear that consideration of boiling curve and flow regimes in parallel micro-channels should take into account such wide variation of the wall temperature.

Boiling regimes

In Fig. 2.36 the dimensionless heat flux q^* is plotted versus the difference $(T_{max} - T_S)$. The dimensionless heat flux is defined as $q^* = [q/(Gc_p\Delta T)](T_{max}/T_{mean})$, where q is the heat flux, G is the mass flux, c_p is the specific heat, $\Delta T = T_S - T_{in}$, T_S is the saturation temperature, T_{in} is the water temperature supplied to the inlet collector, and T_{max} and T_{mean} are maximum and average temperature of the heated surface, respectively. The term $q/(Gc_p\Delta T)$ reflects the effect of subcooling.

In physical objects involving thermal and fluid flow systems, the dual problem of how the heater is represented, and boiling as a local-instantaneous effect, should be considered. The temperature variations on the chip surface are a key characteristic of

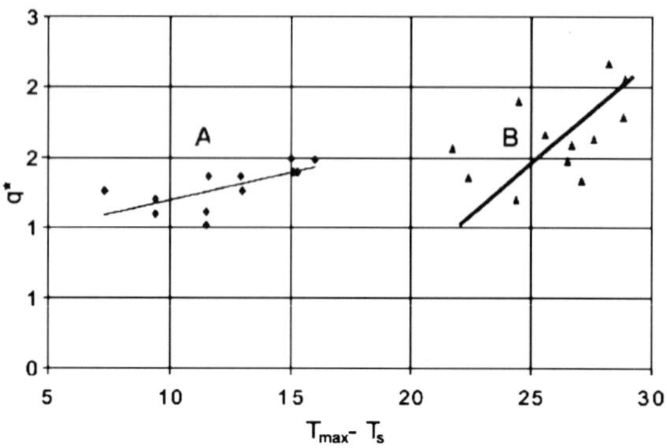

Fig. 2.36 Flow regime map for parallel micro-channels. Region A is the low-heat flux region, and region B the high-heat flux region. Reprinted from Hetsroni et al. (2003b) with permission

such a problem. In Fig. 2.36 two regions of convective boiling may be distinguished. In region A the value of $(T_{max} - T_S) < 20\,K$ and boiling was observed in part of the parallel micro-channels only. It may be referred to as the low-heat flux region. Region B, shown in Fig. 2.36, may be referred to as the high-heat flux region; in this case $(T_{max} - T_S)$ exceeds 20 K.

It should be noted that these results were obtained under conditions of intermittent dryout. The physical implication of such a phenomenon will be discussed in the next section.

Intermittent dryout

Two-phase micro-channel heat sinks generally involve flow boiling in straight, constant cross-section channels. Flow boiling in parallel channels was studied from the subcooled liquid entry at the inlet to a liquid–vapor mixture flow at the outlet. Once nucleation begins, the heat flux causes a sudden release of energy into the vapor bubble, which grows rapidly and occupies the entire channel. The vapor slug may be considered as an elongated bubble. The rapid bubble growth pushes the liquid–vapor interface on both caps of the vapor slug at the upstream and the downstream ends and leads to reversed flow. When in some parallel channels the liquid on the upstream side is pushed back, the other parallel channels carry the resulting excess flow. Figure 2.37 illustrates two-phase flow, driven by the pump, in a certain part of the parallel micro-channels of $d_h = 130\,\mu m$ (Hetsroni et al. 2005b). It is the top view and the flow pattern was observed through the transparent cover. The field of view is 8 mm in the streamwise direction and 4 mm in the spanwise direction, the flow moves from the bottom to the top, the mass flux is $G = 95\,kg/m^2\,s$, and the heat flux is $q = 160\,kW/m^2$. The vapor (the white regions in Fig. 2.37) may be observed at different distances from the inlet. The successive images obtained, using high-speed video, showed that the flow patterns were periodic. The liquid front was seen to pass periodically the region located near the inlet manifold, and the vapor phase occupied most of the channel core; the periodic wetting and rewetting phenomena were also observed. The behavior of long vapor bubbles in a micro-channel was not similar to annular flow with intermittent slugs of liquid between two long vapor

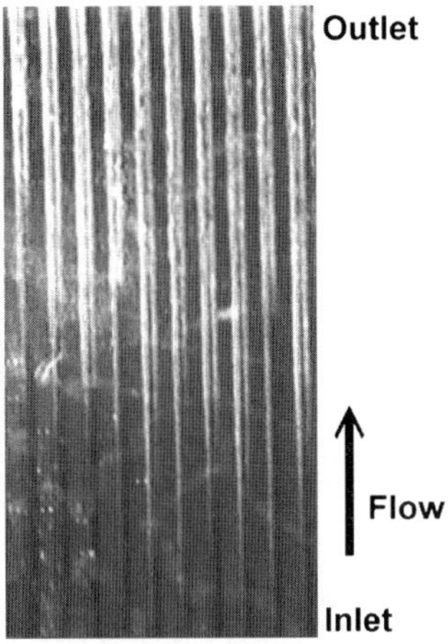

Fig. 2.37 Flow boiling in parallel micro-channels. $G = 95\,\mathrm{kg/m^2 s}$, $q = 160\,\mathrm{kW/m^2}$. Reprinted from Hetsroni et al. (2005b) with permission

trains. The periodic phenomena described above may be regarded as explosive boiling. The trigger mechanism of such a regime is venting of the elongated bubble due to very rapid expansion. Downstream of the onset of nucleate boiling (ONB point) the region of the venting of elongated bubble and dryout was observed.

Bubble growth in micro-channel

The location (along the channel) of appearance of the first bubble at a given flow rate, depends on the local surface temperature variation. Since the heated length of the test section is short (10 mm), at high values of heat flux significant interaction takes place between the steam generated in the micro-channel and the feed water. It is very difficult to capture the behavior of the elongated bubble at high values of heat flux. We investigated this phenomenon at mass flux $G = 95\,\mathrm{kg/m^2\,s}$ and heat flux $q = 80\,\mathrm{kW/m^2}$. Figure 2.38a–h shows a sequence of bubble formation in one of the parallel triangular channels of $d_\mathrm{h} = 130\,\mu\mathrm{m}$. Visualization of the flow pattern was realized near the ONB point. The incipience of boiling is a local phenomenon, which strongly depends on both the hydrodynamic and thermal conditions. The onset of nucleate boiling was established when the first bubbles were observed and the channel wall exceeded the saturation temperature. In general, at a given flow rate the ONB location moved upstream as the heat flux increased, as in ordinary sized channels. The main flow moved from the bottom to the top. Figure 2.38a shows a single bubble, which begins to grow along the channel. Figure 2.38b–d illustrate the bubble growth in the axial direction. Note that at this stage the bubble expands preferentially in the streamwise direction. Figure 2.38e,f show venting of the elongated bubble. Figure 2.38g,h show rewetting. The life time of the elongated bubble (the time between images shown in Fig. 2.38b and Fig. 2.38e) is about 0.01 s. At a given mass flow rate the life time sharply decreases with increasing heat flux. This is why under certain conditions (periodic flow reversal) we consider the mechanism of saturated boiling in micro-channels as an explosive boiling process. Het-

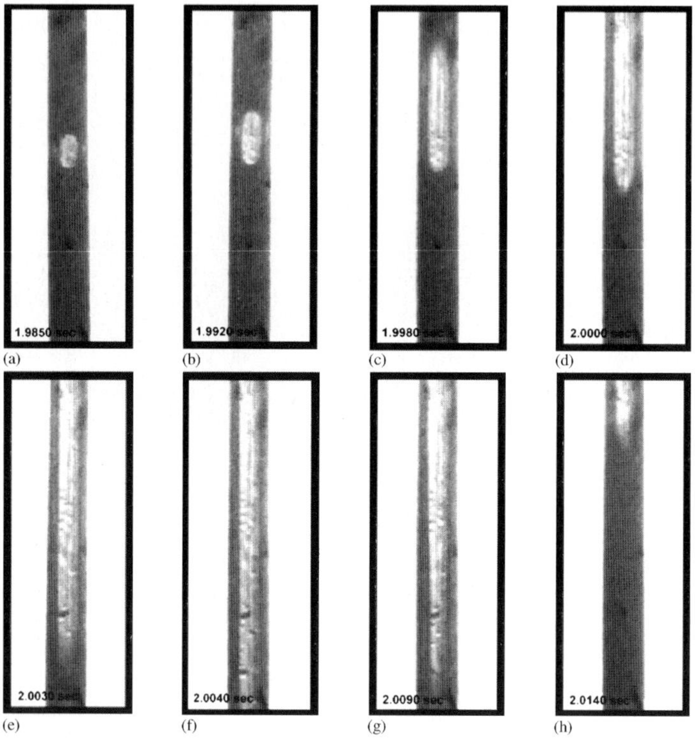

Fig. 2.38a–h Bubble growth at low heat flux. $G = 95\,\mathrm{kg/m^2 s}$, $q = 80\,\mathrm{kW/m^2}$. Reprinted from Hetsroni et al. (2003b) with permission

sroni et al. (2001a, 2002a) observed rapid flow reversal inside an individual channel in parallel triangular micro-channels; Steinke and Kandlikar (2003) reported reverse displacement in a set of six parallel micro-channels, $d_h = 200\,\mu\mathrm{m}$. The system departs from stable operating conditions, and hydrodynamic instability sets in. This boiling mode leads to liquid–vapor alternating flow in the region located near the ONB.

Region along the channel after venting of an elongated bubble

After venting of the elongated bubble, the region of "liquid droplets" begins. The vapor phase occupies most of the channel core. The distinctive feature of this region is the periodic dryout and wetting phenomenon. The duration of the two-phase period, i.e., the presence of a vapor phase and micro-droplet clusters on the heated wall, affects the wall temperature and heat transfer in micro-channels. As the heat flux increases, while other experimental conditions remain unchanged, the duration of the two-phase period decreases, and CHF is closer.

Figure 2.39a–o illustrates a typical example of alternate two-phase flow patterns at a distance of 1,000–1,500 μm downstream from the inlet of the test section. In

Fig. 2.39a–o a single triangular channel of $d_h = 130\,\mu m$ and the walls between the adjacent channels (white strips) are depicted in the central part and in the peripheral parts of each image, respectively. The field of view (including the channel and the walls) is $0.6 \times 0.6\,mm^2$, the flow moves from the bottom to the top, the mass flux is $G = 95\,kg/m^2\,s$, and the heat flux is $q = 160\,kW/m^2$. Figure 2.39a shows

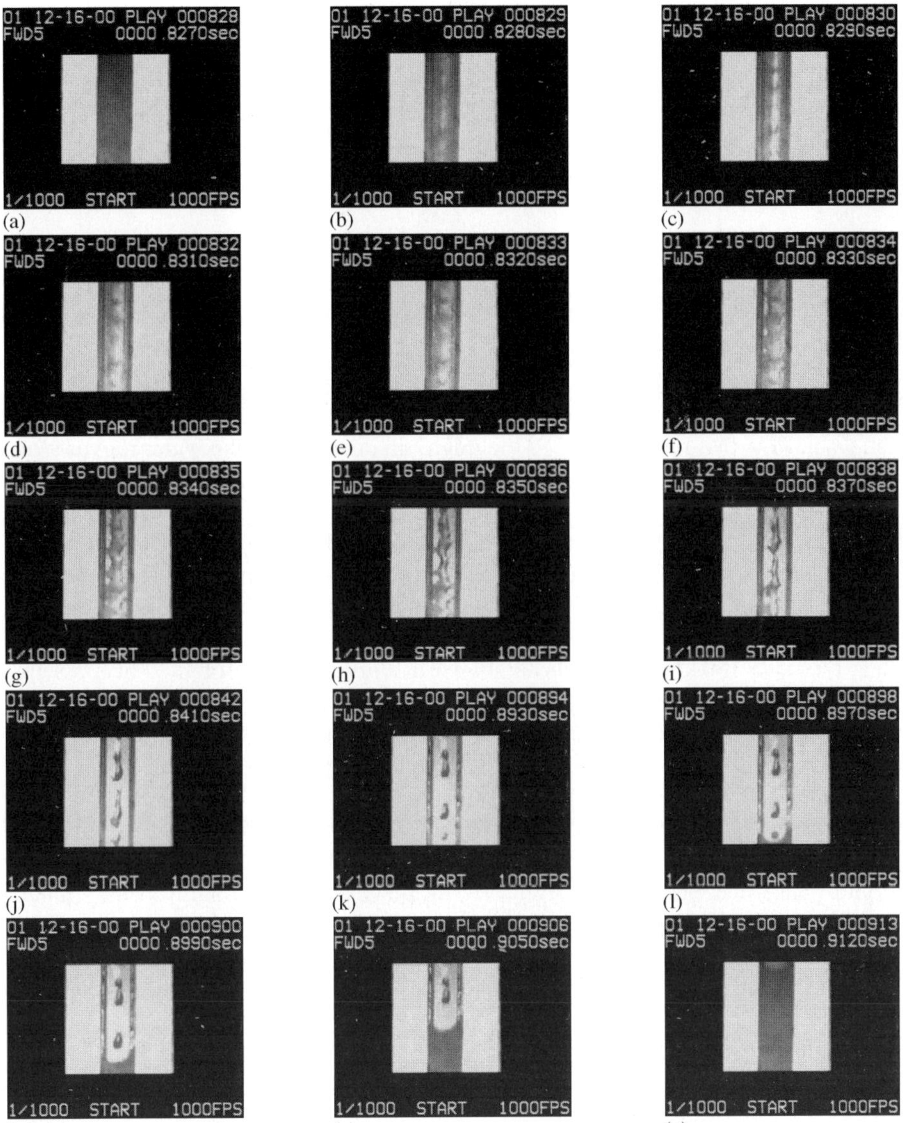

Fig. 2.39a–o Flow pattern at high heat flux. $G = 95\,kg/m^2 s$, $q = 160\,kW/m^2$. Reprinted from Hetsroni et al. (2005b) with permission

Fig. 2.40a,b Flow pattern upstream of the ONB. (**a**) Single-phase water flow. (**b**) Jet that penetrates the bulk of the water. Reprinted from Hetsroni et al. (2006b) with permission

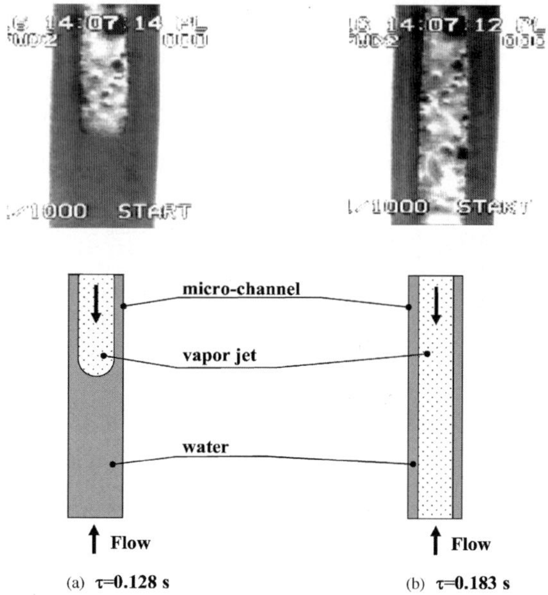

the water flow that moved through the micro-channel. When the water moves along the channel it is heated up to saturation temperature and then the elongated bubble vents in a short time. Figure 2.39b displays the onset of two-phase reversal flow, which evolved due to venting of the elongated bubble downstream from the observation point. From Fig. 2.39a,b one can conclude that at the specified conditions of this experiment, the life time of the elongated bubble did not exceed 0.001 s. Comparison between the present results obtained at mass flux $G = 95\,\mathrm{kg/m^2\,s}$ and heat flux $q = 160\,\mathrm{kW/m^2}$ to those shown in Fig. 2.38 and obtained at mass flux $G = 95\,\mathrm{kg/m^2 s}$ and heat flux $q = 80\,\mathrm{kW/m^2}$ indicates that doubling of the heat flux causes an approximately tenfold decrease of the life time of the elongated bubble.

Figure 2.39c–f show the annular flow, where the liquid film on the wall is symmetrically distributed.

Figure 2.39g,h display the onset of dryout, where liquid droplets are accumulated at the bottom of the triangular channel. Figure 2.39i–k show that the amount of liquid phase decreases, and only a few droplets of water remain on the channel bottom. Figure 2.39l shows the beginning of wetting and the channel is filled with liquid (Fig. 2.39m–o). The flow patterns depicted in Fig. 2.39a–o were also observed by Serizawa et al. (2002) in steam–water flow in a 50 μm silicon tube. The authors reported that at low liquid flow rates, partially continuous liquid film flow changed to rivulets or even to discrete liquid lumps or large liquid droplets.

The periodic phenomenon described above indicates that the entire channel acts like the area beneath a growing bubble, going through periodic drying and rewetting. The cycle was repetitive with venting of the elongated bubble. Such a behavior affects the mean flow characteristics that usually are measured at the manifolds.

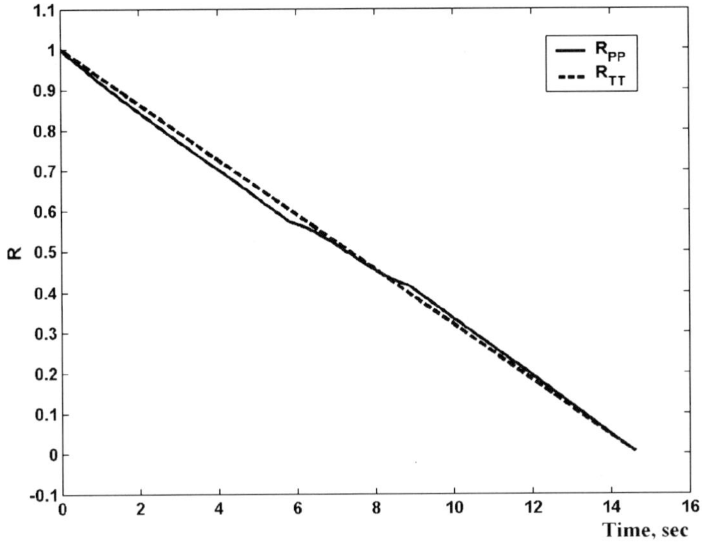

Fig. 2.41 Auto-correlation functions for pressure and temperature fluctuations. Reprinted from Hetsroni et al. (2002b) with permission

Figure 2.40 shows the unsteady flow upstream of the ONB in one of the parallel micro-channels of $d_\mathrm{h} = 130\,\mu\mathrm{m}$ at $q = 228\,\mathrm{kW/m^2}$, $m = 0.044\,\mathrm{g/s}$ (Hetsroni et al. 2001b). In this part of the micro-channel single-phase water flow was mainly observed. Clusters of water appeared as a jet, penetrating the bulk of the water (Fig. 2.40a). The vapor jet moved in the upstream direction, and the space that it occupied increased (Fig. 2.40b). In Fig. 2.40a,b the flow moved from bottom to top. These pictures were obtained at the same part of the micro-channel but not simultaneously. The time interval between events shown in Fig. 2.40a and Fig. 2.40b is 0.055 s. As a result, the vapor accumulated in the inlet plenum and led to increased inlet temperature and to increased temperature and pressure fluctuations.

The auto-correlation functions for the pressure R_pp and temperature R_TT fluctuations are presented in Fig. 2.41. It is clear that the temporal behavior of the temperature fluctuations corresponds to that of the pressure fluctuations (Hetsroni et al. 2002b).

As the heat flux increases, while other experimental conditions remain unchanged, the duration of the two-phase period decreases, and the critical heat flux is closer.

Critical heat flux

As boiling in micro-channel heat sinks is an attractive method for cooling computer CPUs and other high-heat flux devices (such as laser diodes), it is of crucial importance to accurately predict the critical heat flux (CHF) in the small-diameter channels. Critical heat flux or burnout is a limiting value for safe operation of heat dissipation applications and refers to replacement of liquid in contact with the heated surface with a vapor blanket. The thermal conductivity of the vapor is very low compared to the liquid and the surface heat transfer coefficient drops dramatically, resulting in sudden increase of the surface temperature and possible failure of the cooled device. CHF may occur in natural convection boiling, subcooled, as well as in low quality and high-quality saturated boiling conditions.

Natural convection boiling

The effect of annular channel size and length on the CHF was studied under conditions of natural convection boiling by Hetsroni et al. (2007). Restriction of the boiling space was achieved by using unheated glass tubes of different inner diameter, from 2.5 to 9 mm, and length of 24 and 36 mm (typical of micro-channel heat exchangers). The generated vapor rose up through the annular channel between the heated stainless steel tube and the glass tube to the free interface and condensed in a water-cooled condenser.

Data were taken for both increasing and decreasing heat fluxes. Yao and Chang (1983) showed that the squeezing effect of a bubble due to the confinement may be described by the Bond number. For water boiling in gap sizes of $\delta = 0.45$, 1.2, 2.2, and 3.7 mm, the Bond numbers, $\mathrm{Bn} = \delta(\sigma/g(\rho_L - \rho_G))^{-0.5}$, were 0.185, 0.493, 0.9 and 1.52, respectively, where δ is the gap size, σ is the surface tension, g is the acceleration due to gravity, and ρ_L and ρ_G are the liquid and vapor densities.

Effect of channel size

Figure 2.42 shows boiling curves obtained in an annular channel with length 24 mm and different gap size (Bond numbers). The heat flux q is plotted versus the wall excess temperature $\Delta T = T_W - T_S$ (the natural convection data are not shown). The horizontal arrows indicate the critical heat flux. In these experiments we did not observe any signs of hysteresis. The wall excess temperature was reduced as the Bond number (gap size) decreased. One can see that the bubbles grew in the narrow channel, and the liquid layer between the wall and the base of the bubble was enlarged. It facilitates evaporation and increases latent heat transfer.

For low Bond numbers (on the order of or less), the squeezing effect is important since bubbles cannot grow when the channel is narrower than the bubble diameter. In Fig. 2.42 boiling curves 1, 2, and 3 were obtained at Bond numbers 0.185, 0.493, 0.9, respectively. For high Bond numbers, boiling can be considered as almost unconfined. The boiling curve obtained in the present study at $\mathrm{Bn} = 1.52$ did not differ significantly from that obtained by Hetsroni et al. (2002c, 2004b) in pool boiling (unconfined space). For example, at $q = 450\ \mathrm{W/m^2}$ the difference $(T_W - T_S)$ is close to 21 K for boiling at $\mathrm{Bn} = 1.52$ and pool boiling. However, as the gap size decreased, noticeable decrease in CHF occurred.

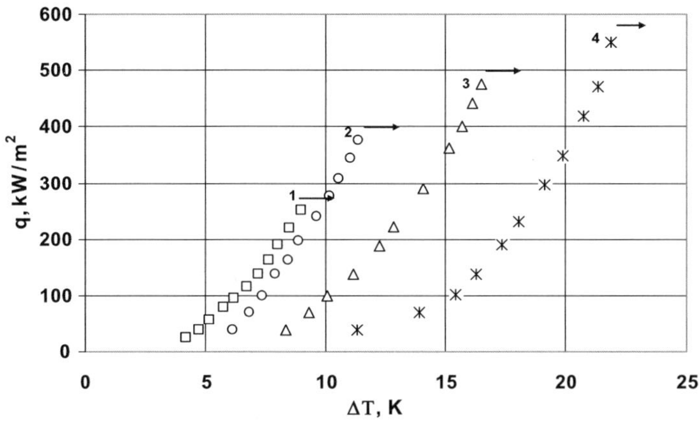

Fig. 2.42 Boiling curves of water in restricted space. *Squares* indicate Bond number Bn = 0.185, *circles* indicate Bond number Bn = 0.493, *triangles* indicate Bond number Bn = 0.9, and last symbol (✳) indicate Bn = 1.52

Effect of channel length

The experimental values of CHF are plotted in Fig. 2.43 as q_{CHF}/q^0_{CHF} versus Bond number, where q^0_{CHF} is the CHF in unconfined pool boiling with the space large enough (Bn > 1.5). The data presented by Hetsroni et al. (2002c, 2004b) were used to calculate the value of q^0_{CHF}. The confined narrow space heavily reduced the CHF as the vapor bubbles have difficulty escaping from the narrow space, especially at larger lengths of the heater surfaces. In a confined space they are able to escape only through the exit. The larger the length of the heater surface, the lower the critical heat flux. Such behavior is in agreement with experimental results described by Katto et al. (1966), Fujita et al. (1988), Bonjour and Lallemand (1988), and Zhao et al. (2002).

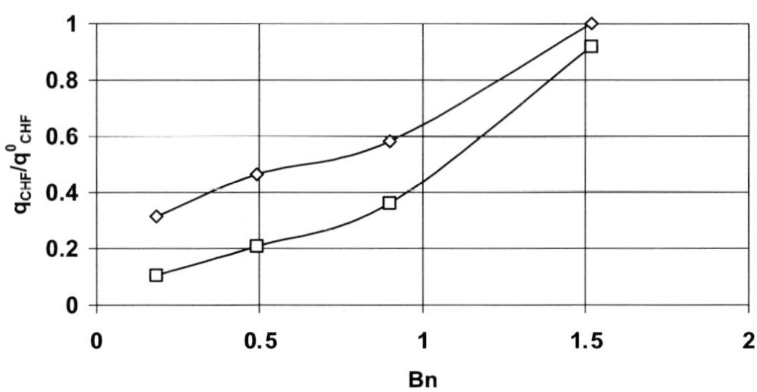

Fig. 2.43 Effect of restriction on CHF. Water boiling. *Diamonds* (◇) indicate $L = 2$ mm, and *squares* indicate $L = 36$ mm. Reprinted from Hetsroni et al. (2007) with permission

Flow in single channel

One of the most widely used correlations for saturated CHF in a single channel is the Katto–Ohno (1984) equation, which in the absence of subcooling reads:

$$\frac{q_{\mathrm{CHF}}}{Gh_{\mathrm{LG}}} = f\left[\frac{\rho_{\mathrm{L}}}{\rho_{\mathrm{G}}}, \frac{\sigma\rho_{\mathrm{L}}}{G^2 L}, \frac{L}{d}\right]. \tag{2.15}$$

For most regimes, they found a linear rise in CHF with increasing liquid subcooling. Accordingly, subcooling was taken into account in the following equation:

$$q_{\mathrm{CHF,sub}} = q_{\mathrm{CHF}}\left(1 + K\frac{\Delta h_{\mathrm{in}}}{h_{\mathrm{LG}}}\right) \tag{2.16}$$

where G is mass flux, $q_{\mathrm{CHF,sub}}$ is the CHF for subcooled conditions, q_{CHF} is the saturated CHF, Δh_{in} is the inlet enthalpy of subcooling with respect to saturation, d is the channel diameter, L is the heated length of the channel, and K is an empirical inlet subcooling parameter.

A series of tests were performed by Wojtan et al. (2006) to determine the saturated critical heat flux in 0.5 and 0.8 mm inner diameter micro-channel tubes as a function of refrigerant mass velocity, heated length, saturation temperature and inlet liquid subcooling. The tested refrigerants were R-134a and R-245fa and the heated length of micro-channel varied between 20 and 70 mm. Figure 2.44 shows the evolution of the CHF as a function of mass velocity. All conditions (heated length and temperature of subcooling of refrigerant) were the same for both diameters. As can be seen, the CHF increased with increasing mass velocity. The CHF for the 0.8 mm micro-channel is higher than that for the 0.5 mm one (Fig. 2.45) and the difference (30–50%) becomes greater as the mass velocity increases. The influence of the heated length on the CHF in both micro-tube diameters at constant mass velocity is depicted in Fig. 2.45. The highest CHF was measured for the shortest heated length.

With the heated length for the 0.8 mm diameter tube enlarged from 20 to 70 mm, the drop in the CHF value is $q_{\mathrm{CHF}} = 400\,\mathrm{kW/m^2}$. It confirms that the heated length, besides the mass velocity, is one of the most important parameters in the design of heat sinks.

Figure 2.46 shows the influence of liquid subcooling on the CHF. Only three experimental points are available as measurements at stronger subcoolings proved impossible. On the basis of the limited data, it can be concluded that CHF does not change significantly for the subcooling range $\Delta T = 4.5$ to 12 K. This observation is in agreement with the results of Qu and Mudawar (2004) for micro-channel heat sinks.

Parallel channels

The analysis of experimental data resulted in version of the Katto–Ohno (1984) equation, namely

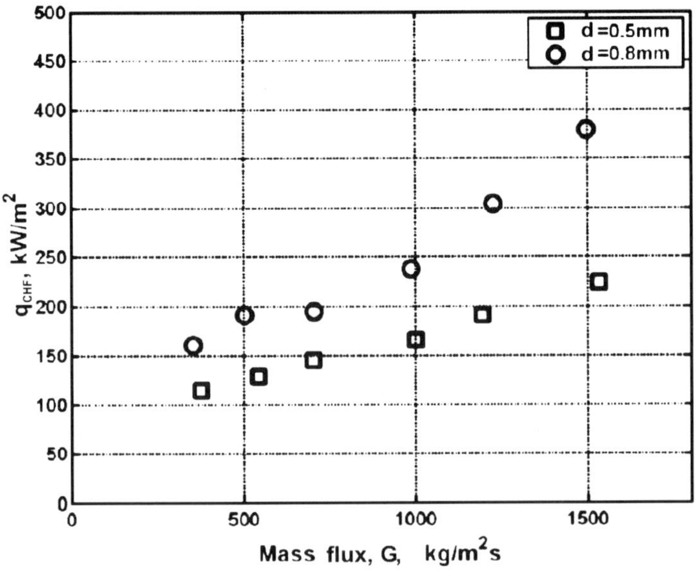

Fig. 2.44 Variation of CHF as a function of the mass velocity in 0.5 and 0.8 mm tubes. R-134a, $\Delta T_{sub} = 8$ K, $T_{sat} = 35\,^{\circ}$C, $L = 70$ mm. Reprinted from Wojtan et al. (2006) with permission

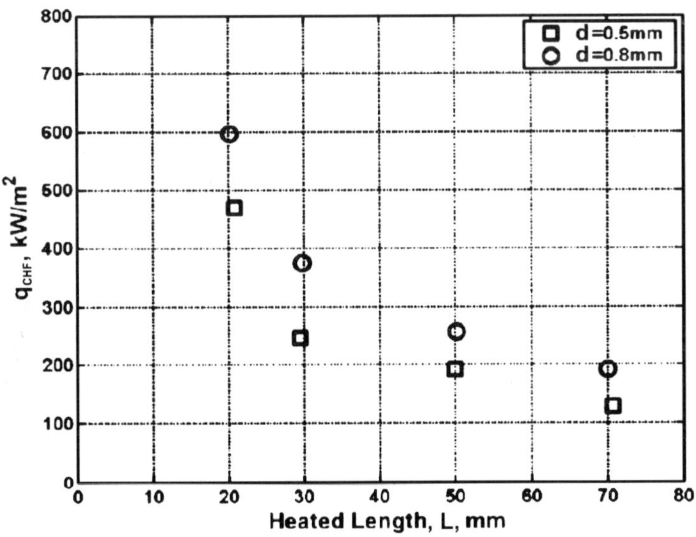

Fig. 2.45 Variation of CHF as a function of the heated length in 0.5 and 0.8 mm tubes. R-134a, $G = 500\ \mathrm{kg/m^2s}$, $\Delta T_{sub} = 8$ K, $T_s = 35\,^{\circ}$C. Reprinted from Wojtan et al. (2006) with permission

$$q_{CHF} = 0.437 \left(\frac{\rho_G}{\rho_L}\right)^{0.073} \mathrm{We}_L^{-0.24} \left(\frac{L}{d}\right)^{-0.72} G h_{LG} \tag{2.17}$$

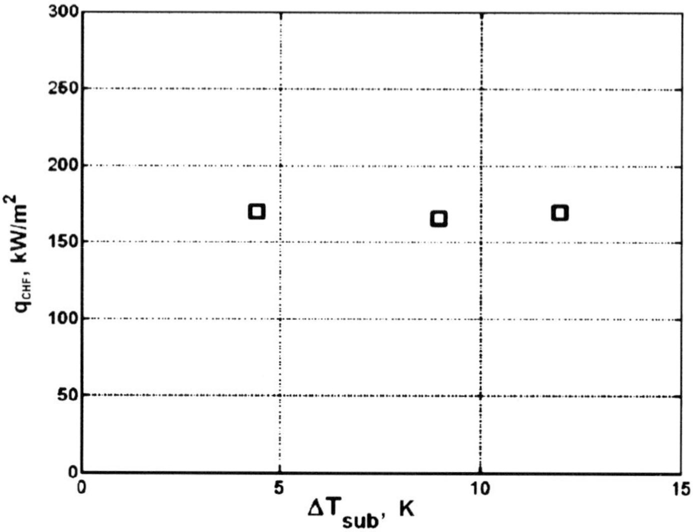

Fig. 2.46 Influence of refrigerant subcooling on CHF. R-134a, $G = 1{,}000\ \text{kg/m}^2\text{s}$, $T_s = 35\,^\circ\text{C}$, $L = 70$ mm, $d = 0.5$ mm. Reprinted from Wojtan et al. (2006) with permission

The first experimental investigation of the CHF in multi-micro-channels was carried out by Bowers and Mudawar (1994), who tested a cooling element with an array of 17 circular channels, 0.51 mm in diameter, 28.6 mm in length, in a 1.59 mm-thick nickel block, heated over a central 10 mm square section. Boiling curves were generated that terminated in well-defined CHF measurements, which were found to be independent of the liquid subcooling and almost directly proportional to the mass flux, for which a dimensionless correlation was proposed. Jiang et al. (1999b) developed two multi-micro-channel heat sinks integrated with a heater and an array of implanted temperature sensors. There were 58 or 34 channels of rhombic shape, having a hydraulic diameter of 0.040 or 0.080 mm, respectively, in their 10×20 mm test section. CHF data were taken for once-through water flow entering subcooled at $\Delta T = 20$ K. The CHF condition was characterized by a rapid rise in the temperature sensors. The critical power limit was found to be proportional to the total volumetric flow rate. In a recent paper, Qu and Mudawar (2004) reported a comprehensive study of the CHF in rectangular micro-channels. Their heated block contained 21 0.215×0.821 mm channels. The heat flux was reported based on the area of the three active sides of the channel. De-ionized, de-aerated water was supplied over a range of $G = 86$ to $368\ \text{kg/m}^2\ \text{s}$, with an inlet temperature of 30–60 °C and outlet pressure of 1.13 bar. The dependence of the heat flux on the mass flux is shown in Fig. 2.47 ($q_{p,m}$ is the CHF based on channel heated inside area, $q_{\text{eff},m}$ is the CHF based on the heat sink's top platform area). One can see that the CHF increases monotonically with increasing G for both inlet temperatures. What is quite surprising is that the inlet temperature T_{in} has an insignificant effect on the CHF. Interestingly, these CHF trends relative to the mass flux G and inlet temperature T_{in} mirror

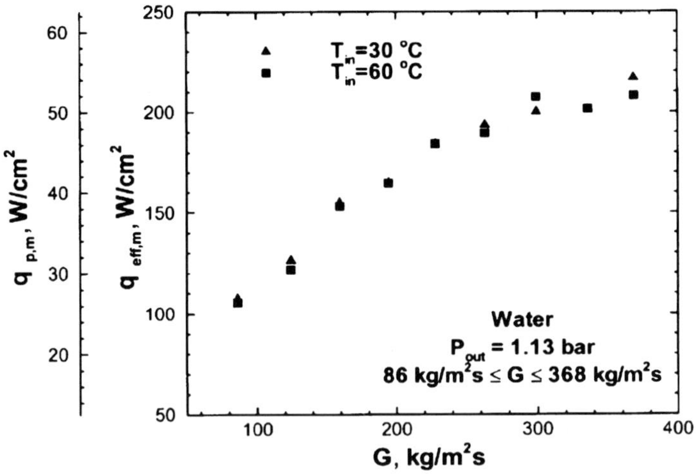

Fig. 2.47 Variation of CHF with mass velocity. Reprinted from Qu and Mudawar (2004)

those of Bowers and Mudawar (1994) for refrigerant R-113 in circular mini- and micro-channel heat sinks. While the trend of increasing CHF with increasing G is quite common, the lack of inlet temperature effect on the CHF seems to be unique to two-phase mini/micro-channel heat sinks, not to single mini/micro-channels. A key difference between these is the aforementioned amplification of parallel channel instability prior to the CHF. As discussed earlier, this amplification causes back flow of vapor into the upstream plenum, which results in strong mixing of the vapor with the incoming liquid. Regardless of how subcooled the incoming liquid, the mixing action appears to increase the liquid temperature close to the local saturation level as it approaches the channel inlet.

The CHF correlation (2.18) was developed by Qu and Mudawar (2004) for water in a rectangular micro-channel heat sink, as well as Bowers and Mudawar's CHF data (1994) for R-113 in a circular mini/micro-channel heat sink.

The following is the CHF correlation:

$$\frac{q_{\mathrm{p},m}}{Gh_{\mathrm{LG}}} = 33.43 \left(\frac{\rho_{\mathrm{G}}}{\rho_{\mathrm{L}}}\right)^{1.11} \mathrm{We}_{\mathrm{L}}^{-0.21} \left(\frac{L}{d_{\mathrm{h}}}\right)^{-0.36}. \tag{2.18}$$

Since the CHF for both mini/micro-channel heat sink databases shows no dependence on inlet subcooling, these databases were correlated without the subcooling multiplier.

Figure 2.48 compares the predictions of this correlation with the flow boiling CHF data for water both in the rectangular micro-channel heat sink (Qu and Mudawar 2004) and in the circular mini/micro-channel heat sinks (Bowers and Mudawar 1994). The overall mean absolute error of 4% demonstrates its predictive capability for different fluids, circumferential heating conditions, channel geometries, channel sizes, and length-to-diameter ratios.

Comparison of the results obtained by Wojtan et al. (2006) and the correlation presented by Qu and Mudawar (2004) is shown in Fig. 2.49. As can be seen there is significant scatter between the respective results. One can conclude that there is a lack of consistency in the reported data.

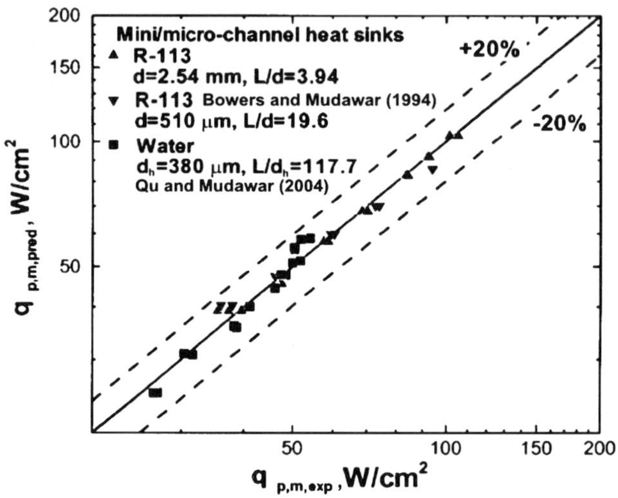

Fig. 2.48 Comparison of CHF data for water and R-113 in mini/micro-channel heat sinks correlation (2.18). Reprinted from Qu and Mudawar (2004) with permission

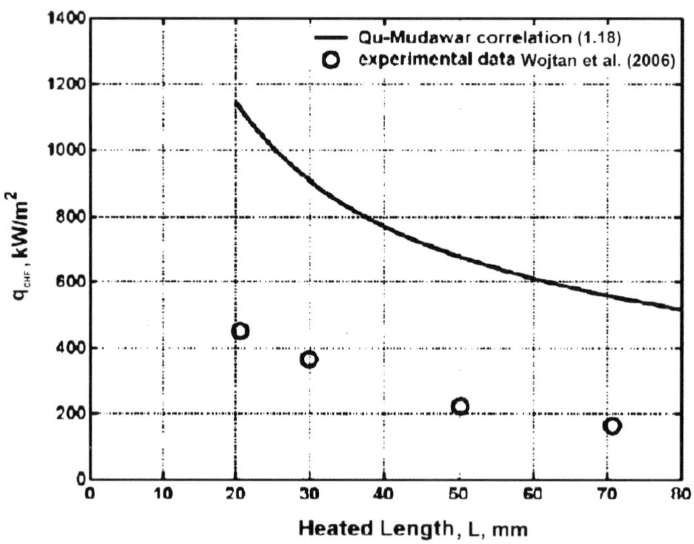

Fig. 2.49 Comparison of the experimental results by Wojtan et al. (2006) to the correlation (2.18) of Qu and Mudawar (2004). R-134a, $G = 500\ \mathrm{kg/m^2s}$, $\Delta T_{\mathrm{sub}} = 8\ \mathrm{K}$, $T_s = 35\,°\mathrm{C}$, $d = 0.5\ \mathrm{mm}$. Reprinted from Wojtan et al. (2006)

2.5 Surfactant Solutions

Properties of surfactant solutions: microstructure

The mechanisms that affect heat transfer in single-phase and two-phase aqueous surfactant solutions is a conjugate problem involving the heater and liquid properties (viscosity, thermal conductivity, heat capacity, surface tension). Besides the effects of heater geometry, its surface characteristics, and wall heat flux level, the bulk concentration of surfactant and its chemistry (ionic nature and molecular weight), surface wetting, surfactant adsorption and desorption, and foaming should be considered.

Surfactants have a unique long-chain molecular structure composed of a hydrophilic head and hydrophobic tail. Based on the nature of the hydrophilic part surfactants are generally categorized as anionic, non-ionic, cationic, and zwitterionic. They all have a natural tendency to adsorb at surfaces and interfaces when added in low concentration in water. Surfactant absorption/desorption at the vapor–liquid interface alters the surface tension, which decreases continually with increasing concentrations until the critical micelle concentration (CMC), at which micelles (colloid-sized clusters or aggregates of monomers) start to form is reached (Manglik et al. 2001; Hetsroni et al. 2003c).

Micro-structure of cationic Habon G surfactant solution at CMC concentration ($C = 530\,\text{ppm}$) is presented in Fig. 2.50 (Hetsroni et al. 2003c). The surfactant solution was examined by direct imaging cryogenic temperature transmission electron microscopy (cryo-TEM). The most common structures are thread-like micelles, shown at high magnification in Fig. 2.50a. Micrographs like this one show quite clearly inner structural details of the micelles (arrowhead). It should be noted that thread-like micelles have been suggested as the one, possibly the most important, micro-structural feature that modifies flow patterns in a flowing fluid and reduces drag (Lu et al. 1998). The drag reduction takes place at values of shear velocity u^* higher than the onset wall shear. The shear velocity is calculated from the pressure drop $u^* = (\tau/\rho)^{0.5}$ (τ shear stress, ρ density). Thus, the drag reducing effect starts with wall shear stresses larger than a threshold value, which depends on the nature of additive and its concentration.

Another feature visible in surfactant solutions is vesicles, denoted by "V" in Fig. 2.50b. These are balloon-like structures made of a double-layer membrane. Such structures are found in many biological and synthetic amphiphiles (Evans and Wennerstrom 1999). In this micrograph we see vesicles co-existing with thread-like micelles (arrows). We have detected junctions between vesicles and thread-like micelles. In fact, three such junctions, one denoted by an arrowhead are seen in the lower left part of the field of view in Fig. 2.50b. The three micelles connecting the three vesicles are connected by a threefold junction seen above the arrowhead. Zheng et al. (2000) hypothesize that the straining actions of flow disrupt vesicles and thus include structural instability of the fragments that leads to their reconstruction into networks of branching thread-like micelles.

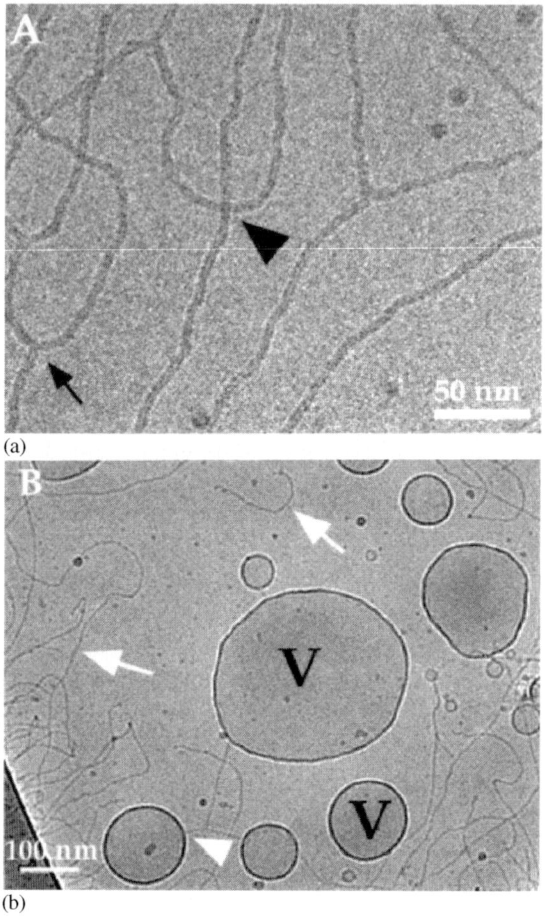

Fig. 2.50a,b Microstructure of the 530 ppm Habon G solution. (**a**) Thread-like micelles and (**b**) vesicles. Reprinted from Hetsroni et al. (2003c) with permission

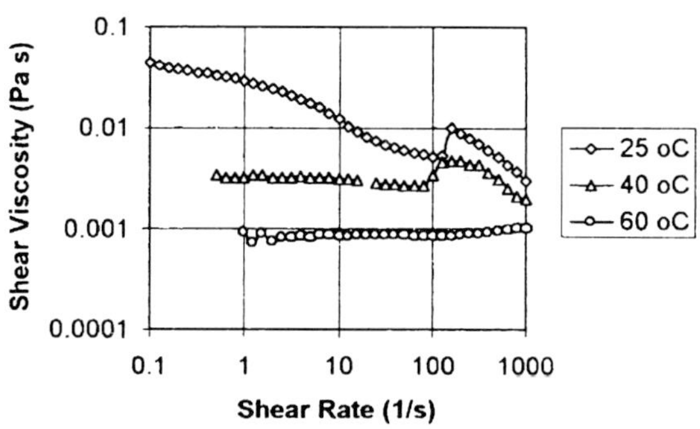

Fig. 2.51 The shear viscosity of 530 ppm Habon G solution vs. shear rate. Reprinted from Hetsroni et al. (2001b) with permission

Viscosity

Viscosity of surfactant solutions depends on the kind of solution, shear rate, temperature and concentration. Figure 2.51 shows the effect of shear rate ϖ on shear viscosity η for a 530 ppm (parts per million weight) Habon G (the cationic surfactant of molecular weight 500) mixture at different temperatures. The magnitude of the shear viscosity as a function of the shear rate decreases, when the temperature of the solution increases. Moreover, the shear viscosity does not change significantly in the range of the shear rate ϖ=1–1000 1/s at T = 60 °C. Based on this result the kinematic viscosity behavior of Habon G solutions was studied at high shear rates for temperatures of T = 55 °C and above. Figure 2.52 shows the effect of temperature on kinematic viscosity of solutions at various Habon G concentrations. One can see the tendency of viscosity curves to approach that of pure water near the saturation temperature. The saturation temperature of aqueous surfactant solutions in the range of concentration 130 to 1060 ppm did not differ from that of pure water (Hetsroni et al. 2001b).

Thermal conductivity and capacity

Thermal conductivity and capacity of aqueous surfactant solutions in the concentration range 130 to 1,060 ppm did not differ from that of pure water (Hetsroni et al. 2001b). Figure 2.53 shows the dependence of thermal conductivity k on the temperature for $C = 530$ ppm Habon G solution. The value of the thermal conductivity agrees well with that for pure water within the standard deviation of 2%.

Contact angle

The change in surface wettability (measured by the contact angle) with concentration for the three surfactants is plotted in Fig. 2.54 (Zhang and Manglik 2005). The contact angle reaches a lower plateau around the CMC where bilayers start to form on the surface. Wettability of non-ionic surfactants in aqueous solutions shows that the contact angle data attains a constant value much below CMC. Direct interactions of their polar chain are generally weak in non-ionics, and it is possible for them to build and rebuild adsorption layers below CMC. The reduced contact an-

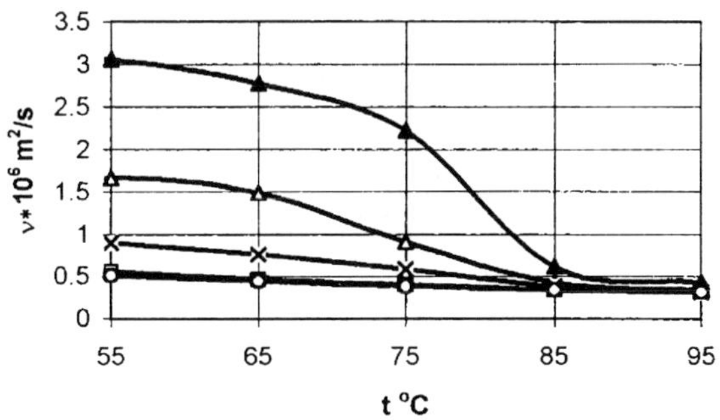

Fig. 2.52 Kinematic viscosity of solution vs. temperature at various Habon G concentrations. *Circles* (○) indicate water; Habon G, *boxes* (□) represent 130 ppm, *crosses* (×) represent 260 ppm, *empty triangles* (△) represent 530 ppm, *filled triangles* (▲) represent 1,060 ppm. Reprinted from Hetsroni et al. (2001b) with permission

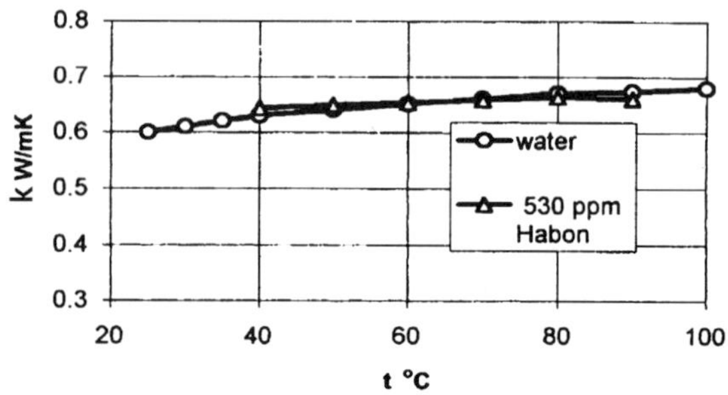

Fig. 2.53 Thermal conductivity vs. temperature. Reprinted from Hetsroni et al. (2001b) with permission

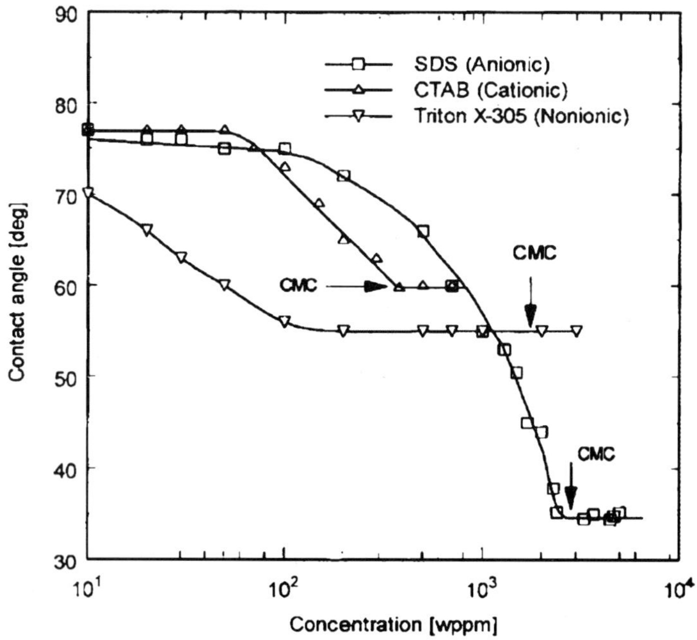

Fig. 2.54 Measured contact angle θ for aqueous surfactant solutions. Reprinted from Zhang and Manglik (2005) with permission

gle at lower concentration ($C <$ CMC) can also be attributed to the absence of any electrical repulsion that could oppose molecular aggregation unlike that associated with ionic surfactants (Zhang and Manglik 2005). Bubble shape is the governing factor in mechanism of bubble growth and motion in nucleate pool boiling. The rapidity of the bubble growth allows the contact angle to deviate from the static value. Figure 2.55 (Hetsroni et al. 2006c) shows time behavior of contact angle for water and non-ionic alkyl (8-16) glucoside aqueous surfactant solution of concentration $C = 600$ ppm.

The results were obtained at heat flux $q = 10\,\text{kW/m}^2$. For both liquids at $t = 1$ ms the contact angle is approximately of $\theta = 60°$, which is very close to the equilibrium surface tension of water. Throughout bubble growth this value decreases approxi-

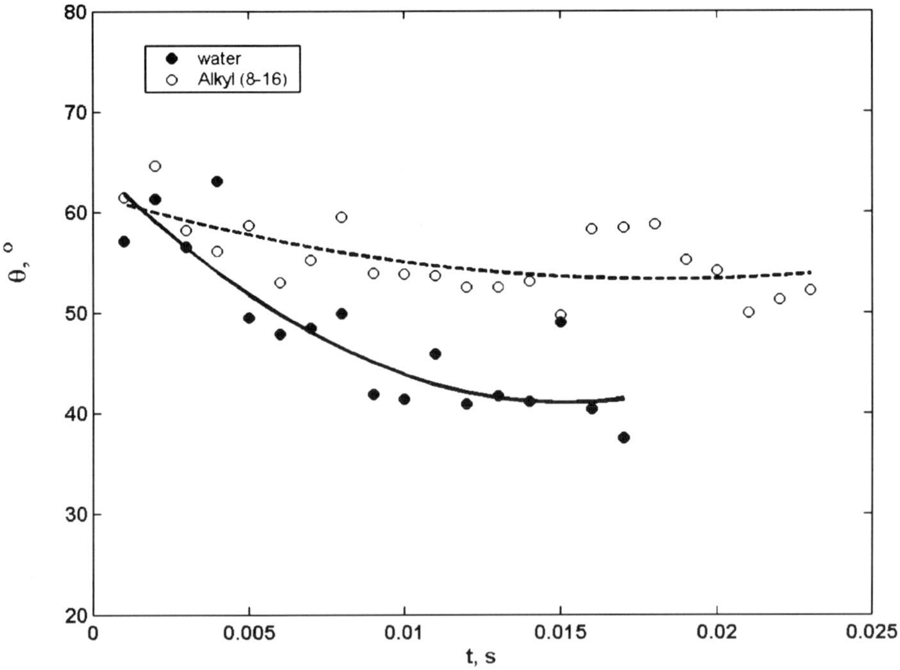

Fig. 2.55 Dependence of contact angle on time, $q = 10\,\mathrm{kW/m^2}$. Reprinted from Hetsroni et al. (2006c) with permission

mately up to 41° and 51° for water and surfactant solution, respectively. The time behavior of the contact angle of a bubble growing in surfactant solution is qualitatively similar to that of water. This in essence explains the time variation of bubble volume for water and surfactant solution shown in Fig. 2.56 (Hetsroni et al. 2006c). A surfactant solution cannot be expected to boil in the same way as a pure liquid even with exactly the same equilibrium surface tension value.

Bubble growth depends on heat flux. With an increase in heat flux, boiling in surfactant solution, when compared with that in pure water, was observed to be more vigorous. Surfactant solution promotes activation of nucleation sites in a clustered mode. This may be due to the retarded bubble coalescence caused by the Marangoni effect of elasticity described by Yang and Maa (2003). The cluster contains a number of small bubbles; the location of the nucleation sites and time behavior cannot be traced exactly. Dependence of a single bubble in water and a single bubble located in the cluster of non-ionic alkyl (8-16) glucoside aqueous surfactant solution of 600 ppm concentration at heat flux $q = 50\,\mathrm{kW/m^2}$ is shown in Fig. 2.57 (Hetsroni et al. 2006c). Analysis of the curves in Fig. 2.57 shows that the growth of bubbles in surfactant solution at high heat flux occurs slower than in water.

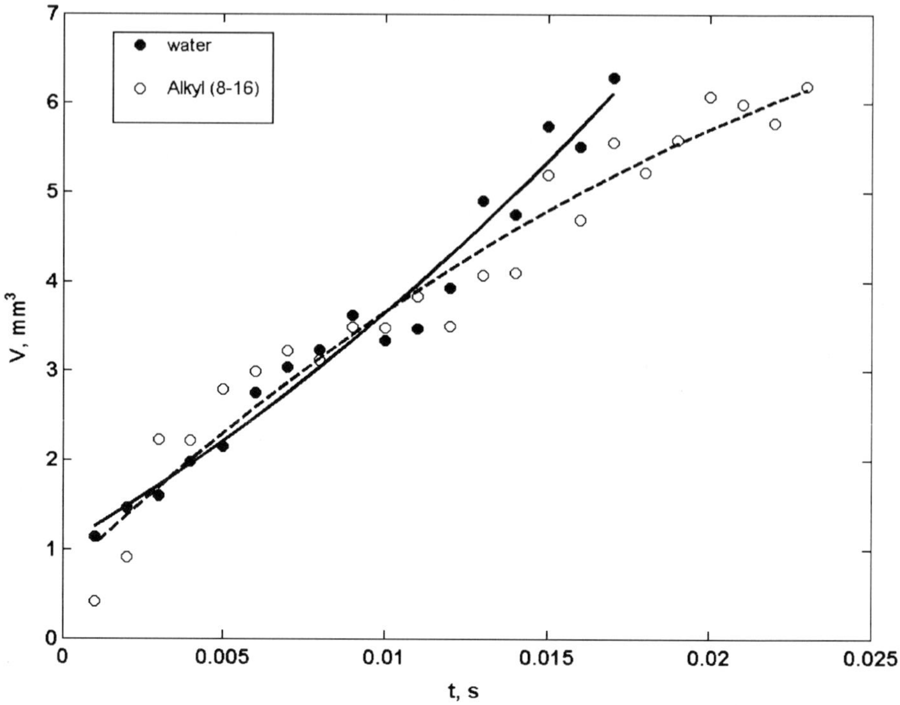

Fig. 2.56 Dependence of bubble volume on time, $q = 10\,\mathrm{kW/m^2}$. Reprinted from Hetsroni et al. (2006c) with permission

Surface tension

In Fig. 2.58 (Hetsroni et al. 2001b) the dependencies of the surface tension of the various surfactants σ divided on the surface tension of water σ_{w} are shown. One can see that beginning from some particular value of surfactant concentration (which depends on the kind of surfactant), the value of the relative surface tension almost does not change with further increase in the surfactant concentration. It should be emphasized that the variation of the surface tension as a function of the solution concentration shows the same behavior for anionic, non-ionic, and cationic surfactants at various temperatures.

For example, for alkyl (8-16) glycoside (Plantacare 818 UP) non-ionic surfactant solution of molecular weight 390 g/mol, an increase in surfactant concentration up to 300 ppm (CMC concentration) leads to a significant decrease in surface tension. In the range $300 \leq C \leq 1{,}200$ ppm the surface tension was almost independent of concentration. In all cases an increase in liquid temperature leads to a decrease in surface tension. This surface tension relaxation is a diffusion rate-dependent process, which typically depends on the type of surfactant, its diffusion/absorption kinetics, micellar dynamics, and bulk concentration levels. As the CMC is approached the absorption becomes independent of the bulk concentration, and the surfactant

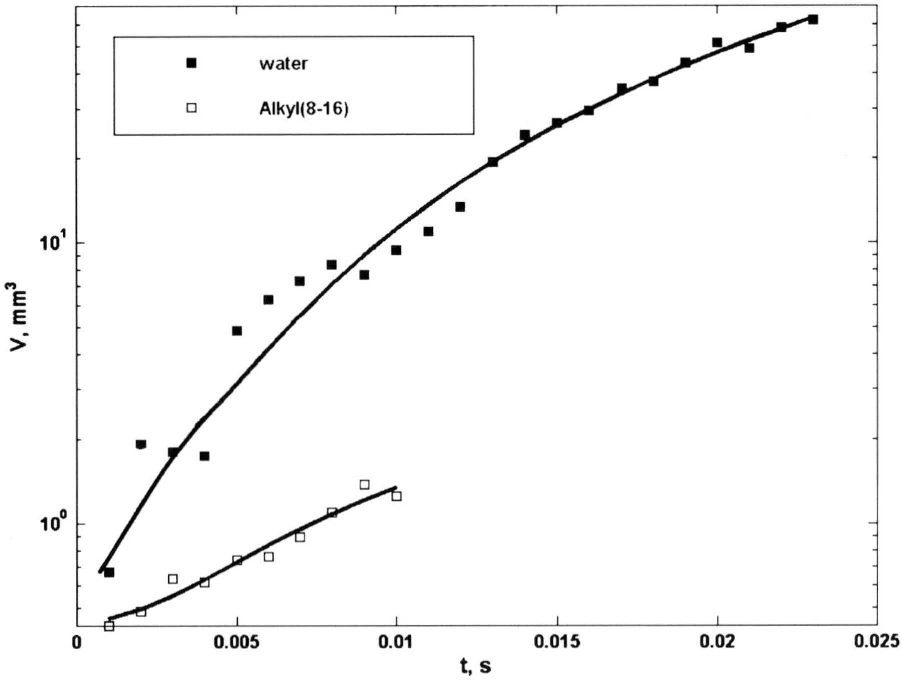

Fig. 2.57 Dependence of bubble volume on time, $q = 50\,\mathrm{kW/m^2}$. Reprinted from Hetsroni et al. (2006c) with permission

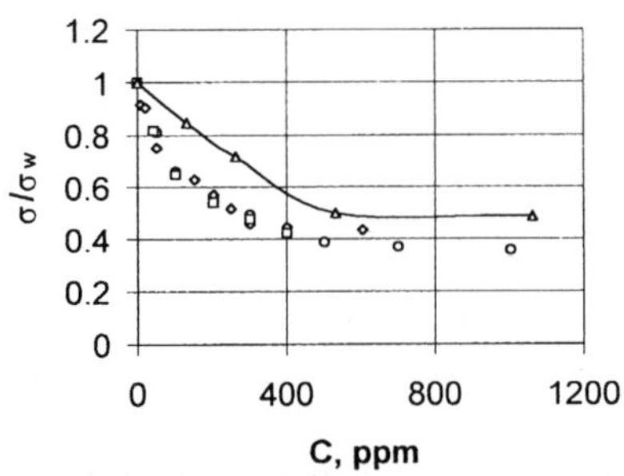

Fig. 2.58 The non-dimensional surface tension of various surfactants vs. solution concentration. *Diamonds* (◇) represent SDS (Wu et al. 1995), $T = 25\,°\mathrm{C}$. Circles (○) represent SDS (Tzanand and Yang 1990), $T = 25\,°\mathrm{C}$. Squares (□) represent SDS (Wu and Yang 1992), $T = 100\,°\mathrm{C}$. Triangles (△) represent Habon G (Hetsroni et al. 2001b), $T = 70\,°\mathrm{C}$. Reprinted from Hetsroni et al. (2001b) with permission

molecules form a bilayer on the surface to make it strongly hydrophilic (Zhang and Manglik 2005). The heat transfer in saturated nucleate boiling of aqueous surfactant solutions generally increases with increase in heat flux and additive concentration up to $C \leq \mathrm{CMC}$. With $C > \mathrm{CMC}$, the enhancement decreases and the heat transfer can even deteriorate. It should be emphasized that the variation of the non-dimensional surface tension as a function of the surfactant concentration shows the same behavior for various temperatures. The normalized nucleate boiling heat transfer coefficient may be related to normalized surface tension of the surfactant solution. We used the magnitude of surfactant characteristic concentration C_0 where the change of relative surface tension reaches 90% of the complete change to normalize the concentration scale. The values $C_0 = 530$ and 700 ppm were chosen for Habon G and SDS solutions, respectively.

For normalization of the value of the heat transfer enhancement, we used its magnitude at the maximum for each curve. The result of such normalization is shown in Fig. 2.59. In this figure, C is the solution concentration, C_0 is the characteristic concentration, h is the heat transfer coefficient at given values of the solution concentration and the heat flux q, h_{max} is the maximum value of the heat transfer coefficient at the same heat flux, and h_{w} is the heat transfer coefficient for pure water at the same heat flux q. Data from all the sources discussed reach the same value of 1.0 at the magnitude of relative surfactant concentration equal to 1.0.

Thus, the enhancement of heat transfer may be connected to the decrease in the surface tension value at low surfactant concentration. In such a system of coordinates, the effect of the surface tension on excess heat transfer $(h - h_{\mathrm{w}})/(h_{\mathrm{max}} - h_{\mathrm{w}})$ may be presented as the linear fit of the value C/C_0. On the other hand, the decrease in heat transfer at higher surfactant concentration may be related to the increased viscosity. Unfortunately, we did not find surfactant viscosity data in the other studies. However, we can assume that the effect of viscosity on heat transfer at surfactant boiling becomes negligible at low concentration of surfactant only. The surface tension of a rapidly extending interface in surfactant solution may be different from the static value, because the surfactant component cannot diffuse to the absorber layer promptly. This may result in an interfacial flow driven by the surface tension gradi-

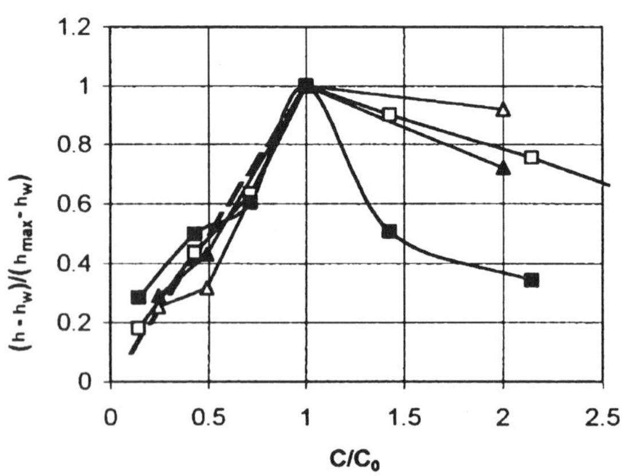

Fig. 2.59 The excess heat transfer coefficient vs. the surfactant concentration. For Habon G *empty triangles* (△) represent $q = 400\,\mathrm{kW/m^2}$, and *filled triangles* (▲) represent $q = 800\,\mathrm{kW/m^2}$. For SDS (Wu and Yang 1992) *empty squares* (□) represent $q = 350\,\mathrm{kW/m^2}$, and *filled squares* (■) represent $q = 400\,\mathrm{kW/m^2}$. Reprinted from Hetsroni et al. (2001b) with permission

ent (known as Marangoni flow). Phase change during boiling is a local phenomenon that is associated with the heat transfer and hydrodynamics in the vicinity of a growing vapor bubble. The concentration sublayer also determines the dynamic surface tension at the interface, which in turn directly influences the bubbly dynamics.

For pool boiling Wasekar and Manglik (2002) presented the results of a study that investigates the dependence of nucleate boiling heat transfer coefficients of aqueous surfactant solutions of different molecular weight and ionic nature. It was reported that the maximum enhancement increased with decreasing surfactant molecular weight. The criteria for nucleate boiling enhancement by surfactant additives was proposed by Yang and Maa (2003). As the first criterion, it was postulated that the surfactant should be soluble in water. As the second criterion, it was postulated that the surfactant should depress the equilibrium surface tension of solution significantly. As the third criterion, it was postulated that the surfactant should not depress the equilibrium contact angle significantly. According to Yang and Maa (2003), boiling heat transfer by addition of surfactant is enhanced by the depression of the equilibrium surface tension but suppressed by the depression of the equilibrium contact angle. Sher and Hetsroni (2002) developed a model of nucleate pool boiling with surfactant additives. Solid–vapor, solid–liquid and liquid–vapor surface tensions were assumed to be surfactant diffusion controlled, and the boiling curves were calculated. To the author's best knowledge no discussion on effect of channel gap size on natural convection boiling was found.

Additive of surfactant leads to enhancement of heat transfer compared to water boiling in the same gap size; however, this effect decreases with decreasing channel size. For the same gap size CHF in surfactant solutions is significantly lower than that in water. At high values of heat flux some foaming patches began to occur; this process increased with decrease in gap size and led to decrease in CHF, Hetsroni et al. (2007).

2.6 Design and Fabrication of Micro-Channel Heat Sinks

Typical channels studied in literature are rectangular, with one side unheated (usually a transparent cover plate in the experimental apparatus) and heat is applied near the opposite wall. Bergles et al. (2003) considered the general geometry, permitting relative simplicity in the vertical dimension, shown in the sketch of Fig. 2.60: a right parallelepiped heated (uniformly) on one side, and perforated by one or more rows of identical channels parallel to the heated surface. The system can be treated as a stack of plates, some with slots for the channels and some solid. Several possible patterns and channel sizes (d, d_i, ...) are shown for illustration, although in the following we shall assume that only one channel pattern is used in any particular case. The channels may be connected at the ends to form flow lengths of $L = 2W$, $3W$, or more, or divided at mid-block to form lengths of $L = W/2$, $W/3$, or less. Let the ratio of channel length to block width be $r = L/W$.

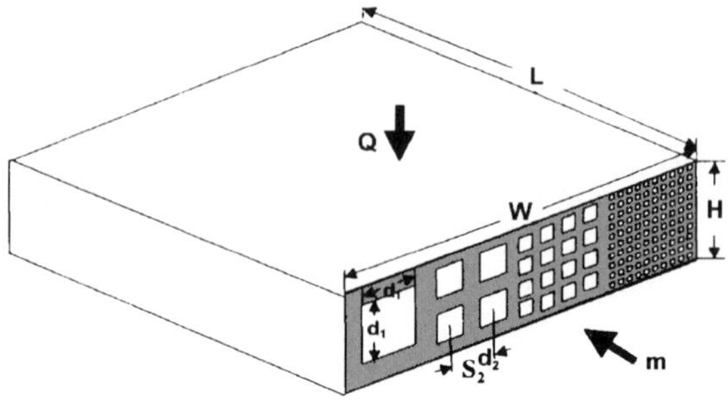

Fig. 2.60 Geometry of micro-channel heat sink. Reprinted from Bergles et al. (2003) with permission

General problem

The design question is: given the heat rate Q, the length L, and width W, select a working fluid with mass flow rate m, channel dimension d, channel spacing $S \times d$ (for S is a number > 1), number of channels n, and material and thickness of the block H. We will be particularly interested in the pumping power P.

The width and thickness of the block, and the number, length, diameter, and spacing of the channels are not independent. The geometric configuration requires that the number of channels in a row, n_c, be

$$n_c = \frac{W}{Sd} \tag{2.19}$$

and that the number of rows be

$$n_r = \frac{H}{Sd}. \tag{2.20}$$

The total number of holes in a cross-section, which is the number of channels, is

$$n = n_c n_r = \frac{WH}{(Sd)^2} \tag{2.21}$$

and the total length of all channels

$$nL = nrW = r\frac{HW^2}{(Sd)^2}. \tag{2.22}$$

The number n must be an integer (i.e., $n = 1, 2, 3, \ldots$) and channel flow is characterized by lengths that are at least several diameters in magnitude. The latter constraint does not preclude operation in the "entry length" mode.

In a phase-change controlled thermal management system, the energy associated with phase change will account for most of the heat removal

$$Q = m\Delta\chi h_{LG} \tag{2.23}$$

where $\Delta\chi$ is the change in quality over the length of the channel and h_{LG} is the latent heat of vaporization for the coolant. For applications where the entire channel is in the subcooled boiling mode, there would be no change in equilibrium quality. In this case, the total heat removal might be on the order of that needed to bring the coolant to saturation temperature:

$$Q \approx mc_p(T_s - T_{sub}) \tag{2.24}$$

where C_p is the specific heat of the liquid coolant, T_s the saturation temperature, and T_{sub} the inlet temperature of the subcooled coolant. Given a desired value of $\Delta\chi$ (low or high) or of $(T_s - T_{sub})$, Eq. (2.23) or (2.24) can be used to get a first estimate for the mass flow rate of the coolant at the system pressure. From the overall energy balance for square channels with sides of length $d_1 = d_2 = d$, we obtain

$$Q = 4dnLq = \frac{4rHW^2}{S^2d}q \tag{2.25}$$

where q is the average heat flux at the channel wall. In the case of circular channels of the same diameter d, π replaces the number 4 in this equation.

The average heat flux must be less than the critical value q_{CHF} for the flow, in order to avoid excessive temperatures in the solid

$$q < q_{CHF}\,. \tag{2.26}$$

It is important to note here that the CHF depends on the flow conditions, including the mass flux and the vapor quality. Using Eq. (2.25) to eliminate q, we find

$$q_{CHF} > \frac{QS^2d}{4rHW^2}\,. \tag{2.27}$$

For the lower heat transfer surfaces in Fig. 2.60 to contribute to the energy transport, the solid should be an effective conductor of heat through its thickness. In other words, conjugate heat transfer effects should not create a more significant resistance to heat flow than that of the fluid in the channel. Since the heat transfer coefficient is generally a maximum at CHF, this leads to

$$\frac{k_{solid}(S-1)}{4H} \gg \frac{q_{CHF}}{(T_{CHF} - T_s)} > \frac{q}{(T_{CHF} - T_s)} \tag{2.28}$$

where k_{solid} is the thermal conductivity of the solid, and T_{CHF} and T_s are the wall temperature at critical heat flux and the saturation temperature of the fluid, respectively. Substituting Eq. (2.25) in Eq. (2.28), we obtain

$$\frac{k_{solid}(S-1)}{H} \gg \frac{Q}{rnLd\,(T_{CHF} - T_s)} \tag{2.29}$$

Note that a good design will be one in which the pumping power P is small compared to the heat removal rate Q:

$$P \ll Q \tag{2.30}$$

Non-uniform temperature distribution

One drawback of a micro-channel heat sink is a relatively high temperature rise along the micro-channel compared to that for the traditional heat sink designs. In the direction of the flow, the wall temperature rises in a single-phase flow even when the wall heat flux is uniform. In a micro-channel heat sink, the large amount of heat generated by the electronic device is removed from the package by a relatively small amount of coolant, which exits at a relatively high temperature. The large temperature rise produces thermal stresses in the elements and packages due to the differences in the coefficient of thermal expansion, thus undermining the device's reliability. This temperature rise may be accompanied by a complex pattern of spatial variations that can produce potentially destructive thermal stresses along the interface between the chip and the substrate or heat sink. This is one of the key justifications for seeking a nearly isothermal heat sink. Furthermore, a large temperature gradient is undesirable for the electronic performance, since many electronic parameters are adversely affected by it. One example is electrical-thermal instability because the basic elements of electronic circuits have a switching time that decreases with increasing temperature.

The surface of ultra large-scale integrated circuits (ULSI), from which the heat should be transferred, may be heated by a uniform heat flux, and more often by a non-uniform one. Even in the former case the temperature of the cooled surface is not uniform, but is determined by the heat transfer coefficients along the surface and in the spanwise direction.

We have designed, manufactured and tested a prototype that may be applied in thermal control of electronic devices. It was fabricated from a silicon substrate and a Pyrex cover, serving as both an insulator and a window through which flow patterns and boiling phenomena could be observed. A number of parallel triangular micro-channels were etched in the substrate. The heat transferred from the device was simulated by different types of electrical heaters that provided uniform and non-uniform heat fluxes, defined here respectively as constant and non-constant values

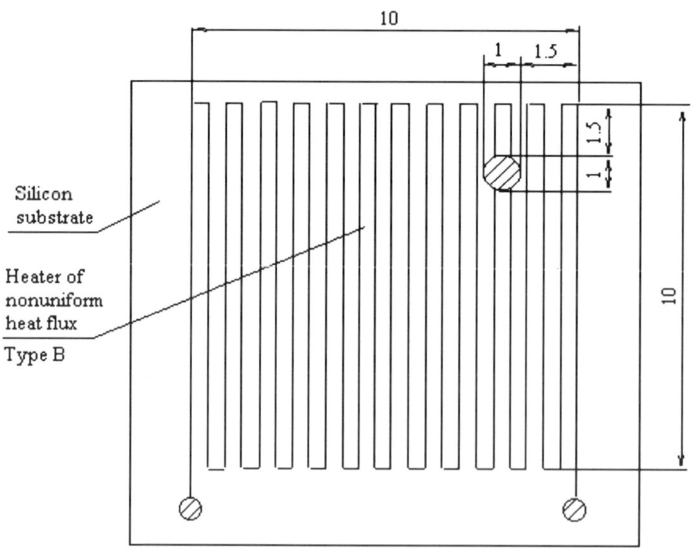

Fig. 2.61 Type of heater providing non-uniform heat flux. Reprinted from Hetsroni et al. (2001a) with permission

on the heated wall (HW). (In reality no heater can generate a really uniform heat flux, because of the concentration effects of the micro-channel geometry.)

In Fig. 2.61 the part of the heater, with a diameter of 1 mm, has the electrical resistance of 20% less than the average one (Hetsroni et al. 2001a).

For triangular micro-channels of $d_h = 130\,\mu m$ results based on experimental measurements at the same entrance flow rate $m = 0.046\,g/s$ are presented in Figs. 2.62 and 2.63.

The measurements performed in the heat flux range from 90 to $220\,kW/m^2$ showed that for all module types tested, the irregularity of $T_{w,max}$ across the heated bottom, increases with the increase of heat flux. This effect is more marked under boiling conditions. At uniform heat flux the irregularity of $T_{w,max}$ exceeds 20 K (Fig. 2.62). In the non-uniform heat flux case (average value $q = 220\,kW/m^2$), the irregularity exceeds 60 K (Fig. 2.63).

The temperature distribution in the flow direction for a fixed flow rate differs for different devices. This suggests that the heat transfer mechanism in these devices is not identical. The non-uniform (of about 20%) heat flux leads to conditions at which the wall temperature increases sharply. Idealizing the heat flux as uniform can result in a significant error in prediction of the temperature distribution.

Manifolds

Several studies (Klein et al. 2005; Mishan et al. 2007) showed that manifold design plays an important role in the liquid distribution among parallel micro-channels, which can lead to spanwise temperature gradients on the device surface, increase the thermal stresses and reduce reliability. To study the effect of entrance conditions

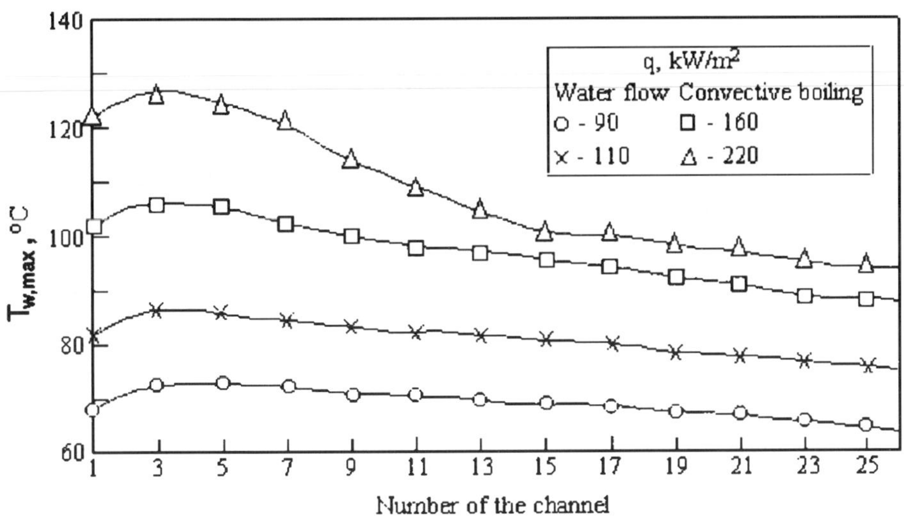

Fig. 2.62 Spanwise irregularity of maximum temperature on HW. Uniform heat flux. Reprinted from Hetsroni et al. (2001a) with permission

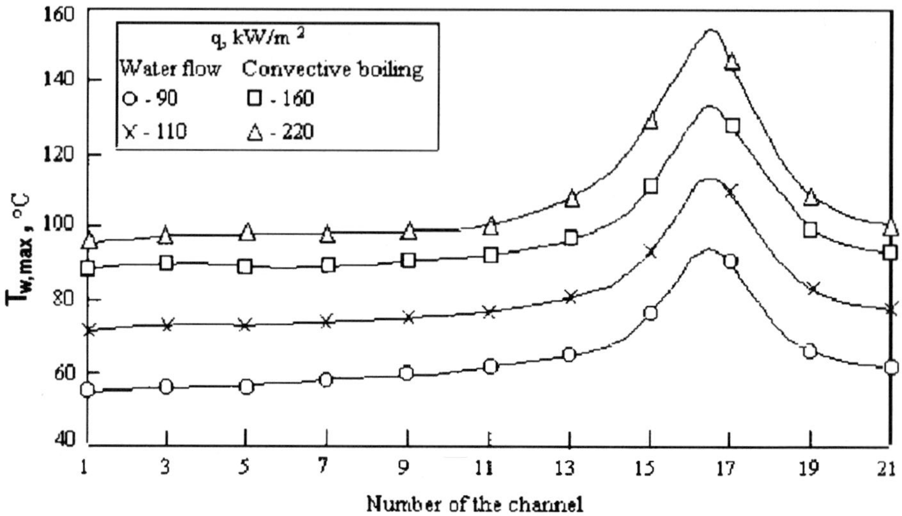

Fig. 2.63 Spanwise irregularity of maximum temperature on HW. Non-uniform heat flux. Reprinted from Hetsroni et al. (2001a) with permission

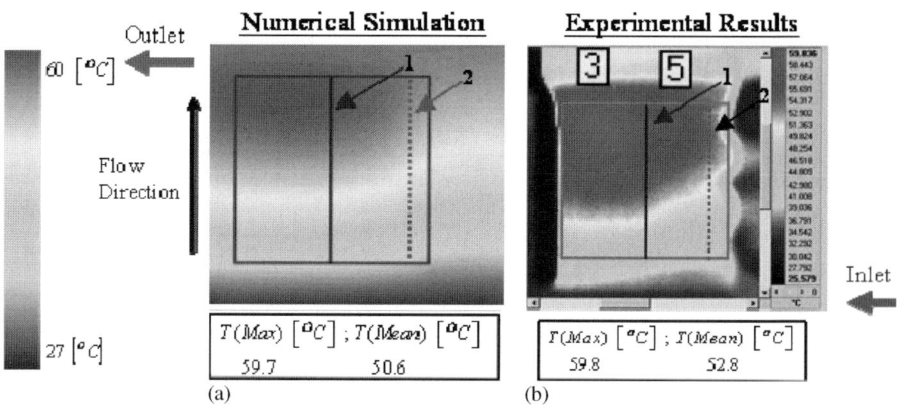

Fig. 2.64a,b Temperature field on heater. Reprinted from Mishan et al. (2007) with permission

on the temperature of the heater calculations using CFD software were conducted (Tiselj et al. 2004; Mishan et al. 2007). Developing heat transfer and fluid flow were investigated experimentally on rectangular micro-channels of $D_h = 440\,\mu m$, with water as working fluid. The experimental results were compared with theoretical predictions from literature and data obtained by numerical modeling of the present experiment. The experimental results of pressure drop and heat transfer confirm that including the entrance effects, the conventional theory is applicable for water flow through micro-channels. Figure 2.64 shows a typical temperature distribution on the heater. The fluid moves from the bottom to the top, the heat flux is $q = 4.4\,W/cm^2$ and the mass flux is $G = 8.33\,kg/m^2\,s$. The figure presents a comparison between the temperature field of the heater obtained from the numerical simulation (A) and that obtained from IR measurements (B). The square enclosed area of $10 \times 10\,\text{mm}^2$ represents the heater. Both of the pictures show very close temperature distributions. Figure 2.65 shows the temperature distribution along the central and side lines, marked as lines 1, 2, respectively, in Fig. 2.64. Numerical and experimental results agree quite well.

To optimize the design of the manifold configuration, a number of numerical calculations were conducted for three types of manifolds shown in Fig. 2.66. Figure 2.67 shows the velocity distribution at 2 mm from the inlet to the micro-channels, from which one can conclude that configuration 1 and 2 ensure uniform velocity distribution at the entrance.

It was shown that data presented by other researchers can be carried over to entrance effects. The present results highlight the importance of accounting for com-

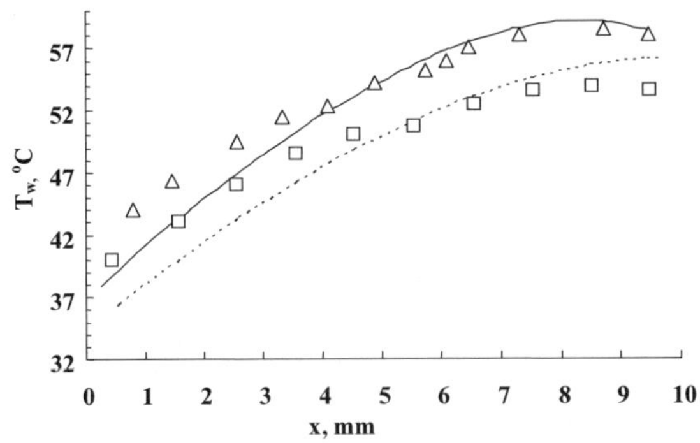

Fig. 2.65 Comparison between numerical simulation and experimental results for the temperature field on the heater surface. The *solid line* represents simulation, and *triangles* (△) experimental results (line 1 in Fig. 2.64); *dotted line* represents simulation, and *squares* experimental results (line 2 in Fig. 2.64). Reprinted from Mishan et al. (2007) with permission

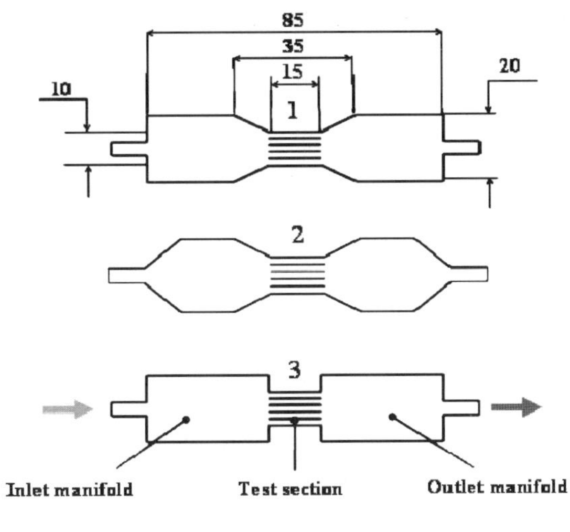

Fig. 2.66 Types of manifolds used for calculations. Reprinted from Mishan et al. (2007) with permission

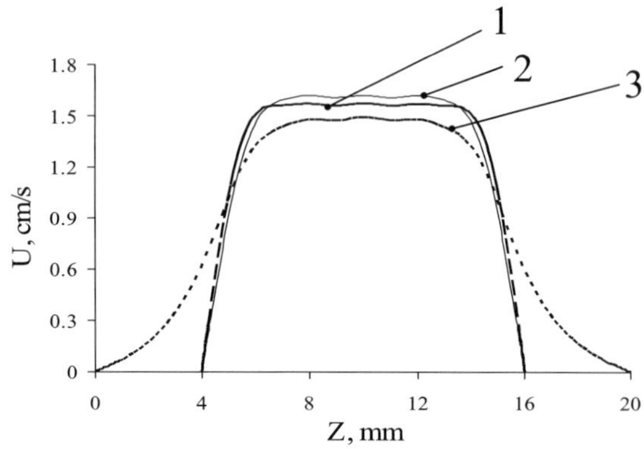

Fig. 2.67 Velocity profile at micro-channel entrance. Manifolds: *1* type 1, *2* type 2, *3* type 3. Reprinted from Mishan et al. (2007) with permission

mon phenomena that are often negligible in standard flows, such as profile of inlet velocity, axial heat conduction, and design inlet and outlet.

Optimization

The object of this optimization is minimization of the unit's overall thermal resistance, at a fixed pumping power per unit area. For a micro-channel heat sink, the thermal resistance consists of conduction resistance, convection resistance and bulk resistance due to the bulk temperature rise of the coolant. Since the micro-channel heat sink is typically thin and the materials used, such as silicon, have very high thermal conductivity, the conduction resistance is normally a small part of the overall resistance. As the channel size decreases, for a fixed pressure drop or pumping power, the convection resistance also decreases because the Nusselt number remains constant. However, the bulk resistance increases because now the flow rate decreases. Therefore there is an optimum channel dimension. If the pumping power is not limited, there is no optimum aspect ratio since the larger the aspect ratio, the better the convective heat transfer. With the constraint of pumping power, however, the aspect ratio cannot be indefinitely increased because the fluid flowing near the fin tip does not remove any heat at very high aspect ratios. Assuming fully developed conditions, Bau (1998) conducted an optimization study to minimize the temperature gradient and the overall thermal resistance. It was demonstrated that further reductions in overall thermal resistance and temperature gradient could be achieved by varying the cross-sectional dimensions of the micro-channel. Another optimization study by Knight et al. (1992) reported that at the optimal configuration, the thermal resistance for the Tuckerman and Pease device (1981) could be reduced by 35% if turbulent flow is allowed. However, the required pumping power is almost five times higher. Phillips (1990) provided an analytical model to estimate the thermal resistance of a micro-channel, whose results were compared with measurements and showed very good agreement. Weisberg et al. (1992) analyzed a micro-channel heat exchanger by numerical simulation of the heat transfer in the fluid and the solid substrate. Several assumptions made in the previous research were examined and found to be valid in this study. The manifold micro-channel heat sink was first proposed by Harpole and Eninger (1991) and numerically studied in Copeland et al. (1997) and Ng and Poh (1999). Unlike a conventional micro-channel heat sink, the manifold type features many inlet and outlet manifolds, alternating at a periodic distance along the length of the micro-channel. It is expected to have a smaller pressure drop compared with its conventional counterpart for a fixed flow rate. Gillot et al. (1998) evaluated thermal performance of micro-channel heat sink for a multi-chip power module. A prototype demonstrated the capability of dissipating 230–350 W/cm with a temperature rise of 35 °C. The overall thermal resistance can be reduced by increasing the pumping power. However, this can be costly in terms of the micro-pump requirement. Another way to reduce thermal resistance is to increase the heat transfer area. For a substrate with fixed area, one way to increase the heat transfer area is to use a heat spreader and then attach a heat sink to the spreader. This is commonly used in cooling of the central processing unit in a computer. The area increase in this case is achieved in the plane of the substrate. The stacked micro-channel analyzed here increases the heat transfer area in the out-of-plane direction. The uniqueness of such a design is that a heat spreader is not required. Wei and Joshi (2000) evaluated the thermal performance of stacked micro-channel heat sinks (schematic shown in Fig. 2.68). At each layer, a number of parallel micro-channels are machined in the surface of a substrate, e.g., copper, silicon or diamond. These layers are then bonded into a stacked micro-heat exchanger.

It was demonstrated that for a fixed pumping power, the overall thermal resistance for a two-layered micro-channel stack is 30% less than for a one-layered micro-channel due to doubling of the heat transfer area. Methods for optimizing the thermal performance of micro-channel heat sinks were discussed by Kim (2004).

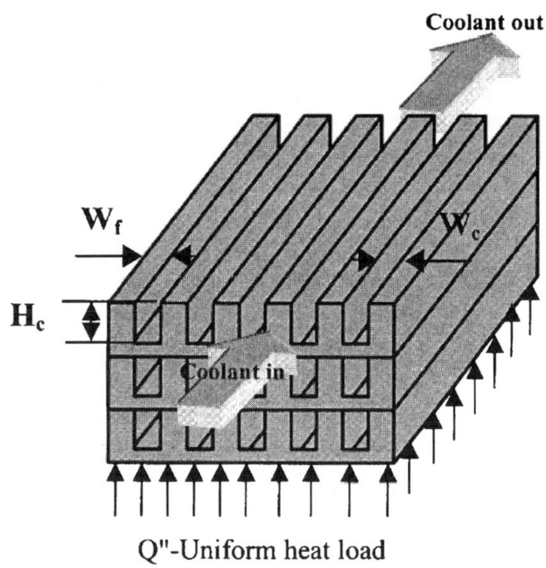

Fig. 2.68 Three-dimensional stack of micro-channels. Reprinted from Wei and Joshi (2000) with permission

These include analytical models (fin and porous medium) and a three-dimensional numerical approach. The fin model is based on the assumption of unidirectional conduction, constant heat transfer coefficient, and uniform fluid temperature, while the porous medium model is based on volume averaging of the velocities and temperatures in the direction perpendicular to the flow. From the optimization of a micro-channel heat sink, which minimizes the total thermal resistance under the constraint of a maximum pumping power, the optimal values of channel height, channel width, and fin thickness are presented. In order to check the validity of the analytical models, the temperature distributions from the two analytical models are compared with those from the numerical simulation. The assumption of unidirectional conduction, and the local volume-averaging technique, are shown to be justified, whereas the assumption of constant heat transfer coefficient with reference to the bulk mean temperature is invalid for large values of the aspect ratio. This latter assumption, combined with incorrect values of fin efficiency, may result in errors in the thermal design if the aspect ratio is larger than eight. Because of this defect, the fin model fails to provide the design variables.

It is noteworthy that several studies exhibit contradictory results for both the mechanical and thermal characteristics of the flow. This is generally due to differences in the many parameters that characterize these studies such as the geometry, shape and surface roughness of the channels, the fluid, the boundary conditions and the measuring methodology itself. These discrepancies indicate the need for extension of the experimental base to provide the necessary background to the theoretical model.

In the past few years, Bar-Cohen and co-workers documented an extended series of heat sink design and optimization studies aimed at achieving the highest energy efficiency for high-performance heat sinks (Bahadur and Bar-Cohen 2005; Bar-Cohen et al. 2006). In their work they introduced an approach involving a new metric, called the total coefficient of performance (COPT), which namely is the ratio of the heat removed by the sink to the total invested work, including both the pumping power and energy content of the heat sink. They successfully demonstrated the suitability of this approach for air-cooled heat sinks, and found that for a wide parametric range, the most energy efficient heat sink design involves use of "least material" optimum fins. This technique has yet to be applied to liquid-cooled heat sinks, where much higher heat transfer coefficients and heavier pumping power losses than experienced in air-cooled heat sinks will be encountered. Extension of this design and optimization methodology to high-performance micro-channel coolers can be expected to dramatically improve the energy efficiency of these thermal management devices and establish the broad applicability of the COPT approach.

A typical example of calculation of the thermal coefficient of performance for forced convection of air is shown in Fig. 2.69. COP surfaces are presented for the maximum and least material aluminum heat sink configurations in the design flow space between 0.01–0.04 m^3/s and 20–80 Pa.

These upper-bound COP values for maximum thermal performance aluminum arrays are seen to approach remarkably high values of nearly 1000 and provide

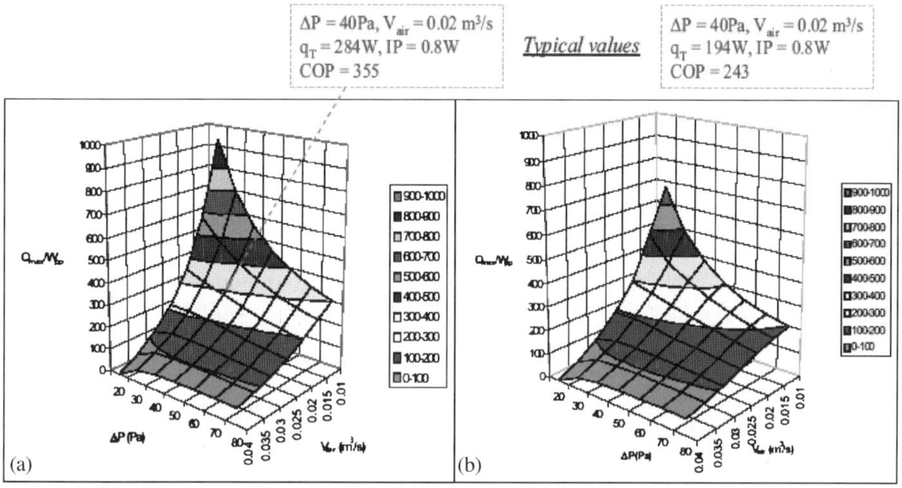

Fig. 2.69a,b Coefficient of performance for aluminum plate fin array in forced convection (**a**) maximum heat transfer design, and (**b**) least material design. Reprinted from Bar-Cohen et al. (2006) with permission

a broad range of flow rate/pressure head combinations with COPs above 300. Predictably, the least material configurations – in which some thermal performance is sacrificed in favor of significant mass reduction – offer lower, though still remarkably high COP values, peaking at nearly 700, with a broad range of options for attaining COP values above 200. It should be noted that for the maximum, as well as least material heat sinks, the peak COPs occur at the lowest values of pressure drop and flow rate and then decrease rather steeply with increasing volumetric flow rate and heat sink pressure drop.

Fabrication technique

The earliest micro-channels were built using anisotropic wet chemical etching techniques based on alkaline solutions. Recently, anisotropic dry processes have been developed. These are significantly faster and, from the manufacturing standpoint, involve fewer contamination and waste treatment problems. Some fabrication areas that may spur advances include new materials, high-aspect ratio patterning techniques other than dry etching, active fluid flow elements, and micro-molding. Miniaturized traditional techniques are in some respects the most straightforward approach to creating micro-channels. Saw cuts on the order of 25 μm width with accuracy of 4 μm can be obtained with commercially available equipment. Micro-discharge-electromachining has been demonstrated, using very fine wires as electrodes. Other cutting techniques such as ultrasonic and water jet machining are especially effective on hard brittle materials. Laser machining has become a powerful tool that can handle a wide variety of difficult materials. Focused ion beam machining offers many similar benefits and can operate in the submicron regime.

A review of micro-channel fabrication technology was published by Kandlikar and Grande (2002). In many cases silicon is the substrate of choice for reasons of cost and process compatibility. The etching can be carried out in either a wet chemical or dry plasma format, and both techniques have isotropic and anisotropic variants (Bean 1978; Bhardwaj et al. 1997). Anisotropic wet chemical etching (WCE) (using potassium hydroxide or ethylene diamine pyrocatechol) can produce specific micro-channel device designs. In most cases triangular micro-channels may be created with a 54.74° angle to the surface. Isotropic wet chemical etching (using hydrofluoric acid, nitric acid, and acetic acid) results in hemispherical etch profiles. Another constraint of anisotropic WCE is that features formed by the intersection of planes are stable only where the corner angle is less than 180°. A recent advance in etch technology are deep reactive ion etch processes (DRIE), which can produce vertical etch profiles in silicon. Figure 2.70 shows a set of micro-channels with 42 μm trenches etched to more than 100 μm. Bulk micro-machining etch processes are summarized in Fig. 2.71. Etch techniques can be combined in unique ways to create complex micro-channel structures. Figure 2.72 schematically illustrates a process that can create micro-channels buried inside a wafer. It is also possible to build sets of micro-channels at different depths and have them cross over one another. The micro-channels with hydraulic diameters in the range of 1.01 to 35.91 μm and length of 10–11 cm were fabricated using photolithography and wet etching techniques (Harley et al. 1995).

A class of very high-aspect ratio fabrication processes is based on the lost wax molding technique (LIGA) (lithography, electroforming, molding). As shown in Fig. 2.73, LIGA uses highly collimated X-rays projected through a special X-ray mask to provide near diffraction-free exposure of a thick photoresist. The technique can create structures with aspect ratios in excess of 100:1 and can hold submicron tolerances over many hundreds of microns of vertical height (Mohr et al. 1988).

Fabrication of micro-channels in dielectrics, using femtosecond lasers, has also been used (Hwang et al. 2004). In these applications amplified pulsed lasers produc-

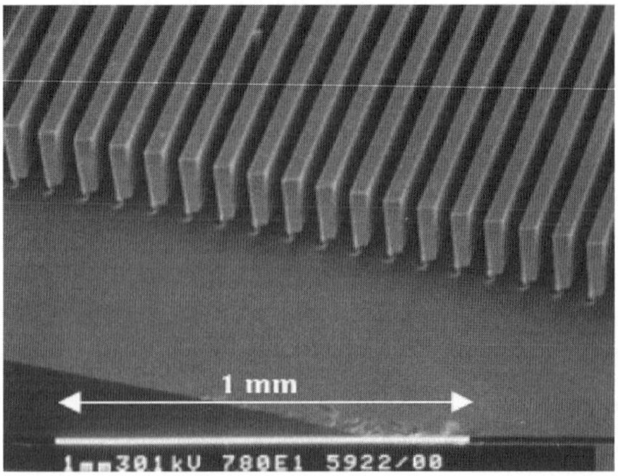

Fig. 2.70 Micro-channel array formed by silicon DRIE. Reprinted from Kandlikar and Grande (2002) with permission

Fig. 2.71a–c Bulk micro-machining etch profiles. Reprinted from Kandlikar and Grande (2002) with permission

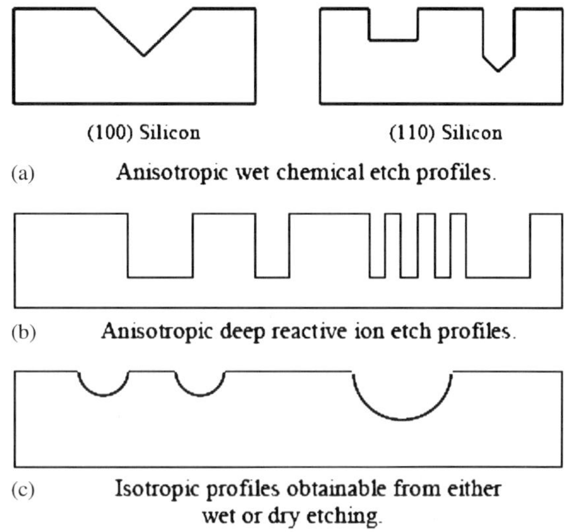

Fig. 2.72a–c Construction of buried micro-channels by combined anisotropic and isotropic etching. Reprinted from Kandlikar and Grande (2002) with permission

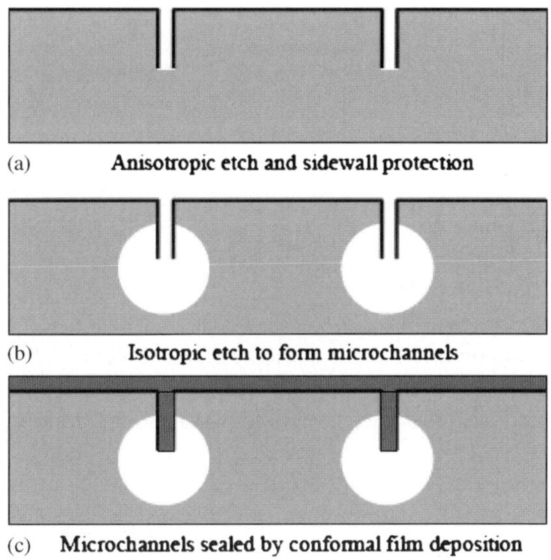

ing micro-joules per pulse were used to overcome the optical breakdown threshold of the substrate, typically glass or fused silica. Based on this method, cylindrical micro-channels with diameters ranging from 8 to 20 μm can be fabricated in a water-immersed polymethyl methacrylate (PMMA) substrate (Fig. 2.74). The laser beam is focused onto the sample by a long working distance objective. Water immersion is used to reduce aberrations and improve the performance and consistency of the fabrication method.

None of the technologies described above can individually produce a complete micro-system. Hybridization is the means of combining all the necessary disparate

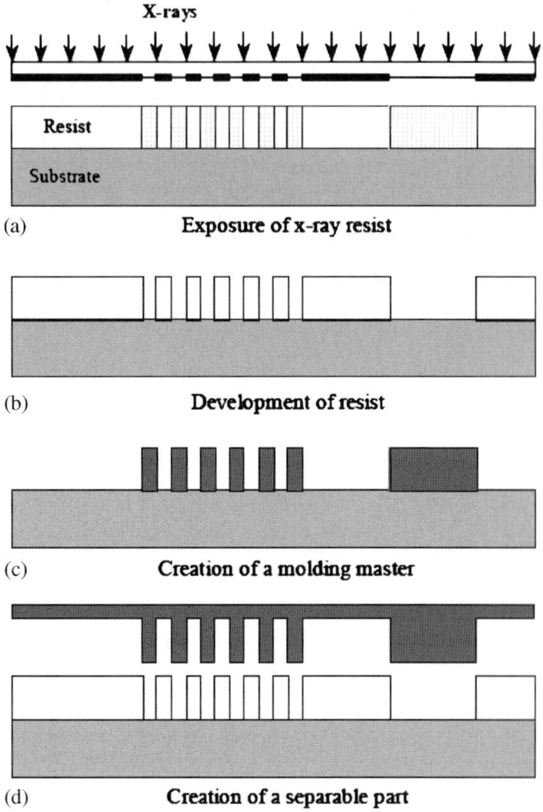

Fig. 2.73a–d The LIGA process. Reprinted from Kandlikar and Grande (2002) with permission

components. Direct wafer bonding can generally be used to obtain a wide range of adhesive bond strengths (Plosl and Krauter 1999). Three bonding techniques of particular interest are fusion bonding, anodic bonding, and adhesive bonding. In fusion bonding two wafers whose surfaces are silicon or silicon compounds can be bonded through a combination of chemical surface treatments, pressure, and annealing at elevated temperature. In anodic bonding silicon and ionic glass surfaces are jointed through a combination of pressure, temperature, and electric field. While both fusion and anodic bonding can produce interfaces of high strength, they are quite material-specific. For generic hetero-bonding, adhesive techniques are the solution. Figure 2.75 shows how micro-channel devices can be formed by wafer bonding.

A method that creates patterned micro-structures distributed on the bottom wall of the micro-channel was proposed by Yang et al. (2006). A roughened bottom wall was created using the crystal orientation characteristics of the wafers.

Wet etching procedures were used in fabrication of the micro-channel heat sink. The substrates were covered with thermally grown silicon dioxide and patterned in a photolithography process. Hexagonal cavities distributed over the channel bottom wall were created using a mask containing parallelograms in the micro-channel. Figure 2.76 shows a typical example of created micro-structures distributed along

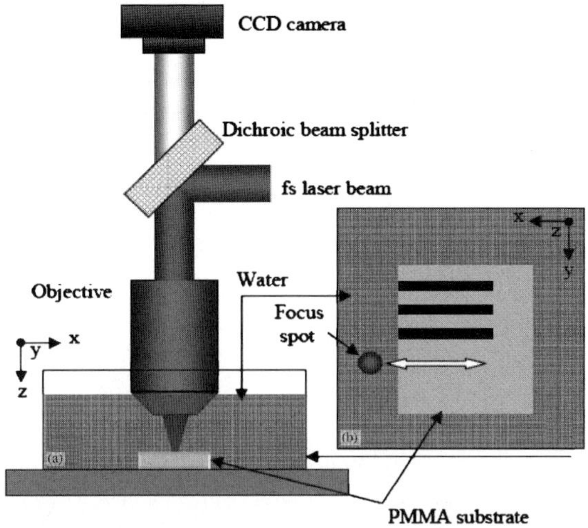

Fig. 2.74a,b Schematic diagram of (**a**) the experimental set-up for fabrication of micro-channels and (**b**) the fabrication geometry in the sample. Reprinted from Hwang et al. (2004) with permission

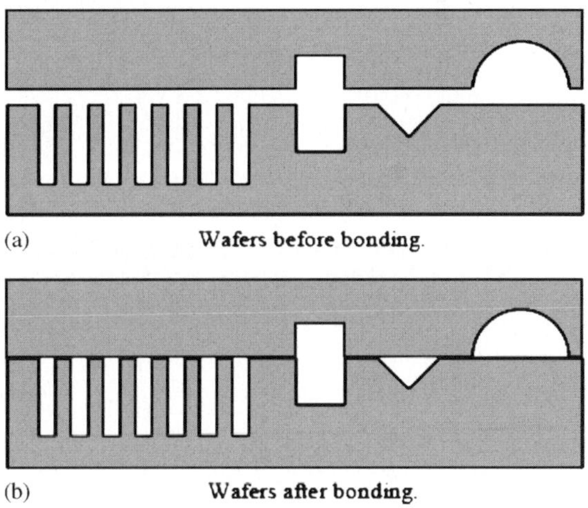

Fig. 2.75 The wafer bonding process. Reprinted from Kandlikar and Grande (2002) with permission

the bottom walls. The channel width is 500 μm and the cavity size is 200 × 200 × 55 μm. The microstructures are expected to induce fluid flow and thermal boundary layer disturbances, leading to an enhanced heat transfer coefficient.

A completely different way involves use of metal foams, or metal made porous otherwise (North and Cho 2003; Hetsroni et al. 2006a).

Aluminum foam can be used as a porous medium in the model of a heat sink with inner heat generation (Hetsroni et al. 2006a). Open-cell metal foam has a good effective thermal conductivity and a high specific solid–fluid interfacial surface area.

Depending on the metal foam configuration, its specific surface area varies from 500 for original foam to 10,000 m^2/m^3 for compressed foam. Aluminum foam of 40 pores per inch (ppi) was studied. The structure of the porous material is presented in Fig. 2.77.

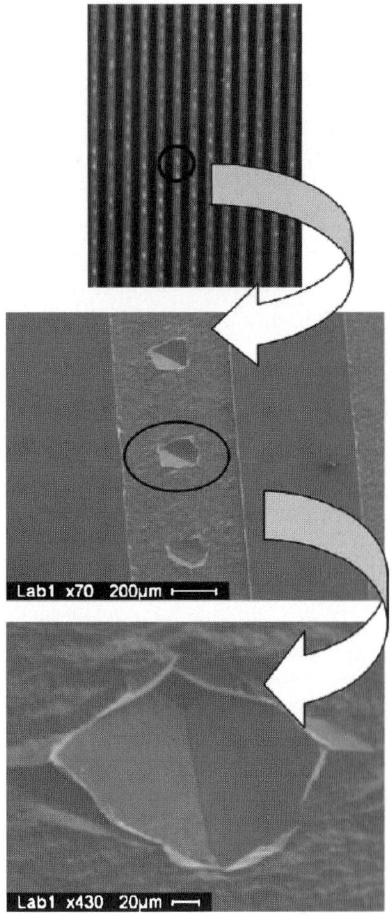

Fig. 2.76 SEM micrograph of roughened micro-channel bottom wall with distributed hexagonal micro-structures. Reprinted from Yang et al. (2006) with permission

Summary

1. The increased power density in electronic and other devices leads to higher operational temperature, which limits the device performance. Cooling problems are becoming acute for such devices, and an efficient cooling system is required to maintain an isothermal temperature of the device. Direct liquid cooling has emerged as one of the most promising thermal management techniques for micro-systems where the control of both the operating temperature and the temperature cycling is still a challenging task.

The active micro-channel cooling contains a cooling system centered on a micro-heat collector that is fabricated much in the same way as the chip. Heat collector usually has a dense amount of micro-channels etched into the surface that works to transfer heat to a fluid, which is circulated through the entire package.

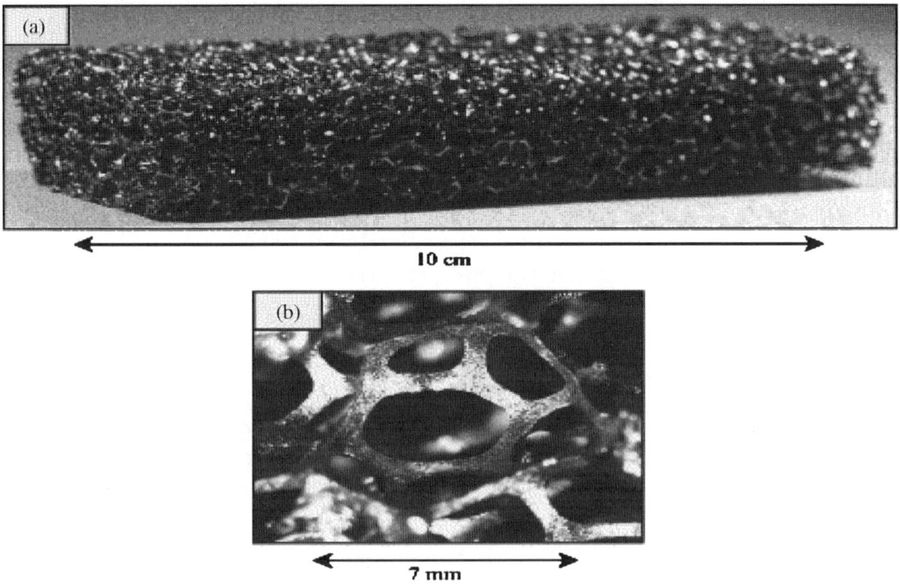

Fig. 2.77a,b Sample of aluminum foam. Reprinted from Hetsroni et al. (2006a) with permission

There are now companies engaged in the development and commercialization not only of micro-channels, but also spray cooling, synthetic jets, liquid metal cooling, and sintered porous inserts. For heat flux densities up to 10–50 W/cm^2 air cooling may remain the cooling option of choice. For heat fluxes over 100–500 W/cm^2 some form of liquid cooling appears to be the most viable option. Several investigations demonstrated that using water as cooling fluid has the potential for industrial application in the range between 500 and 1,000 W/cm^2. There are also devices with cooling heat flux requirements on the order of 10^3–10^4 W/cm^2. Fusion reactors, for example, contain components that require continuous cooling on the order of 10^4 W/cm^2. Other examples are directed energy devices such as high efficiency, multi-megawatt continuous-wave magnetrons used for short-pulse lasers and radars, optical devices that deliver high brilliance beams. The enormous cooling requirements of these high-energy devices preclude the use of refrigerants. For these applications, water and liquid metals remain the coolants of choice. The high cost of the systems, the stringent material and compatibility requirements constrain the application of liquid metals to cooling. Water may be chosen as a cooling fluid because of its outstanding thermophysical properties: very high specific heat and latent heat of vaporization.

The high-performance supercomputers, certain power devices, electric vehicles, and advanced avionics need to maintain their temperature in the range of 30–60 °C at the high-flux range. However, because of its very low vapor pressure at the expected design saturation temperatures of 40–65 °C, water is not a candidate fluid for the flow boiling coolant. The most promising are liquid refrigerants, which at atmospheric pressure have saturation temperatures in the range of 30–60 °C. For example, FC-72 may be taken as a possible candidate of various dielectric fluids. It has at atmospheric pressure the saturation temperature of 56.6 °C, and the latent heat 94.8 kJ/kg. It should be taken into account that there is some concern over the global warming potential of all refrigerants, as well as the danger of chemical breakdown and production of harmful substances upon exposure to high temperatures.

2. In this book we will consider the channels with hydraulic diameters ranging roughly from 5 to 500 μm as micro-channels and the channels with hydraulic diameters $d_h > 500$ μm as conventional size channels. Heat transfer in micro-channels has been studied in a number of investigations, and has been compared with the behavior of conventional (i.e., large-sized) length scales. However, there have been wide discrepancies between different sets of published results. Measured heat transfer coefficients in single-phase and two-phase micro-channel flows have either well exceeded, or fallen far below, those predicted for conventional channels. The Reynolds number at which the flow behavior indicates a transition from laminar to turbulent flow has also differed widely in these studies We considered these problems with regard to micro-channels to gain a better understanding of the distinctive properties of the measurement techniques and uncertainties, the conditions, under which the experimental results should be compared to analytical or numerical predictions, boiling phenomenon, as well as different types of micro-channel heat sinks.

3. Pressure drop measurements. For the majority of experiments the instrumentation was relatively similar. Due to limitations associated with the small size of the channels, pressures were not measured directly inside the micro-channels. To obtain the channel entrance and exit pressures, measurements were taken in a plenum or supply line prior to entering the channel. It is insufficient to assume that the friction factor for laminar compressible flow can be determined by means of analytical predictions for incompressible flow.

The method of the measurement pressure drop inside the channel itself should be used to check the overall agreement between experimental and predicted results. The internal pressure measurements should be used also to validate the entrance and exit losses.

4. Temperature measurements. Reliable measurement and control of temperature in the micro-scale are highly required to develop various micro-devices. Many temperature measurement methods traditionally applied to macro-devices are evolving into more advanced techniques applicable to micro-devices taking into consideration enhanced spatial, temporal, and temperature resolution. The thermochromic liquid crystal may be employed for full-field mapping of temperature fields. The good results obtained by the widespread use of infrared thermography (IR) in experimental studies of convective heat transfer and boiling in micro-channels have proved this method to be an effective tool in overcoming several limitations of the standard sensors originating both from the measurement and the visualization techniques. Recently IR has been developed to measure the temperature of the fluid and wall in a micro-channel, using a transparent cover. Measurement of the temperature field of a micro-object by an infrared camera has a number of problems. The small size of the object causes a substantial amount of infrared radiation from the

background. The problem of background influence on the object temperature measurement should be taken into account.

5. In general, the result of measurement is only an approximation or estimate of the value of the specific quantity subject to measurement, and thus the result is complete only when accompanied by a quantitative statement of its uncertainty. Because the reliability of evaluations of components of uncertainty depends on the quality of the information available, it is recommended that all parameters upon which the measurand depends be varied to the fullest extent practicable so that the evaluations are based as much as possible on observed data.

6. Pressure drop and heat transfer in a single-phase incompressible flow. According to conventional theory, continuum-based models for channels should apply as long as the Knudsen number is lower than 0.01. For air at atmospheric pressure, Kn is typically lower than 0.01 for channels with hydraulic diameters greater than 7 μm. From descriptions of much research, it is clear that there is a great amount of variation in the results that have been obtained. It was not clear whether the differences between measured and predicted values were due to determined phenomenon or due to errors and uncertainties in the reported data. The reasons why some experimental investigations of micro-channel flow and heat transfer have discrepancies between standard models and measurements will be discussed in the next chapters.

7. Steam–liquid flow. Two-phase flow maps and heat transfer prediction methods which exist for vaporization in macro-channels and are inapplicable in micro-channels. Due to the predominance of surface tension over the gravity forces, the orientation of micro-channel has a negligible influence on the flow pattern. The models of convection boiling should correlate the frequencies, length and velocities of the bubbles and the coalescence processes, which control the flow pattern transitions, with the heat flux and the mass flux. The vapor bubble size distribution must be taken into account.

The flow pattern in parallel micro-channels is quite different from that found in a single micro-channel. At the same values of heat and mass flux, different flow regimes exist in a given micro-channel depending on the time. Moreover, at the same time, different flow regimes may exist in various parallel micro-channels. At low vapor quality heat flux causes a sudden release of energy into the vapor bubble, which grows rapidly and occupies the entire channel. The rapid bubble growth pushes the liquid–vapor interface on both caps of the vapor bubble, at the upstream and the downstream ends, and leads to a reverse flow. This phenomenon may be regarded as explosive boiling. The CHF phenomenon is different from that observed in annular flow. A key difference is the amplification of flow and wall temperature instabilities prior to CHF.

Confined boiling of water and surfactant solutions under condition of natural convection causes a heat transfer enhancement. Additive of surfactant leads to enhancement of heat transfer compared to water boiling in the same gap size; however, this effect decreased with decreasing gap size. For the same gap size, CHF decreases with an increase in the channel length. CHF in surfactant solutions is significantly lower than in water.

It is noteworthy that several studies presented very different results for both the heat transfer at flow boiling and CHF in micro-channels. This is generally due to differences in many parameters that characterize these studies such as the geometry, the hydraulic diameter, the shape and surface roughness of the channels, the fluid nature, the boundary conditions, the flow regimes and the measuring technique. Such a large variety of experimental conditions often makes it difficult to apply the results of a given study to other investigations. The present chapter gives only a general insight on the problem. In the next chapters the main categories of effects will be discussed to study the boiling in micro-channels.

8. For micro-channel heat sink there is an optimum channel dimension. If the pumping power is not limited, there is no optimum aspect ratio since the larger the aspect ratio, the better the heat transfer. With the constraint on pumping power, however, the aspect ratio cannot be indefinitely increased because the fluid flowing near the fin does not remove any heat at very high aspect ratios. For an actual micro-channel cooling element, the micro-channels often have a rectangular shape in its cross-section, with aspect ratio in the range of 5–10 (with the short side at the base of the cooling element and the long sides forming fins between channels).

9. Fabrication technique. The micro-channels on the order of 25 μm width with accuracy of 4 μm can be manufactured with commercially available equipment. The micro-channels with diameters ranging from 8 to 20 μm can be fabricated using femtosecond lasers. In some experiments on the effect of rarefaction on the friction factor the channels of $d_\mathrm{h} = 4.688\,\mu\mathrm{m}$ with surface roughness ranging from 0.002 to 0.06 μm were used. Using photolithography and wet etching technique it was possible to fabricate micro-channels with hydraulic diameters from 1.01 to 35.91 μm and lengths from 10 to 11 cm.

References

Acikalin T, Wait S, Garimella S, Raman A (2004) Experimental investigation of the thermal performance of piezoelectric fans. Heat Transfer Eng 25:4–14

Adams TM, Abdel-Khalik SI, Jeter SM, Qureshi ZH (1998) An experimental investigation of single-phase forced convection in micro-channels. Int J Heat Mass Transfer 41:851–857

Adams TM, Dowling MF, Abdel-Khalik SI, Jeter SM (1999) Applicability of traditional turbulent single phase forced convection correlations to non-circular micro-channels. Int J Heat Mass Transfer 42:4411–4415

American Society Of Mechanical Engineers (ASME) (2000) Policy on reporting uncertainties in experimental measurements and results. J Heat Transfer 122:411–413

Azar K, Benson JR, Manno VP (1991) Liquid crystal imaging for temperature measurement of electronic devices. In: Proceedings of 7th Annual IEEE Semiconductor Thermal Measurement and Management Symposium, Phoenix, 12–14 February 1991, pp 23–33

Bahadur R, Bar-Cohen A (2005) Thermal design and optimization of polymer-based pin fin natural convection heat sinks. IEEE Trans Comp Packag Technol 28(2):238–246

Bar-Cohen A, Bahadur R, Iyengar M (2006) Least-energy optimization of air-cooled heat sinks for sustainability-theory, geometry and material selection. Energy 31:579–619

Bau HH (1998) Optimization of conduits' shape in micro heat exchangers. Int J Heat Mass Transfer 41:2717–2723

Bean KE (1978) Anisotropic etching of silicon. IEEE Trans Electron Dev 25(10):1185–1193

Beratlis N, Smith M (2003) Optimization of synthetic jet cooling for micro-electronics applications. In: Proceedings of 19th SemiTherm Symposium, San Jose, 11–13 March 2003, pp 66–73

Bergles AE, Lienhard JH, Gall V, Kendall GE, Griffith P (2003) Boiling and evaporation in small diameter channels. Heat Transfer Eng 24:18–40

Bhardwaj J, Ashraf H, McGuarrie A (1997) Dry silicon etching for MEMS. In: Proceedings of the 191st Meeting of the Electrochemical Society, Microstructures and Microfabricated Systems, III Symposium, Montreal, 4–9 May 1997, vol 97-5, pp 118–130

Bonjour J, Lallemand M (1998) Flow patterns during boiling in a narrow space between two vertical surfaces. Int J Multiphase Flow 24:947–960

Bowers MB, Mudawar I (1994) High flux boiling in low flow rate, low pressure drop mini-channel and micro-channel heat sinks. Int J Heat Mass Transfer 37:321–332

Boyd RD (1985) Subcooled flow boiling critical heat flux and its application to fusion energy components. Part 1. A review of fundamentals of CHF and related data base. Fusion Tech 7:7–31

Calame JP, Myers RE, Binari SC, Wood FN, Garven M (2007) Experimental investigation of micro-channel coolers for the high heat flux thermal management of GaN-on-SiC semiconductor devices. Int J Heat Mass Transfer 50: 4767-4779

Celata GP, Cumo M, Zummo G (2004) Thermal-hydraulic characteristics of single- phase flow in capillary pipes. Exp Thermal Fluid Sci 28:87–95

Celata GP (2004). Heat transfer and fluid flow in micro-channels. Begell House, N.Y.

Celata GP, Cumo M, McPhail SJ, Tesfagabir L, Zummo G (2005) Experimental study on compressibility effects in micro-tubes, in Proceedings of the XXIII UIT Italian National Coference, 2005:53-60

Chaudhari AM, Woudenberg TM, Albin M, Goodson KE (1998) Transient liquid crystal thermometry of microfabricated PCR vessel arrays. J Microelectromech Sys 7:345–355

Cheng P, Wu WY (2006) Mesoscale and microscale phase heat transfer. In: Greene G, Cho Y, Hartnett J, Bar-Cohen A (eds) Advances in heat transfer, vol 39. Elsevier, Amsterdam

Choi SB, Barron RF, Warrington RQ (1991) Fluid flow and heat transfer in micro- tubes. ASME DSC 40:89–93

Chung PM-Y, Kawaji M (2004) The effect of channel diameter on adiabatic two-phase flow characteristics in micro-channels. Int J Multiphase Flow 30:735–761

Colgan E (2005) A practical implementation of silicon microchannel coolers for high power chips. In: Proceedings of 21st SemiTherm Symposium, San Jose, 15–17 March 2005, pp 1–7

Copeland D, Behnia M, Nakayama W (1997) Manifold micro-channel heat sinks: isothermal analysis. IEEE Trans Comp Packag Manuf Technol A 20:96–102

Dupont V, Thome JR, Jacobi AM (2004) Heat transfer model for evaporation in microchannels. Part II. Comparison with the database. Int J Heat Mass Transfer 47:3387–3401

Evans FE, Wennerstrom H (1999) The colloidal domain, 2nd edn. VCH, New York

Fabbri M, Jiang S, Dhir VK (2005) A comparative study of cooling of high power density electronics using sprays and microjets. J Heat Transfer 127:38–48

Fujita Y, Ohta H, Uchida S, Nishikawa K (1988) Nucleate boiling heat transfer and critical heat flux in narrow space between rectangular surface. Int J Heat Mass Transfer 31:229–239

Furjes P, Vízvary Zs, Adam M, Barsony I, Morrissey A, Ducso Cs (2002) Materials and processing for realization of micro-hotplates operated at elevated temperature. J Micromech Microeng 12:425–429

Gad-el-Hak M (1999) The fluid mechanics of micro-devices. The Freemen Scholar Lecture. J Fluid Eng 121:5–33

Garimella S, Sobhan C (2003) Transport in microchannels: a critical review. Ann Rev Heat Transfer 13:1–50

Gillot C, Schaeffer C, Bricard A (1998) Integrated micro heat sink for power multichip module. IEEE Trans Ind Appl 36(1):217–221

Guo ZY, Li ZX (2003) Size effect on single-phase channel flow and heat transfer at microscale. Int J Heat Mass Transfer 24:284–298

Harley JC, Huang Y, Bau H, Zemel JN (1995) Gas flow in micro-channels. J Fluid Mech 284:257–274

Harpole GM, Eninger JE (1991) Micro-channel heat exchanger optimization. In: Proceedings of 7th IEEE SemiTherm Symposium, Scottsdale, 12–14 February 1991, pp 59–63

Hetsroni G, Gurevich M, Mosyak A, Pogrebnyak E, Rozenblit R, Yarin LP (2003a) Boiling in capillary tubes. Int J Multiphase Flow 29:1551–1563

Hetsroni G, Gurevich M, Mosyak A, Rozenblit R (2004a) Drag reduction and heat transfer of surfactants flowing in a capillary tube. Int J Heat Mass Transfer 47:3797–3809

Hetsroni G, Gurevich M, Mosyak A, Rozenblit R, Yarin LP (2002c) Subcooled boiling of surfactant solutions. Int J Multiphase Flow 28:347–361

Hetsroni G, Gurevich M, Rozenblit R (2006a) Sintered porous medium heat sink for cooling of high-power mini-devices. Int J Heat Fluid Flow 27:259–262

Hetsroni G, Mosyak A, Segal Z (2001a) Nonuniform temperature distribution in electronic devices cooled by flow in parallel micro-channels. IEEE Trans Comp Packag Technol 24:16–23

Hetsroni G, Zakin JL, Lin Z, Mosyak A, Pancallo EA, Rozenblit R (2001b) The effect of surfactants on bubble growth, wall thermal patterns and heat transfer in pool boiling. Int J Heat Mass Transfer 44:485–497

Hetsroni G, Mosyak A, Pogrebnyak E, Segal Z (2005b) Explosive boiling of water in parallel micro-channels. Int J Multiphase Flow 31:371–392

Hetsroni G, Mosyak A, Pogrebnyak E, Segal Z (2006b) Periodic boiling in parallel micro-channels at low vapor quality. Int J Multiphase Flow 32:1141–1159

Hetsroni G, Mosyak A, Pogrebnyak E, Sher I, Segal Z (2006c) Bubble growth in saturated pool boiling in water and surfactant solution. Int J Multiphase Flow 32:159–182

Hetsroni G, Mosyak A, Pogrebnyak E, Segal Z (2007) Natural convection boiling of water and surfactants in narrow horizontal annular channels. Int J Multiphase Flow 33:469–483

Hetsroni G, Mosyak A, Pogrebnyak E, Yarin L (2005a) Fluid flow in micro-channels. Int J Heat Mass Transfer 48:1982–1998

Hetsroni G, Mosyak A, Pogrebnyak E, Yarin LP (2005c) Heat transfer in micro-channels: comparison of experiments with theory and numerical results. Int J Heat Mass Transfer 48:5580–5601

Hetsroni G, Mosyak A, Segal Z, Pogrebnyak E (2003b) Two-phase flow patterns in parallel micro-channels. Int J Multiphase Flow 29:341–360

Hetsroni G, Mosyak A, Bernheim-Groswasser A, Talmon Y, Zakin JL (2003c) The effect of cationic surfactant on turbulent flow patterns. J Heat Transfer ASME 125:947–950

Hetsroni G, Mosyak A, Segal Z, Pogrebnyak E (2002a) Two-phase flow patterns in parallel micro-channels. Int J Multiphase Flow 9:341–260

Hetsroni G, Mosyak A, Segal Z, Ziskind G (2002b) A uniform temperature heat sink for cooling of electronic devices. Int J Heat Mass Transfer 45:3275–3286

Hetsroni G, Zakin JL, Gurevich M, Mosyak A, Pogrebnyak E, Rozenblit R (2004b) Saturated flow boiling heat transfer of environmentally acceptable surfactants. Int J Multiphase Flow 30:717–734

Hwang DJ, Choi TY, Grigoropoulos CP (2004) Liquid assisted femtosecond laser drilling of straight and three-dimensional micro-channels in glass. Appl Phys A 79:605–612

Jiang L, Wang Y, Wong M, Zohar Y (1999a) Fabrication and characterization of a microsystem for a micro-scale heat transfer study. J Micromech Microeng 9:422–428

Jiang L, Wong M, Zohar Y (1999b) Phase change in microchannel heat sinks with integrated temperature sensors. J Microelectromech Syst 8:358–365

Jiang L, Wong M, Zohar Y (2000) Unsteady characteristics of a thermal microsystem. Sens Actuators A 82:108–113

Kakac S, Vasiliev LL, Bayazitoglu Y, Yener Y (2005). Micro-scale heat transfer. Springer, Berlin Heidelberg.

Kandlikar SG, Grande W (2002) Evolution of micro-channel flow passages – thermo-hydraulic performance and fabrication technology. In: Proceedings of IMECE ASME International Mechanical Engineering Congress and Exposition, New Orleans, 17–22 November 2002, IMECE 2002-32043, pp 1–13

Kandilikar SG, Upadhye H (2005) Extending the heat flux limit with enhanced microchannels in direct single-phase cooling of computer chips. In: Proceedings of 21st SemiTherm Symposium, San Jose, 15–17 March 2005, pp 8–15

Kandlikar SG, Garimella S, Li D, Colin S, King MR (2005) Heat Transfer And Fluid Flow In Minichannels And Microchannels. Elsevier Science & Technology (Netherlands)

Karniadakis GE, Beskon A (2002) Fundamentals and simulation. Springer, Berlin Heidelberg New York

Katto Y, Ohno H (1984) An improved version of the generalized correlation of critical heat flux for the forced convective boiling in uniformly heated vertical tubes. Int J Heat Mass Transfer 27(9):1641–1648

Katto Y, Yokoya S, Teraoka K (1966) Experimental study of nucleate boiling in case of making interference-plate approach to the heating surface. In: Proceedings of 3rd International Heat Transfer Conference, 1966, vol 3, pp 219–227

Kim SJ (2004) Methods for thermal optimization of microchannel heat sinks. Heat Transfer Eng 25(1):37–49

Kim HJ, Kihm KD (2001) Application of a two-color laser induced fluorescence (LIF) technique for temperature mapping. In: Proceedings of ASME International Mechanical Engineering Congress and Exposition, IMECE2001/HTD-24411, New York, 11–16 November 2001

Kim J, Golliher E (2002) Steady state model of a micro loop heat pipe. Proceedings of 18th SemiTherm Symposium, San Jose, 15–17 March 2005, pp 137–144

Klein D, Hetsroni G, Mosyak A (2005) Heat transfer characteristics of water and APG surfactant solution in a micro-channel heat sink. Int J Multiphase Flow 31:393–415

Kline SJ, McClintock FA (1953) Describing uncertainties in single-sample experiments. Mech Eng 75(Jan):3–8

Knight RW, Hall DJ, Goodling JS, Jaeger RC (1992) Heat sink optimization with application to micro-channels. IEEE Trans Comp Hybrids Manuf Technol 15:832–842

Kohl MJ, Abdel-Khalik SI, Jeter SM, Sadowski DL (2005) An experimental investigation of microchannel flow with internal pressure measurements. Int J Heat Mass Transfer 48:1518–1533

Lasance CJM, Simons RE (2005) Advances in high performance cooling for electronics. http://electronics-cooling.com/html/2005nov.article2.html. Accessed 2007

Le Berre M, Pandraud G, Morfouli P, Lallemand M (2006) The performance of micro heat pipes measured by integrated sensors. J Micromech Microeng 16:1047–1050

Lee DY, Vafai K (1999) Comparative analysis of jet impingement and microchannel cooling for high heat flux applications. Int J Heat Mass Transfer 42:1555–1568

Lee PC, Tseng FG, Pan C (2004) Bubble dynamics in micro-channels. Part 1. Single microchannel. Int J Heat Mass Transfer 47:5575–5589

Li X, Lee WY, Wong M, Zohar Y (2000) Gas flow in constriction microdevices. Sens Actuators A 83:277–283

Lin Q, Jiang F, Wang X-Q, Han Z, Tai Y-C, Lew J, Ho C-M (2000) MEMS Thermal Shear-Stress Sensors: Experiments, Theory and Modeling, Technical Digest, Solid State Sensors and Actuators Workshop, Hilton Head, SC, 4–8 June 2000, pp 304–307

Lin TY, Yang CY (2007) An experimental investigation of forced convection heat transfer performance in micro-tubes by the method of liquid crystal thermography. Int. J. Heat Mass Transfer 50: 4736-4742

Liu J, Enzelberger M, Quake S (2002) A nanoliter rotary device for polymerase chain reaction. Electrophoresis 23:1531–1536

Lu B, Zheng Y, Scriven LE, Davis HT, Talmon Y, Zakin JL (1998) Effect of variation counterion-to-surfactant ratio on rheology and micro-structures of drag reducing cationic surfactant systems. Rheol Acta 37:528–548

Luo K, Shi Z, Varesi J, Majumdar A (1997) Sensor nanofabrication, performance, and conduction mechanisms in scanning thermal microscopy. J Vac Sci Technol B 15:349–360

Majumdar A (1999) Scanning thermal microscopy. Annu Rev Mater Sci 29:505–585

Manglik RM, Wasekar VM, Zhang J (2001) Dynamic and equilibrium surface tension of aqueous surfactant and polymeric solutions. Exp Thermal Fluid Sci 25:55–64

Maynes D, Webb AR (2002) Velocity profile characterization in sub-millimeter diameter tubes using molecular tagging velocimetry. Exp Fluids 32:3–15
Mehendale SS, Jacobi AM, Shah RK (1999) Heat exchangers at micro- and meso-scales. In: Proceedings of International Conference on Compact Heat Exchangers and Enhance Technology for the Process Industries, Banff, 18–23 July 1999, pp 55–74
Miner A, Ghoshal U (2004) Cooling of high-power density microdevices using liquid metal coolants. Appl Phys Lett 85:506–508
Mishan Y, Mosyak A, Pogrebnyak E, Hetsroni G (2007) Effect of developing flow and thermal regime on momentum and heat transfer in micro-scale heat sink. Int J Heat Mass Transfer 50:3100–3114
Mohapatra S, Loikitis D (2005) Advances in liquid coolant technologies for electronics cooling. In: Proceedings of 21st SemiTherm Symposium, San Jose, 15–17 March 2005, pp 354–360
Mohr J, Ehrfeld W, Munchmeyer D (1988) Requirements on resist layers in deep-etch synchrotron radiation lithography. J Vac Sci Technol B6:2264–2267
Morini GL (2004) Single phase convective heat transfer in micro-channels: overview of experimental results. Int J Thermal Sci 43:631–651
Morini Gl, Lorenzini M, Salvigini S (2006) Friction characteristics of compressible gas flows in micro-tubes. Exp. Thermal and Fluid Science 30:733-744
Mudawar I (2001) Assessment of high-heat-flux thermal management schemes. IEEE CPT Trans 24:122–141
Ng EYK, Poh ST (1999) Investigative study of manifold microchannel heat sinks for electronic cooling design. J Electron Manuf 9(2):155–166
North M, Cho W (2003) High heat flux liquid-cooled porous metal heat sink. In: Proceedings of IPAK 2003, Maui, 6–11 July 2003, Paper 35320
Palm R (2001) Heat transfer in micro-channels. Micro-scale Thermophys Eng 5:155–175
Peng XF, Peterson GP (1995) The effect of thermo-fluid and geometric parameters on convection of liquid through rectangular micro-channels. Int J Heat Mass Transfer 38:755–758
Peng XF, Peterson GP (1996) Convective heat transfer and flow friction for water flow in microchannel structures. Int J Heat Mass Transfer 39:2599–2608
Peng XF, Peterson GP, Wang BX (1994a) Frictional flow characteristics of water flowing through micro-channels. Exp Heat Transfer 7:249–264
Peng XF, Peterson GP, Wang BX (1994b) Heat transfer characteristics of water flowing through microchannels. Exp Heat Transfer 7:265–283
Peng XF, Wang BX (1998) Forced convection and boiling characteristics in micro-channels. In: Proceedings of 11 IHTC, 1, Kyonji, Korea, 23–28 August 1998, pp 371–390
Peterson M, Fisher T, Garimella S, Schlitz D (2003) Experimental characterization of low voltage field emission from carbon-based cathodes in atmospheric air. In: Proceedings of IMECE'03, Washington DC, 15–23 November 2003, Paper 41775
Pfund D, Rector D, Shekarriz A, Popescu A, Welty J (2000) Pressure drop measurements in a microchannel. AIChE J 46:1496–1507
Phillips RJ (1990) Micro-channel heat sinks. In: Bar-Cohen A, Kraus AD (eds) Advances in thermal modeling of electronic components and systems, vol 2, pp 109–184
Plosl A, Krauter G (1999) Wafer direct bonding: tailoring adhesion between brittle materials. Mater Sci Eng R25:92–98
Prasher RS, Chang JY, Sauciuc I, Narasimhan S, Chau D, Chrysler G, Myers A, Prstic S, Hu C (2005) Nano and micro technology-based next-generation package-level cooling solutions. Intel Technol J 4(4):285–296
Qu W, Mudawar I (2002a) Experimental and numerical study of pressure drop and heat transfer in a single-phase micro-channel heat sink. Int J Heat Mass Transfer 45:2549–2565
Qu W, Mudawar I (2004) Measurement and correlation of critical heat flux in two-phase microchannel heat sinks. Int J Heat Mass Transfer 47:2045–2059
Qu W, Mudawar I (2002b) Prediction and measurement of incipient boiling heat flux in microchannel heat sinks. Int J Heat Mass Transfer 45:3933–3945

Revellin R, Dupont V, Ursenbacher T, Thome JR, Zun I (2006) Characterization of diabatic two-phase flows in microchannels: flow parameter results for R-134a in a 0.5 mm channel. Int J Multiphase Flow 32:755–774

Ross D, Gaitan M, Locascio LE (2001) Temperature measurement in microfluidic systems using a temperature-dependent fluorescent dye. Anal Chem 73:4117–4123

Sammarco TS, Burns MA (1999) Thermocapillary pumping of discrete drops in microfabricated analysis devices. AIChE J 45:350–366

Schlicting H, Gersten K (2000) Boundary layer theory, 8th rev and England edn. Springer, Berlin Heidelberg New York

Serizawa A, Feng Z, Kawara Z (2002) Two-phase flow in microchannels. Exp Thermal Fluid Sci 26:703–714

Shah RK, London AL (1978) Laminar flow forced convection in ducts. Academic, New York

Sher I, Hetsroni G (2002) An analytical model for nucleate pool boiling with surfactant additives. Int J Multiphase Flow 28:699–706

Shih JC, Ho C-M, Liu J, Tai Y-C (1996) Monatomic and polyatomic gas flow through uniform microchannels. Microelectromech Syst (MEMS) DSC 59:197–203

Simons RE (1996) Direct liquid immersion cooling for high power density microelectronics. Electron Cooling 2(2):7–12

Sobhan CB, Garimella SV (2001) A comparative analysis of studies on heat transfer and fluid flow in micro-channels. Microscale Thermophys Eng 5:293–311

Steinke M, Kandlikar SG (2003) Flow boiling and pressure drop in parallel flow micro-channels. In: Kandlikar SG (ed) Proceedings of 1st International Conference on Micro-channels and Mini-channels, Rochester, 24–25 April 2003, pp 567–579

Thome JR (2006) State-of-the-art overview of boiling and two-phase flows in microchannels. Heat Transfer Eng 27(9):4–19

Thome JR, Dupont V, Jacobi AM (2004) Heat transfer model for evaporation in microchannels. Part I. Presentation of the model. Int J Heat Mass Transfer 47:3375–3385

Tiselj I, Hetsroni G, Mavko B, Mosyak A, Pogrebnyak E, Segal Z (2004) Effect of axial conduction on the heat transfer in micro-channels Int J Heat Mass Transfer 47:2551–2565

Triplett KA, Ghiaasiaan SM, Abdel-Khalik SI, Sadowski DL (1999) Gas–liquid two-phase flow in microchannels. Part I. Two-phase flow patterns. Int J Multiphase Flow 25:377–394

Tsai J-H, Lin L (2002) Transient thermal bubble formation on polysilicon micro-resisters. J Heat Transfer 124:375–382

Tuckerman D, Pease RFW (1981) High performance heat sinking for VLSI. IEEE Electron Device Lett 2:126–129

Turner SE, Lam LC, Faghri M, Gregory OJ (2004) Experimental investigation of gas flow in microchannels. J Heat Transfer 126:753–763

Turner SE, Sun H, Faghri M, Gregory OJ (2002) Gas flow through smooth and rough microchannels. In: Proceedings of the 12th International Heat Transfer Conference, Grenoble, 18–23 August 2002

Tzanand YL, Yang YM (1990) Experimental study of surfactant effects on pool boiling heat transfer J Heat Transfer 112:207–212

Varesi J, Majumdar A (1998) Scanning joule expansion microscopy at nanometer scales. Appl Phys Lett 72:37–39

Wang BX, Peng XF (1994) Experimental investigation of liquid forced-convection heat transfer through micro-channels. Int J Heat Mass Transfer 37:73–82

Wasekar VM, Manglik RM (2002) The influence of additive molecular weight and ionic nature on the pool boiling performance of aqueous surfactant solutions. Int J Heat Mass Transfer 45:483–493

Webb RL, Zhang M (1998) Heat transfer and friction in small diameter channels. Microscale Thermophys Eng 2:189–202

Wei XJ, Joshi Y (2002) Optimization study of stacked micro-channel heat sinks for micro-electronic cooling. IEEE Trans Comp Packag Technol 26(1):55–61

Weisberg A, Bau HH, Zemel JN (1992) Analysis of micro-channels for integrated cooling. Int J Heat Mass Transfer 35(10):2465–2474

Wojtan L, Revellin R, Thome JR (2006) Investigation of saturated critical heat flux in a single, uniformly heated microchannel. Exp Thermal Fluid Sci 30:765–774

Wu PY, Little WA (1984) Measurement of the heat transfer characteristics of gas flow in fine channel heat exchangers used for micro-miniature refrigerators. Cryogenics 23:415–420

Wu WT, Yang YM (1992) Enhanced boiling heat transfer by surfactant additives. In: Pool and External Flow Boiling. ASME, New York, pp 361–366

Wu WT, Yang YM, Maa JR (1995) Enhancement of nucleate boiling heat transfer and depression of surface tension by surfactant additives. J Heat Transfer 117:526–529

Yang H, Lee F, Chein R (2006) Micro-channel heat sink fabrication with roughened bottom walls. Microsyst Technol 12:760–765

Yang YM, Maa JR (2003) Boiling heat transfer enhancement by surfactant additives. In: Proceedings of the 5th International Conference Boiling Heat Transfer, ICBHT, Montego Bay, Jamaica, 4–8 May 2003

Yao SC, Chang Y (1983) Pool boiling heat transfer in a narrow space. Int J Heat Mass Transfer 26:841–848

Yen T-H, Shoji M, Takemura F, Suzuki Y, Kasagi N (2006) Visualization of convective boiling heat transfer in single micro-channels with different shaped cross-sections. Int J Heat Mass Transfer 49:3884–3894

Yoo JY (2006) Recent studies on fluid flow and heat transfer in thermal microdevices. Nanoscale Microscale Thermophys Eng 10:67–81

Yoon DS, Lee Y-S, Lee Y, Cho HJ, Sung SW, Oh KW, Cha J, Lim G (2002) Precise temperature control and rapid thermal cycling in a micromachined DNA polymerase chain reaction chip. J Micromech Microeng 12:813–823

Yu DL, Warrington RO, Barron RF, Ameel T (1995) An experimental and theoretical investigation of fluid flow and heat transfer in micro-tubes. ASME/JSME Thermal Eng Conf 1:523–530

Zhang J, Manglik RM (2005) Additive adsorption interfacial characteristics of nucleate pool boiling in aqueous surfactant solutions. J Heat Transfer ASME 127:684–690

Zhang L, Goodson KE, Kenny TW (2004). Silicon micro-channel heat sinks. Springer-Verlag Berlin Heidelberg

Zhang Z (2007) Nano/Micro-scale Heat Transfer. McGraw-Hill

Zhao YH, Masuoka T, Tsuruta T (2002) Unified theoretical prediction of fully developed nucleate boiling and critical heat flux based on a dynamic microlayer model. Int J Heat Mass Transfer 45:3189–3197

Zheng Y, Lin Z, Zakin JL, Talmon Y, Davis HT, Scriven LE (2000) Cryo-TEM imaging the flow induced transition from vesicles to threadlike micelles. J Phys Chem B 104(22):5263–5271

Nomenclature

A	Surface area
B	Bias limit
C	Concentration
C_0	Characteristic concentration
c_p	Heat capacity at constant pressure
d	Diameter
G	Mass flux, mass velocity of liquid plus vapor
g	Acceleration due to gravity
h	Heat transfer coefficient

h_{LG}	Latent heat
H	High, depth
K	Subcooling parameter
k	Thermal conductivity
k_s	Surface roughness
L	Channel length
m	Mass flow rate
N	Power, number of sample
N	Number of channels
P	Pressure, precession limit
Q	Heat rate
q	Heat flux
R	Thermal resistance, gas constant, auto-correlation function, radius
r	Radius, ration of channel length to block width
S	Distance between channels
T	Temperature
U	Average flow velocity in micro-channel, uncertainty
U_{LS}	Superficial liquid velocity
U_{GS}	Superficial gas velocity
u^*	Shear velocity
V	Volume
w	Width
x, y, z	Cartesian coordinates
$\mathrm{Bn} = \delta[\frac{g(\rho_L - \rho_G)}{\sigma}]^{0.5}$	Bond number
$\mathrm{Bo} = \frac{q}{Gh_{LG}}$	Boiling number
$\mathrm{Kn} = \frac{\overline{\lambda}}{d_h}$	Knudsen number
$\mathrm{Ma} = \frac{U}{U_{sound}}$	Mach number
$\mathrm{Nu} = \frac{hd_h}{k}$	Nusselt number
$\mathrm{Po} = \lambda \cdot Re$	Poiseuille number
Pr	Prandtl number
$\mathrm{Re} = \frac{Ud_h}{\nu}$	Reynolds number
$\mathrm{We} = \frac{G^2 d_h}{\sigma\rho}$	Weber number

Greek symbols

δ	Gap size, thickness
ΔP	Pressure drop
ΔT	Temperature difference
φ	Ratio of total heated area to circuit area
$\overline{\Phi}$	Pressure drop ratio
$\overline{\lambda}$	Wave length, mean free path
$\lambda = 2\Delta P \frac{d_h}{L} \frac{1}{\rho U^2}$	Friction factor
μ	Dynamic viscosity
ν	Kinematic viscosity
ρ	Density
σ	Surface tension
τ	Shear stress
χ	Vapor quality
θ	Contact angle
η	Shear viscosity
ϖ	Shear rate

Subscripts

c	Channel
cond	Conduction
conv	Convection
CHF	Critical heat flux
crit	Critical
dry	Dryout
eff,m	Heated platform area
exp	Experimental
film	Film
G	Gas
h	Hydraulic, heated
heat	Heat
in	Inlet
L	Liquid
LG	Liquid/gas
max	Maximum
mean	Average
min	Minimum
out	Outlet
p	Liquid-bubble slug

pm	Heated inside area
pred	Predicted
r	Row
s	Saturation
sound	Sound
sub	Subcooled
theory	Theory
w	Wall

Chapter 3
Velocity Field and Pressure Drop in Single-Phase Flows

The available experimental data on the flow of incompressible fluids are generalized. These data encompass a wide range of Reynolds numbers that correspond to laminar, transient and turbulent regimes of the flow. Thermal effects due to energy dissipation are estimated. Laminar drag reduction in micro-channels using hydrophobic surfaces is discussed. Data of experimental investigations related to the flow in smooth and rough micro-channels are compared with predictions of the conventional theory. Possible sources of divergence of experimental and theoretical results are also discussed.

3.1 Introduction

Flow in small tubes has been studied by many researchers over the years. Schlichting (1979) documents the then-available theories and experimental data, starting with the pioneering works by Hagen (1839) and Poiseuille (1840). Rapid development of micro-mechanics stimulated during the last decades numerous investigations in the field of fluid mechanics of micro-devices (Ho and Tai 1998; Gad-el-Hak 1999; Bayraktar and Pidugu 2006). Research in this field is important for different applications in micro-system technology, in particular, micro-scaled cooling systems of electronic devices, which generate high power (Tuckerman 1984; Incropera 1999).

The problems of micro-hydrodynamics were considered in different contexts: (1) drag in micro-channels with a hydraulic diameter from 10^6 m to 10^3 m at laminar, transient and turbulent single-phase flows, (2) heat transfer in liquid and gas flows in small channels, and (3) two-phase flow in adiabatic and heated micro-channels. The studies performed in these directions encompass a vast class of problems related to flow of incompressible and compressible fluids in regular and irregular micro-channels under adiabatic conditions, heat transfer, as well as phase change.

In spite of the existence of numerous experimental and theoretical investigations, a number of principal problems related to micro-fluid hydrodynamics are not well-studied. There are contradictory data on the drag in micro-channels, transition from laminar to turbulent flow, etc. That leads to difficulties in understanding the essence of this phenomenon and is a basis for questionable discoveries of special "micro-effects" (Duncan and Peterson 1994; Ho and Tai 1998; Plam 2000; Herwig 2000; Herwig and Hausner 2003; Gad-el-Hak 2003). The latter were revealed by comparison of experimental data with predictions of a conventional theory based on the Navier–Stokes equations. The discrepancy between these data was interpreted as a display of new effects of flow in micro-channels. It should be noted that actual conditions of several experiments were often not identical to conditions that were used in the theoretical models. For this reason, the analysis of sources of disparity between the theory and experiment is of significance.

We attempt here to reveal the actual reasons of disparity between the theoretical predictions and measurements obtained for single-phase flow in micro-channels. For this purpose, we consider the effect of different factors (roughness, energy dissipation, etc.) on flow characteristics. Some of these factors were also discussed by Sharp et al. (2001), and Sharp and Adrian (2004).

We consider the problem of liquid and gas flow in micro-channels under the conditions of small Knudsen and Mach numbers that correspond to the continuum model. Data from the literature on pressure drop in micro-channels of circular, rectangular, triangular and trapezoidal cross-sections are analyzed, whereas the hydraulic diameter ranges from 1.01 to 4,010 μm. The Reynolds number at the transition from laminar to turbulent flow is considered. Attention is paid to a comparison between predictions of the conventional theory and experimental data, obtained during the last decade, as well as to a discussion of possible sources of unexpected effects which were revealed by a number of previous investigations.

This chapter has the following structure: in Sect. 3.2 the common characteristics of experiments are discussed. Conditions that are needed for proper comparison of experimental and theoretical results are formulated in Sect. 3.3. In Sect. 3.4 the data of flow of incompressible fluids in smooth and rough micro-channels are discussed. Section 3.5 deals with gas flows. The data on transition from laminar to turbulent flow are presented in Sect. 3.6. Effect of measurement accuracy is estimated in Sect. 3.7. A discussion on the flow in capillary tubes is given in Sect. 3.8.

3.2 Characteristics of Experiments

For the analysis of flow in micro-channels we use the following experimental data:

1. Smooth micro-channels by Li et al. (2003), Yang et al. (2003), Pfund et al. (2000), Xu et al. (2000), Wu and Cheng (2003), Maynes and Webb (2002), Judy et al. (2002), Sharp and Adrian (2004), and Celata et al. (2006)

2. Micro-channels with roughness by Peng and Peterson (1996), Peng and Wang (1998), Mala and Li (1999), Qu et al. (2000), Pfund et al. (2000), Li et al. (2003), and Kandlikar et al. (2003)
3. Transition from laminar to turbulent flow by Peng and Peterson (1996), Peng and Wang (1998), Pfund et al. (2000), Li et al. (2003) and Sharp and Adrian (2004)
4. Gas flow in micro-channels by Harley et al. (1995), and Hsieh et al. (2004)

Brief characteristics of these experiments are given in Table 3.1. The flow characteristics (flow rate, pressure gradient, average and fluctuating velocities) were measured for flow in micro-channels, both smooth and with roughness, of different geometry. The physical properties of the fluids used in the experiments, as well as the material of the micro-channel walls varied widely, which allows for analyzing the effect of fluid composition, in particular, the ionic component, as well as the influence of fluid–wall interaction, on the flow characteristics. The Reynolds number in the experiments varied in the range $10^3 < \mathrm{Re} < 4\times10^3$, which covers the regimes of laminar, transition and turbulent flow. An extremely large relative length of micro-channels is characteristic for the considered experiments: $10^2 < L/d_* < 15\times10^4$, where L is the length of the micro-channel and d_* is the characteristic dimension of diameter or depth for circular or rectangular micro-channels, respectively.

In general, conventional theory has been tested for flow in micro-channels by comparing the experimental and theoretical data on pressure drop as a function of flow rate. During the last few years, better methods have been used for measurement of the mean velocity, as well as rms of the velocity fluctuations (Maynes and Webb 2002; Sharp and Adrian 2004).

Table 3.1 Flow characteristics in micro-channels

Geometry of the tube		Circular, Rectangular, Trapezoidal, Triangular
Size	d_h μm	$3–4\times10^3$
	h μm	0.51–250
$\frac{L}{d_*}$		$90–15\times10^4$
Wall	Material	Glass, Silica, Stainless steel
	Surface	Smooth, Rough
Fluid	Liquid	Tap water, Dist. water, De-ion. water, R-134a, Methanol, Isopropanol, Carbon tetrachloride solutions with several ionic compositions
	Gas	Air, N_2, He, Ar
Re		$10^3–4\times10^3$
U	m/s	$10^4–30$
Measured parameter		Flow rate, Pressure gradient, Mean velocity, rms of velocity fluctuations

3.3 Comparison Between Experimental and Theoretical Results

The classical solution of the problem of steady laminar flow in straight ducts is based on a number of assumptions on flow conditions, Hetsroni et al. (2005)

1. The flow is generated by a force due to a static pressure in the fluid.
2. The flow is stationary and fully developed, i.e., it is strictly axial.
3. The flow is laminar.
4. The Knudsen number is small enough so that the fluid is a continuous medium.
5. There is no slip at the wall.
6. The fluids are incompressible Newtonian fluids with constant viscosity.
7. There is no heat transfer to/from the ambient.
8. The energy dissipation is negligible.
9. There is no fluid/wall interaction (except purely viscous).
10. The channel walls are straight.
11. The channel walls are smooth.

In this case the problem of developed laminar flow in a straight duct reduces to integrating the equation

$$\mu\left(\frac{\partial^2 u}{\partial y^2}+\frac{\partial^2 u}{\partial z^2}\right)=\frac{\mathrm{d}P}{\mathrm{d}x} \tag{3.1}$$

with no-slip condition on the contour, which determines the channel shape.

In Eq. (3.1) $u = u(y,z)$ is the velocity component in the longitudinal x-direction, P is the pressure, μ is the dynamic viscosity, and x,y,z are the Cartesian coordinates.

The solution of Eq. (3.1) leads to the following expression for the friction factor

$$\lambda=\frac{\mathrm{const}}{\mathrm{Re}} \tag{3.2}$$

or

$$\mathrm{Po}=\mathrm{const} \tag{3.3}$$

where $\lambda=\frac{2\Delta P}{L}\frac{d_*}{\rho U^2}$ is the friction factor, $\mathrm{Po}=\lambda\cdot\mathrm{Re}$ is the Poiseuille number, and ΔP is the pressure drop on a channel length L, d_* is the characteristic size.

The constant in Eqs. (3.2) and (3.3) is determined by the micro-channel shape and does not depend on the flow parameters, as shown in Table 3.2 (Loitsianskii 1996). In Table 3.2, d is the diameter of the tube, α and β are the semi-axes of the ellipse, a is the side of an equilateral triangle, H is the side of a rectangle, and $f(\xi)$

Table 3.2 Parameters of micro-channels (Loitsianskii 1996)

Shape of micro-channel	Characteristic size	Constant in Eqs. (3.2), (3.3)
Circular	$d_*=d$	64
Elliptical	$d_*=\frac{1}{2\sqrt{2}}\frac{\alpha\beta}{(\alpha^2+\beta^2)^{1/2}}$	64
Equilateral triangular	$d_*=a$	160
Rectangular	$d_*=H$	$128/f(\xi)$

is a tabulated function of height-to-width ratio; $f(\xi)$ changes from 2.253 to 5.333 when ξ varies from 1 to ∞.

The data on pressure drop in irregular channels are presented by Shah and London (1978) and White (1994). Analytical solutions for the drag in micro-channels with a wide variety of shapes of the duct cross-section were obtained by Ma and Peterson (1997). Numerical values of the Poiseuille number for irregular micro-channels are tabulated by Sharp et al. (2001). It is possible to formulate the general features of Poiseuille flow as follows:

$$\frac{d\mathrm{Po}}{d\mathrm{Re}} = 0 \tag{3.4}$$

$$C^* = \frac{\mathrm{Po}_{\mathrm{exp}}}{\mathrm{Po}_{\mathrm{theor}}} \tag{3.5}$$

where C^* is the ratio of the Poiseuille number defined experimentally to the same one defined from theory. $\mathrm{Po}_{\mathrm{exp}}$ and $\mathrm{Po}_{\mathrm{theor}}$ are the experimental and theoretical Poiseuille numbers, respectively.

Equation (3.4) reflects the dependence of the friction factor on the Reynolds number, whereas Eq. (3.5) shows conformity between actual and calculated shapes of a micro-channel. Condition (3.5) is the most general since it testifies to an identical form of the dependencies of the experimental and theoretical friction factor on the Reynolds number.

Basically, there may be three reasons for the inconsistency between the theoretical and experimental friction factors: (1) discrepancy between the actual conditions of a given experiment and the assumptions used in deriving the theoretical value, (2) error in measurements, and (3) effects due to decreasing the characteristic scale of the problem, which leads to changing correlation between the mass and surface forces (Ho and Tai 1998).

3.4 Flow of Incompressible Fluid

3.4.1 Smooth Micro-Channels

We begin the comparison of experimental data with predictions of the conventional theory for results related to flow of incompressible fluids in smooth micro-channels. For liquid flow in the channels with the hydraulic diameter ranging from 10^6 m to 10^3 m the Knudsen number is much smaller than unity. Under these conditions, one might expect a fairly good agreement between the theoretical and experimental results. On the other hand, the existence of discrepancy between those results can be treated as a display of specific features of flow, which were not accounted for by the conventional theory. Bearing in mind these circumstances, we consider such experiments, which were performed under conditions close to those used for the theoretical description of flows in circular, rectangular, and trapezoidal micro-channels.

Glass and silicon tubes with diameters of 79.9–166.3 μm, and 100.25–205.3 μm, respectively, were employed by Li et al. (2003) to study the characteristics of friction factors for de-ionized water flow in micro-tubes in the Re range of 350 to 2,300. Figure 3.1 shows that for fully developed water flow in smooth glass and silicon micro-tubes, the Poiseuille number remained approximately 64, which is consistent with the results in macro-tubes. The Reynolds number corresponding to the transition from laminar to turbulent flow was Re = 1,700–2,000.

Yang et al. (2003) obtained the friction characteristics for air, water, and liquid refrigerant R-134a flow in tubes with inside diameters from 173 to 4,010 μm. The test results showed that the pressure drop correlations for large tubes might be adequately used for water, refrigerant, and low-speed air flow in micro-tubes. The laminar-turbulent transition Reynolds number varied from 1,200 to 3,800 and increased with decreasing tube diameters. The test friction factors agree very well with the Poiseuille equation for the laminar flow regime.

Pfund et al. (2000) studied the friction factor and Poiseuille number for 128–521 μm rectangular channels with smooth bottom plate. Water moved in the channels at Re = 60–3,450. In all cases corresponding to Re < 2,000 the friction factor was inversely proportional to the Reynolds number. A deviation of Poiseuille number from the value corresponding to theoretical prediction was observed. The deviation increased with a decrease in the channel depth. The ratio of experimental to theoretical Poiseuille number was 1.08 ± 0.06 and 1.12 ± 0.12 for micro-channels with depths 531 and 263 μm, respectively.

Xu et al. (2000) investigated de-ionized water flow in micro-channels with hydraulic diameter ranging from 30 to 344 μm at Reynolds numbers ranging from 20 to 4,000. Two test modules were used. The first test module consisted of a cover and an aluminum plate, into which a micro-channel, inlet and outlet sumps were machined. A Plexiglas plate was used to cover the channel. The second module was fabricated from a silicon wafer, and a 5 mm thick Pyrex glass was utilized to

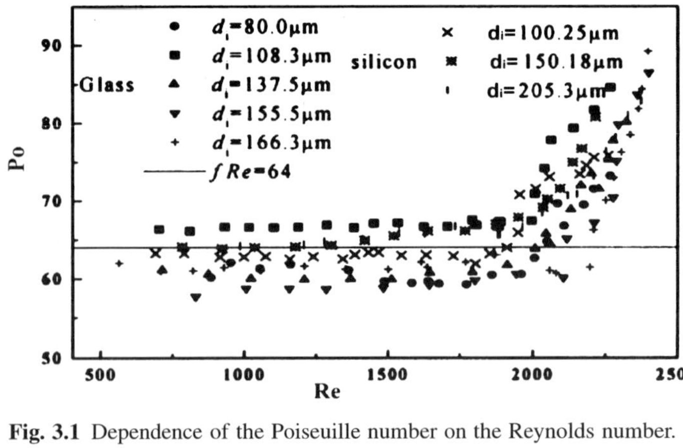

Fig. 3.1 Dependence of the Poiseuille number on the Reynolds number. Reprinted from Li et al. (2003) with permission

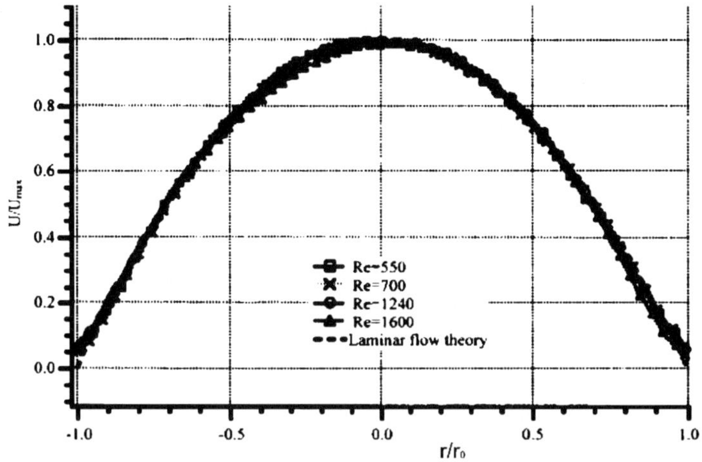

Fig. 3.2 Velocity profiles vs. r/r_0. Reprinted from Maynes and Webb (2002) with permission

cover the channel by using anodic bonding. The experimental data obtained in both modules showed that the trend of water flow in micro-channels is similar to the prediction of the conventional theory as a whole, i.e., that Po = const. for flow in the laminar region. However, some discrepancy between the results obtained in these modules occurred. For flow in the first test module, in which the hydraulic diameter of the channels varied from 50 to 300 μm, the Po values in the channels smaller than 100 μm were lower than those predicted by the theory. This phenomenon did not occur in the second test module. Experimental results there almost agree with the theory for flow in micro-channels varying from 30 to 60 μm. The authors explained the disagreement and showed that the actual dimension of the first module would be inaccurate when the cover and micro-channel plate were sealed by bonding with glue due to the thickness of the layer of the cured material.

The transition to turbulent flow occurred at Re of about 1,500. The authors noted that for smaller micro-channels, the flow transition would occur at lower Re. The early transition phenomenon might be affected by surface roughness and other factors.

Wu and Cheng (2003) measured the friction factor of laminar flow of de-ionized water in smooth silicon micro-channels of trapezoidal cross-section with hydraulic diameters in the range of 25.9 to 291.0 μm. The experimental data were found to be in agreement within ±11% with an existing theoretical solution for an incompressible, fully developed, laminar flow in trapezoidal channels under the no-slip boundary condition. It is confirmed that Navier–Stokes equations are still valid for the laminar flow of de-ionized water in smooth micro-channels having hydraulic diameter as small as 25.9 μm. For smooth channels with larger hydraulic diameters of 103.4–103.4–291.0 μm, transition from laminar to turbulent flow occurred at Re = 1,500–2,000.

Maynes and Webb (2002) presented pressure drop, velocity and rms profile data for water flowing in a tube 0.705 mm in diameter, in the range of Re $= 500-5{,}000$. The velocity distribution in the cross-section of the tube was obtained using the molecular tagging velocimetry technique. The profiles for Re $= 550, 700, 1{,}240$, and 1,600 showed excellent agreement with laminar flow theory, as presented in Fig. 3.2. The profiles showed transitional behavior at Re $> 2{,}100$. In the range Re $= 550-2{,}100$ the Poiseuille number was Po $= 64$.

Lelea et al. (2004) investigated experimentally fluid flow in stainless steel micro-tubes with diameter of 100–500 μm at Re $= 50-800$. The obtained results for the Poiseuille number are in good agreement with the conventional theoretical value Po $= 64$. Early transition from laminar to turbulent flow was not observed within the studied range of Reynolds numbers.

Papautsky et al. (1999) investigated the flow friction characteristics of water flowing through rectangular micro-channels with width varying from 150 to 600 μm, height (depth) ranging from 22.71 to 26.35 μm, and relative roughness of 0.00028. The experiments were conducted in the range of extremely low Reynolds numbers, $0.001 <$ Re < 10. The measurements showed that the ratio C^* was independent of Re. This value was about 1.2, i.e., nearly 20% above the value of C^* corresponding to the conventional theory.

The frictional pressure drop for liquid flows through micro-channels with diameter ranging from 15 to 150 μm was explored by Judy et al. (2002). Micro-channels fabricated from fused silica and stainless steel were used in these experiments. The measurements were performed with a wide variety of micro-channel diameters, lengths, and types of working fluid (distilled water, methanol, isopropanol), and showed that there were no deviations between the predictions of conventional theory and the experiment. Sharp and Adrian (2004) studied the fluid flow through micro-channels with the diameter ranging from 50 to 247 μm and Reynolds number from 20 to 2,300. Their measurements agree fairly well with theoretical data.

Cui et al. (2004) studied the flow characteristics in micro-tubes driven by high pressure ranging from 1 to 30 MPa. The diameters of the micro-tubes were from 3 to 10 μm and de-ionized water, isopropanol and carbon tetrachloride were used as the working fluid. The Reynolds number ranged from 0.1 to 24. The measurements showed that the ratio $C^* = \mathrm{Po}_{\mathrm{exp}}/\mathrm{Po}_{\mathrm{theor}}$ varied slightly with the pressure for de-ionized water. But for the other two liquids, isopropanol and carbon tetrachloride, C* increased markedly with pressure, as shown in Fig. 3.3. It should be noted that the viscosity is extremely sensitive to the flow temperature. The viscosity of water in the pressure range from 1 to 30 MPa can be approximately regarded as constant. Therefore C^* varied very slightly with pressure. The results of isopropanol and carbon tetrachloride were corrected taking into account the viscosity–pressure relationship to obtain the revised theoretical flow rate and revised normalized friction coefficient. Two revised results for $d = 10$ μm for isopropanol and carbon tetrachloride are shown in Fig. 3.4. It can be seen that the experimental data agree well with theoretical prediction. For $d = 5$ and 3 μm micro-tubes, similar tendencies were observed.

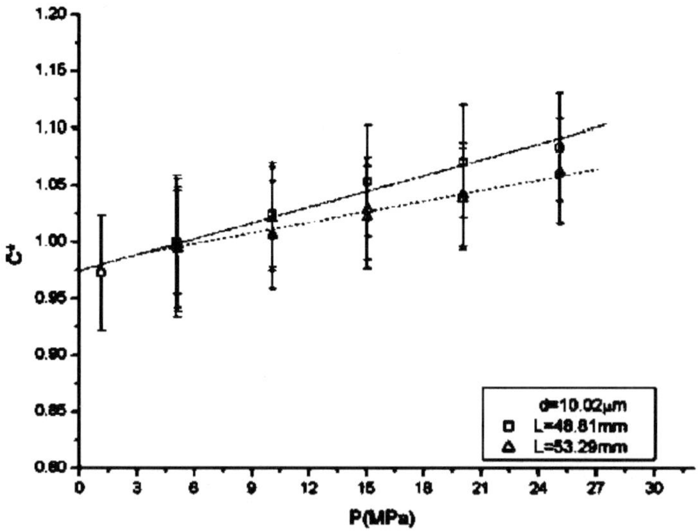

Fig. 3.3 The normalized Poiseuille number as a function of pressure for carbon tetrachloride in 10 μm micro-tube. Reprinted from Cui et al. (2004) with permission

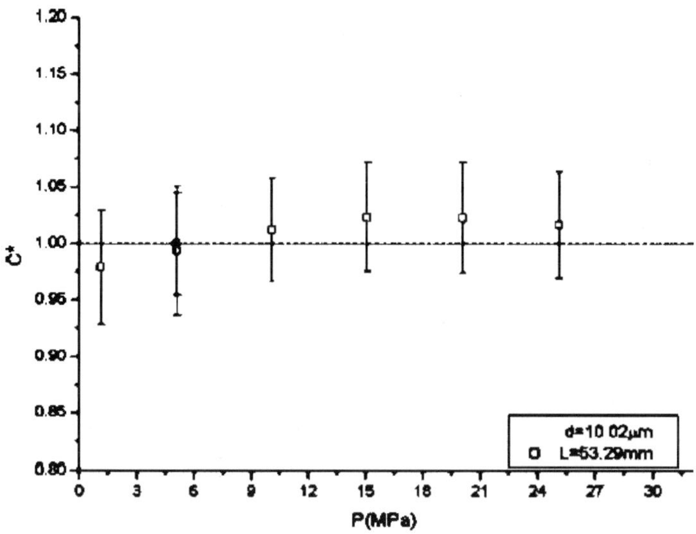

Fig. 3.4 The revised normalized Poiseuille number as a function of pressure for carbon tetrachloride in 10 μm micro-tube. Reprinted from Cui et al. (2004) with permission

Celata et al. (2006) studied experimentally the drag in glass/fused silica micro-tubes with inner diameter ranging from 31 to 259 μm for water flow with Re > 300. The drag measurements show that the friction factor for all diameters agrees well with predictions of conventional theory: $\lambda = 64/\mathrm{Re}$ (for the smallest diameter 31 μm, the deviations of experimental points from the line $\lambda = 64/\mathrm{Re}$ do not exceed

Table 3.3 Experimental results of single-phase fluid flow in smooth micro-channels

Author	Micro-channel		Fluid	Re	Poiseuille number		Re_{cr}
	Shape	d_h [μm]			$\frac{Po_{exp}}{Po_{theor}}$	Dependence on Re ($Re < Re_{cr}$)	
Sharp, Adrian (2004)	Circular tube	50–245	De-ionized water, 1-propanol and 20% weigh of glycerol solution	20–400 400–2,900	1	Independent	1,800–2,300
Li et al. (2003)	Circular tube	79.9–166.3 100.25–205.5	De-ionized water	350–2,300	1	Independent	1,700–2,000
Yang et al. (2003)	Circular tube	173, 4,010	Water R-134a	1,200–3,800	1	Independent	1,200–3,800
Maynes, Webb (2002)	Circular tube	705	Water	500–5,000	1	Independent	2,200
Judy et al. (2002)	Circular tube square micro-channel	15–150	Distilled water Methanol Isopropanol	8–2,300	1	Independent	~ 2,000
Celata et al. (2006)	Circular tube	31–251	Water	300–3,000	1	Independent	2,000–3,000
Lelea et al. (2004)	Circular tube	100, 300, 500	Distilled water	50–800	1	Independent	2,800
Cui et al. (2004)	Circular tube	3–10	De-ionized Isopropanol Carbon tetrachloride	0.1–24	~ 1 (For revised data)		
Hwang, Kim (2006)	Circular tube	244–792	R-134a	200–10,000	1	Independent	~ 2,000
Pfund et al. (2000)	Rectangular	128–1,050	Water	60–3,450	1.08–1.12	Independent	1,700–2,200
Xu et al. (2000)	Rectangular	30–344	Clean water	20–4,000	1	Independent	1,500
Papautsky et al. (1999)	Rectangular	Width 150–600 Height 22.71–26.35	Water	0.001–10	1.195	Independent	–
Wu, Chen (2003)	Trapezoidal	25.9–291	De-ionized water		1±0.11	Weak dependent	1,500–2,000

±19%). The transition from laminar to turbulent flow occurs at $2 \times 10^3 < \mathrm{Re} < 3 \times 10^3$.

The experimental results of single-phase flow in smooth micro-channels are summarized in Table 3.3.

3.4.2 Micro-Channels with Rough Walls

Several investigators obtained friction factors in micro-channels with rough walls that were greater than those in smooth wall channels. These observations should be considered taking into account the entrance effects, losses from change in channel size, etc.

The existence of roughness leads also to decreasing the value of the critical Reynolds number, at which transition from laminar to turbulent flow occurs. The character of the dependence of the friction factor on the Reynolds number in laminar flow remains the same for both smooth and rough micro-channels, i.e., $\lambda = \mathrm{const}/\mathrm{Re}$.

The general characteristics of experimental investigations of pressure drop in micro-channels with roughness are presented in Table 3.4. The experiments involve a wide range of flow conditions as related to shape of micro-channels (circular, rectangular, trapezoidal), their sizes ($50\,\mu\mathrm{m} < d_\mathrm{h} < 10^3\,\mu\mathrm{m}$), and the Reynolds numbers ($10^2 < \mathrm{Re} < 4 \times 10^3$). The relative roughness of these micro-channels k_s/r_0 varied from 0.32 to 7%. It is known that hydraulically smooth flow regime occurs when the Reynolds number, which is defined by height of roughness k_s and friction velocity u_*, varies in the range $0 < k_\mathrm{s} u_*/\nu < 5$ (Tani 1969; Schlichting 1979). The upper limit of this inequality determines the maximum value of the velocity at which laminar flow is possible. Taking into account that $u_* = \sqrt{\tau/\rho}$, $\tau = \mu(\mathrm{d}u/\mathrm{d}y)_\mathrm{w}$, $u = U_\mathrm{max}(1-\eta^2)$, $\eta = r/r_0$ and $U_\mathrm{max} = 2U$, where τ is the shearing stress at a wall, U_max and U are the maximum and average velocities, r_0 is the micro-channel radius, and subscript w refers to the wall, we arrive at the following estimate of the relative roughness, corresponding to the boundary that subdivides the laminar flow in smooth and rough channels, Hetsroni et al. (2005)

$$\frac{k_\mathrm{s}}{r_0} < \frac{5}{1.41\mathrm{Re}^{1/2}}. \tag{3.6}$$

For $\mathrm{Re} \sim 2 \times 10^3$, the relative roughness that corresponds to the boundary between the smooth and rough channels is $k_\mathrm{s}/r_0 \sim 0.08$. The latter shows that the hydraulic characteristics of the present micro-channels (Table 3.4) are close to characteristics of smooth micro-channels.

The deviation of the data related to flow in smooth (Table 3.3) and rough (Table 3.4) micro-channels may be a result of heterogeneity of the actual roughness, where height of some roughness peaks significantly exceeds its mean value. For example, under conditions of experiment by Pfund et al. (2000) the mean heigh of

Table 3.4 Experimental results of single-phase flow in rough micro-channels

Author	Micro-channel		Fluid	Re	Poiseuille number			Re_{cr}
	Shape	d_h [mm]			Po_{exp}/Po_{theor}	Dependence on Re	Relative roughness % $[k_s/r_0]$	
Peng, Peterson (1996)	Rectangular	133–367	Water	2×10^2–4×10^3	Inversely proport. to $Re^{0.98}$	Inversely proport. to $Re^{0.98}$		300–700
Peng, Wang (1998)	Rectangular	133–343	Water	2×10^2–4×10^3	Inversely proport. to $Re^{0.98}$	Inversely proport. to $Re^{0.98}$		300–700
Mala, Li (1999)	Circular tube	50–254	De-ion. water	10^2–2.1×10^3	1 for $Re < 500$	Independ. on Re at $Re < 500$	1.36–7	300–900
Qu et al. (2000)	Trapezoidal	51–169	De-ion. ultra filtered water	10^2–1.6×10^3	1.15–1.3	Weakly dependent on Re	3.6–5.7	1,000
Pfund et al. (2000)	Rectangular	Depth 257	De-ion. water	60–3,450	1.25	Independ. on Re at $Re < 1,700$	Mean 0.74 Max 5.7	1,700–2,000
Li et al. (2003)	Circular tube	128.76–179.8	De-ion. water	350–2,300	1.15–1.37	Weak depend. on Re at $Re < 1,500$	3–4	1,700
Kandlikar et al. (2003)	Circular tube	620–1,067	Water	500–2,600	~ 1	Independ. on Re at $Re < 2\times10^3$	0.32–0.71	$\sim 2,300$

the roughness wall was $\pm 1.9\,\mu$m, with the maximum peak value height of approximately 14.67 μm.

A study of forced convection characteristics in rectangular channels with hydraulic diameter of 133–367 μm was performed by Peng and Peterson (1996). In their experiments the liquid velocity varied from 0.2 to 12 m/s and the Reynolds number was in the range 50–4,000. The main results of this study (and subsequent works, e.g., Peng and Wang 1998) may be summarized as follows: (1) friction factors for laminar and turbulent flows are inversely proportional to $\mathrm{Re}^{1.98}$ and $\mathrm{Re}^{1.72}$, respectively; (2) the Poiseuille number is not constant, i.e., for laminar flow it depends on Re as $\mathrm{Po} \sim \mathrm{Re}^{-0.98}$; (3) the transition from laminar to turbulent flow occurs at Re about 300–700. These results do not agree with those reported by other investigators and are probably incorrect.

Mala and Li (1999) investigated experimentally the pressure losses in microchannels with diameters ranging from 50 to 254 μm. The micro-tubes were fabricated from two different materials, silica and stainless steel, and had mean surface roughness of $\pm 1.75\,\mu$m. Thus, the relative roughness changed from 1.36 to 7.0% for the pipes with $d = 254\,\mu$m and $d = 50\,\mu$m, respectively. The measurements indicate the existence of significant divergence between experimental values of pressure gradient ΔP_{exp} and values predicted by the conventional theory, $\Delta P_{\mathrm{theor}}$. The difference $\Delta P_{\mathrm{exp}} - \Delta P_{\mathrm{theor}}$ depends on the diameter of the micro-tube, as well as on the Reynolds number. At relatively large Re the dependence $\Delta P(\mathrm{Re})$ is close to linear, whereas at $\mathrm{Re} \sim 10^3$ and $d < 100\,\mu$m it shows significant deviation from the dependence predicted by the Poiseuille equation. It is worth noting that there is some difference between pressure losses corresponding to flow in silica and stainless steel micro-tubes: the pressure gradient in a silica micro-tube is slightly higher than that for stainless steel. Under conditions corresponding to flow in micro-tubes with the same roughness, such a difference points to an existence of some non-hydrodynamic interaction between the fluid and micro-tube wall. Based on the dependence $C^* = f(\mathrm{Re})$, it is possible to select two branches corresponding to relatively low ($\mathrm{Re} < 10^3$) and high ($\mathrm{Re} > 10^3$) Reynolds numbers. Within the first branch the ratio is about of $C^* = 1.12$. At $\mathrm{Re} > 10^3$ the ratio C^* increases with Re. The latter result means that within the ranges of Reynolds number of $0 < \mathrm{Re} < 10^3$ and $10^3 < \mathrm{Re} < 2 \times 10^3$ the friction factor is inversely proportional to Re and $(\mathrm{Re})^n$ ($n < 1$), respectively. Thus, the present results correspond (qualitatively) to the known data on the drag in macro-tubes at laminar and developed turbulent flows. The difference between the results corresponding to micro- and macro-rough tubes manifests itself as shift of the transition to low Reynolds number regions where $\lambda \sim 1/\mathrm{Re}$ and $\lambda = \mathrm{const}$. We suggest two possible reasons of the effects mentioned above: the first is an earlier transition from laminar to turbulent flow, and the second is the direct effect of roughness on the momentum changes in the liquid layer adjacent to the solid wall.

The hypothesis on the earlier transition from laminar to turbulent flow in micro-tubes is based on analysis of the dependence of pressure gradient on Reynolds number. As shown by the experimental data by Mala and Li (1999), this dependence may be approximated by three power functions: $\Delta P \sim \mathrm{Re}^{1.072}$ ($\mathrm{Re} < 600$),

$\Delta P \sim \mathrm{Re}^{1.3204}$ $(600 < \mathrm{Re} < 1{,}500)$ and $\Delta P \sim \mathrm{Re}^{2.0167}$ $(1{,}500 < \mathrm{Re} < 2{,}200)$ corresponding to laminar, transition and developed turbulent flow, respectively. Naturally, such phenomenological analysis does not reveal the actual reasons of the shift of the boundary transition to the low Reynolds numbers region.

One of the possible ways to account for the effect of roughness on the pressure drop in a micro-tube is to apply a modified-viscosity model to calculate the velocity distribution. Qu et al. (2000) performed an experimental study of the pressure drop in trapezoidal silicon micro-channels with the relative roughness and hydraulic diameter ranging from 3.5 to 5.7% and 51 to 169 μm, respectively. These experiments showed significant difference between experimental and theoretical pressure gradient.

It was found that the pressure gradient and flow friction in micro-channels were higher than that predicted by the conventional laminar flow theory. In a low Re range, the measured pressure gradient increased linearly with Re. For $\mathrm{Re} > 500$, the slope of the P_x–Re relationship increases with Re. The ratio C^* was about 1.3 for micro-channels of hydraulic diameter 51.3–64.9 μm and 1.15–1.18 for micro-channels of hydraulic diameter 114.5–168.9 μm. It was also found that the ratio of C^* depends on the Reynolds number.

A roughness-viscosity model was proposed to interpret the experimental data. An effective viscosity μ_{ef} was introduced for this purpose as the sum of physical μ and imaginary $\mu_M = \mu_M(r)$ viscosities. The momentum equation is

$$\frac{1}{r}\frac{\partial}{\partial r}\left(\mu_{\mathrm{ef}} r \frac{\partial u}{\partial r}\right) = \frac{\mathrm{d}P}{\mathrm{d}x} \tag{3.7}$$

with no-slip conditions on the walls, where μ_M is determined by a semi-empirical correlation $\mu_M = \mu f(r/k_{\mathrm{s}}, \mathrm{Re}, \mathrm{Re}_k)$, $\mathrm{Re}_k = u_k k_{\mathrm{s}}/\nu$, $\mathrm{Re} = Ud/\nu$, u_k is the velocity at the top of the roughness element, U is the average velocity, and d is the micro tube diameter. The correlation for μ_M also contains an empirical constant that has to be determined by using the experimental data.

Friction factors and the friction constant for a 257 μm deep channel with rough bottom plate were measured by Pfund et al. (2000). The width of the channel cut into each spacer was fixed at 1 cm. The micro-channel was 10 cm long, so that the flow profile would be fully developed within at least the middle third of the channel length. The wide channel gave an approximately 2D flow, thereby simplifying the theoretical description of the flow. Water entered the system through a pressured pulsation damper. The mean amplitude of roughness on the rough bottom was of ±1.90 μm with a maximum peak-valley height of approximately 14.67 μm. The value of C^* reached, at low Re, approximately 1.25. These results were compared to those obtained in the smooth micro-channels with depths ranging from 128 to 521 μm. The authors conclude that a possible reduction in channel depth at constant roughness, and an increase in roughness at constant depth, produced no significant changes in the friction constant. Friction factors in laminar flow were proportional to $1/\mathrm{Re}$ to the degree of accuracy of the experiments. High relative roughness might cause the Poiseuille number to vary with Re. Uncertainties in the measured constant obscured this effect.

Li et al. (2003) studied the flow in a stainless steel micro-tube with the diameter of 128.76–179.8 μm and relative roughness of about 3–4%. The Poiseuille number for tubes with diameter 128.76 and 171.8 μm exceeded the value of Po corresponding to conventional theory by 37 and 15%, respectively. The critical value of the Reynolds number was close to 2,000 for 136.5 and 179.8 μm micro-tubes and about 1,700 for micro-tube with diameter 128.76 μm.

The effect of roughness on pressure drop in micro-tubes 620 and 1,067 μm in diameter, with relative roughness of 0.71, 0.58 and 0.321% was investigated by Kandlikar et al. (2003). For the 1,067 μm diameter tube, the effect of roughness on pressure drop was insignificant. For the 620 μm tube the pressure drop results showed dependence on the surface roughness.

3.4.3 Surfactant Solutions

Drag reduction in a surfactant solution flowing in a micro-channel was the subject of experimental study by Hetsroni et al. (2004). The pressure drop in fully developed laminar flow in a micro-tube of inner diameter 1.07 mm was measured in the range of $10 \leq \mathrm{Re} \leq 450$. The study was performed for water surfactant solution of 350 and 1,060 ppm in adiabatic and diabatic flows. The cationic surfactant Habon G (molecular weight 500, trade name Hoe S4089, Hoechst AG) was used. The cation of the surfactant is hexadecyldimethyl hydroxyethyl ammonium and the counter-ion is 3-hydroxy-z-naphthoate. It was shown by Zakin et al. (1996, 2002) that although micro-structure of Habon G was mechanically degraded under high-shear conditions, it recovered quickly, no matter how many times it was broken up by shear.

The measurements of physical properties were carried out over a wide range of temperatures and for various concentrations. All solutions were prepared by dissolving the powdered surfactant in de-ionized water with gentle stirring. The shear viscosity of all surfactant solutions was determined in the temperature range of 25 to 60 °C (Hetsroni et al. 2001). Figure 3.5 shows the effect of shear rate on shear viscosity for 1,060 ppm Habon G solutions at different temperatures. One can see that at low-shear rates, the shear viscosity of Habon G solution is significantly higher than that of clear water. The curves come closer to one another for higher shear rates. The magnitude of the shear viscosity as a function of the shear rate decreases when the temperature of the solution increases.

The plot of the pressure drop depending on the bulk velocity in adiabatic and diabatic flows is shown in Fig. 3.6a,b. The data related to the adiabatic flow correspond to constant temperature of the fluids $T_{\mathrm{in}} = 25\,^{\circ}\mathrm{C}$, whereas in the diabatic flow the fluid temperature increased along micro-channel approximately from 40 to 60 °C. It is seen that in both cases the pressure drop for Habon G increases compared to clear water. The difference between pressure drop corresponding to flows of a surfactant solution and solvent increases with increasing bulk velocity.

The experimental data on micro-channel drag are presented in Figs. 3.7 and 3.8 in the form of f versus Re where $f = (\Delta P d)/(2L\rho U^2)$, i.e., $f = \lambda/4$. The Reynolds

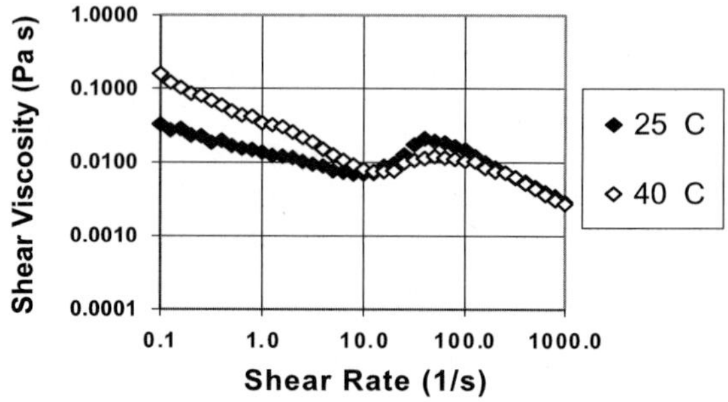

Fig. 3.5 Dependence of shear viscosity of Habon G solution on shear rate. Reprinted from Hetsroni et al. (2004) with permission

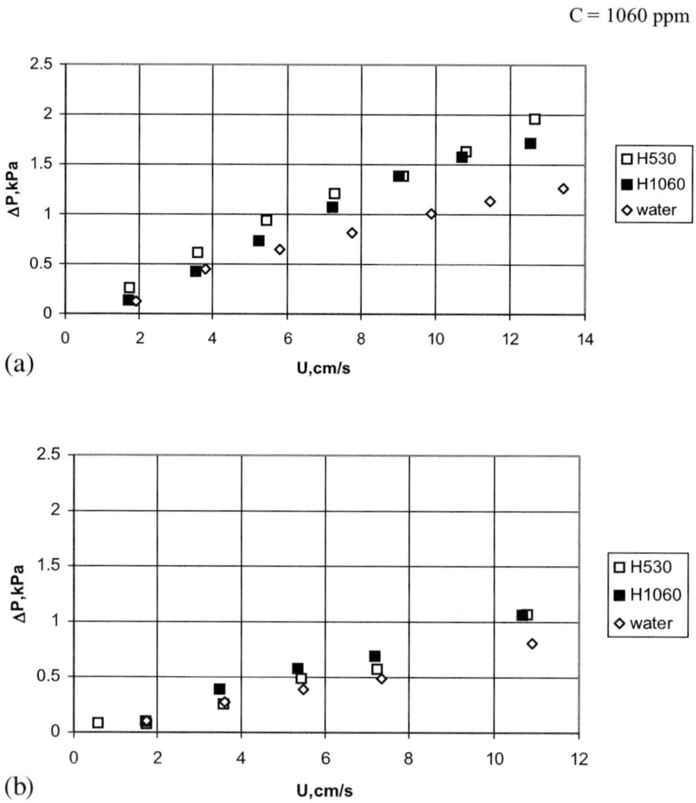

Fig. 3.6a,b Dependence of pressure drop on fluid bulk velocity in (**a**) adiabatic flow, and (**b**) diabatic flow. Reprinted from Hetsroni et al. (2004) with permission

numbers are defined based on the solvent (Re_{wat}) or shear viscosity of Habon G (Re_{sh}), where the shear viscosity, ν_{sh}, determines for a given value of shear rate that defined as $8U/d$ (Cogswell 1981). Figure 3.7a,b shows that the friction factor corresponding to the Habon G flow exceeds (at the same Re_{wat}) the friction factor in the pure solvent flow. In contrast to that, the dependences for both 530 and 1,060 ppm Habon G solutions $f(Re_{sh})$ are located significantly lower than that for a Newtonian

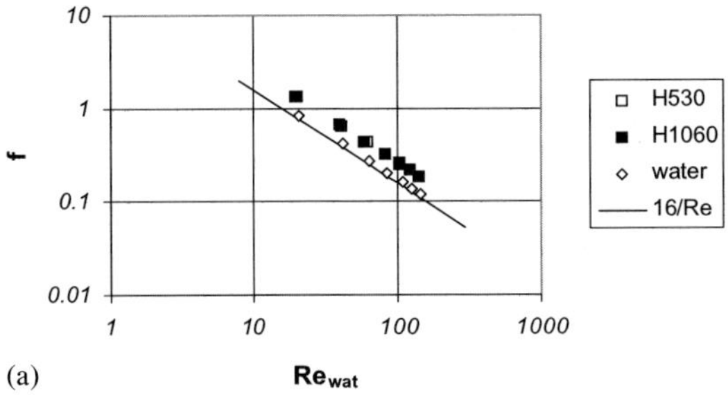

(a)

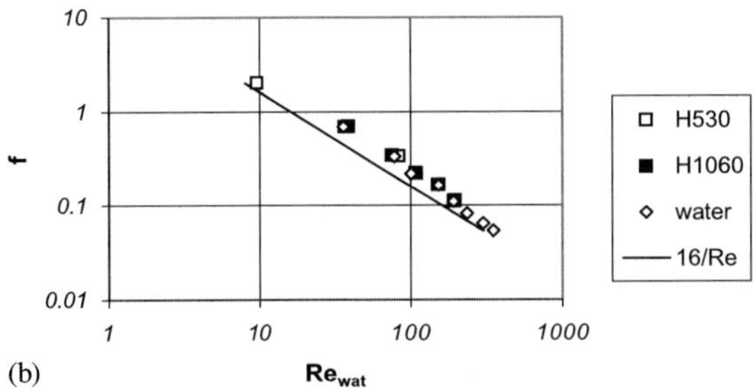

(b)

Fig. 3.7a,b Friction coefficients as function of solvent Reynolds number Re_{wat} in (**a**) adiabatic flow, and (**b**) diabatic flow. Reprinted from Hetsroni et al. (2004) with permission

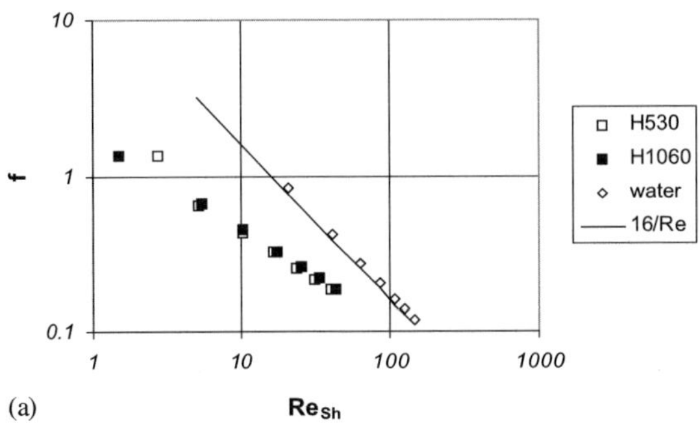

(a)

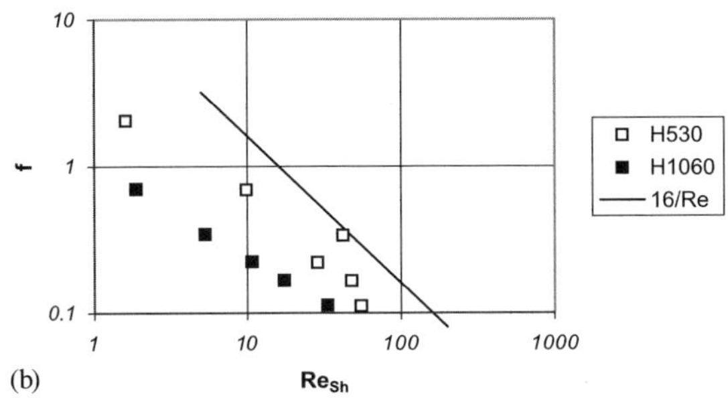

(b)

Fig. 3.8a,b Friction coefficients as a function of solution Reynolds number Re_{sh} in (**a**) adiabatic flow, and (**b**) diabatic flow. Reprinted from Hetsroni et al. (2004) with permission fluid in a pipe with smooth walls, where the friction factor is $f = 16/\mathrm{Re}_{sh}$. These results are somewhat unexpected. As can be concluded from Fig. 3.8a,b, the Habon G solutions are not drag reducing in laminar flow. The dependence of f versus Re_{sh} does not reflect this phenomenon. The Reynolds number based on shear viscosity is not valid to describe drag reduction in laminar flow. According to Lumley (1969) definition the flows have to be compared using the same viscosity, given by the solution. It has become customary to use the kinematic viscosity of the solvent with respect to the definition of the friction factor as a function of the Reynolds number (Cho and Hartnett 1982; Zakin et al. 1996; Virk et al. 1970; Warholic et al. 1999).

3.5 Gas Flows

Gas flows are flows of compressible fluids that flow under conditions of gas expansion and changes its density, pressure, velocity and temperature along the channel length (Shapiro 1953). There is a paucity of theoretical study of laminar gas flows in micro-channels. Berg et al. (1993) considered this problem in conjunction with calculation of the viscosity from the data on mass flow rate. An experimental and theoretical investigation of low Reynolds number, high subsonic Mach number compressible gas flow in channels was presented by Harley et al. (1995). Nitrogen, helium, and argon gases were used. By means of analytical and numerical solutions of the problem the detailed data on velocity, density and temperature distributions along micro-channel's axis were obtained. The effect of the Mach number on profiles of axial and transverse velocities and temperature was revealed. The friction factor in trapezoidal micro-channels typically 100 μm wide, 10^4 μm long and ranging from 0.5 to 20 μm in depth, was measured. The measurements of the pressure drop for the flow of nitrogen, helium and argon were carried out by varying the Knudsen number from 10^{-3} to 0.4, the inlet and exit Mach numbers 0.07 and 0.15, and 0.22, and 0.84, respectively and $\mathrm{Re} < 2 \times 10^3$. Figure 3.9 shows that under conditions which correspond to continuum flow, the measured value of friction factor is close to that which is predicted by conventional theory: the average friction constant was within 3% of the theoretical value for fully developed incompressible flow.

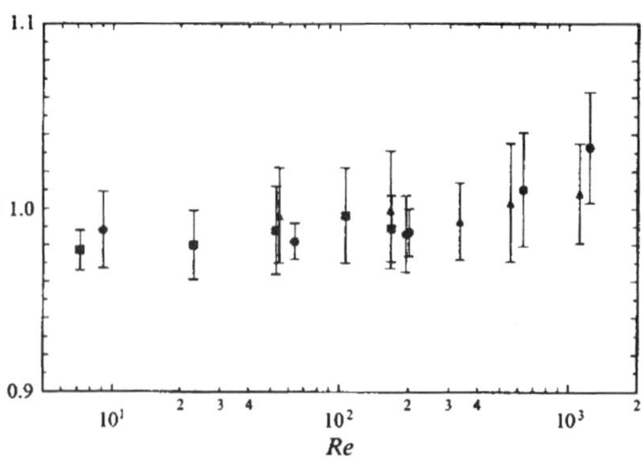

Fig. 3.9 Normalized Poiseuille number as a function of the Reynolds number in 11.04 μm deep channel. *Circles* (●) represent nitrogen, *squares* (■) represent helium, and *triangles* (▲) represent argon. Reprinted from Harley et al. (1995) with permission

3.6 Transition from Laminar to Turbulent Flow

The data on critical Reynolds numbers in micro-channels of circular and rectangular cross-section are presented in Tables 3.5 and 3.6, respectively. We also list geometrical characteristics of the micro-channels and the methods used for determination of the critical Reynolds number.

For the most part of the experiments one can conclude that transition from laminar to turbulent flow in smooth and rough circular micro-tubes occurs at Reynolds numbers about $\mathrm{Re}_{cr} = 2{,}000$, corresponding to those in macro-channels. Note that other results were also reported. According to Yang et al. (2003) Re_{cr} derived from the dependence of pressure drop on Reynolds number varied from $\mathrm{Re}_{cr} = 1{,}200$ to $\mathrm{Re}_{cr} = 3{,}800$. The lower value was obtained for the flow in a tube 4.01 mm in diameter, whereas the higher one was obtained for flow in a tube of 0.502 mm diameter. These results look highly questionable since they contradict the data related to the flow in tubes of diameter $d > 1$ mm. Actually, the 4.01 mm tube may be considered

Table 3.5 Critical Reynolds number in circular micro-channels

Author	Micro-channel				Re_{cr}	Remarks: considered characteristic
	Smooth/ rough	d [μm]	L [mm]	L/d		
Li et al. (2003)	Smooth	79.9–449	14.56–118.9	182–714	2,000	Friction factor
Yang et al. (2003)	Smooth	502–4,010	200–1,000	244–567	2,200	Friction factor ($\mathrm{Re}_{cr} \sim 1200$–3800 – pressure gradient)
Maynes, Webb (2002)	Smooth	705	141.9	201.2	2,200	Friction factor
Mala, Li (1999)	Smooth	50–254	55–88		300–900	Pressure gradient
Li et al. (2003)	Rough	128–179.8	39.30–84.26	305–470	1,700–1,900	Friction factor
Kandlikar et al. (2003)	Rough	620–1,032	–	–	2,000	Friction factor
Sharp, Adrian (2004)	Smooth	50–247	–	–	1,800–2,200	Friction factor, centerline velocity, rms of centerline velocity fluctuations
Hwang, Kim (2006)	Smooth	244–792	400–462	583–1,639	∼ 2,000	Friction factor

Table 3.6 Critical Reynolds number in rectangular and trapezoidal micro-channels

Author	Micro-channel				Re_{cr}	Remarks: considered characteristic
	Smooth/rough	H [μm]	L [mm]	L/H		
Xu et al. (2000)	Smooth	15.4	10–50	367–1,700	1,500	Friction factor
Pfund et al. (2000)	Smooth	128–1,050	100	95–781	1,700 for $H = 263$ μm 2,000–2,100 for $H = 521$ μm	Friction factor
Pfund et al. (2000)	Rough	257	100	389	1,700	Friction factor
Qu et al. (2000)	Rough (trapezoidal)	28–113.84	30	263–1,071	<2,300	Pressure gradient
Peng, Peterson (1996)	Rough	200–300	45	150–225	200–700	Friction factor

to be a macro-tube, in which the critical Reynolds number is about 2,000. A possible reason for this inconsistency arises from the method of determination of Re_{cr}. The data discussed above were obtained by analysis of the dependence of the pressure gradient on the Reynolds number. The same experimental data, presented by Yang et al. (2003) in the form of dependence of the friction factor on the Reynolds number, clearly showed that Re_{cr} is about 2,000. Mala and Li (1999), using the method based on dependence of the pressure drop gradient on Reynolds number, obtained $Re_{cr} = 300-900$.

The transition from laminar to turbulent flow in micro-channels with diameters ranging from 50 to 247 μm was studied by Sharp and Adrian (2004). The transition to turbulent flow was studied for liquids of different polarities in glass micro-tubes having diameters between 50 and 247 μm. The onset of transition occurred at the Reynolds number of about 1,800–2,000, as indicated by greater-than-laminar pressure drop and micro-PIV measurements of mean velocity and rms velocity fluctuations at the centerline.

In the laminar region the rms of streamwise velocity fluctuations was expected to be zero (Sharp et al. 2001). Figure 3.10 shows that the first evidence of transition, in the form of an abrupt increase in the rms, occurs at $1{,}800 \leq Re \leq 2{,}200$, in full agreement with the flow resistance data. There was no evidence of transition below these values. Thus, the behavior of the flow in micro-tubes, at least down to a 50 μm diameter, shows no perceptible differences with the macro-scale flow.

Hwang and Kim (2006) investigated the pressure drop in circular stainless steel smooth micro-tubes ($k_s/d < 0.1\%$) with inner diameters of 244 μm, 430 μm and 792 μm. The measurements showed that the onset of flow transition from laminar to turbulent motion occurs at the Reynolds number of slightly less than 2,000. It

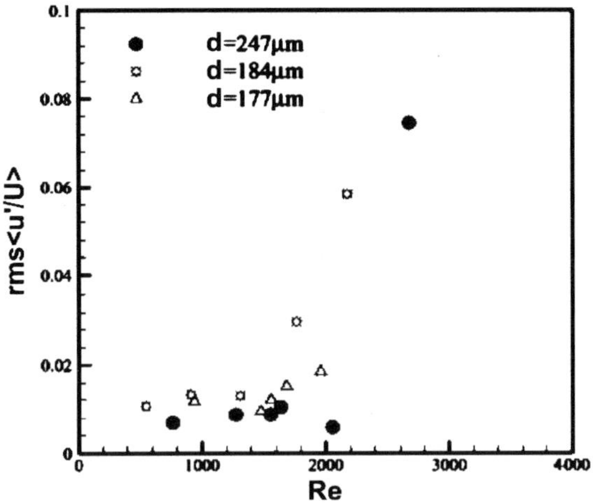

Fig. 3.10 Measured rms of the centerline velocity, divided by measured average velocity, vs. the Reynolds number. Reprinted from Sharp et al. (2001) with permission

also showed that the conventional theory predicted the friction factor well within an absolute average deviation of 8.9%.

Hao et al. (2007) investigated the water flow in a glass tube with diameter of 230 μm using micro particle velocimetry. The streamwise and mean velocity profile and turbulence intensities were measured at Reynolds number ranging from 1,540 to 2,960. Experimental results indicate that the transition from laminar to turbulent flow occurs at Re = 1,700–1,900 and the turbulence becomes fully developed at Re > 2,500.

Thus, the available data related to transition in circular micro-tubes testify to the fact that the critical Reynolds number, which corresponds to the onset of such transition, is about 2,000. The evaluation of critical Reynolds number in irregular micro-channels will entail great difficulty since this problem contains a number of characteristic length scales. This fact leads to some vagueness in definition of critical Reynolds number that is not a single criterion, which determines flow characteristics.

Let us explain this assertion by an example of the developed laminar flow in a rectangular micro-channel. As is well known (Loitsianskii 1966) this problem reduces to integrating the momentum equation

$$\mu\left(\frac{\partial^2 u}{\partial y^2}+\frac{\partial^2 u}{\partial z^2}\right)=\frac{\mathrm{d}P}{\mathrm{d}x}=-\frac{\Delta P}{L} \tag{3.8}$$

with the following boundary conditions

$$u=0 \text{ at } y=\pm H,\ |z|<W; u=0 \text{ at } z=\pm W,\ |y|<H \tag{3.9}$$

where u is the longitudinal velocity component, $\Delta P/L$ is the pressure drop per unit length, μ is dynamic viscosity, and $2H$ and $2W$ are the depth and width of the micro-channel.

Bearing in mind the conditions of the problem, we can assume that pressure drop per unit length is determined by viscosity, average velocity, as well as the depth and width of the micro-channel:

$$\frac{\Delta P}{L} = f(\mu, U, H, W) \tag{3.10}$$

where U is the average velocity.

Applying the Π-theorem to Eq. (3.10), we arrive at the following expression for the friction factor

$$\lambda = \frac{\varphi(\varepsilon)}{\mathrm{Re}_*} \tag{3.11}$$

where $\mathrm{Re}_* = U2H/\nu$ is the Reynolds number determined by micro-channel depth, λ is the friction factor, $\varphi(\varepsilon)$ is some unknown function of channel aspect ratio, $\varepsilon = W/H$, and ν is the kinematic viscosity of fluid.

It appears that the Poiseuille number in a rectangular channel depends on the aspect ratio, ε. In order to reveal an explicit form of the dependence $\varphi(\varepsilon)$, it is necessary to solve the problem defined by Eqs. (3.8) and (3.9) to obtain

$$\varphi(\varepsilon) = \frac{128}{\psi(\varepsilon)} \tag{3.12}$$

where $\psi(\varepsilon) = \frac{16}{3} - \frac{1024}{\pi^5 \varepsilon}\left(th\frac{\pi\varepsilon}{2} + \frac{1}{3^3}th\frac{3\pi\varepsilon}{2} + \ldots\right)$.

Equations (3.11) and (3.12) show that the friction factor of a rectangular micro-channel is determined by two dimensionless groups: (1) the Reynolds number that is defined by channel depth, and (2) the channel aspect ratio. It is essential that the introduction of a hydraulic diameter as the characteristic length scale does not allow for the reduction of the number of dimensionless groups to one. We obtain

$$\lambda = \frac{\varphi'(\varepsilon)}{\mathrm{Re_h}} \tag{3.13}$$

where $\varphi'(\varepsilon) = (\varphi(\varepsilon)2\varepsilon)/(1+\varepsilon)$. Accordingly, the Poiseuille number that is defined by hydraulic diameter depends also on channel aspect ratio.

The results related to the laminar-to-turbulent transition can be generalized by using the Obot–Jones model (Jones 1976; Obot 1988). A detailed discussion of this model is found in the paper by Morini (2004).

An experimental study of the laminar-turbulent transition in water flow in long circular micro-tubes, with diameter and length in the range of 16.6–32.2 μm and 1–30 mm, respectively, was carried out by Rands et al. (2006). The measurements allowed to estimate the effect of heat released by energy dissipation on fluid viscosity under conditions of laminar and turbulent flow in long micro-tubes.

The data on the drag for micro-tube diameters of 16.6, 19.7, 26.3 and 32.2 μm are presented in Fig. 3.11 in the form of the dependence of the Poiseuille number on Re. The latter was determined by an average of the mixed-mean temperature at the inlet and outlet of the micro-tube. The data of Fig. 3.11 show that the Poiseuille number practically shows no dependence on Re in the range $500 < \mathrm{Re} < 2{,}000$. The

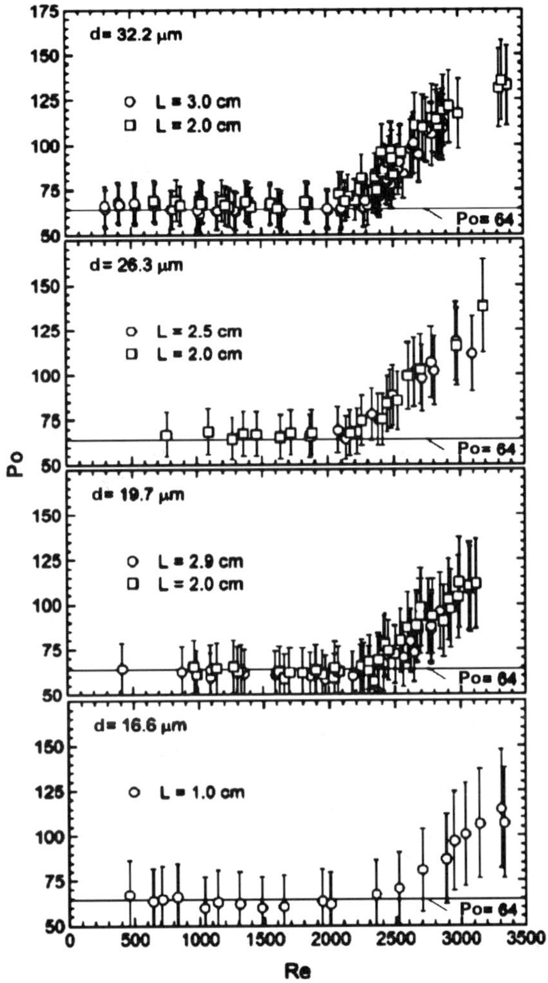

Fig. 3.11 Dependence of the Poiseuille number on Reynolds number. Reprinted from Rands et al. (2006) with permission

absolute of Po is close to 64, corresponding to the predictions of the conventional theory. The transition from laminar to turbulent flow occurs at Re = 2,100–2,500.

The dependence of the measured rise in fluid mixed-cup temperature on Reynolds number is illustrated in Fig. 3.12. The difference between outlet and inlet temperatures increases monotonically with increasing Re at laminar and turbulent flows. Under conditions of the given experiments, the temperature rise due to energy dissipation is very significant: $\Delta T = 15{-}35$ K at $L/d = 900{-}1{,}470$ and Re = 2,500. The data on rising temperature in long micro-tubes can be presented in the form of the dependence of dimensionless viscous heating parameter $\mathrm{Re}/[\mathrm{Ec}(L/d)]$ on Reynolds number (Fig. 3.13).

In the latter figure, the theoretical dependence

$$\mathrm{Re}/[\mathrm{Ec}(L/d)] = 32 \tag{3.14}$$

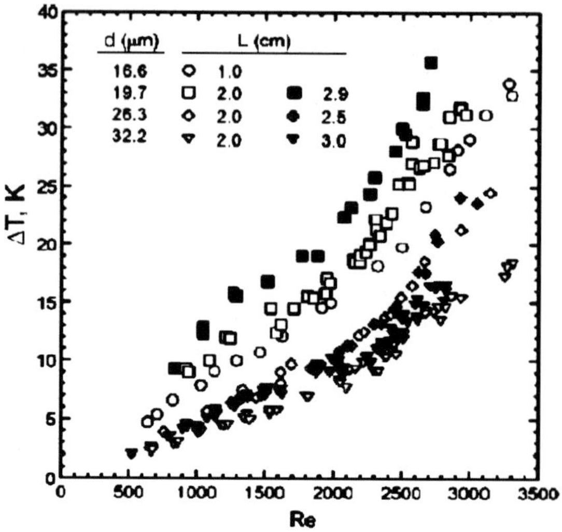

Fig. 3.12 Increase in the fluid average temperature. Reprinted from Rands et al. (2006) with permission

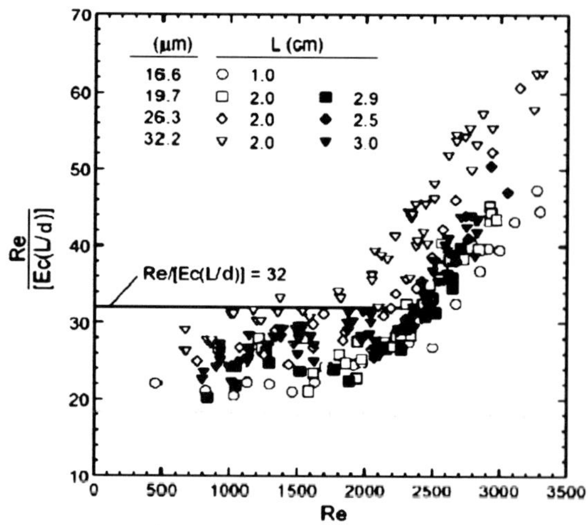

Fig. 3.13 Variation of the viscous heating parameter $\mathrm{Re}/[\mathrm{Ec}(L/d)]$ with Reynolds number. Reprinted from Rands et al. (2006) with permission

corresponding to laminar flow in an adiabatic micro-tube is also presented. It is seen that at relatively low Reynolds numbers that correspond to laminar flow, the viscous heating parameter does not depend on Re. At $\mathrm{Re} > 2{,}000$ the shape of the dependence $\mathrm{Re}/[\mathrm{Ec}(L/d)] = f(\mathrm{Re})$ sharply changes. An increase in the Reynolds number leads to an increase in the viscous heating parameter. Thus, both characteristic parameters of flow Po and $\mathrm{Re}/[\mathrm{Ec}(L/d)]$ that account for the hydraulic resistance and energy dissipation change identically with the Reynolds number. In the range of relatively low Reynolds numbers, $\mathrm{Re} < 2{,}000$, they are constant, whereas at $\mathrm{Re} > 2{,}000–2{,}500$ the Poiseuille number and viscous heating parameter monotonically increase with increasing Re. The latter testifies to the fundamental transformation of the flow structure, which occurs in this range of Re, and the Reynolds numbers in the range of 2,000–2,500 can be considered as critical corresponding to the transition from laminar to turbulent flow in micro-tubes.

3.7 Effect of Measurement Accuracy

In experiments related to flow and heat transfer in micro-channels, some parameters, such as the flow rate and channel dimensions are difficult to measure accurately because they are very small. For a single-phase flow in micro-channels the uncertainty of $\lambda \mathrm{Re}$ is (Guo and Li 2002, 2003)

$$\frac{\delta(\lambda \mathrm{Re})}{\lambda \mathrm{Re}} = \left\{ \left(\frac{\delta(\Delta P)}{\Delta P}\right)^2 + \left(\frac{4(\delta d_\mathrm{h})}{d_\mathrm{h}}\right)^2 + \left(\frac{\delta(L)}{L}\right)^2 + \left(\frac{\delta(m)}{m}\right)^2 \right\}^{1/2} \tag{3.15}$$

where m, ΔP, d_h, L are the mass flow rate, pressure drop, hydraulic diameter and channel length, respectively. Equation (3.15) shows that the channel hydraulic diameter measurement error may play a very important part in the resulting uncertainty of the product $\lambda \mathrm{Re}$. For example, in experiments measuring the friction factor in a circular glass micro-tube by Guo and Li (2002), they initially measured the diameter as 84.7 μm using a 40× microscope. The data reduction with this diameter showed that the friction factors were larger than that predicted by the conventional theory. The averaged value of the diameter measured using a 400× microscope and scanning electron microscope for the same micro-tube was only 80.0 μm. With this more accurate value of the diameter, the friction factors obtained from the experimental data were in good agreement with the conventional values.

3.8 Specific Features of Flow in Micro-Channels

3.8.1 General Remarks

The data presented in the previous chapters, as well as the data from investigations of single-phase forced convection heat transfer in micro-channels (e.g., Bailey et al. 1995; Guo and Li 2002, 2003; Celata et al. 2004) show that there exist a number of principal problems related to micro-channel flows. Among them there are: (1) the dependence of pressure drop on Reynolds number, (2) value of the Poiseuille number and its consistency with prediction of conventional theory, and (3) the value of the critical Reynolds number and its dependence on roughness, fluid properties, etc.

All available experimental data (except the data by Peng and Peterson 1996; Peng and Wang 1998) show that the friction factor is inversely proportional to the Reynolds number, i.e., $\lambda = \mathrm{const}/\mathrm{Re}$. The constant depends on the micro-channel shape only and agrees fairly well with the result of a dimensional analysis carried out by Sedov (1993) and the analytical solution of the problem (Loitsianskii 1966). The qualitative difference between the data by Peng and Peterson (1996) and Peng and Wang (1998) and the experimental and theoretical data of other researchers is, probably, due to experimental uncertainties.

Concerning the critical Reynolds number, several groups of experiments related to laminar-to-turbulent flow can be set apart (Tables 3.5 and 3.6):

- Li et al. (2003), Yang et al. (2003), Maynes and Webb (2002), Kandlikar et al. (2003) and Sharp and Adrian (2004) obtained the critical Reynolds number in the range $\mathrm{Re}_{\mathrm{cr}} \approx 1{,}700-2{,}300$.
- Li et al. (2003), Xu et al. (2000), and Pfund et al. (2000) also obtained the critical Reynolds number $\mathrm{Re}_{\mathrm{cr}} \approx 1{,}500-1{,}900$.
- Peng and Peterson (1996), Peng and Wang (1998), and Mala and Li (1999) obtained the anomalously low critical Reynolds number $\mathrm{Re}_{\mathrm{cr}} \approx 200-900$.

The critical Reynolds numbers of 1,700–2,300 agree fairly well with the well-known values of $\mathrm{Re}_{\mathrm{cr}}$ corresponding to flow in macro-channels (Lindgren 1958; Leite 1959; Wygnanskii and Champagne 1973; Schlichting 1979). This result is not unexpected since from a simple physical consideration it is obvious that under conditions of the continuous model with viscous fluid/wall interaction and moderate flow velocity, when energy dissipation is negligible, there are no reasons for a decrease in the critical Reynolds number. Taking into consideration the dimension of the characteristic parameters, we determine in this case that the lifetime of vortices is inversely proportional to the fluid viscosity

$$t \sim \frac{d^2}{\nu}. \tag{3.16}$$

The characteristic hydrodynamic time, t_{h} is

$$t_{\mathrm{h}} = \frac{L}{U} \tag{3.17}$$

and

$$\frac{t}{t_{\mathrm{h}}} \sim \frac{\mathrm{Re}}{\overline{L}} \tag{3.18}$$

where $\overline{L} = L/d$.

The ratio of t/t_{h}, which is characteristic of the possibility of vortices, does not depend on the micro-channel diameter and is fully determined by the Reynolds number and L/d. The lower value of Re at which $t/t_{\mathrm{h}} \geq 1$ can be treated as a threshold. As was shown by Darbyshire and Mullin (1995), under conditions of an artificial disturbance of pipe flow, a transition from laminar to turbulent flow is not possible for $\mathrm{Re} < 1{,}700$, even with a very large amplitude of disturbances.

A direct study of the transition from laminar to turbulent flow in micro-tubes was performed by Sharp and Adrian (2004). The measurements of mean velocity and rms of velocity fluctuations showed that the value of the Reynolds number, at which the transition from laminar to turbulent flow occurs, is about $\mathrm{Re}_{\mathrm{cr}} \approx 1{,}800-2{,}200$.

Thus, the measurements of integral flow characteristics, as well as mean velocity and rms of velocity fluctuations testify to the fact that the critical Reynolds number is the same as Re_{cr} in the "macroscopic" Poiseuille flow. Some decrease in the critical Reynolds number down to $Re \sim 1{,}500-1{,}700$, reported by the second group above, may be due to energy dissipation. The energy dissipation leads to an increase in fluid temperature. As a result, the viscosity would increase in gas and decrease in liquid. Accordingly, in both cases the Reynolds number based on the inlet flow viscosity differs from that based on local viscosity at a given point in the micro-channel.

There is a significant scatter between the values of the Poiseuille number in micro-channel flows of fluids with different physical properties. The results presented in Table 3.1 for de-ionized water flow, in smooth micro-channels, are very close to the values predicted by the conventional theory. Significant discrepancy between the theory and experiment was observed in the cases when fluid with unknown physical properties was used (tap water, etc.). If the liquid contains even a very small amount of ions, the electrostatic charges on the solid surface will attract the counter-ions in the liquid to establish an electric field. Fluid-surface interaction can be put forward as an explanation of the Poiseuille number increase by the fluid ionic coupling with the surface (Brutin and Tadrist 2003; Ren et al. 2001; Papautsky et al. 1999).

The results obtained by Brutin and Tadrist (2003) showed a clear effect of the fluid on the Poiseuille number. Figure 3.14 shows results of experiments that were done in the same experimental set-up for hydraulic diameters of 152 and 262 μm, using distilled water and tap water. The ion interactions with the surface can perhaps explain such differences. Tap water contains more ions such as Ca^{2+}, Mg^{2+}, which are 100 to 1,000 times more concentrated than H_3O^+ or OH^-. In distilled water only H_3O^+ and OH^- exist in equal low concentrations. The anion and cation interactions with the polarized surface could modify the friction factor. This is valid only in the case of a non-conducting surface.

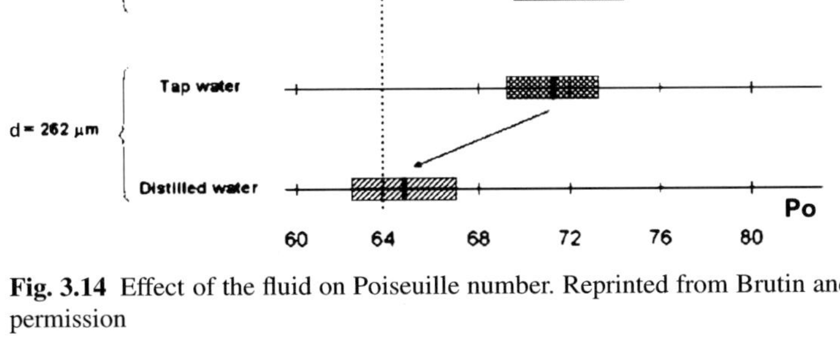

Fig. 3.14 Effect of the fluid on Poiseuille number. Reprinted from Brutin and Tadrist (2003) with permission

3.8.2 Thermal Effects

Gruntfest et al. (1964) showed that the thermal effects due to energy dissipation of liquid flow in a pipe lead to significant transformation of the flow field. It occurs due to the dependence of liquid density on temperature, distortion of velocity profile and development of flow instability and its transition to turbulence. The influence of energy dissipation on thermo- and hydrodynamic characteristics of the liquid flow in micro-channels was considered by Tso and Mahulikar (1998, 1999, 2000). A detailed analysis of viscous dissipation effects in micro-tubes and micro-channels was performed by Koo and Kleinstreuer (2004). It was shown that viscous dissipation becomes significant for fluids with low specific heat capacities and high viscosities, even in relatively low Reynolds number flows. The channel size, aspect ratio of the channel diameter to its length, as well as the Reynolds and Brinkman numbers are important factors that determine the effect of viscous dissipation.

Celata et al. (2005) evaluated the effect of viscous heating on friction factor for flow of an incompressible fluid in a micro-channel. By integrating the energy equation over the micro-channel length, a criterion that determines conditions when viscous dissipation effect is significant was obtained:

$$\frac{\mathrm{Ec}}{\mathrm{Re}}\left(\lambda \mathrm{Re}\overline{L}\right) \geq 1\,. \tag{3.19}$$

The simplest evaluation value of the complex $(\mathrm{Ec}/\mathrm{Re})\left(\lambda \mathrm{Re}\overline{L}\right)$ shows that it is essentially smaller than unity for the realistic conditions typical for water flow in micro-channels: $\mathrm{Re} \sim 10^3$, $d_\mathrm{h} \sim 100\,\mu\mathrm{m}$, $\overline{L} \sim 500$.

The behavior of liquid flow in micro-tubes and channels depends not only on the absolute value of the viscosity but also on its dependence on temperature. The non-linear character of this dependence is a source of an important phenomenon – hydrodynamic thermal explosion, which is a sharp change of flow parameters at small temperature disturbances due to viscous dissipation. This is accompanied by radical changes of flow characteristics. Bastanjian et al. (1965) showed that under certain conditions the steady-state flow cannot exist, and an oscillatory regime begins.

We can estimate the effect of energy dissipation on liquid heating and values of flow parameters corresponding to arising oscillations in the flow. We assume that the density of the fluid and its thermal conductivity are constant. Then, the energy equation attains the form

$$\rho(v\nabla)h = k\nabla^2 T + \Phi \tag{3.20}$$

where h is the enthalpy, k is the thermal conductivity, and Φ is the energy dissipation. The present analysis is restricted to estimation of the maximum heating of the liquid that corresponds to adiabatic flow in a micro-channel.

For laminar flow

$$u = U_\mathrm{max}(1-\eta^2) \tag{3.21}$$

where $U_\mathrm{max} = 2U$ or $U_\mathrm{max} = 3U/2$ for axisymmetric and plane flows, respectively. Hence, we obtain from Eq. (3.20) the following estimation for adiabatic rise of the liquid temperature:

$$\frac{\Delta T}{T_{\rm in}} = 2\frac{\nu^2}{r_0^2}\left(\frac{L}{r_0}\right)\frac{\rm Re}{c_{\rm p}T_{\rm in}} \tag{3.22}$$

for circular micro-channel, and

$$\frac{\Delta T}{T_{\rm in}} = 12\frac{\nu^2 L{\rm Re}}{H^2 d_{\rm h} c_{\rm p} T_{\rm in}} \tag{3.23}$$

for plane micro-channel, where $\Delta T = T_{\rm out} - T_{\rm in}$ is the difference between outlet and inlet temperatures. The difference between outlet and inlet temperatures essentially depends on the heat capacity of the fluid and sharply increases as $c_{\rm p}$ decreases.

Estimation of adiabatic increase in the liquid temperature in circular micro-tubes with diameter ranging from 15 to 150 μm, under the experimental conditions reported by Judy et al. (2002), are presented in Table 3.7. The calculations were carried out for water, isopropanol and methanol flows, respectively, at initial temperature $T_{\rm in} = 298$ K and $\nu = 8.7\times10^{-7}\,{\rm m^2/s}$, $2.5\times10^{-6}\,{\rm m^2/s}$, $1.63\times10^{-6}\,{\rm m^2/s}$, and $c_{\rm p} = 4{,}178$ J/kg K, 2,606 J/kg K, 2,531 J/kg K, respectively. The lower and higher values of $\Delta T/T_{\rm in}$ correspond to limiting values of micro-channel length and Reynolds numbers. Table 3.7 shows adiabatic heating of liquid in micro-tubes can reach ten degrees: the increase in mean fluid temperature $(T_{\rm in}+T_{\rm out})/2$ is about 9 °C, 121 °C, 38 °C for the water ($d = 20$ μm), isopropanol ($d = 20$ μm) and methanol ($d = 30$ μm) flows, respectively.

Energy dissipation also significantly affects the temperature of gas flow. In order to estimate adiabatic increase in gas temperature in micro-channels we used the approximate expression for $\Delta T/T_{\rm in}$, which follows from Eq. (3.23) assuming that $k\nabla^2 T \sim 0$, $h = c_{\rm p}T$, $\Phi = \mu\left(\frac{{\rm d}u}{{\rm d}y}\right)^2 \sim \mu\frac{U^2}{\delta^2}$ and $\frac{{\rm d}T}{{\rm d}x} = \frac{T_{\rm out}-T_{\rm in}}{L}$:

$$\frac{\Delta T}{T_{\rm in}} = \frac{\nu^2}{2\delta^2 c_{\rm p} T_{\rm in}}\left(\frac{L}{\delta}\right){\rm Re} \tag{3.24}$$

where $\delta = d_{\rm h}/2$ and U are characteristic size and average velocity, ${\rm Re} = U2\delta/\nu$.

Table 3.7 Estimation of adiabatic increase in the liquid temperature in a circular micro-tube. Experimental conditions correspond to those of Judy et al. (2002)

d [μm]	L [m]	Re	$(\Delta T/T_{\rm in})\cdot 10^2$ Water	Isopropanol	Methanol
15	0.036	34–41	0.35–0.43	4.7–5.6	2.0–2.5
20	0.03–0.05	18–989	0.066–6.0	0.87–80	0.38–35
30	0.04–0.07	8–1,716	0.012–4.3	0.15–57	0.067–25
40	0.05–0.37	17–769	0.013–4.3	0.17–57	0.075–25
50	0.07–0.29	44–1,451	0.024–3.3	0.32–43	0.14–19
75	0.3–0.39	146–1,883	0.10–1.7	1.3–22	0.59–9.8
100	0.39	109–1,858	0.041–0.70	0.55–9.3	0.24–4.1
150	0.2–0.3	137–1,540	0.0079–0.13	0.11–1.8	0.046–0.77

For parameters presented in Table 3.8, which correspond to conditions in the experiments by Harley et al. (1995), the mean increase in the temperature of nitrogen, helium and argon ($\mathrm{Re}_{N_2} = 60$, $\mathrm{Re}_{He} = 20$, $\mathrm{Re}_{Ar} = 50$, $T_{in} = 300$ K, $M \sim 0.1$) is about 50–200 K (Table 3.9).

Energy dissipation leads to a significant increase in gas viscosity and decrease in the actual Reynolds number Re_{ac} defined by the mean viscosity $\nu_{ac} = (\nu_{out} + \nu_{in})/2$ compared to that defined by inlet parameters.

Under conditions of real experiments, the thermal regime of the flow determines not only the energy dissipation, but also the heat losses to the micro-channel walls. In this case increase in the fluid temperature depends on the relation between the rate of heat release by energy dissipation and heat losses due to heat transfer (Koo and Kleinstreuer 2004). This does not distort the qualitative picture of the phenomenon – in all cases energy dissipation leads to temperature growth, changing viscosity of fluid and actual Reynolds number. Moreover, in certain cases, energy dissipation leads to radical transformation of flow and transition from a stable to an oscillatory regime.

3.8.3 Oscillatory Regimes

To estimate the parameters resulting in such transitions we use the approach by Bastanjian et al. (1965) and Zel'dovich et al. (1985). The momentum and energy equations for a steady and fully developed flow in a circular tube are

$$\frac{1}{r}\frac{\partial}{\partial r}\left(r\mu\frac{\partial u}{\partial r}\right) - \frac{\mathrm{d}P}{\mathrm{d}x} = 0 \tag{3.25}$$

$$\frac{\partial^2 T}{\partial r^2} + \frac{1}{2}\frac{\partial T}{\partial r} + \frac{\mu}{kT}\left(\frac{\partial u}{\partial r}\right)^2 = 0 \tag{3.26}$$

Table 3.8 Characteristics of a micro-channel in the experiments by Harley et al. (1995)

Top width [μm]	Bottom width [μm]	Depth [mm]	Hydraulic diameter d_h [μm]	Length L [mm]	L/d_h	Gases tested
94.4	80.0	11.04	19.18	10.18	992.1	N_2, He, Ar

Table 3.9 Estimation of adiabatic increase in average gas temperature in experiments by Harley et al. (1995)

Gases	$\Delta T/T_{in}$	$\Delta T_{av}/T_{in}$	ν_{ac}/ν_{in}	R_{ac}/R_{in}
Nitrogen	0.37	0.18	∼1.35	∼0.74
Helium	1.522	0.761	∼2.87	∼0.35
Argon	0.096	0.048	∼1.09	∼0.919

where $\mu = \mu(T)$.

The dependence of the liquid viscosity on the temperature is given by Frenkel (1946):

$$\mu = \mu_0 \exp\left(\frac{E}{RT}\right). \tag{3.27}$$

Using the Frank-Kamenetskii (1969) transformation we present the relation in Eq. (3.27) in the following form (Zel'dovich et al. 1985):

$$\mu = \mu_0 \exp\left(\frac{E}{RT_0}\right) \exp\left(-\frac{E(T-T_0)}{RT_0^2}\right) \tag{3.28}$$

where E is activation energy, R is the universal gas constant, $\mu_0 = f(T_0)$, T_0 is the characteristic temperature.

Introducing new variables

$$\xi = \left(\frac{r}{r_0}\right)^2, \quad \theta = \frac{E(T-T_0)}{RT_0^2} \tag{3.29}$$

we reduce the energy equation (3.20) to the following form (Zel'dovich et al. 1985):

$$\frac{\partial^2 \theta}{\partial \xi^2} + \frac{1}{\xi}\frac{\partial \theta}{\partial \xi} + \chi e^{\theta} = 0 \tag{3.30}$$

where the dimensionless parameter χ is

$$\chi = \frac{(\partial_x P)^2 r_0^4}{16\mu_0 k}\frac{E}{RT_0^2} \exp\left(-\frac{E}{RT_0}\right). \tag{3.31}$$

The boundary conditions corresponding to constant temperature are

$$\theta(1) = 0, \quad \left(\frac{\partial \theta}{\partial \xi}\right)_{\xi=0} = 0. \tag{3.32}$$

A steady-state solution of Eq. (3.30) exists only for $\chi \le 2$. At $\chi > 2$ hydrodynamic thermal explosion occurs and oscillatory flow takes place.

Bearing in mind that $\Delta P = \lambda \frac{L}{d}\frac{\rho U^2}{2}$, $\lambda = \frac{64}{\mathrm{Re}}$, $U = \frac{\mathrm{Re} v}{d}$, $\partial_x P = \frac{\Delta P}{L} = 32\frac{\rho v^2}{d^3}\mathrm{Re}$, and assuming that $\frac{\mu_0}{\rho} \simeq v_0 \approx v$ we obtain

$$\chi = \frac{\mathrm{Re}^2}{r_0^2}\frac{v_0^2}{c_p T_0}\frac{E}{RT_0}\exp\left(-\frac{E}{RT_0}\right)\mathrm{Pr} \tag{3.33}$$

or

$$\mathrm{Re}_{\mathrm{cr}} = \left(\frac{\chi_{\mathrm{cr}}}{N}\right)^{1/2} r_0 \tag{3.34}$$

where $\chi_{\mathrm{cr}} = 2$, $N = \mathrm{Pr}\frac{v_0^2}{C_p T_0}\frac{E}{RT_0}\exp\left(-\frac{E}{RT_0}\right)$.

From approximation of actual values $\mu(T)$ and E/R in the range of $T = 273-323$ K it was found $E/R = 2{,}090$ K and $\mu_0 = 0.845 \times 10^{-6} Pas$ at $T_0 = 300$ K.

We estimate the critical Reynolds number for flows of water and transformer oil, using the physical properties presented by Vargaftik et al. (1996) to be

$$\mathrm{Re}_{\mathrm{cr}} = 1.94 \times 10^{10} r_0 \tag{3.35}$$

for water, and

$$\mathrm{Re}_{\mathrm{cr}} = 2.5 \times 10^{8} r_0 \tag{3.36}$$

for transformer oil, where r_0 is the micro-channel radius in meters.

The critical Reynolds number depends significantly on physical properties of the liquid (kinematic viscosity, heat capacity and the Prandtl number), and micro-channel radius. For flow of highly viscous liquids, e.g., transformer oil, in micro-channels of $r_0 < 10^{-5}$ m, the critical Reynolds number is less than 2,300. Under these conditions, oscillatory regime occurs at all Re corresponding to laminar flow. The existence of velocity fluctuations does not indicate a change of the flow regime from laminar to turbulent. This phenomenon shows only the occurrence of oscillatory laminar flow. In water flow (small kinematic viscosity, large heat capacity) rising of velocity oscillations is not possible at any realistic r_0.

Thus, the comparison of experimental results to those obtained by conventional theory is correct when the experimental conditions were consistent with the theoretical ones. The experimental results corresponding to these requirements agree quite well with the theory:

- For single-phase fluid flow in smooth micro-channels of hydraulic diameter from 15 to 4,010 μm, in the range of the Reynolds numbers $\mathrm{Re} < \mathrm{Re}_{\mathrm{cr}}$, the Poiseuille number, Po, is independent of the Reynolds number, Re.
- For single-phase gas flow in micro-channels of hydraulic diameter from 101 to 4,010 μm, in the range of Reynolds numbers $\mathrm{Re} < \mathrm{Re}_{\mathrm{cr}}$, the Knudsen number $0.001 \leq \mathrm{Kn} \leq 0.38$, and the Mach number $0.07 \leq \mathrm{Ma} \leq 0.84$, the experimental friction factor agrees quite well with the theoretical one predicted for fully developed laminar flow.
- The behavior of the flow in micro-channels, at least down to 50 μm in diameter, shows no difference with macro-scale flow. For smooth and rough micro-channels with relative roughness $0.32\% \leq k_s \leq 7\%$, the transition from laminar to turbulent flow occurs between $1{,}800 \leq \mathrm{Re}_{\mathrm{cr}} \leq 2{,}200$, in full agreement with flow visualization and flow resistance data. In the articles used for the present study there was no evidence of transition below these results.
- The relation of hydraulic diameter to channel length and the Reynolds number are important factors that determine the effect of the viscous energy dissipation on flow parameters.
- Under certain conditions the energy dissipation may lead to an oscillatory regime of laminar flow in micro-channels. The oscillatory flow regime occurs in micro-channels at Reynolds numbers less that $\mathrm{Re}_{\mathrm{cr}}$. In this case the existence of velocity fluctuations does not indicate change from laminar to turbulent flow.

3.8.4 Laminar Drag Reduction in Micro-Channels Using Ultrahydrophobic Surfaces

The question of the conditions to be satisfied by a moving fluid in contact with a solid body was one of considerable difficulty for quite some time, as pointed out by Goldstein (1965), and the assumption of no-slip is now generally accepted for practical purposes. On the other hand, if we can make an artificial solid surface where there is very little interaction between the surface and the liquid in contact with it, slip would be appreciable for liquid flow. The analysis of the phenomenon was presented by Watanabe et al. (1999).

Because fluid slip occurs at highly water-repellent walls when the contact angle is about 150°, Watanabe et al. (1999) analyzed the friction factor of slip flow in a circular pipe. For a fully developed steady flow in a pipe, the Navier–Stokes equation can be written as

$$\frac{\mu}{r}\left[\frac{\mathrm{d}}{\mathrm{d}r}\left(r\frac{\mathrm{d}u}{\mathrm{d}r}\right)\right]=\left(\frac{\mathrm{d}P}{\mathrm{d}z}\right). \tag{3.37}$$

By integrating this equation, and owing to the physical consideration that the velocity must be finite at $r=0$,

$$u=\frac{r^2}{4\mu}\left(\frac{\mathrm{d}P}{\mathrm{d}z}\right)+C_1. \tag{3.38}$$

The constant C_1 is evaluated under the boundary conditions at the pipe wall: $r=a$, $u=u_\mathrm{s}$. Then, the slip velocity u_s is determined (Goldstein 1965) from a macroscopic point of view:

$$\tau_w=\mu\left(-\frac{\mathrm{d}u}{\mathrm{d}r}\right)_{r=r_0}=\beta u_\mathrm{s} \tag{3.39}$$

where β is the sliding coefficient. For the case of $\beta\to\infty$, Eq. (3.39) agrees with the no-slip condition. Consequently, this gives

$$C_1=\left(\frac{r_0}{2\beta}+\frac{r_0^2}{4\mu}\right)\left(\frac{-\mathrm{d}P}{\mathrm{d}z}\right) \tag{3.40}$$

and hence,

$$u=\left[\frac{r_0^2}{4\mu}\left(1-\frac{r^2}{r_0^2}\right)+\frac{r_0}{2\beta}\right]\left(\frac{-\mathrm{d}P}{\mathrm{d}z}\right). \tag{3.41}$$

The volume flow rate is

$$Q=\int_0^a 2\pi r u\,\mathrm{d}r=\frac{\pi r_0^4\Delta P}{8\mu L}\left(1+\frac{4\mu}{r_0\beta}\right) \tag{3.42}$$

where $(\Delta P/L)$ is the pressure gradient in a fully developed flow, and equals $(-\mathrm{d}P/\mathrm{d}z)$. The friction factor for laminar flow, λ, is

$$\lambda=\frac{64}{\mathrm{Re}}\frac{1}{1+\frac{4\mu}{r_0\beta}}. \tag{3.43}$$

Thus, in laminar flow with fluid slip, the friction factor is a function of not only the Reynolds number Re, but also the non-dimensional parameter $(\mu/a\beta)$. If the flow does not exhibit fluid slip, Eq. (3.43) gives $\lambda = 64/\text{Re}$ on substituting $\beta \to \infty$ into the equation.

A series of experiments was presented by Ou et al. (2004), which demonstrate significant drag reduction for the laminar flow of water through micro-channels using hydrophobic surfaces with well-defined micron-sized surface roughness.

The difference between a hydrophobic surface and an ultrahydrophobic surface lies not in the surface chemistry, but in the micro-scale surface roughness. Ultrahydrophobic surfaces are actually very rough with large, micron-sized protrusions coming out of the surface. An optical micrograph of an ultrahydrophobic lithographically etched silicon surface is shown in Fig. 3.15. The equilibrium contact angles for each of these surfaces were found to be greater than 160°.

A schematic diagram of the physical model is shown in Fig. 3.16.

Average pressure drop reduction as a function of flow rate for a series of different surfaces in a micro-channel is shown in Fig. 3.17, where $\text{Dr} = \frac{(\Delta P_{\text{no-slip}} - \Delta P)}{\Delta P_{\text{no-slip}}}$, ΔP is the experimentally measured pressure drop and $\Delta P_{\text{no-slip}}$ is the theoretical pressure drop prediction for flow over a no-slip surface at the same flow rate.

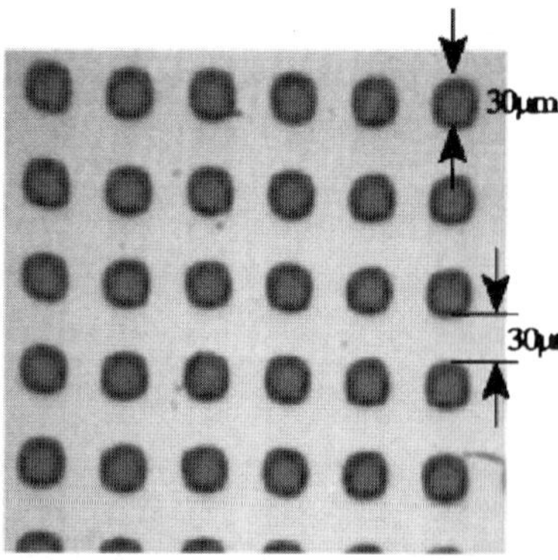

Fig. 3.15 Examples of ultrahydrophobic surfaces. Lithographically etched silicon surface patterned with 30 μm tall cubic micro-posts. Reprinted from Ou et al. (2004) with permission

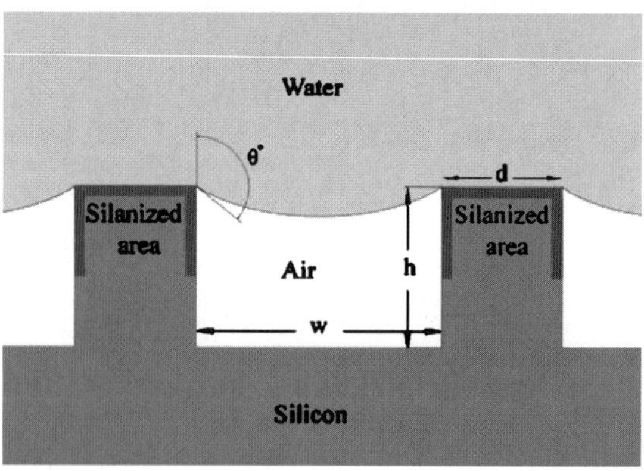

Fig. 3.16 Schematic diagram of a model for ultrahydrophobic drag reduction. A combination of surface hydrophobicity and roughness combine to allow water to stand away from the solid surface. Reprinted from Ou et al. (2004) with permission

The paper by Davies et al. (2006) reports results of a numerical investigation of the laminar, periodically repeating flow in a parallel-plate micro-channel with superhydrophobic walls. In particular, the influence of the Reynolds number and the vapor cavity size on the overall flow dynamics was explored. A schematic of the near-wall and cavity regions is shown in Fig. 3.18.

The walls exhibit micro-rib and cavity structures oriented perpendicular (transverse) to the flow direction and the walls are treated with a hydrophobic coating. Re-

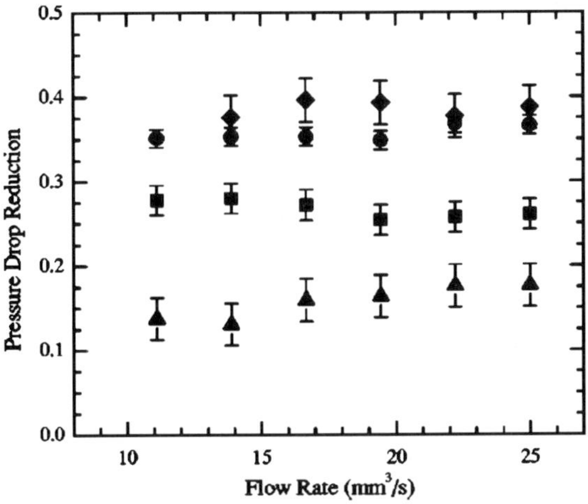

Fig. 3.17 Average pressure drop reduction as a function of flow rate for a series of different surfaces in a micro-channel having dimensions $W = 2.54$ mm, $H = 127$ μm, and $L = 50$ mm. The experimental data include a series of ultrahydrophobic surfaces with a regular array of square micro-posts with $d = 30$ μm with a spacing between micro-posts of $w = 15$ μm represented by *triangles* (▲), $d = 30$ μm and $w = 30$ μm represented by *squares* (■), $d = 30$ μm and $w = 60$ μm represented by *circles* (●), and $d = 30$ μm and $w = 150$ μm represented by *diamonds* (♦). Reprinted from Ou et al. (2004) with permission

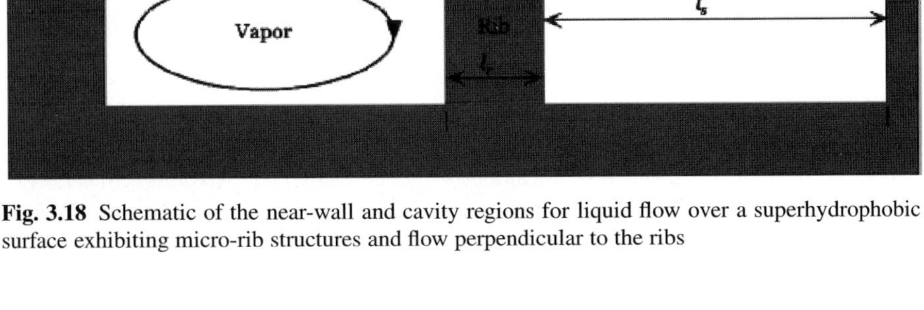

Fig. 3.18 Schematic of the near-wall and cavity regions for liquid flow over a superhydrophobic surface exhibiting micro-rib structures and flow perpendicular to the ribs

ductions in the frictional drag are greater as the cavity-to-rib length ratio is increased (increasing shear-free fraction) and as the channel hydraulic diameter is decreased.

The results also show that the normalized slip length and the average friction factor–Reynolds number product exhibit Reynolds dependence. Furthermore, the predictions reveal that the impact of the vapor cavity depth on the overall frictional resistance is minimal provided the depth of the vapor cavity is greater than 25% of its width.

Summary

Because most applications for micro-channel heat sinks deal with liquids, most of the former studies were focused on micro-channel laminar flows. Several investigators obtained friction factors that were greater than those predicted by the standard theory for conventional size channels, and, as the diameter of the channels decreased, the deviation of the friction factor measurements from theory increased. The early transition to turbulence was also reported. These observations may have been due to the fact that the entrance effects were not appropriately accounted for. Losses from change in tube diameter, bends and tees must be determined and must be considered for any piping between the channel plenums and the pressure transducers. It is necessary to account for the loss coefficients associated with single-phase flow in micro-channels, which are comparable to those for large channels with the same area ratio.

Internal pressure measurements would eliminate these effects. In addition to accounting for the losses outside the channel, it is also necessary to consider the pressure drop associated with developing flow in the entrance region of the channel.

The uncertainty of calculating the Poiseuille number from the measurements must be taken into account. The viscosity–pressure relationship of certain liquids (e.g., isopropanol, carbon tetrachloride) must be kept in mind to obtain the revised theoretical flow ratc. Thc cffcct of evaporation from the collection dish during the mass flow rate measurement must be taken into consideration. The effect of evaporation of collected water into the room air may not be negligible, and due to the extremely low mass flow rates through the micro-channel this effect can become significant.

The main aim of the present chapter is to verify the capacity of conventional theory to predict the hydrodynamic characteristics of laminar Newtonian incompressible flows in micro-channels in the hydraulic diameter range from $d_{\mathrm{h}} = 15$ to $d_{\mathrm{h}} = 4{,}010\,\mu\mathrm{m}$, Reynolds number from $\mathrm{Re} = 10^3$ up to $\mathrm{Re} = \mathrm{Re}_{\mathrm{cr}}$, and Knudsen number from $\mathrm{Kn} = 0.001$ to $\mathrm{Kn} = 0.4$. The following conclusions can be drawn from this study:

The comparison of experimental results accounting for effects discussed above to those obtained by conventional theory is correct when the experimental conditions were consistent with theoretical ones, and agrees well with Poiseuille flow predictions.

For single-phase fluid flow in smooth micro-channels of hydraulic diameter the Poiseuille number is independent of the Reynolds number. For single-phase gas flow in micro-channels of hydraulic diameter 16.6 to 4,010 μm, Knudsen number of $Kn = 0.001-0.38$, Mach number of $Ma = 0.07-0.84$, the experimental friction factor agrees quite well with theoretical one predicted for fully developed laminar flow.

The problem of transition to turbulence has been studied. The data for all channels showed no indication that a transition to turbulence began for Re up to 2,000. The behavior of the flow in micro-channels, at least down to 50 μm diameter, shows no difference with macro-scale flow. For smooth and rough micro-channels with relative roughness in the range of $k_s = 0.32-7\%$ the transition from laminar to turbulent flow occurs in the range of Reynolds number $Re = 1{,}800-2{,}200$, in full agreement with flow visualization and flow resistance data. These results support the standard findings that laminar flow is maintained for $Re < 2{,}300$. It is important to note that when the data from a given paper is examined independently, it often shows a consistent deviation from the theoretical predictions. However, when the data from many papers are summarized they are randomly scattered both above and below the standard predictions for large channels.

For flow of some kind of surfactant solutions (Habon G solutions at concentration 530 and 1,060 ppm) in the tube of $d = 1.07$ mm in the range of Reynolds number based on solvent viscosity $Re = 10-450$, the increase of pressure drop in adiabatic and diabatic conditions was observed compared to that of pure water.

Under certain conditions the energy dissipation may lead to an oscillatory regime of laminar flow in micro-channels. The relation of hydraulic diameter to channel length and the Reynolds number are important factors that determine the effect of viscous energy dissipation on flow parameters. The oscillatory flow regime occurs in micro-channels at Reynolds numbers less than Re_{cr}. In this case the existence of velocity fluctuations does not indicate change from laminar to turbulent flow.

One approach recently proposed reducing the pressure drop of liquids flowing in micro-channels by creating micro-ribs and cavities on the micro-channel walls. If these micro-rib/cavity structures are treated with a hydrophobic coating and the cavity size is small enough, the liquid flowing in the micro-channels wets only the surface of the ribs, and does not penetrate the cavities. The liquid thus forms a free surface meniscus in the cavity regions between the micro-ribs. The result is a reduction in the surface contact area between the channel walls and the flowing liquid, and a reduction in pressure drop up to 40% may be achieved.

References

Bailey DK, Ameel TA, Warrington RO, Savoie TI (1995) Single phase forced convection heat transfer in micro-geometries. In: Proceedings the 13th of Intersociety Energy Conversion Engineering Conference, San Diego, 20–25 August 1978. American Society of Mechanical Engineers, New York, pp 301–310

Bastanjian SA, Merzhanov AG, Xudiaev SI (1965) On hydrodynamic thermal explosion. Sov Phys Docl 163:133–136

Bayraktar T, Pidugu SB (2006) Characterization of liquid flows in micro-fluidic systems. Int J Heat Mass Transfer 49:815–824

Berg van den HR, Seldom ten CA, Gulik van der PS (1993) Compressible laminar flow in a capillary. J Fluid Mech 246:1–20

Brutin D, Tadrist L (2003) Experimental friction factor of a liquid flow in micro-tubes. Phys Fluids 15:653–661

Celata GP, Gumo M, Zummo G (2004) Thermal-hydraulic characteristics of single-phase flow in capillary pipes. Exp Thermal Fluid Sci 28:87–95

Celata GP, Morini GL, Marconi V, McPhail SS, Zummo G (2005) Using viscous heating to determine the friction factor in micro-channels: an experimental validation. In: Proceedings of ECI International Conference on Heat Transfer and Fluid Flow in Microchannel, Caste/Vecchio Pascoli, Italy, 25–30 September 2005

Celata GP, Cumo M, McPhail S, Zummo G (2006) Characterization of fluid dynamics behavior and channel wall effects in micro-tube. Int J Heat Fluid Flow 27:135–143

Cho YI, Hartnett JP (1982) Non-Newtonian fluids in circular pipe flows. Adv Heat Transfer 15:60–141

Cogswell FN (1981) Polymer melt rheology: a guide for industrial practice. Woodhead, Cambridge

Cui HH, Silber-Li ZH, Zhu SN (2004) Flow characteristics of liquids in micro-tubes driven by high pressure. Phys Fluids 16:1803–1810

Darbyshire AG, Mullin T (1995) Transition to turbulence in constant-mass-flux pipe flow. J Fluid Mech 289:83–114

Davies J, Maynes D, Webb BW, Woolford B (2006) Laminar flow in a microchannel with super hydrophobic walls exhibiting transverse ribs. Phys Fluids 18:087110

Duncan AB, Peterson GP (1994) Review of microscale heat transfer. Appl Mech Rev 47:397–428

Frank-Kamenetskii DA (1969) Diffusion and heat transfer in chemical kinetics, 2nd edn. Plenum, New York

Frenkel L (1946) Kinetic theory of liquids. Clarendon, Oxford

Gad-el-Hak M (1999) The fluid mechanics of microdevices. The Freeman Scholar Lecture. J Fluid Eng 121:5–33

Gad-el-Hak M (2003) Comments or critical view on new results in micro-fluid mechanics. Int J Heat Mass Transfer 46:3941–3945

Goldstein S (1965) Modern developments in fluid dynamics, vol 2. Dover, New York, pp 676–680

Gruntfest IJ, Young JP, Jhonson NL (1964) Temperatures generated by the flow of liquids in pipes. J Appl Phys 35:18–23

Guo ZY, Li ZX (2002) Size effect on micro-scale single phase flow and heat transfer. In: Proceedings of the 12th International Heat Transfer Conference, Grenoble, France, 18–23 August 2002

Guo ZY, Li ZX (2003) Size effect on micro-scale single-phase flow and heat transfer. Int J Heat Mass Transfer 46:149–159

Hagen G (1839) Über die Bewegung des Wassers in engen zylindrischen Rohren. Pogg Ann 46:423–442

Hao PF, Zhang XW, Yao FHe (2007) Transitional and turbulent flow in circular micro-tube. Exp. Thermal and Fluid Science 32:423-431

Harley JC, Huang Y, Bau HH, Zewel JN (1995) Gas flow in micro-channels. J Fluid Mech 284:257–274

Herwig H (2000) Flow and heat transfer in micro systems. Is everything different or just smaller. ZAMM 82:579–586

Herwig H, Hausner O (2003) Critical view on new results in micro-fluid mechanics: an example. Int J Heat Mass Transfer 46:935–937

Hetsroni G, Gurevich M, Mosyak A, Rozenblit R (2004) Drag reduction and heat transfer of surfactants flowing in a capillary tube. Int J Heat Mass Transfer 47:3797–3869

Hetsroni G, Zakin JL, Lin Z, Mosyak A, Pancallo EA, Rozenblit R (2001) The effect of surfactants on bubble grows, wall thermal patterns and heat transfer in pool boiling. Int J Heat Mass Transfer 44:485–497

Hetsroni G, Mosyak A, Pogrebnyak E, Yarin LP (2005) Fluid flow in micro-channels. Int J Heat Mass Transfer 48:1982-1998

Ho C-M, Tai Y-C (1998) Micro-electro-mechanical systems (MEMS) and fluid flows. Ann Rev Fluid Mech 30:579–612

Hsieh SS, Tsai HH, Lin CY, Huang CF, Chien CM (2004) Gas flow in long micro-channel. Int J Heat Mass Transfer 47:3877–3887

Hwang YW, Kim MS (2006) The pressure drop in microtubes and correlation development. Int J Heat Mass Transfer 49:1804–1812

Incropera FP (1999) Liquid cooling of electronic devices by single-phase convection. Wiley, New York

Jones OC (1976) An improvement in the calculation of turbulent friction factor in rectangular ducts. Trans ASME J Fluid Eng 98:173–181

Judy J, Maynes D, Webb BW (2002) Characterization of frictional pressure drop for liquid flows through micro-channels. Int J Heat Mass Transfer 45:3477–3489

Kandlikar SG, Joshi S, Tian S (2003) Effect of surface roughness on heat transfer and fluid flow characteristics at low Reynolds numbers in small diameter tubes. Heat Transfer Eng 24:4–16

Koo J, Kleinstreuer C (2004) Viscous dissipation effects in microtubes and microchannels. Int J Heat Mass Transfer 47:3159–3169

Leite RJ (1959) An experimental investigation of the stability of Poiseuille flow. J Fluid Mech 5:81–96

Lelea D, Nishio S, Takano K (2004) The experimental research on micro-tube heat transfer and fluid flow of distilled water. Int J Heat Mass Transfer 47:2817–2830

Li ZX, Du DX, Guo ZY (2003) Experimental study on flow characteristics of liquid in circular micro-tubes. Microscale Thermophys Eng 7:253–265

Lindgren ER (1958) The transition process and other phenomena in viscous flow. Arkiv für Physik 12:1–169

Loitsianskii LG (1966) Mechanics of liquid and gases. Pergamon, Oxford

Lumley JL (1969) Drag reduction by additives. Ann Rev Fluid Mech 1:367–384

Ma HB, Peterson GP (1997) Laminar friction factor in microscale ducts of irregular cross section. Microscale Thermophys Eng 1:253–265

Mala GM, Li D (1999) Flow characteristics of water in micro-tubes. Int J Heat Fluid Flow 20:142–148

Maynes D, Webb AR (2002) Velocity profile characterization in sub-diameter tubes using molecular tagging velocimetry. Exp Fluids 32:3–15

Morini GL (2004) Laminar-to-turbulent transition in microchannels. Microscale Thermophys Eng 8:15–30

Obot NT (1988) Determination of incompressible flow friction in smooth circular and noncircular passages. A generalized approach including validation of the century old hydraulic diameter concept. Trans ASME J Fluid Eng 110:431–440

Ou J, Perot B, Rothstein JP (2004) Laminar drag reduction in microchannels using ultrahydrophobic surfaces. Phys Fluids 16(12):4635–4643

Papautsky I, Brazzle J, Ameel T, Frazier B (1999) Laminar fluid behavior in micro-channels using micro-polar fluid theory. Sens Actuators 73:101–108

Peng XF, Peterson GP (1996) Convective heat transfer and friction for water flow in micro-channel structures. Int J Heat Mass Transfer 39:2599–2608

Peng XF, Wang BX (1998) Forced convection and boiling characteristics in micro-channels. In: Heat Transfer 1998 Proceedings of the 11th IHTC, Kyongju, Korea, 23–28 August 1998, vol 11, pp 371–390

Pfund D, Rector D, Shekarriz A (2000) Pressure drop measurements in a micro-channel. AIChE J 46:1496–1507

Plam B (2000) Heat transfer in microchannels. In: Heat Transfer and Transport Phenomena in Microscale. Banff Oct:54–64

Poiseuille JLM (1840) J Recherches experimentelles tubes de tris petits diameters. Comptes Rendus 11:961–967, 1041–1048

Qu W, Mala GM, Li D (2000) Pressure driven water flows in trapezoidal silicon micro-channels. Int J Heat Mass Transfer 43:353–364

Rands C, Webb BW, Maynes D (2006) Characterization of transition to turbulence in micro-channels. Int J Heat Mass Transfer 49:2924–2930

Ren L, Qu W, Li D (2001) Interfacial electrokinetic effects on liquid flow in micro-channels. Int J Heat Mass Transfer 44:3125–3134

Schlichting H (1979) Boundary layer theory. McGraw-Hill, New York

Sedov LI (1993) Similarity and dimensional methods in mechanics, 10th edn. CRC, Boca Raton

Shah RK, London AL (1978) Laminar flow forced convection in duct. Academic, New York

Shapiro AK (1953) The dynamics and thermodynamics of compressible fluid flow. Wiley, New York

Sharp KV, Adrian RJ (2004) Transition from laminar to turbulent flow in liquid filled microtubes. Exp Fluids 36:741–747

Sharp KV, Adrian R, Santiago J, Molho JI (2001) Liquid flows in micro-channels. In: CRC Handbook of MEMS. CRC, Boca Raton, pp 6.1–6.38

Tani I (1969) Boundary layer transition. Ann Review of Fluid Mech 1:169–196

Tso CP, Mahulikar SP (1998) The use of the Brinkman number for single phase forced convective heat transfer in micro-channels. Int J Heat Mass Transfer 41:1759–1769

Tso CP, Mahulikar SP (1999) The role of the Brinkman number in analysis flow transition in micro-channel. Int J Heat Mass Transfer 42:1813–1833

Tso CP, Mahulikar SP (2000) Experimental verification of the role of Brinkman number in micro-channels using local parameters. Int J Heat Mass Transfer 43:1837–1849

Tuckerman DB (1984) Heat transfer micro-structure for integrated circuits. Dissertation, Department of Electrical Engineering, Stanford University

Vargaftik NB, Vinogradov YK, Vargin VS (1996) Handbook of physical properties of liquids and gases. Pure substances and mixtures, 3rd augm. rev. edn. Begell House, New York

Virk PS, Mickley HS, Smith KA (1970) The ultimate asymptote and mean flow structure in Toms' phenomenon. ASME J Appl Mech 37:488–493

Warholic MD, Schmidt GM, Hanratty TJ (1999) The influence of a drag-reducing surfactant on a turbulent velocity field. J Fluid Mech 388:1–20

Watanabe K, Udagawa Y, Udagawa H (1999) Drag reduction of Newtonian fluid in a circular pipe with a highly water-repellent wall. J Fluid Mech 381:225–238

White FM (1994) Fluid mechanics, 3rd edn. McGraw-Hill, New York

Wu HY, Cheng P (2003) Friction factors in smooth trapezoidal silicon micro-channels with different aspect ratio. Int J Heat Mass Transfer 46:2519–2525

Wygnanskii IJ, Champagne FH (1973) On transition in a pipe. Part 1. The origin of puffs and slugs and the flow in a turbulent slug. J Fluid Mech 59:281–351

Xu B, Ooi KT, Wong NT, Choi WK (2000) Experimental investigation of flow friction for liquid flow in micro-channels. Int Comm Heat Transfer 27(8):1165–1176

Yang CY, Wu JC, Chien HT, Lu SR (2003) Friction characteristics of water, R-134a, and air in small tubes. Microscale Thermophys Eng 7:335–348

Zakin JL, Myska J, Chara Z (1996) New limiting drag reduction and velocity profile asymptotes for nonpolymeric additives systems. AIChE J 42:3544–3546

Zakin JL, Qi Y, Zhang Y (2002) In: Proceedings of 15th International Congress of Chemical and Process Engineering, CHISA 2002, Prague, Czech Republic, 25–29 August 2002

Zel'dovich JaB, Barenblatt GI, Librovich VB, Makhviladze GM (1985) Mathematical theory of combustion and explosion. Plenum, New York

Nomenclature

a	Side of a equilateral triangle
C^*	Ratio of $\mathrm{Po}_{\mathrm{exp}}/\mathrm{Po}_{\mathrm{theor}}$
c_p	Specific heat
d	Diameter
d_*	Characteristic size
E	Activation energy
$f = \dfrac{\lambda}{4}$	Fanning friction factor
h	Enthalpy, heat transfer coefficient
H	Side of a rectangle; depth of plane micro-channel
k	Thermal conductivity
k_s	Average height of surface roughness
L	The length of the micro-channel
$\overline{L} = L/d_*$	Dimensionless length
m	Mass flow rate
P	Pressure
r	Current radius
r_0	Inner radius
R	Universal gas constant
T	Temperature
t	Time
U	Average velocity
u	Longitudinal component of velocity
u'	rms velocity fluctuation
u_s	Slip velocity
u_*	Friction velocity
u_k	Velocity at the top of the roughness element
x, y, z	Cartesian coordinates
$\mathrm{Ec} = \frac{U^2}{2C_p\Delta T}$	Eckert number
$\mathrm{Kn} = \frac{\overline{\lambda}}{d_h}$	Knudsen number
$\mathrm{Ma} = \frac{U}{U_{\mathrm{sound}}}$	Mach number
$\mathrm{Po} = \lambda \mathrm{Re}$	Poiseuille number
$\mathrm{Pr} = \frac{\nu}{\alpha}$	Prandtl number
$\mathrm{Re} = \frac{Ud_h}{\nu}$	Reynolds number
$\mathrm{Re}_k = \frac{u_k k_s}{\nu}$	Reynolds number based on average height of surface roughness

Greek Symbols

α	Thermal diffusivity
β	Sliding coefficient
α, β	Semi-axis of the ellipse
ΔT	Temperature difference
ΔP	Pressure drop
δ	Uncertainty, characteristic size
ε	Channel aspect ratio
η	Dimensionless radius
θ	Dimensionless temperature, variable
λ	Friction factor
$\overline{\lambda}$	Mean free path
μ	Dynamic viscosity
μ_m	Viscosity determined by roughness – viscosity model
ν	Kinematic viscosity
ξ	Height to width ratio, variable
ρ	Density
τ	Shearing stress at a wall
Φ	Energy dissipation
χ	Variable

Subscript

ac	Actual
av	Average
cr	Critical
ef	Effective
exp	Experimental
h	Hydraulic
in	Inlet
lam	Laminar
max	Maximum
out	Outlet
sh	Shear
theor	Theoretical
w	Wall
wat	Water

Chapter 4
Heat Transfer in Single-Phase Flows

The subject of this chapter is single-phase heat transfer in micro-channels. Several aspects of the problem are considered in the frame of a continuum model, corresponding to small Knudsen number. A number of special problems of the theory of heat transfer in micro-channels, such as the effect of viscous energy dissipation, axial heat conduction, heat transfer characteristics of gaseous flows in micro-channels, and electro-osmotic heat transfer in micro-channels, are also discussed in this chapter.

4.1 Introduction

Heat transfer in straight tubes and channels has been the subject of much research for the last century. Theoretical predictions in this field agree fairly well with known experimental data related to heat transfer in the conventional size channels (cf. Petukhov 1967; Kays and Crawford 1993; Baehr and Stephan 1998; Schlichting 2000). The development of micro-mechanics during the last decades stimulated a great interest in heat transfer studies in micro-channels (cf. Ho and Tai 1998; Gad-el-Hak 1999). A number of theoretical and experimental investigations devoted to this problem were performed from 1994 to 2006, including Wang and Peng (1994), Peng and Peterson (1995), Peng et al. (1995), Peng and Peterson (1996), Ma and Peterson (1997), Mala et al. (1997a,b), Mala and Li (1999), Qu et al. (2000), Qu and Mudawar (2002a,b), Gao et al. (2002), Zhao and Lu (2002), Wu and Cheng (2003), Weigand and Lauffer (2004), Male et al. (2004), Lelea et al. (2004), Gamart et al. (2005), Reynaud et al. (2005), Lelea (2005), and Yoo (2006). Data on heat transfer in laminar and turbulent flows in micro-channels of different geometry were obtained. Several special problems related to heat transfer in micro-channels were discussed, including the effect of axial conduction in the wall and viscous dissipation effect (Tso and Mahulikar 1998; Tso and Mahulikar 1999; Tso and Mahulikar 2000; Tunc and Bayazitoglu 2001; Koo and Kleinstreuer 2004; Maranzana et al. 2004). Comprehensive surveys may be found in Bailey et al. (1995), Sobhan and Garimella (2001), Kandlikar and Grande (2002), Gua and Li (2003), Garimella and Sobhan (2003), Celata et al. (2004), Hassan et al. (2004), and Morini (2004).

We note that different methods of heating are used by various researchers, as shown in Fig. 4.1.

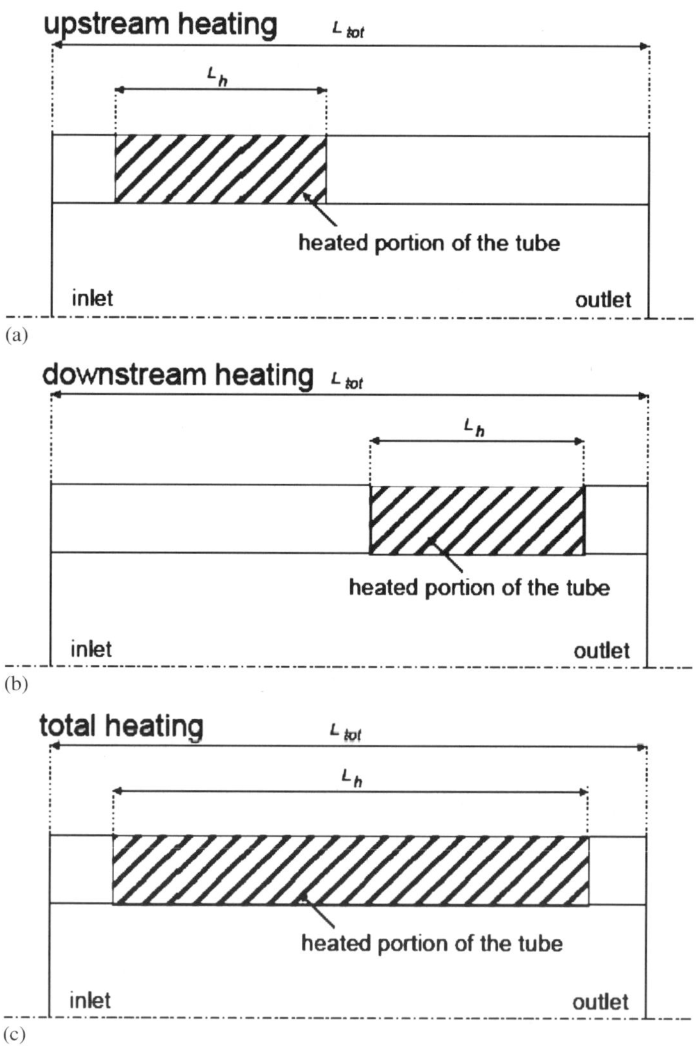

Fig. 4.1a–c Different heating cases in experiments and numerical modeling. Reprinted from Lelea (2005) with permission

Considering the available literature on the experimental research, one can conclude that there is a large scatter in the results on heat transfer. There is no convincing explanation of the difference between experimental and theoretical results for laminar flow, and between experimental and semi-empirical results for turbulent flow. On the one hand, several researchers argue that some new effects exist in micro-channels, e.g., Ho and Tai (1998), Tso and Mahulikar (1998, 1999, 2000), Gad-el-Hak (2003). On the other hand, the phenomenon can be related to the discrepancy between the actual conditions of a given experiment, and theoretical or numerical solution obtained in the frame of conventional theory (Herwig 2000; Herwig and Hausner 2003). The aim of the present chapter is to address this issue. The problem of heat transfer is considered here in the frame of a continuum model, corresponding to small Knudsen number. The data on heat transfer in circular, triangular, rectangular, and trapezoidal micro-channels with hydraulic diameters ranging from 10 to 2,000 μm are analyzed. The effects of geometry, axial heat flux due to thermal conduction through the working fluid and channel walls, and energy dissipation are also discussed. We focus on comparing experimental data, obtained by a number of investigators, to the conventional theory of heat transfer. The analysis includes a discussion of possible sources of unexpected effects reported in some experimental investigations.

In our analysis, we discuss experimental results of heat transfer obtained by previous investigators and related to incompressible fluid flow in micro-channels of different geometry. The basic characteristics of experimental conditions are given in Table 4.1. The studies considered herein were selected to reveal the physical basis of scale effect on convective heat transfer and are confined mainly to consideration of laminar flows that are important for comparison with conventional theory.

Table 4.1 Basic characteristics of micro-channels and experimental conditions

Cross-section	Micro-channel size			Working fluid	Walls		Re
	d_h [μm]	L [mm]	L/d_h		Material	Surface	
Circular	125.4–1,070	53–335	72–500	Distilled water, de-ionized water	Stainless steel	Smooth Rough	10–2,600
Rect-angular	133–2,000	25–325.125	13–433	De-min-eralized water, de-ionized water	Silicon, copper	Smooth Rough	40–9,000
Trapez-oidal and triangular	62.3–168.9	30	180–500	R-134a FC-84	Silicon	Smooth Rough	15–1,450

4.2 Experimental Investigations

4.2.1 Heat Transfer in Circular Tubes

Laminar flow

The schemes of the test sections used by some investigators are shown in Fig. 4.2. The geometrical parameters are presented in Tables 4.2 and 4.3.

Figure 4.2a shows the experimental set-up used by Lelea et al. (2004). The inner diameters of smooth micro-tubes were 125.4, 300, and 500 μm and the flow regime was laminar with Reynolds number Re = 95–774. The micro-tube was placed inside a vacuum chamber to eliminate heat loss to the ambient. It was heated by Joule heating with an electrical power supply. Distilled water was used and the measurements of heat transfer coefficient were performed under the thermal boundary condition of a constant heat flux on the wall. In the experimental set-up, there were two electrodes at both ends of the test tube. The insulated parts were included in the test section. Thus, for the heating length of $L_h = 250$, 95, and 53 mm the total length of the test section was $L = 600$, 123, and 70 mm, respectively. The experimental results have been compared both with theoretical predictions from the literature, and the results obtained by numerical modeling under the same thermal boundary conditions at the inlet and outlet of the tube. The experimental results confirm that the conventional theories are applicable for water flow through micro-channels, including the entrance effects (see Fig. 4.3a).

Table 4.2 Smooth circular micro-channels: experimental conditions

Author	Number of channels in the test section	Inner diameter d_{in} [μm]	Outer diameter d_{out} [μm]	Heating length L_h [mm]	Dimensionless length L_h/d_{in}	Reynolds number Re
Lelea et al. (2004)	1	500	700	250	500	95–774
		300	500	95	317	
		125.4	300	53	424	
Hetsroni et al. (2004)	1	1,070	1,500	335	313	10–450

Table 4.3 Rough circular micro-channels: experimental conditions

Author	Number of channels in the test section	Inner diameter d_{in} [μm]	Relative roughness k_s/d_{in}	Heating length L_h [mm]	Dimensionless length L_h/d_{in}	Reynolds number Re
Kandlikar et al. (2003)	1	1,067	0.00178–0.00281	76.5	72	500–2,600
		620	0.00161–0.00355	67	108	

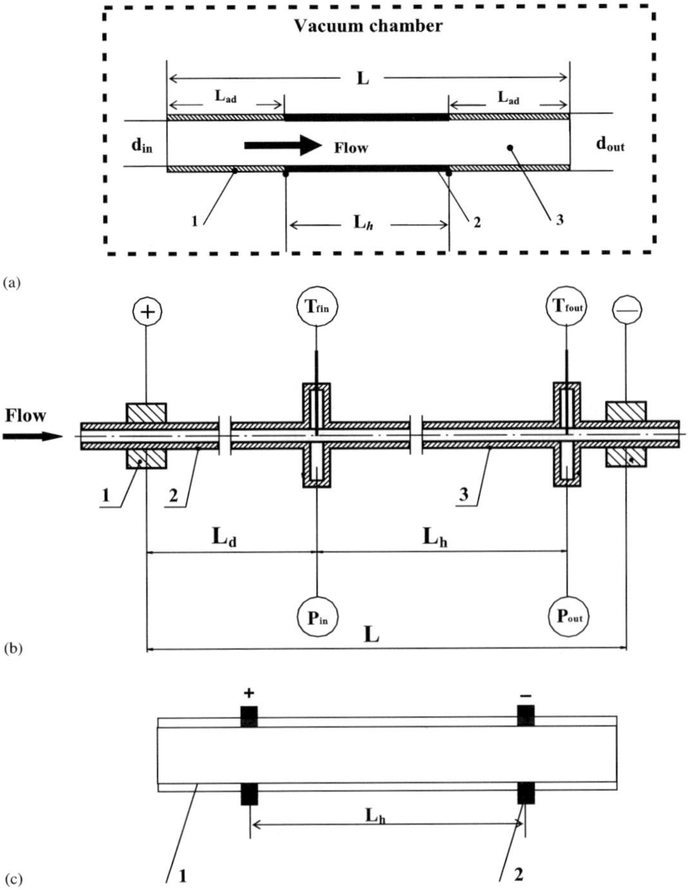

Fig. 4.2a–c Circular micro-channels. (**a**) $d_{\text{in}} = 125.4{-}500\,\mu\text{m}$. Test section used by Lelea et al. (2004) (schematic view): *1* adiabatic section, *2* heated section, *3* micro-channel. (**b**) $d_{\text{in}} = 1{,}070\,\mu\text{m}$. Test section used by Hetsroni et al. (2004): *1* electrical contact, *2* thermal developing section, *3* measurement section. (**c**) Test section of rough circular micro-channel used by Kandlikar et al. (2003) (schematic view). $d_{\text{in}} = 1{,}067\,\mu\text{m}$, $k_s/d_{\text{in}} = 0.00178{-}0.00281$; $d_{\text{in}} = 620\,\mu\text{m}$, $k_s/d_{\text{in}} = 0.00161{-}0.00355$: *1* steel tube, *2* electrical contact. Reprinted from Lelea et al. (2004), Hetsroni et al. (2004), and Kandlikar et al. (2003) with permission

The micro-channels utilized in engineering systems are frequently connected with inlet and outlet manifolds. In this case the thermal boundary condition at the inlet and outlet of the tube is not adiabatic. Heat transfer in a micro-tube under these conditions was studied by Hetsroni et al. (2004). They measured heat transfer to water flowing in a pipe of inner diameter 1.07 mm, outer diameter 1.5 mm, and 0.600 m in length, as shown in Fig. 4.2b. The pipe was divided into two sections. The development section of $L_{\text{d}} = 0.245$ m was used to obtain fully developed flow and thermal fields. The test section proper, of heating length $L_{\text{h}} = 0.335$ m, was used for collecting the experimental data.

DC current was supplied through the development and test sections for direct heating. The outer temperature on the heated wall was measured by means of an infrared radiometer. Experiments were carried out in the range of Re = 10−450. The average Nusselt number was calculated using the average temperature of the inner tube wall and mean temperature of the fluid at the inlet and outlet of the tube.

The dependence of the local Nusselt number on non-dimensional axial distance X^+ is shown in Fig. 4.3a. The dependence of the average Nusselt number on the Reynolds number is presented in Fig. 4.3b. The Nusselt number increased drastically with increasing Re at very low Reynolds numbers, 10 < Re < 100, but this increase became smaller for 100 < Re < 450. Such a behavior was attributed to the effect of axial heat conduction along the tube wall. Figure 4.3c shows the dependence of the relation N_a/N on the Peclet number Pe, where N_a is the power conducted axially in the tube wall, and N is total electrical power supplied to the tube. Comparison between the results presented in Fig. 4.3b and those presented in Fig. 4.3c allows one to conclude that the effect of thermal conduction in the solid wall leads to a decrease in the Nusselt number. This effect decreases with an increase in the

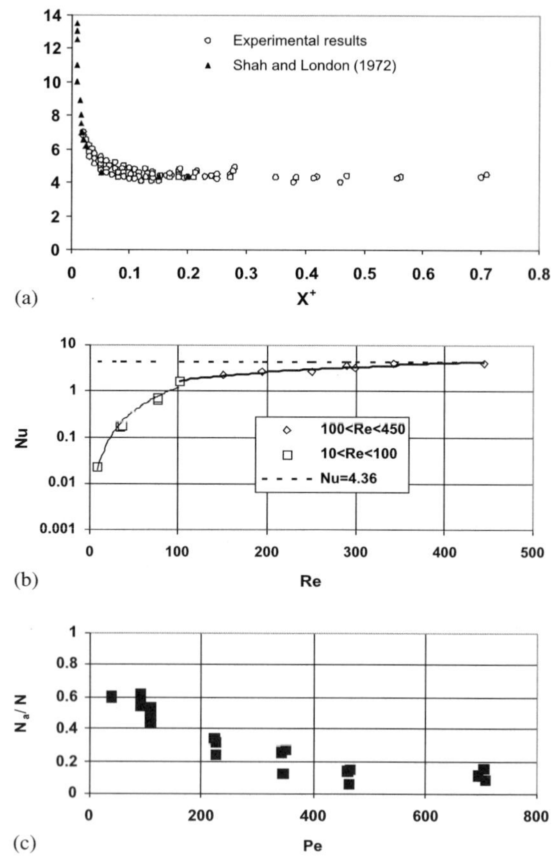

Fig. 4.3a–c Experimental results for smooth circular tubes. (**a**) Dependence of the Nusselt number on non-dimensional axial distance d_{in} = 125.4, 300 and 500 μm, Re = 95−774. Reprinted from Lelea et al. (2004) with permission. (**b**) d_{in} = 1,070 μm. Dependence of average Nusselt number on Reynolds number. Reprinted from Hetsroni et al. (2004) with permission. (**c**) d_{in} = 1,070 μm. Dependence of the relation of the power conducted axially through the heated wall to the power supplied to the heat section. Reprinted from Hetsroni et al. (2004) with permission

Reynolds number. It should be stressed that the heat transfer coefficient depends on the character of the wall temperature and the bulk fluid temperature variation along the heated tube wall. It is well known that under certain conditions the use of mean wall and fluid temperatures to calculate the heat transfer coefficient may lead to peculiar behavior of the Nusselt number (see Eckert and Weise 1941; Petukhov 1967; Kays and Crawford 1993). The experimental results of Hetsroni et al. (2004) showed that the use of the heat transfer model based on the assumption of constant heat flux, and linear variation of the bulk temperature of the fluid at low Reynolds number, yield an apparent growth of the Nusselt number with an increase in the Reynolds number, as well as underestimation of this number.

Turbulent flow

Adams et al. (1998) investigated turbulent, single-phase forced convection of water in circular micro-channels with diameters of 0.76 and 1.09 mm. The Nusselt numbers determined experimentally were higher than those predicted by traditional Nusselt number correlations such as the Gnielinski correlation (1976). The data suggest that the extent of enhancement (deviation) increases as the channel diameter decreases. Owhaib and Palm (2004) investigated the heat transfer characteristics

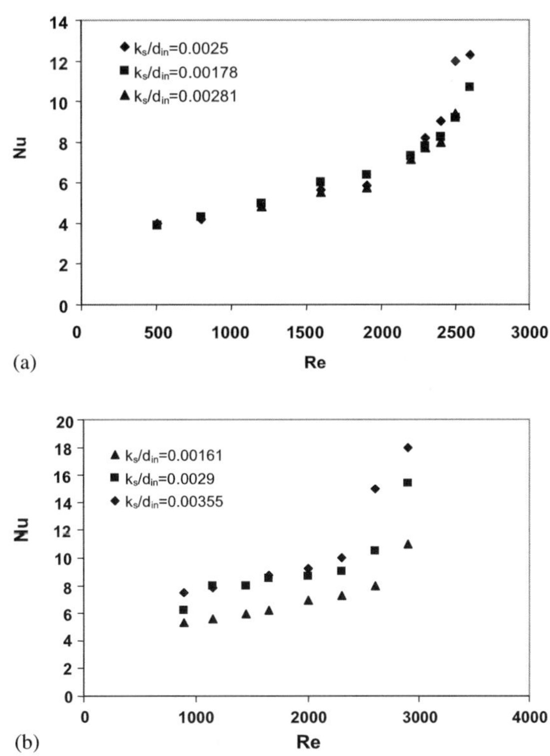

Fig. 4.4a,b Plots of local Nusselt number for different k_s/d_{in} ratios. (**a**) $d_{in} = 1{,}067\,\mu m$ (**b**) $d_{in} = 620\,\mu m$. Reprinted from Kandlikar et al. (2003) with permission

for single-phase forced convection of R-134 through single circular micro-channels. The test sections consisted of stainless steel tubes with 1.7, 1.2, and 0.8 mm inner diameters, and 325 mm in length. The results show good agreement between the classical correlations (Dittus-Boelter 1930; Petukhov et al. 1973; Gnielinski 1976) and the experimentally measured data in the turbulent region. Yang and Lin (2007), Lin and Yang (2007) measured the heat transfer coefficients for water flow through stainless steel micro-tubes with inner diameters ranging from 123 to 962 μm by the method of liquid crystal thermography. In the range of Re = 200–20,000, roughness surface of 1.16–1.48 μm the transfer correlations for laminar and turbulent flow can be well applied for predicting the fully developed heat transfer performances in micro-tubes. The transition occurs at Reynolds numbers from 2,300 to 3,000. This is also the same range as that for convectional tubes. On the contrary, correlations suggested for micro-channels (Wu and Little 1984; Choi et al. 1991; Adams et al. 1998) do not agree with this test. Kandlikar et al. (2003) studied experimentally the effect of surface roughness on heat transfer in circular tubes 1.067 and 0.62 mm in diameter. Brief details of these experiments are given in Fig. 4.2c and Table 4.3. The results are presented in Fig. 4.4a,b. They concluded that tubes above $d_{in} = 1.067$ mm with relative roughness k_s/d_{in} about 0.003 may be considered as smooth tubes. However, for small diameter tubes ($d_{in} < 0.62$ mm), the same relative roughness increases the heat transfer.

4.2.2 Heat Transfer in Rectangular, Trapezoidal and Triangular Ducts

The schemes of the test sections reported in the literature are shown in Fig. 4.5. The geometrical parameters are presented in Tables 4.4 and 4.5. Peng and Peterson (1996) investigated experimentally the single-phase forced convective heat transfer

Table 4.4 Rectangular micro-channels

Author	Number of channels in the test section	Heating length L_h [mm]	Hydraulic diameter d_h [μm]	Dimensionless length L_h/d_h	Reynolds number Re
Peng and Peterson (1996)	Not reported	45	133–367	123–338	90–9,000
Harms et al. (1999)	1 68	25	1,923 404	13 62	173–12,900
Qu and Mudawar (2002b)	21	44.764	348	129	139–1,672
Warrier et al. (2002)	5	325.125	750	433.5	557–1,552
Gao et al. (2002)	1	62	200–2,000	31–310	40–8,000
Lee et al. (2005)	10	25.4	318–903	28–80	300–3,500

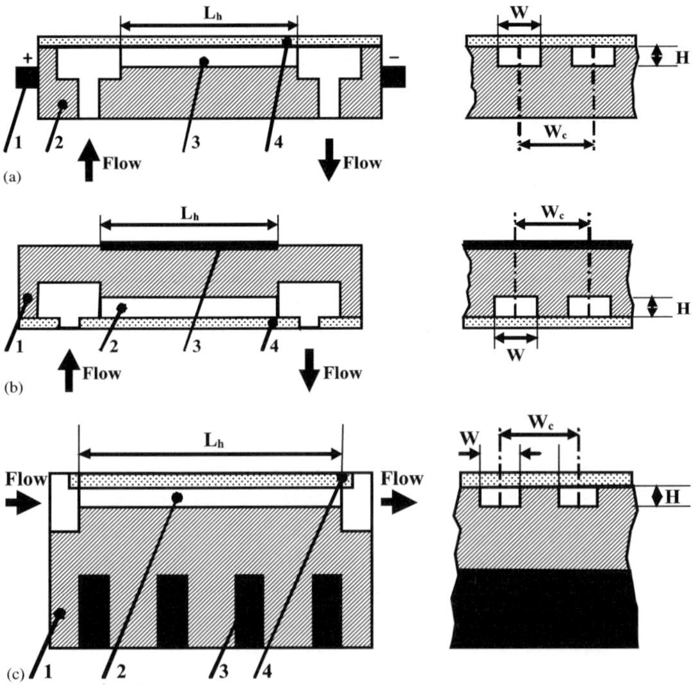

Fig. 4.5a–c Rectangular micro-channels. (**a**) $d_h = 133-367\ \mu m$. Test section used by Peng and Peterson (1996) (schematic view): *1* electrical contact, *2* heated stainless steel block, *3* micro-channel, *4* cover plate. (**b**) $d_h = 404-1{,}923\ \mu m$. Test section used by Harms et al. (1999) (schematic view): *1* silicon wafer, *2* micro-channel, *3* heater, *4* cover plate. (**c**) $d_h = 348\ \mu m$. Test section used by Qu and Mudawar (2002a) (schematic view): *1* copper block, *2* micro-channel, *3* heater, *4* cover plate. Reprinted from Peng and Peterson (1996), Harms et al. (1999), Warrier et al. (2002), Qu and Mudawar (2002a), Gao et al. (2002), and Lee et al. (2005) with permission

Table 4.5 Trapezoidal and triangular micro-channels

Author	Number of channels in the test section	Length L [mm]	Hydraulic diameter d_h [μm]	Dimension-less length L/d_h	Relative roughness k_s/d_h	Reynolds number Re
Qu et al. (2000)	5	30	62.3 168.9	482 178	1.12×10^{-2} 1.75×10^{-2}	100–1,450
Wu and Cheng (2003)	13			195.34–453.79	3.26×10^{-5}–1.09×10^{2}	15–1,500
Tiselj et al. (2004)	17	10	160	63		3.2–84

of water in micro-channel structures with small rectangular channels having a hydraulic diameter of 0.133–0.367 mm and the distinct geometric configuration shown in Fig. 4.5a. The heat flux of the micro-channel structure was based on the micro-

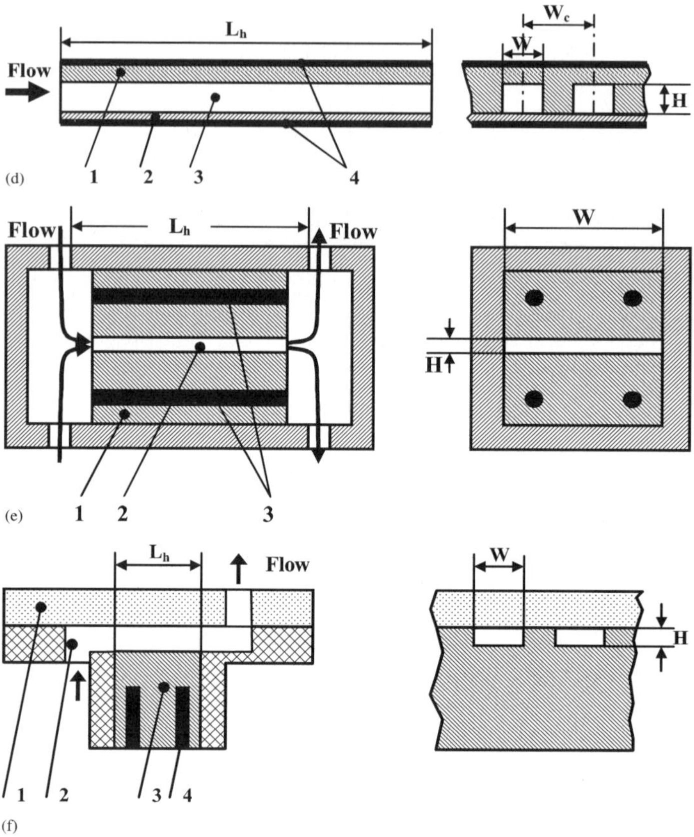

Fig. 4.5d–f Rectangular micro-channels. (**d**) $d_h = 750\,\mu m$. Test section used by Warrier et al. (2002) (schematic view): *1* upper aluminum plate, *2* down aluminum plate, *3* micro-channel, *4* heater. (**e**) $d_h = 200–2{,}000\,\mu m$. Test section used by Gao et al. (2002) (schematic view): *1* brass block, *2* micro-channel, *3* heater. (**f**) Thermally developing flow in rectangular micro-channel (Lee et al. 2005) (schematic view): *1* cover plate, *2* micro-channel, *3* copper block, *4* heater. Reprinted from Peng and Peterson (1996), Harms et al. (1999), Warrier et al. (2002), Qu and Mudawar (2002a), Gao et al. (2002), and Lee et al. (2005) with permission

channel plate area. The heat transfer coefficient was evaluated using the log-mean temperature difference. Thus, the heat transfer coefficient corresponds to some integral value of heat flux. The dependence of the heat transfer coefficient on the flow and geometric parameters was presented.

Harms et al. (1999) obtained experimental results for single-phase forced convection in deep rectangular micro-channels (Fig. 4.5b). Two configurations were tested, a single-channel system and a multiple-channel system. All tests were performed with de-ionized water, and Reynolds number ranging from 173 to 12,900. For the single channel design, the experimental Nusselt number was higher than theoretically predicted for heat transfer in laminar flow. For example, at $Re = 1{,}383$

the Nusselt number was Nu $= 40.9$. Harms et al. (1999) concluded that this enhancement may be due to the effect of the inlet bend. The results for the multiple channel design in the range Re $= 173-1{,}188$ showed an increase in Nusselt number with increasing Re. For example, at Re $= 173$ the Nusselt number was equal to 2.65, and at Re $= 1{,}188$ the Nusselt number was 8.41. This deviation from theoretical prediction was attributed to flow bypass in the manifold. The authors believe that in systems with a small heater area compared to the projected channel area, three-dimensional conduction occurs. However, for multiple channel design three-dimensional conduction also occurs when the heater area covers the entire projected channel area.

Qu and Mudawar (2002b) studied both experimentally and numerically heat transfer characteristics of a single-phase micro-channel heat sink. The heat sink was made from oxide-free copper and fitted with a polycarbonate plastic cover plate (Fig. 4.5c). The heat sink consisted of an array of rectangular micro-channels 231 μm wide and 713 μm deep (Table 4.4). The Reynolds number ranged from 139 to 1,672. The three-dimensional heat transfer characteristics of the heat sink were analyzed numerically by solving the conjugate heat transfer problem. The measured temperature distributions showed good agreement with the corresponding numerical predictions (Fig. 4.6a). These findings demonstrate that conventional Navier–Stokes and energy equations can adequately predict the fluid flow and heat transfer characteristics of micro-channel heat sinks.

Warrier et al. (2002) conducted experiments of forced convection in small rectangular channels using FC-84 as the test fluid. The test section consisted of five parallel channels with hydraulic diameter $d_{\mathrm{h}} = 0.75$ mm and length-to-diameter ratio $L_{\mathrm{h}}/d_{\mathrm{h}} = 433.5$ (Fig. 4.5d and Table 4.4). The experiments were performed with uniform heat fluxes applied to the top and bottom surfaces. The wall heat flux was calculated using the total surface area of the flow channels. Variation of single-phase Nusselt number with dimensionless axial distance is shown in Fig. 4.6b. The numerical results presented by Kays and Crawford (1993) are also shown in Fig. 4.6b. The measured values agree quite well with the numerical results.

The study of Gao et al. (2002) is devoted to investigations of the flow and associated heat transfer in micro-channels of large-span rectangular cross-section and adjustable height in the range of $H = 0.1-1$ mm (Fig. 4.5e and Table 4.4). The fluid used was de-mineralized water. The active channel walls were two plane brass blocks of heating length $L_{\mathrm{h}} = 62$ mm, which were separated by a foil with a hollowed out central part of width $W = 25$ mm. The thickness of this foil fixed the channel height, H. A set of foils allowed variations of the hydraulic diameter d_{h} from 200 to 2,000 μm. A typical temperature distribution along the channel for the case of a very narrow micro-channel ($d_{\mathrm{h}} = 200$ μm) is shown in Fig. 4.6c, where TC1 through TC5 denote thermocouples. The local Nusselt number Nu_x expressed as a function of $X^+ = x/(d_{\mathrm{h}}\mathrm{Pe})$, was compared with the theoretical solution of Shah and London (1978). For $H = 1$ mm the results demonstrate good agreement with the theoretical solution. However, for $H = 0.1$ mm, the plot $\mathrm{Nu}(X^+)$ shows departure from the theoretical heat transfer law, as the Nusselt number is smaller than the conventional value for large-scale channels. This trend is in agreement with results

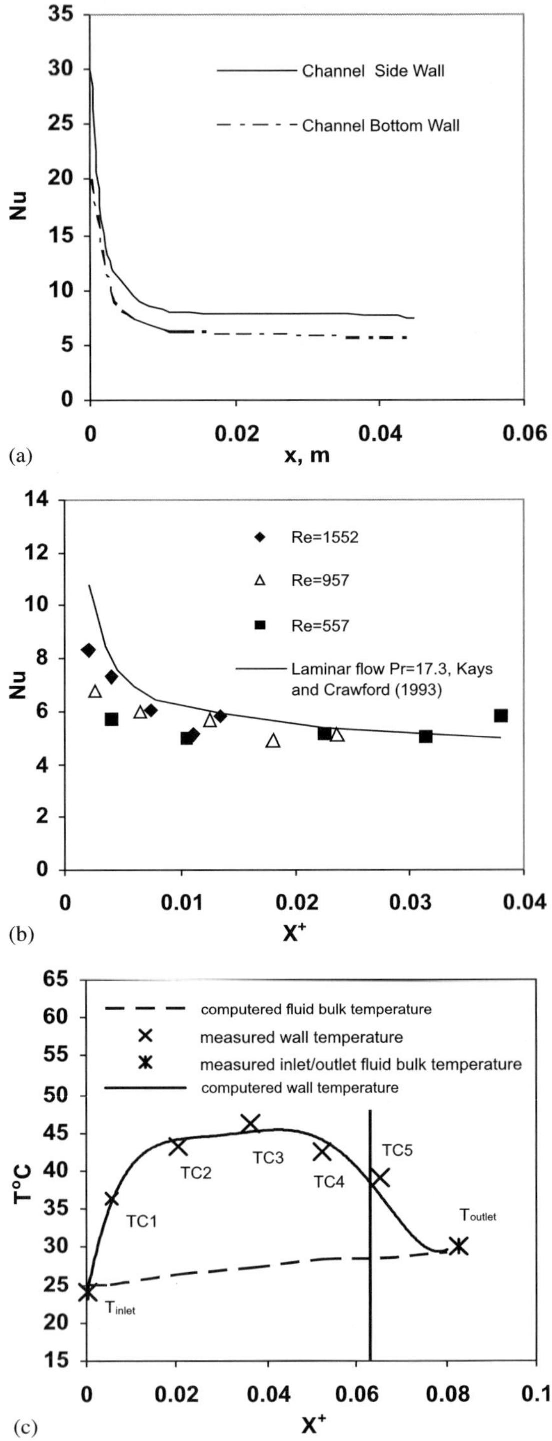

Fig. 4.6a–c Rectangular micro-channels. Calculation and experimental data for (**a**) $d_h = 348\ \mu m$. Numerical predictions of average Nusselt number, Re = 864 (Qu and Mudawar 2002). (**b**) $d_h = 750\ \mu m$. Variation of Nusselt number with axial distance (Warrier et al. 2002). (**c**) $d_h = 200\ \mu m$, Re = 1,780. Temperature distribution along the channel (Gao et al. 2002). Reprinted from Qu and Mudawar (2002), Warrier et al. (2002), Goa et al. (2002), and Lee et al. (2005) with permission

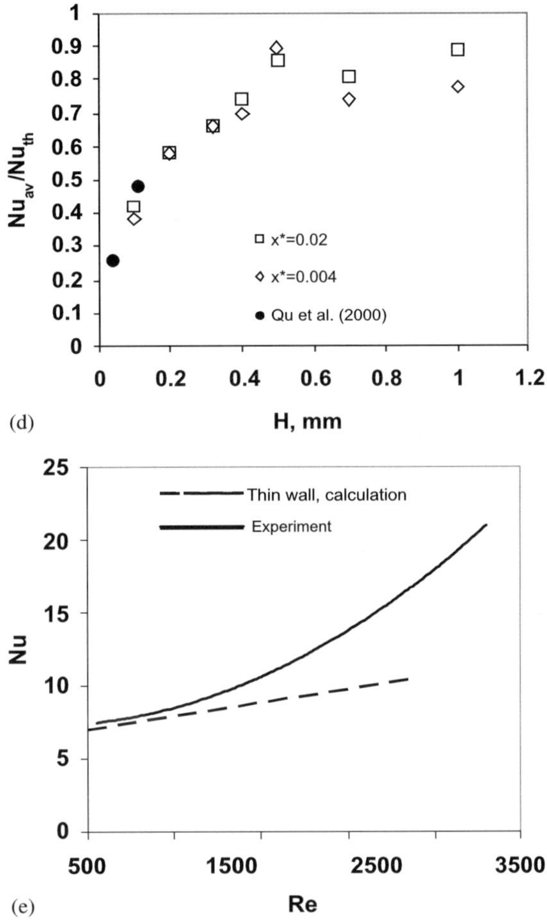

Fig. 4.6d,e Rectangular micro-channels. (**d**) d_h = 200–2,000 μm. Effect of channel size on Nusselt number (Gao et al. 2002). (**e**) Comparison of the average Nusselt number obtained from numerical analyses for the 194 μm wide micro-channels (Lee et al. 2005). Reprinted from Qu and Mudawar (2002), Warrier et al. (2002), Goa et al. (2002), and Lee et al. (2005) with permission

reported by Qu et al. (2000). Figure 4.6d shows the dependence of Nu_{av}/Nu_{th} as a function of the channel height, where Nu_{th} is the theoretical value of the Nusselt number for the same value of X^+. The significant reduction in the Nusselt number cannot be explained by roughness effects. The modification of heat transfer laws by electrokinetic effects also should be discarded, due to the large difference of scales between the channel height and the double-diffusive layer thickness.

An experimental investigation was conducted by Lee et al. (2005) to explore the validity of classical correlations based on conventional sized channels for predicting the thermal behavior in single-phase flow through rectangular micro-channels. The micro-channels ranged in width from 194 to 534 μm, with the channel depth being nominally five times the width in each case (Fig. 4.5f and Table 4.4). Each test piece was made of copper and contained ten micro-channels in parallel. The experiments were conducted with de-ionized water, with Reynolds number ranging from approximately 300 to 3,500. The tests were carried out either at the hydrodynamically developed but thermally developing (TD) regime or a simultaneously developing (SD) regime. The average heat transfer coefficient was determined by using the area available for convection of channels, average temperature of the channel wall and mean fluid temperature at the inlet and outlet of the micro-channels. In Fig. 4.6e the numerical calculations that did not include the conduction in the substrate (thin wall) are compared with the experimental results.

Qu et al. (2000) carried out experiments on heat transfer for water flow at $100 < \mathrm{Re} < 1{,}450$ in trapezoidal silicon micro-channels, with the hydraulic diameter ranging from 62.3 to 168.9 μm. The dimensions are presented in Table 4.5. A numerical analysis was also carried out by solving a conjugate heat transfer problem involving simultaneous determination of the temperature field in both the solid and fluid regions. It was found that the experimentally determined Nusselt number in micro-channels is lower than that predicted by numerical analysis. A roughness-viscosity model was applied to interpret the experimental results.

Wu and Cheng (2003) investigated laminar convective heat transfer of water in a single trapezoidal silicon micro-channel. They used a set of 13 micro-channels having different dimensions and different relative roughness. The geometrical parameters of micro-channels are presented in Table 4.5. A silicon chip was anodically bonded with a thin Pyrex glass plate from the top. Figure 4.7a shows the effect of relative surface roughness, a, on the Nusselt number, where W_t and W_b are the width (μm) of the channels top and bottom, respectively, and H is the height of the channel (in microns). One can see that at the same Reynolds numbers the Nusselt number increases with increasing relative surface roughness. The laminar convective heat transfer showed two different characteristics at low and high Reynolds number ranges. Figure 4.7b shows the effect of geometric parameters on the Nusselt number, in the range of relative surface roughness $a = 3.62\times10^{-5}$–9.85×10^{-5}. From Fig. 4.7b it can be observed that for very low Reynolds number flow, $\mathrm{Re} = 0$–100, the Nusselt number increased acutely with the increase in the Reynolds number. However, the increase in the Nusselt number when $\mathrm{Re} > 100$ is gentle with an increase in the Reynolds number.

4.2.3 *Heat Transfer in Surfactant Solutions Flowing in a Micro-Channel*

The dependence of the average heat transfer coefficient on the bulk fluid velocity is plotted in Fig. 4.8. These data are related to laminar flow of Habon G aqueous solutions of $C = 530$ ppm (H530) and $C = 1{,}200$ ppm (H1200) in a micro-channel with the inner diameter of 1.07 mm (Hetsroni et al. 2004). The figure shows that the heat transfer coefficient in surfactant solutions is higher than in laminar pipe water flow. Figure 4.9 shows relative enhancement in heat transfer coefficient $\mathrm{DR}(h) = (1 - h_{\mathrm{sur}}/h_{\mathrm{wat}})$ versus fluid bulk velocity, where h_{sur} and h_{wat} are the heat transfer coefficients in the tube at the same bulk velocity of surfactant solution, Habon G and water, respectively. Note that data for $C = 530$ ppm and $C = 1{,}060$ ppm

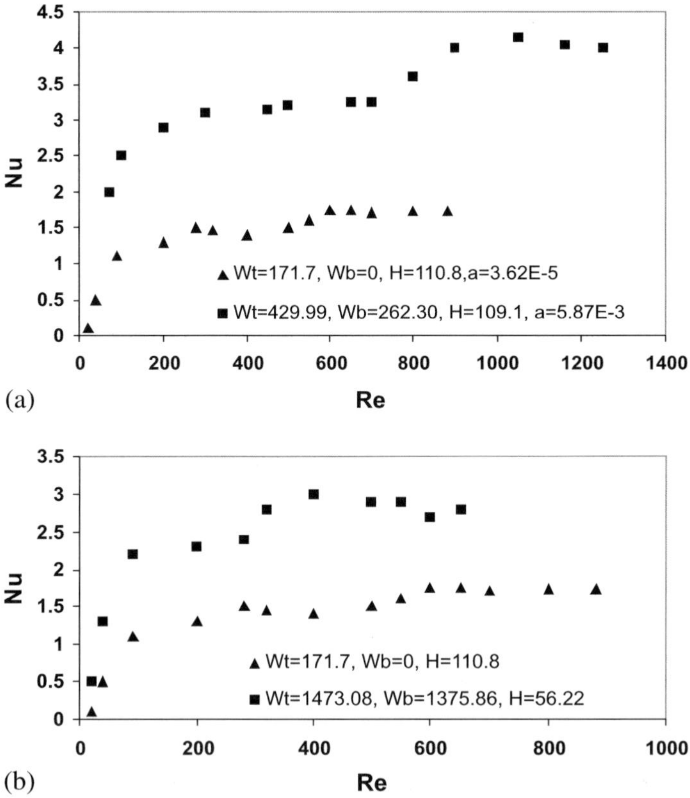

Fig. 4.7a,b Trapezoidal micro-channels. (**a**) Effect of surface roughness on Nusselt number (Wu and Cheng 2003). (**b**) Effect of geometric parameters on Nusselt number (Wu and Cheng 2003). Reprinted from Wu and Cheng (2003) with permission

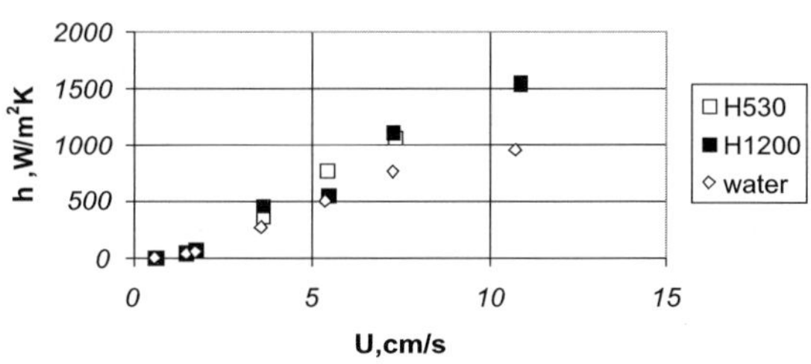

Fig. 4.8 Average heat transfer coefficient dependence on flow velocity. Reprinted from Hetsroni et al. (2004) with permission

are expressed as the same symbols. The negative values of DR(h) correspond to an increase in the heat transfer coefficient. It can be seen from Fig. 4.9 that the increase in the heat transfer coefficient does not depend on the solution bulk velocity. With the addition of surfactant macromolecules to water, the viscosity and pressure drop in laminar pipe flows increase. It is thus quite unexpected to find that for heat transfer the opposite can sometimes take place. Kostic (1994) observed that some drag-reducing solutions augmented the heat transfer in laminar flow in a non-circular

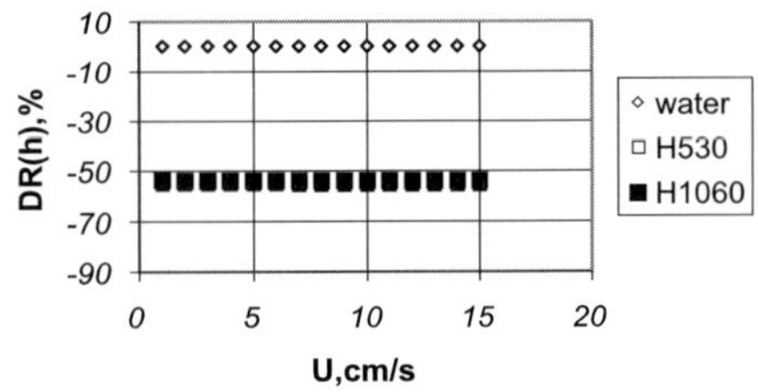

Fig. 4.9 Relative heat transfer coefficient on liquid velocity. Reprinted from Hetsroni et al. (2004) with permission

duct. They suggested that the fluid elasticity may lead to secondary flows, which are responsible for the increase in the heat transfer. The secondary flows increase the pressure drops in both adiabatic and diabatic flows. However, in diabatic flow the total pressure drop (due to friction and due to secondary flows) may be smaller compared to that in the adiabatic flow. In this case the common effect includes also some decrease in the fluid viscosity with an increase in the temperature of the surfactant solution. On the other hand, it may be assumed that macromolecules of surfactant change the flow structure in the near-wall region. It also may be responsible for the increase in pressure drop and heat transfer.

The average Nusselt number, Nu, is presented in Fig. 4.10a,b versus the shear Reynolds number, Re_{sh}. This dependence is qualitatively similar to water behavior for all surfactant solutions used. At a given value of Reynolds number, Re_{sh}, the Nusselt number, Nu, increases with an increase in the shear viscosity. As discussed in Chap. 3, the use of shear viscosity for the determination of drag reduction is not a good choice. The heat transfer results also illustrate the need for a more appropriate physical parameter. In particular, Fig. 4.10a shows different behavior of the Nusselt number for water and surfactants. Figure 4.10b shows the dependence of the Nusselt number on the Peclet number. The Nusselt numbers of all solutions are in agreement with heat transfer enhancement presented in Fig. 4.8. The data in Fig. 4.10b show

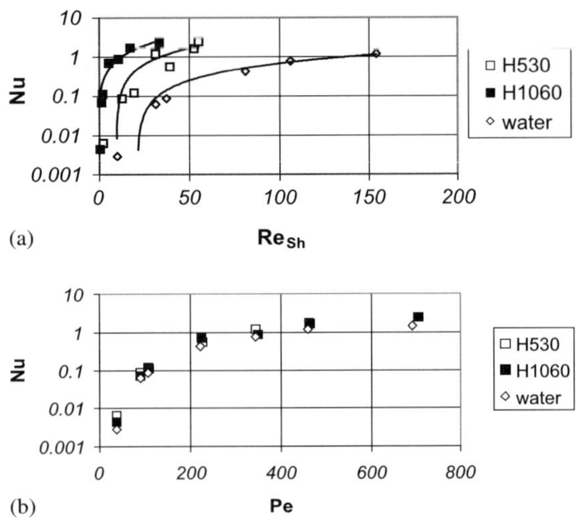

Fig. 4.10a,b The dependence of the Nusselt number on the Reynolds and Peclet numbers. (**a**) Dependence of the average Nusselt number on the solution Reynolds number. (**b**) Dependence of the Nusselt number on the Peclet number. Reprinted from Hetsroni et al. (2004) with permission

that the use of the Peclet number may be considered for description of experimental data of laminar pipe flow of certain non-Newtonian fluids.

4.3 Effect of Viscous Energy Dissipation

Under some conditions, the heat released due to viscous dissipation leads to a drastic change of flow and temperature field: in particular, it leads to flow instability, transition to turbulence, oscillatory motions, etc. (Gruntfest et al. 1964; Bastanjian et al. 1965; Zel'dovich et al. 1985; Hetsroni et al. 2005). This problem was also discussed by Tso and Mahulikar (1998, 1999, 2000). To reveal the effect of viscous dissipation, the experimental data by Wang and Peng (1994) and Tso and Mahulikar (2000) were used. Experiments were performed using water flow in microchannels. The experimental data in the laminar flow regime were found to correlate well with the Brinkman number. As a result, a semi-empirical equation for the Nusselt number was suggested and the dependence of $\mathrm{Nu}/\mathrm{Re}^{0.62}\,\mathrm{Pr}^{0.33}$ on the Brinkman number was demonstrated. The Brinkman number, $\mathrm{Br} = \mu U^2/k\Delta T$, is the ratio of the heat production due to viscous forces, to heat transferred from the wall to the fluid. However, most of the data used for that correlation were obtained under conditions in which the Reynolds number and the Prandtl number also varied, so that it was difficult to separate the effect of the Brinkman number from the effects of the Reynolds and Prandtl numbers. For instance, at $0.4559 \times 10^{-5} \leq \mathrm{Br} \leq 2.8333 \times 10^{-5}$ the Reynolds and Prandtl numbers varied as $80 \leq \mathrm{Re} \leq 107$, and $4.80 \leq \mathrm{Pr} \leq 6.71$ (Tso and Mahulikar 1998). To estimate the real effect of viscous dissipation on heat transfer it is necessary to determine the dependence of the Nusselt number on the Brinkman number at fixed values of the Reynolds and the Prandtl numbers. This is done for the data reported by Tso and Mahulikar (1998) as shown in Table 4.6. One can see that the effect of the Brinkman number on the Nusselt number is negligible. The same conclusion may be derived also from experiments performed by Tso and Mahulikar (2000). According to their measurements (Table 4.7) variation of the Brinkman number from 1.1195×10^{-8} to 2.3048×10^{-8} did not affect the Nusselt number, when the Reynolds and the Prandtl numbers did not change significantly. It should be stressed that the effect of viscous dissipation on heat transfer in microchannels at extremely small values of the Brinkman number, $\mathrm{Br} \sim 10^{-8}$–10^{-5}, is

Table 4.6 Experimental data in the laminar regime presented by Tso and Mahulikar (1998) at approximately fixed values of the Reynolds and the Prandtl numbers

$\mathrm{Br} \times 10^5$	Re	Pr	Nu
0.4559	107	4.80	0.35
1.4541	124	4.82	0.37
2.8333	80	6.71	0.46
4.4115	93	6.62	0.49

Table 4.7 Experimental data in the laminar regime obtained by Tso and Mahulikar (2000) at approximately fixed values of the Reynolds and the Prandtl numbers

$Br \times 10^8$	Re	Pr	Nu
1.1195	17.5	3.50	0.3031
2.3048	22.4	3.82	0.3068

not realistic from the physical point of view. At small Br the contribution of the heat released due to viscous forces is negligible, consequently the effect on the heat transfer is also negligible. It should also be noted that evaluation of the role of the Brinkman number performed by Tso and Mahulikar (1998, 1999, 2000) was based on the experiments that were carried out under some specific conditions (the heat transfer characteristics were found to be affected by the channel geometry, axial heat conduction in the channels walls, liquid velocity and temperature, etc.). Some aspects of this problem were discussed by Herwig and Hausner (2003) and Gad-el-Hak (2003).

The effect of viscous heating was investigated by Tunc and Bayazitoglu (2001) when the fluid was heated ($T_{in} < T_w$) or cooled ($T_{in} > T_w$). In the range of $0 < Kn < 0.12$ the Nusselt number decreased as the Knudsen number increased. The viscous dissipation significantly affected heat transfer. Tunc and Bayazitoglu (2001) showed that the decrease was greater when viscous dissipation occurred. The effect of viscous dissipation on the temperature field was investigated by Koo and Kleinstreuer (2004) for three working fluids: water, methanol and isopropanol. Channel size, the Reynolds number and the Prandtl number are the key factors, which determine the impact of viscous dissipation. Viscous dissipation effects may be very important for fluids with low specific heats and high viscosities, even in relatively low Reynolds number flows. For water the relative magnitude of the ratio, A_d, of convective heat transfer to dissipation term is given in Table 4.8.

Experimental and numerical analyses were performed on the heat transfer characteristics of water flowing through triangular silicon micro-channels with hydraulic diameter of 160 μm in the range of Reynolds number $Re = 3.2–84$ (Tiselj et al. 2004). It was shown that dissipation effects can be neglected and the heat transfer may be described by conventional Navier–Stokes and energy equations as a common basis. Experiments carried out by Hetsroni et al. (2004) in a pipe of inner diameter of 1.07 mm also did not show effect of the Brinkman number on the Nusselt number in the range $Re = 10–100$.

Table 4.8 Ratio A_d of convective heat transfer to dissipation term in tubes

Tube radius r_0 [m]	$Re = 20A_d$	$Re = 200A_d$	$Re = 2{,}000A_d$
10^{-3}	3.45×10^4	3.45×10^3	3.45×10^2
10^{-4}	3.45×10^2	3.45×10^1	3.45×10^0
10^{-5}	3.45×10^0	3.45×10^{-1}	3.45×10^{-2}

Hetsroni et al. (2005) evaluated the effect of inlet temperature, channel size and fluid properties on energy dissipation in the flow of a viscous fluid. For fully developed laminar flow in circular micro-channels, they obtained an equation for the adiabatic increase of the fluid temperature due to viscous dissipation:

$$\frac{\Delta T}{T_{\text{in}}} = 2\frac{v^2}{r_0^2}\left(\frac{L}{r_0}\right)\frac{\text{Re}}{c_{\text{p}}T_{\text{in}}}. \tag{4.1}$$

For an incompressible fluid, the density variation with temperature is negligible compared to the viscosity variation. Hence, the viscosity variation is a function of temperature only and can be a cause of radical transformation of flow and transition from stable flow to the oscillatory regime. The critical Reynolds number also depends significantly on the specific heat, Prandtl number and micro-channel radius. For flow of high-viscosity fluids in micro-channels of $r_0 < 10^{-5}$ m the critical Reynolds number is less than 2,300. In this case the oscillatory regime occurs at values of Re $< 2{,}300$.

We can estimate the values of the Brinkman number, at which the viscous dissipation becomes important. Assuming that the physical properties of the fluid are constant, the energy equation for fully developed flow in a circular tube at $T_{\text{w}} = \text{const.}$ is:

$$\rho u c_{\text{p}}\frac{\partial T}{\partial x} = \frac{1}{r}\frac{\partial}{\partial r}\left(kr\frac{\partial T}{\partial r}\right) + \mu\left(\frac{\partial u}{\partial r}\right)^2 \tag{4.2}$$

where ρ, μ and k are the density, dynamic viscosity and thermal conductivity, respectively, c_{p} is the specific heat, and u is the actual streamwise velocity.

The actual velocity, u, may be expressed as:

$$u = 2U(1 - R^2) \tag{4.3}$$

where $U = \frac{1}{r_0^2}\int_0^{r_0} ur\,\mathrm{d}r$ is the average velocity, and $R = \frac{r}{r_0}$, r_0 is the micro-channel radius.

From Eqs. (4.2) and (4.3) one can obtain:

$$\frac{2\int_0^{r_0}\frac{\partial}{\partial x}\left\{\rho u c_{\text{p}}(T - T_{\text{w}})\right\} r\,\mathrm{d}r}{k(\overline{T} - T_{\text{w}})} = -\text{Nu} \pm 8\,\text{Br} \tag{4.4}$$

where $\overline{T}$ is the average fluid temperature, T_{w} is the wall temperature, $\text{Nu} = -\left(k\left(\partial T/\partial r\right)_{r=r_0} d\right) / \left(k(\overline{T} - T_{\text{w}})\right)$ is the Nusselt number, and Br is the Brinkman number. The minus or plus sign in front of the last term in Eq. (4.4) corresponds to cooling of fluid when $\overline{T}_0 > T_{\text{w}}$ or its heating when $\overline{T}_0 < T_{\text{w}}$, respectively, where $\overline{T}_0$ is the average fluid temperature at the inlet of the micro-channel.

A detailed study of the influence of viscous heating on the temperature field in micro-channels of different geometries (rectangular, trapezoidal, double-trapezoidal) has been performed by Morini (2005). The momentum and energy conservation equations for flow of an incompressible Newtonian fluid were used to estimate the effect of viscous dissipation on bulk temperature. For steady two-dimensional fully developed laminar flow with constant physical properties the following relation for longitudinal gradient of the bulk temperature was derived

$$\frac{\mathrm{d}T_b}{\mathrm{d}x} = \left\{4\frac{\mathrm{Ec}}{\mathrm{Re}}(\lambda \mathrm{Re})\right\}\frac{\Delta T}{d_\mathrm{h}} \tag{4.5}$$

where $T_b = \frac{1}{A}\int_A u(y,z)T(x,y,z)\,\mathrm{d}A$ is the bulk temperature, A is the cross-section area, $u(y,z)$ and $T(x,y,z)$ are the longitudinal components of velocity and temperature, respectively, $\mathrm{Re} = Ud_\mathrm{h}/\nu$ and $\mathrm{Ec} = U^2/\left(2c_\mathrm{p}\Delta T\right)$ are the Reynolds and Eckert numbers, λ is the friction factor, ΔT is the temperature difference, x, y, z are the Cartesian coordinates, and x is directed along the micro-channel axis.

The effect of viscous dissipation on temperature change along the micro-channel axis is illustrated in Fig. 4.11, where the dependences $\mathrm{d}T_b/\mathrm{d}x$ on d_h that correspond to water and isopropanol flows are presented. One can see that under the conditions corresponding to the Judy et al. (2002) experiments ($d_\mathrm{h} = 74.1\,\mu\mathrm{m}$, $L = 114\,\mathrm{mm}$, $L/d = 1{,}543$), the rise of bulk temperature due to viscous dissipation is small enough. So, at $d_\mathrm{h} \geq 100\,\mu\mathrm{m}$ the temperature gradient is $\mathrm{d}T_b/\mathrm{d}x \leq 1\,\mathrm{K/m}$. In this case, the difference between outlet and inlet temperature is about 0.1 K. Under conditions that are typical for micro-channels of electronic devices ($L/d_\mathrm{h} \sim 10^2$) this difference is about 0.01 K. The rise of temperature due to viscous dissipation is small enough even at water flow in micro-channels with $d_\mathrm{h} \sim 20\,\mu\mathrm{m}$. Thus, for micro-channels with $d_\mathrm{h} = 20\,\mu\mathrm{m}$ and $L/d_\mathrm{h} = 10^2$, we have $T_\mathrm{out} - T_\mathrm{in} \approx 0.8\,\mathrm{K}$.

To estimate the value of the term on the left-hand side of Eq. (4.4), we use an approximate expression for the local and average fluid temperature in the tube. We use

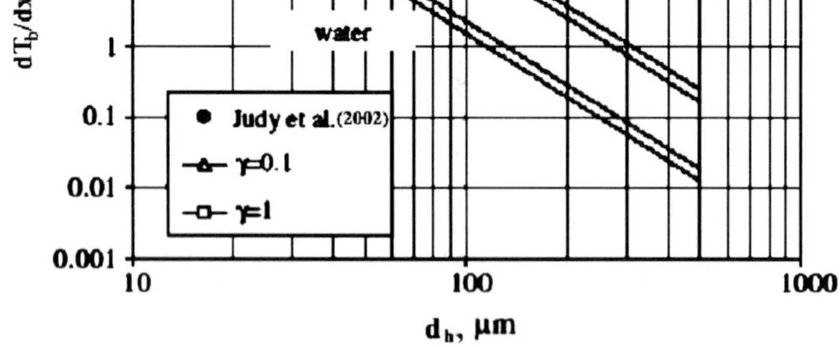

Fig. 4.11 Temperature gradients due to viscous dissipation at Re = 300. Flow in rectangular ($\gamma = 0.1$), and square ($\gamma = 1$) micro-channel. Reprinted from Morini (2005) with permission

as the initial guess the expressions for $T(x,r)$ and $\overline{T}(x)$ that correspond to negligible viscous dissipation. For $T_{\mathrm{w}} = \text{const.}$ they are:

$$T - T_{\mathrm{w}} = (T_0 - T_{\mathrm{w}}) \sum_{n=0}^{\infty} A_n \varphi_n(R) \exp\left(-2\varepsilon_n^2 \frac{1}{\mathrm{Pe}} \frac{x}{d}\right) \tag{4.6}$$

$$\overline{T} - T_{\mathrm{w}} = 8(\overline{T}_0 - T_{\mathrm{w}}) \sum_{n=1}^{\infty} \frac{B_n}{\varepsilon_n^2} \exp\left(-2\varepsilon_n^2 \frac{1}{\mathrm{Pe}} \frac{x}{d}\right) \tag{4.7}$$

where $\mathrm{Pe} = Ud/\alpha$ is the Peclet number, α is the thermal diffusivity, A_n and B_n are constants, ε_n and $\varphi_n(R)$ are the corresponding eigen functions (Petukhov 1967).

For large values of dimensionless axial distance, $X^+ = x/(\mathrm{Pe}\, d)$ the following relations were obtained from Eqs. (4.4), (4.6), and (4.7):

$$T - T_{\mathrm{w}} = (T_0 - T_{\mathrm{w}}) A_0 \varphi_0(R) \exp\left(-2\varepsilon_0^2 \frac{1}{\mathrm{Pe}} \frac{x}{d}\right) \tag{4.8}$$

$$\overline{T} - T_{\mathrm{w}} = 8(\overline{T}_0 - T_{\mathrm{w}}) \frac{B_0}{\varepsilon_0^2} \exp\left(-2\varepsilon_0^2 \frac{1}{\mathrm{Pe}} \frac{x}{d}\right) \tag{4.9}$$

and:

$$-2\frac{\frac{\mathrm{d}}{\mathrm{d}x}\int_0^{r_0} \rho u c_{\mathrm{p}} T r\,\mathrm{d}r}{k(\overline{T} - T_{\mathrm{w}})} = \frac{1}{4}\varepsilon_0^4 \frac{A_0}{B_0} \int_0^1 (1 - R^2)\varphi_0(R) R\,\mathrm{d}R. \tag{4.10}$$

Substitution of the values of ε_0, A_0, B_0, as well as the expression for $\varphi_0(R)$ in the right hand side of Eq. (4.10) gives:

$$\frac{1}{4}\varepsilon_0^4 \frac{A_0}{B_0} \int_0^1 (1 - R^2)\varphi_0(R) R\,\mathrm{d}R = 3.64 = \mathrm{Nu}_0 \tag{4.11}$$

where Nu_0 is the Nusselt number that corresponds to negligible viscous dissipation.

Equation (4.4) is rewritten as:

$$\mathrm{Nu} = \mathrm{Nu}_0 \pm 8\,\mathrm{Br} \tag{4.12}$$

where the plus and minus signs correspond to the cooling $\overline{T}_0 > T_{\mathrm{w}}$ or heating $\overline{T}_0 < T_{\mathrm{w}}$ regimes, respectively.

Equation (4.12) indicates the effect of viscous dissipation on heat transfer in micro-channels. In the case when the inlet fluid temperature, $\overline{T}_0$, exceeds the wall temperature, viscous dissipation leads to an increase in the Nusselt number. In contrast, when $\overline{T}_0 < T_{\mathrm{w}}$, viscous dissipation leads to a decrease in the temperature gradient on the wall. Equation (4.12) corresponds to a relatively small amount of heat released due to viscous dissipation. Taking this into account, we estimate the lower boundary of the Brinkman number at which the effect of viscous dissipation may be observed experimentally. Assuming that $(\mathrm{Nu} - \mathrm{Nu}_0)/\mathrm{Nu}_0 \geq 10^{-2}$ the following evaluation of the Brinkman number was obtained: $Br \geq 5 \times 10^{-3}$. This estimation shows that the conclusions of Tso and Mahulicar (1998, 1999, 2000) cannot be derived from experiments performed at extremely low Brinkman numbers of $Br \sim 10^{8}-10^{5}$.

It should be noted that for some fluids, viscous dissipation can affect the development length in tubes at this physical scale, particularly at high average fluid velocities. The influence of viscous dissipation was explored by Judy et al. (2002). Viscous heating has the effect of increasing the temperature of the flowing fluid along the tube axis, yielding continuously varying thermophysical properties. The Nahme number is used by Judy et al. (2002) to estimate the importance of viscous dissipation. For pipe flow the Nahme number is defined as $Na = -(4\beta\mu U^2)/k$ where β is the temperature sensitivity of viscosity defined as $-(1/\mu)(\partial\mu/\partial T)$, μ is the viscosity, k is the thermal conductivity, and U is the average fluid velocity in the tube.

Viscous dissipation effects become significant for increasing Na. For the liquids and Reynolds number ranges used in the study by Judy et al. (2002), the largest Nahme number for the presented data was 0.02, and was found for isopropanol. This small Na suggests that for the data presented in that study the viscous dissipation effects were small. Despite this small Na, viscous heating of the fluid can affect the results. Figure 4.12 illustrates the Poiseuille number Po as a function of Reynolds number for a fused silica square micro-channel of length $L = 11.4$ cm and hydraulic diameter $d_h = 74.1$ μm ($L/d_h = 1{,}543$) with isopropanol as the working fluid. The maximum rise in liquid temperature for this case was 6.2 K, found at the maximum Reynolds number tested, $Re \approx 300$. The figure shows Po calculated in two ways: (1) using a viscosity based on the tube inlet temperature, and (2) using a viscosity based on the average of tube inlet and exit temperatures. Note that when the average temperature is used, Po is effectively independent of the Reynolds number. By contrast, when the temperature at the tube inlet is used to evaluate the viscosity, the friction factor drops with increasing Reynolds number, as the effect of viscous heating becomes more pronounced at higher fluid velocities. Even for the small Nahme

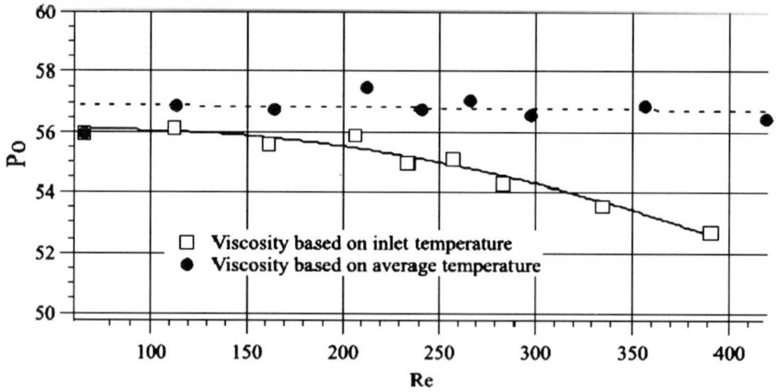

Fig. 4.12 Measured Po vs. Reynolds number with viscosity based on average tube fluid temperature represented by *circles*, and inlet temperature represented by *squares* for a fused silica square micro-channel with isopropanol. Reprinted from Judy et al. (2002) with permission

number characterizing this data set (Na = 0.02), the viscous heating and associated viscosity variation can result in a 7–10% drop in Po. For the small Nahme numbers and associated small rise in mean temperature over tube length found in tests reported here, the average of the fluid temperatures at the tube inlet and exit appears to be an appropriate reference condition for thermophysical properties.

It should be emphasized that under conditions of energy dissipation the definition of the heat transfer coefficient as $k\left(\partial T/\partial r\right)_{r=r_0}/(\overline{T}-T_{\mathrm{w}})$, where $\overline{T}$ is the average fluid temperature and T_{w} is the wall temperature, does not characterize the actual heat transfer properly (Kays and Crawford 1993; Schlichting 2000).

The factors that determine the temperature distribution of the fluid are (1) convective heat transfer, and (2) heat released due to viscous dissipation. For a cooling regime when the fluid temperature at the inlet $T_0 > T_{\mathrm{w}}$, convective heat transfer from the fluid to the wall leads to a decrease in the fluid temperature along a relatively small dimensionless distance X^+. The heat released due to viscous dissipation causes an increase in the fluid temperature. The contribution of each component to the behavior of the fluid temperature depends on X^+. At small X^+ the dominant role belongs to convective heat transfer, whereas at large X^+ the effect of viscous dissipation becomes significant. These two factors determine the specific shape of the temperature distribution along the micro-channel: the fluid temperature variation along the micro-channel has a minimum. It is worth noting that under the condition of viscous dissipation the fluid temperature does not reach the wall temperature at any value of X^+.

For the heating regime at small X^+, the heat transferred from the wall to the cold fluid and the heat released due to viscous dissipation lead to an increase in

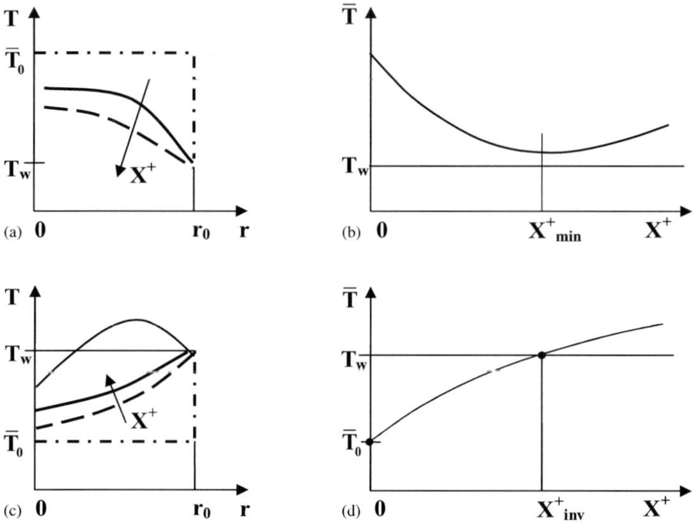

Fig. 4.13a–d Temperature distribution (**a**) and (**b**) $\overline{T}_0 > T_{\mathrm{w}}$, (**c**) and (**d**) $\overline{T}_0 < T_{\mathrm{w}}$. The *dotted line* corresponds to temperature distribution at $X^+ = 0$; the *arrows* show the direction of the increase of X^+. Reprinted with permission from Hetsroni et al. (2005)

fluid temperature. At some value of X^+ the fluid temperature may exceed the wall temperature (Fig. 4.13). The difference $T - T_w$ decreases up to the inversion point $X^+ = X^+_{inv}$. This leads to physically unrealistic results like infinite growth of the Nusselt number in the vicinity of the inversion point.

The analysis of the behavior of the fluid temperature and the Nusselt number performed for a circular tube at the thermal wall boundary condition $T_w = \text{const.}$ also reflects general features of heat transfer in micro-channels of other geometries.

4.4 Axial Conduction

4.4.1 Axial Conduction in the Fluid

This problem was the subject of a number of theoretical investigations carried out for conventional size channels, e.g., Petukhov (1967), Hehnecke (1968), Nguyen (1992), Nguyen et al. (1996), and Weigand and Lauffer (2004). We consider the effect of axial conduction in the fluid on heat transfer in micro-channels. The energy equation is formulated for the flow of an incompressible fluid with constant physical properties. Assuming that the energy dissipation is negligible, we obtain:

$$\rho u c_p \frac{\partial T}{\partial x} = k\left\{\frac{1}{r}\frac{\partial}{\partial r}\left(r\frac{\partial T}{\partial r}\right) + \frac{\partial^2 T}{\partial x^2}\right\} \tag{4.13}$$

where x and r are the longitudinal and radial coordinates, ρ, c_p and k are the density, specific heat and thermal conductivity of the fluid, respectively.

For $q = \text{const.}$ (q is the heat flux on the wall), we introduce new variables

$$\tilde{u} = \frac{u}{U_{ax}},\ R = \frac{r}{r_0},\ X^+ = \frac{2}{\text{Pe}}\frac{x}{d},\ \theta = \frac{T - T_0}{\left(\frac{qd}{k}\right)} \tag{4.14}$$

where $\text{Pe} = \frac{Ud}{\alpha}$, $\tilde{u} = (1 - R^2)$, $d = 2r_0$, U_{ax} and $U = U_{ax}/2$ are the axial and average velocities, respectively, T_0 is the fluid temperature at the entrance of the heating section, and α is the thermal diffusivity.

The dimensionless form of Eq. (4.13) is:

$$\tilde{u}\frac{\partial \theta}{\partial X^+} = \frac{1}{R}\frac{\partial}{\partial R}\left(R\frac{\partial \theta}{\partial R}\right) + \frac{1}{\text{Pe}^2}\frac{\partial^2 \theta}{\partial X^{+2}}. \tag{4.15}$$

Transferring Eq. (4.15) to divergent form and integrating this equation through the micro-channel cross-section we obtain:

$$\frac{\partial}{\partial X^+}\left(\int_0^1 \tilde{u}\theta R\,\mathrm{d}R\right) = \left(R\frac{\partial \theta}{\partial R}\right)\Bigg|_0^1 + \frac{1}{\text{Pe}^2}\frac{\partial^2}{\partial X^{+2}}\left(\int_0^1 \theta R\,\mathrm{d}R\right). \tag{4.16}$$

Assuming that the exit of the micro-channel is connected to an adiabatic section the boundary conditions are:

$$X^+ = 0\,, \quad \theta = 0\,; \qquad X^+ > 0 \begin{cases} R = 0, & \dfrac{\partial \theta}{\partial R} = 0 \\ R = 1, & \dfrac{\partial \theta}{\partial R} = \dfrac{1}{2} \end{cases} \;; \qquad X^+ = X_*^+\,, \quad \frac{\partial \theta}{\partial X^+} = 0 \tag{4.17}$$

where X_*^+ corresponds to the micro-channel exit.

Taking into account conditions (4.17) we obtain from (4.16):

$$\frac{\mathrm{d}}{\mathrm{d}X^+}\left(\int_0^1 \tilde{u}\theta R\,\mathrm{d}R\right) = \frac{1}{2} + \frac{1}{\mathrm{Pe}^2}\frac{d^2\overline{\theta}}{\mathrm{d}X^{+2}} \tag{4.18}$$

where $\overline{\theta} = \int_0^1 \theta R\,\mathrm{d}R$ is the average temperature.

Assuming that θ is a weak function of R and $\theta \approx \overline{\theta}$ we can estimate:

$$\int_0^1 \tilde{u}\theta R\,\mathrm{d}R \approx \frac{\overline{\theta}}{4}\,. \tag{4.19}$$

Then Eq. (4.18) becomes

$$\frac{\partial^2 \overline{\theta}}{\partial X^{+2}} - \frac{1}{4}\mathrm{Pe}^2 \frac{\partial \overline{\theta}}{\partial X^+} + \frac{1}{2}\mathrm{Pe}^2 = 0\,. \tag{4.20}$$

The solution of Eq. (4.20) is:

$$\overline{\theta} = C_1 + C_2 \exp\left(\frac{\mathrm{Pe}^2}{4}X^+\right) + 2X^+ \tag{4.21}$$

where C_1 and C_2 are constants.

Using the first and third conditions of (4.17) we find:

$$C_1 = C_2, \quad C_2 = -\frac{2}{(\mathrm{Pe}^2/4)}\exp\left(-\frac{\mathrm{Pe}^2}{4}X_*^+\right)\,. \tag{4.22}$$

The effect of axial conduction on heat transfer in the fluid in the micro-channel can be characterized by a dimensionless parameter

$$M = \frac{|q_{\mathrm{cond}}|}{|q_{\mathrm{conv}}|} \tag{4.23}$$

that expresses the relation between heat fluxes due to conduction and convection. In Eq. (4.23) $q_{\mathrm{conv}} = \rho U c_{\mathrm{p}}(T - T_0)$, $q_{\mathrm{cond}} = k\frac{\mathrm{d}T}{\mathrm{d}x}$. Substitution of the expression for variable $\overline{\theta}$ and $\frac{\partial \overline{\theta}}{\partial X^+}$ in relation (4.23) gives:

$$M = \frac{1}{4}\frac{1 - \exp(\chi^+ - \chi_*^+)}{\chi^+ - \exp(\chi^+ - \chi_*^+) + \exp(-\chi_*^+)} \tag{4.24}$$

where $\chi^+ = \frac{\mathrm{Pe}^2 X^+}{4}$, $\chi_* = \frac{\mathrm{Pe}^2 X_*^+}{4}$.

The limiting cases are:

$$1.\ \chi^+ \ll 1,\ \chi_*^+ \gg 1$$

$$2.\ \chi^+ - \chi_*^+ \ll 1 \quad \begin{cases} \text{a. } \chi_*^+ \gg 1 \\ \text{b. } \chi_*^+ \ll 1 \end{cases}$$

which correspond to heat transfer at the cooling inlet and heat transfer in the vicinity of the adiabatic outlet, respectively. In the first case we obtain the following evaluation of parameter M:

$$M = \frac{1}{4}\frac{1}{\chi^+} \gg 1\,. \tag{4.25}$$

It shows that close to the micro-channel inlet, heat losses to the cooling inlet due to axial conduction in the fluid are dominant. In the second case parameter M is:

a. Long micro-channel

$$M = \frac{\chi_*^+ - \chi^+}{\chi_*^+ - 1} \ll 1 \tag{4.26}$$

b. Short micro-channel

$$M = \frac{\chi_*^+ - \chi^+}{\chi_*^+} \ll 1 \tag{4.27}$$

Thus, M decreases when χ (and the Peclet number) increases. Accordingly, the Nusselt number decreases when the Peclet number increases, and approaches its limiting value Nu_∞ that corresponds to $\mathrm{Pe} \to \infty$.

The existence of heat transfer due to axial conduction in the fluid leads to increasing difference between wall and fluid temperatures and decreasing value of the Nus-

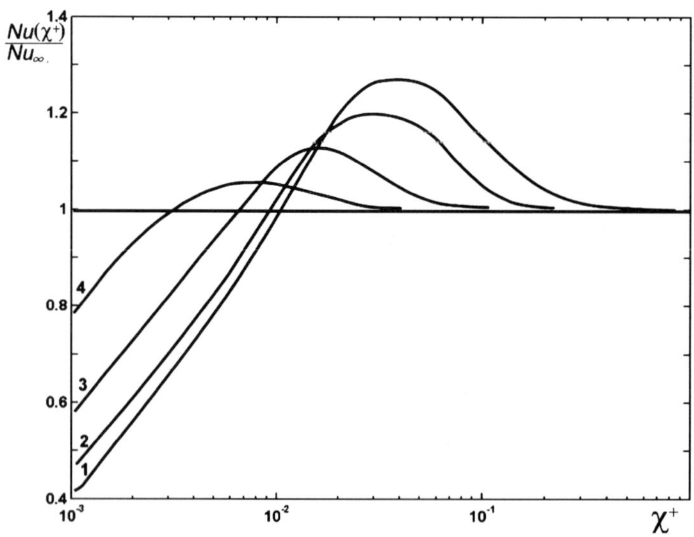

Fig. 4.14 Numerical calculations of the dependences $\mathrm{Nu}(X^+)$: *1* Pe = 1, *2* Pe = 2.5, *3* Pe = 10, *4* Pe = 45

selt number within the entrance section as compared to Nu_∞. These effects are illustrated in Fig. 4.14, which shows the dependence $\text{Nu}(\chi^+)/\text{Nu}_\infty$ on χ^+. It is possible to estimate the critical value of the parameter M that subdivides the states at which the effect of axial conduction dominates ($M > M_{\text{cr}}$) or it is negligible ($M < M_{\text{cr}}$). According to the data of Petukhov (1967), this critical value is $M_{\text{cr}} \simeq 0.01$. In line with these results we obtain the following evaluation of the length where the effect of axial conduction in the fluid should be taken into consideration: $\frac{x}{r_0}\text{Pe} \le 20$.

4.4.2 Axial Conduction in the Wall

The problem of axial conduction in the wall was considered by Petukhov (1967). The parameter used to characterize the effect of axial conduction is $P = \left(1 - d_0^2/d_{\text{in}}^2\right)(k_2/k_1)$. The numerical calculations performed for $q = $ const. and neglecting the wall thermal resistance in radial direction, showed that axial thermal conduction in the wall does not affect the Nusselt number Nu_∞. Davis and Gill (1970) considered the problem of axial conduction in the wall with reference to laminar flow between parallel plates with finite conductivity. It was found that the Peclet number, the ratio of thickness of the plates to their length are important dimensionless groups that determine the process of heat transfer.

The effect of axial conduction in the wall on the heat transfer in micro-channels was recently investigated by Maranzana et al. (2004) and Tiselj et al. (2004). In the first study the thermal structure of laminar flow between parallel plates was investigated. In the second one, heat transfer characteristics of water flowing through triangular silicon micro-channels were analyzed. For the physical interpretation of the effects due to axial conduction in the wall, Maranzana et al. (2004) introduced "axial conduction number" M defined as the ratio of conductive heat flux to convective heat flux. Numerical calculations by Maranzana et al. (2004) showed that the effect of axial conduction in the wall is significant when $M > 10^{-2}$ or estimated using relation (Celata et al. 2005, 2006) suggested the relation $k_{\text{w}}\left(d_{\text{out}}^2 - d_0^2\right)/k_{\text{f}} d_0 L.01\text{Re}\,\text{Pr}$, where k_{w} and k_{f} are the thermal conductivity of wall and liquid, respectively, d_{out} and d_0 are the outer and inner tube diameter, and L is the tube length. According to Tiselj et al. (2004), axial conduction in the wall significantly affects the longitudinal fluid and wall temperature distribution and longitudinal distribution of the normal and axial heat flux. We will discuss these results in Sect. 4.5.

4.4.3 Combined Axial Conduction in the Fluid and in the Wall

The energy equations for the fluid and the wall are (Petukhov 1967):

$$\frac{\partial \tilde{u}\theta_1}{\partial X^+} = \frac{1}{R}\frac{\partial}{\partial R}\left(R\frac{\partial \theta_1}{\partial R}\right) + \frac{1}{\text{Pe}^2}\frac{\partial^2 \theta_1}{\partial X^{+2}} \tag{4.28}$$

$$\frac{1}{R}\frac{\partial}{\partial R}\left(R\frac{\partial\theta_2}{\partial R}\right)+\frac{1}{\mathrm{Pe}^2}\frac{\partial^2\theta}{\partial X^{+2}}=0 \tag{4.29}$$

where $\theta_i = (T_i - T_0)/(qd_1/k_i)$, $i = 1,2$ for the fluid and wall, respectively.

Integration of Eqs. (4.28) and (4.29) through the cross-section of the micro-channel gives:

$$\frac{\partial}{\partial X^+}\left(\int_0^1 \tilde{u}\theta_1 R\,\mathrm{d}R\right)=\left(R\frac{\partial\theta_1}{\partial R}\right)\Bigg|_0^1+\frac{1}{\mathrm{Pe}^2}\frac{\partial^2}{\partial X^{+2}}\left(\int_0^1 \theta_1 R\,\mathrm{d}R\right) \tag{4.30}$$

$$\left(R\frac{\partial\theta_2}{\partial R}\right)\Bigg|_1^{R_*}+\frac{1}{\mathrm{Pe}^2}\frac{\partial^2}{\partial X^{+2}}\left(\int_1^{R_*} \theta_2 R\,\mathrm{d}R\right)=0 \tag{4.31}$$

where

$$R_*=\frac{r_0}{r_{\mathrm{out}}},\quad \frac{\partial\theta_1}{\partial R}\bigg|_0=0,\quad \left(\frac{\partial\theta_1}{\partial R}=\frac{\partial\theta_2}{\partial R}\right)\bigg|_{R=1}=0,\quad \frac{\partial\theta_2}{\partial R}\bigg|_{R=R_*}=\frac{k_1}{k_2}\frac{1}{2}. \tag{4.32}$$

The sum of Eq. (4.30) and Eq. (4.31) is:

$$\frac{\partial}{\partial X^+}\left(\int_0^1 \tilde{u}\theta_1 R\,\mathrm{d}R\right)$$
$$=\left(R\frac{\partial\theta_2}{\partial R}\right)\Bigg|_{R=R_*}+\frac{1}{\mathrm{Pe}^2}\frac{\partial^2}{\partial X^{+2}}\left(\int_0^1 \theta_1 R\,\mathrm{d}R\right)+\frac{1}{\mathrm{Pe}^2}\frac{\partial^2}{\partial X^{+2}}\left(\int_1^{R_*} \theta_2 R\,\mathrm{d}R\right) \tag{4.33}$$

Taking into account that

$$\overline{\theta}_1=\int_0^1 \theta_1 R\,\mathrm{d}R,\quad \overline{\theta}_2=(R_*^2-1)^{-1}\int_1^{R_*}\theta_2 R\,\mathrm{d}R \tag{4.34}$$

we obtain:

$$\frac{\mathrm{d}}{\mathrm{d}X^+}\left(\int_0^1 \tilde{u}\theta_1 R\,\mathrm{d}R\right)=\frac{k_1}{k_2}\frac{1}{2}+\frac{1}{\mathrm{Pe}^2}\frac{\mathrm{d}^2\overline{\theta}_1}{\mathrm{d}X^{+2}}+\frac{(R_*^2-1)}{\mathrm{Pe}^2}\frac{\mathrm{d}^2\overline{\theta}_2}{\mathrm{d}X^{+2}}. \tag{4.35}$$

Assuming that $\mathrm{d}^2\overline{\theta}_1/\mathrm{d}X^{+2}\approx \mathrm{d}^2\overline{\theta}_2/\mathrm{d}X^{+2}$ the following equation for average fluid temperature was obtained

$$\frac{\mathrm{d}^2\overline{\theta}_1}{\mathrm{d}X^{+2}}-\frac{\mathrm{Pe}^2}{4R_*^2}\frac{\mathrm{d}\overline{\theta}_1}{\mathrm{d}X^{+2}}+\frac{1}{2}k_{1,2}\frac{\mathrm{Pe}^2}{R_*^2}=0. \tag{4.36}$$

The longitudinal fluid temperature distribution is:

$$\overline{\theta}_1=C_1+C_2\exp\left(\frac{\mathrm{Pe}^2}{4R_*^2}X^+\right)+2\frac{k_1}{k_2}X^+ \tag{4.37}$$

where $C_1=-C_2$; $C_2=-\frac{2(k_1/k_2)}{(\mathrm{Pe}^2/4R_*^2)}\exp\left(\frac{\mathrm{Pe}^2X_*^+}{4R_*^2}\right)$.

Then, the parameter M is

$$M = \frac{1}{4} \frac{1 - \exp(\tilde{\chi} - \tilde{\chi}_*)}{(k_1/k_2)\tilde{\chi} - \exp(\tilde{\chi} - \tilde{\chi}_*) + \exp(-\tilde{\chi}_*)} \tag{4.38}$$

where $\tilde{\chi} = \frac{\mathrm{Pe}^2 X^+}{4R_*^2}$, $\tilde{\chi}_* = \frac{\mathrm{Pe}^2 X_*^+}{4R_*^2}$.

For the limiting cases $\tilde{\chi} \ll 1$ and $\tilde{\chi}_* \gg 1$ we obtain the following estimation:

$$M = \frac{1}{4k_{1,2}} \frac{1}{\tilde{\chi}} = \frac{\Lambda}{4\chi^+} \tag{4.39}$$

where $\Lambda = (k_2/k_1)(r_0/r_{\mathrm{out}})^2$. For conditions corresponding to flow in micro-channels, this factor is $\Lambda \gg 1$. For example Λ equals 25 and 250 for water flows in micro-channels of $r_0/r_{\mathrm{out}} = 1.1$ made of stainless steel and silicon, respectively. This shows that conduction in the wall has significant effect on the heat transfer in the micro-channels. It can be an important factor that leads to changes in the heat transfer coefficient.

4.5 Micro-Channel Heat Sinks

4.5.1 Three-Dimensional Heat Transfer in Micro-Channel Heat Sinks

The cooling systems fabricated from a large number of rectangular (Qu and Mudawar 2002a; Toh et al. 2002; Li et al. 2004), triangular (Tiselj 2004) or circular (Kroeker et al. 2004) micro-channels were investigated both theoretically and experimentally. The micro-channel heat sinks are highly complicated systems with a non-uniform distribution of thermal characteristics. The existence of a non-uniform temperature field in the liquid and solid substrate leads to a non-uniform distribution of heat fluxes in the streamwise and spanwise direction. Qu and Mudawar (2002a) carried out calculations at Re = 140, 700, and 1,400 for heat sinks. The micro-channels had a width of 57 μm and a depth of 180 μm, and were separated by a 43 μm wall. The major approximations introduced in the classical fin analysis method for micro-channel heat sinks operating in the laminar flow regime are summarized and assessed based on numerical results. The numerical results of that study revealed that the classical fin method could only provide a qualitatively correct picture of the heat transport in a micro-channel heat sink.

Numerical results of the heat transfer inside four 1 cm^2 heat sinks with 150 and 200 channels were presented by Toh et al. (2002). Their calculation predicted the local thermal resistance very well. The micro-heat sink modeled in the numerical investigation by Li et al. (2004) consisted of a 10 mm long silicon substrate. The rectangular micro-channels had a width of 57 μm, and a depth of 180 μm. The heat transfer calculations were performed for Reynolds numbers of 144, 77 and 42. The longitudinal heat conduction along the silicon wafer at different Reynolds number is different. In reality, because it is difficult to achieve an adiabatic boundary at the inlet and outlet of the heat sink as assumed in the numerical model, a significant portion of the heat is transferred to the inlet and outlet manifolds, especially for low fluid flow conditions. Thus, when evaluating the heat transfer in micro-heat sinks with low fluid flow rates, particular attention should be paid to the effects of heat conduction through the wafer.

Kroeker et al. (2004) investigated thermal characteristics of heat sinks with circular micro-channels using the continuum model based on the conventional Navier–Stokes equations and the energy conservation equation. Developing flow (both hydrodynamically and thermally) was assumed in the flow region and three-dimensional conjugate heat transfer was assumed in the solid region. At the inlet and outlet of the solid region (copper or silicon), adiabatic boundary conditions were imposed. The calculations of local Nusselt number performed at Re = 500 and Re = 1,000 follow closely the classical solution reported by Shah and London (1978) for forced convection in tubes with constant wall temperature.

One particular characteristic of conduction heat transfer in micro-channel heat sinks is the strong three-dimensional character of the phenomenon. The smaller the hydraulic diameter, the more important the coupling between wall and bulk fluid temperatures, because the heat transfer coefficient becomes high. Even though the thermal wall boundary conditions at the inlet and outlet of the solid wall are adiabatic, for small Reynolds numbers the heat flux can become strongly non-uniform: most of the flux is transferred to the fluid at the entrance of the micro-channel. Maranzana et al. (2004) analyzed this type of problem and proposed the model of channel flow heat transfer between parallel plates. The geometry shown in Fig. 4.15 corresponds to a flow between parallel plates, the uniform heat flux is imposed on the upper face of block 1; the lower face of block 0 and the side faces of both blocks

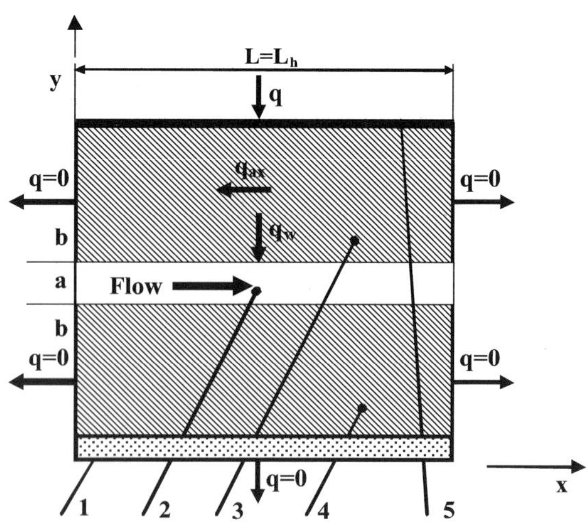

Fig. 4.15 Effect of axial conduction, channel between two parallel plates (Maranzana et al. 2004) (schematic view): *1* cover plate, *2* micro-channel, *3* silicon block 1, *4* silicon block 0, *5* heater

are adiabatic. The two 10 mm long and 500 μm thick blocks are made of silicon, water flows in the 100 μm thick channel. Figure 4.16a–c shows interface temperature, interface heat flux and the Nusselt number. The lower the Reynolds number, the larger the axial conduction effects.

The numerical and experimental study of Tiselj et al. (2004) (see Fig. 4.17) was focused on the effect of axial heat conduction through silicon wafers on heat transfer in the range of Re = 3.2−84. Figure 4.17 shows their calculation model of a triangular micro-channels heat sink. The results of calculations are presented in Fig. 4.18.

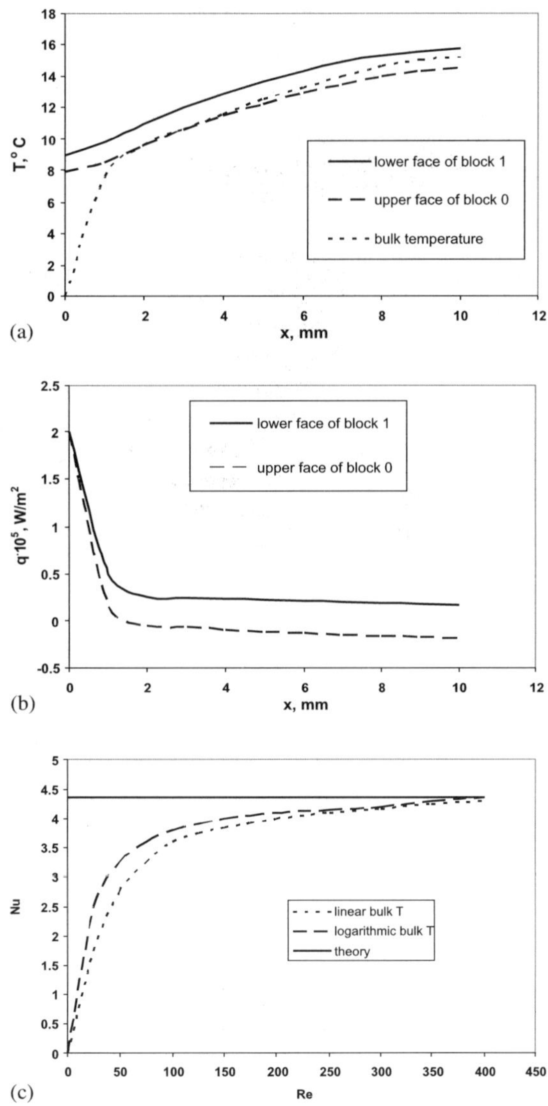

Fig. 4.16a–c Effect of axial conduction. Numerical simulation. (**a**) Interface temperature. (**b**) Interface heat flux. (**c**) The Nusselt numbers. Reprinted from Maranzana et al. (2004) with permission

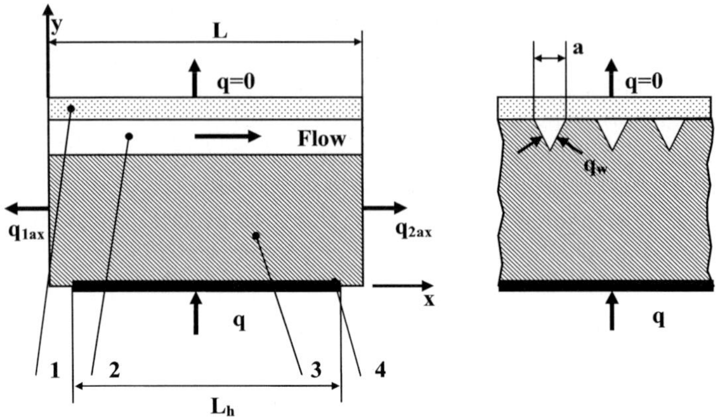

Fig. 4.17 Calculation model of triangular micro-channels heat sink: *1* cover plate, *2* micro-channel, *3* silicon wafer, *4* heater. Reprinted from Tiselj et al. (2004) with permission

The thermal wall boundary conditions used by Tiselj et al. (2004) at the inlet and outlet of the solid wall are not adiabatic (cf. Maranzana et al. 2004). The heat sink was made from a square-shaped silicon substrate of 15 × 15 mm and 530 μm thick, and in the silicon substrate, 17 parallel micro-channels were etched. The cross-section of each channel was an isosceles triangle with a base of 310 μm, the length of the micro-channels was $L = 15$ mm, the heating length was $L_h = 10$ mm. The angles at the base were 55°, the hydraulic diameter was $d_h = 160$ μm. The experimental results were used for numerical calculation. The bulk water temperature and the inner heated wall temperature are shown in Fig. 4.18a. Both temperatures did not change linearly along the longitudinal direction. In fact, a linear temperature rise cannot be regarded as a good approximation for both temperatures. When axial heat flux is directed to both the inlet and the outlet manifolds, the water and heated surface temperatures did not change monotonically. The longitudinal distributions of wall normal heat fluxes are presented in Fig. 4.18b. The heat fluxes are taken to be positive if they are directed from the solid to the fluid, and negative otherwise. The direction and the magnitude of the negative heat flux depend on relation between the two thermal resistances: the first defines the heat transport from the wall through the boundary layer to the fluid core, and the second one quantifies the possibility of heat transport through the silicon wafer. The importance of the second alternative pathway for the heat transfer can be truly appreciated only in a three-dimensional conjugate heat transfer problem. The thermal wall boundary conditions have a dominant role in such a problem. Numerical predictions of the local Nusselt number variation in the streamwise direction are plotted in Fig. 4.18c. As indicated by Incropera and De Witt (1996) the thermal entry length of a circular tube is:

$$L_{\text{therm}} = 0.05\text{Pe} \cdot d_h \,. \tag{4.40}$$

In the study by Tiselj et al. (2004), the thermal entry length was 0.13 and 3.3 mm for Re = 3.2 and 84, respectively.

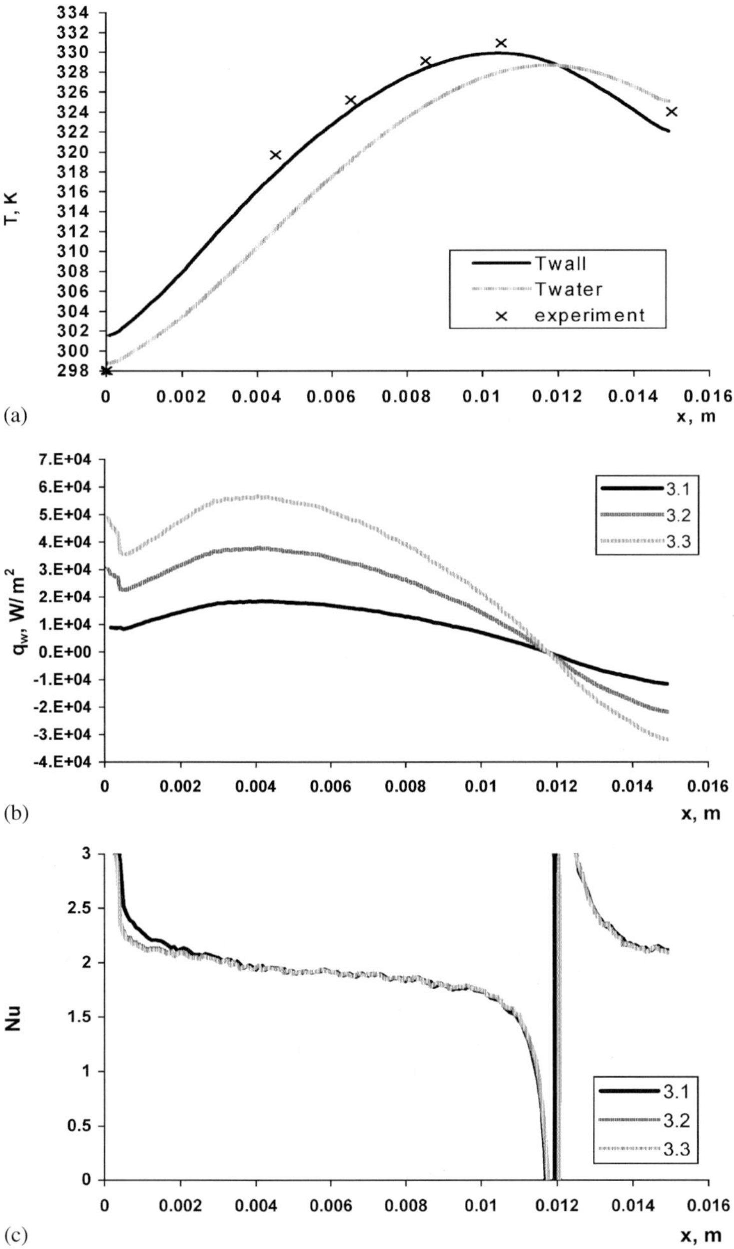

Fig. 4.18a–c Effect of axial conduction. Triangular micro-channels. $\dot{m} = 0.0356$ g/s. (**a**) Average water and silicon temperature distributions. $N = 8.424$ W. (**b**) Wall normal heat flux distribution in the silicon chip. 3-1 $N = 2.816$ W, 3-2 $N = 5.676$ W, 3-3 $N = 8.424$ W. (**c**) Axial distribution of the Nusselt number in the chip. 3-1 $N = 2.816$ W, 3-2 $N = 5.676$ W, 3-3 $N = 8.424$ W. Reprinted from Tiselj et al. (2004) with permission

4.5.2 Entrance Effects

The entrance effects in a single channel were extensively studied in the past (Petukhov 1967; Kays and Crawford 1993; Baehr and Stephan 1998; Schlichting 2000; Toh et al. 2002; Gamart et al. 2005). We restrict our discussion to the effect of inlet and the outlet manifolds on the flow and temperature distributions between the parallel micro-channels. Hetsroni et al. (2001) and Klein et al. (2005) observed an uneven liquid distribution in the parallel micro-channels. Depending on the particular manifold design, the difference between the flow rates into some parallel micro-channels was up to about 20%. Furthermore, due to relatively high thermal conductivity of the manifolds, fluid pre-warming occurred in the inlet manifold and additional warming occurred in the outlet manifold. The behavior of the Nusselt number depends, at least partly, on the entrance effects, which may be important in the laminar regime. The problem was also studied by Gamart et al. (2005). Idealizing the flow rate as uniform can result in a significant error when predicting the temperature distribution of a heated electronic device. Lee et al. (2005) showed that the entrance and boundary conditions imposed in the experiment need to be carefully matched in the predictive approaches. In this case numerical predictions based on a classical, continuum approach were in good agreement with the experimental data.

4.5.3 Characteristic Parameters

Two definitions were considered by Qu and Mudawar (2004) for heat flux to the heat sink. The first is an "effective" heat flux defined as the total electrical power input divided by the top area of the heat sink. The second definition is a mean heat flux averaged over the micro-channel heated inside area. We determine heat flux as the power calculated from the energy balance based on fluid temperature at the inlet and outlet manifolds, divided by the heated area of the micro-channel side walls (Tiselj et al. 2004). Often special effects are proposed to explain unexpected experimental results. A common assumption often made (Wang and Peng 1994; Peng and Peterson 1995; Peng and Peterson 1996) is to consider the wall heat flux to be uniform along the channels. However, according to Fig. 4.16b, the wall heat flux was far from uniform. As shown in Figs. 4.6c and 4.16a, the bulk temperature of the fluid did not vary linearly (Gao et al. 2002; Maranzana et al. 2004; Tiselj et al. 2004). This effect is especially important when the M number is large. For example, Maranzana et al. (2004) utilized exact modeling of heat conduction in the wall. The corresponding simulated estimation of the convective heat transfer coefficient at $M = 0.32$ was equal to $25{,}900\,\mathrm{W/m^2\,K}$. By comparison, the one-dimensional model, assuming that the bulk temperature is linear, yields a mean convective heat transfer coefficient of $6{,}100\,\mathrm{W/m^2\,K}$. Axial conduction can be neglected for M number lower than 0.01.

4.5.4 *Effect of Wall Roughness*

Wall roughness leads to increasing friction factor at the same Reynolds number. Existence of roughness leads also to a decrease in the value of the critical Reynolds number, at which the transition from laminar to turbulent flow occurs. We suggested a following estimation of the relative roughness, corresponding to the hydrodynamic threshold that subdivides the flow in smooth and rough channels (Hetsroni 2005): $k_s/r_0 < 5/1.41\mathrm{Re}^{0.5}$, where k_s is the average height of roughness, r_0 is the channel hydraulic radius, and Re is the Reynolds number. For Re $\sim$ 2,000, the relative roughness that corresponds to the boundary between the smooth and rough channels is about 0.08. Turner et al. (1999, 2000) concluded that micro-channel surfaces with relative surface roughness of 0.06 did not cause any statistical change in the friction factor for laminar flow. The effect of surface roughness on heat transfer depends on the Prandtl number. Kandlikar et al. (2003) reported that for a 1,067 μm diameter tube, the effect of relative roughness of about 0.003 on heat transfer in water flow was insignificant. For 620 μm, the same relative roughness increases the heat transfer. New experiments should be performed to clarify the effect of wall roughness on heat transfer.

4.5.5 *Interfacial Effects*

Turner et al. (1999) reported that for the Knudsen number Kn < 0.04 (ratio of mean free path to channel hydraulic diameter) the continuum-based equations can be used for flow in micro-channels. Because the micro-devices have a large surface-to-volume ratio, factors related to surface effects have more impact on the flow at small scales. Among these are the surface electrostatic charges. If the liquid contains even a small amount of ions, the electrostatic charges on the solid surface will attract the counter-ions in the fluid to establish an electrical field. The arrangement of the electrostatic charges on the solid surface and the balancing charges in the liquid is called the electrical double layer (EDL). Mala et al. (1997b), Yang and Li (1998), and Ren et al. (2001) reported numerical and experimental results for the EDL effect with different liquids. They found that the EDL effect led to higher friction coefficient for pure water and dilute solutions. No results for heat transfer were presented.

4.5.6 *Effect of Measurement Accuracy*

In experiments of flow and heat transfer in micro-channels, some parameters, such as the Reynolds number, heat transfer coefficient, and Nusselt number, are difficult to obtain with high accuracy. The channel hydraulic diameter measurement error may play a very important role in the uncertainty of the friction factor (Hetsroni et al. 2005). The analysis carried out by Hetsroni et al. (2003) for micro-tubes reveals the following values of standard uncertainties: for infrared measurements, systematic error 0.1 K, random error 0.2 K; for thermocouples, systematic error 0.1 K, random error 0.15 K; for temperature acquisition system, systematic error 0.1 K, random error 0.16 K. The 95% confidence uncertainty of the heat transfer coefficient was 13.2%. The uncertainty must be taken into account in the presentation of experimental data and in comparison between experimental results and theoretical predictions.

4.6 Compressibility Effects

Two-dimensional compressible momentum and energy equations were solved by Asako and Toriyama (2005) to obtain the heat transfer characteristics of gaseous flows in parallel-plate micro-channels. The problem is modeled as a parallel-plate channel, as shown in Fig. 4.19, with a chamber at the stagnation temperature T_{stg} and the stagnation pressure P_{stg} attached to its upstream section. The flow is assumed to be steady, two-dimensional, and laminar. The fluid is assumed to be an ideal gas. The computations were performed to obtain the adiabatic wall temperature and also to obtain the total temperature of channels with the isothermal walls. The governing equations can be expressed as

$$\frac{\partial \rho u}{\partial x} + \frac{\partial \rho v}{\partial y} = 0 \tag{4.41}$$

$$\frac{\partial \rho uu}{\partial x} + \frac{\partial \rho uv}{\partial y} = -\frac{\partial P}{\partial x} + \mu\left(\frac{\partial^2 u}{\partial x^2} + \frac{\partial^2 u}{\partial y^2}\right) + \frac{\mu}{3}\frac{\partial}{\partial x}\left(\frac{\partial u}{\partial x} + \frac{\partial v}{\partial y}\right) \tag{4.42}$$

$$\frac{\partial \rho uv}{\partial x} + \frac{\partial \rho vv}{\partial y} = -\frac{\partial P}{\partial y} + \mu\left(\frac{\partial^2 v}{\partial x^2} + \frac{\partial^2 v}{\partial y^2}\right) + \frac{\mu}{3}\frac{\partial}{\partial y}\left(\frac{\partial u}{\partial x} + \frac{\partial v}{\partial y}\right) \tag{4.43}$$

$$\frac{\partial \rho uh}{\partial x} + \frac{\partial \rho vh}{\partial y} = -P\left(\frac{\partial u}{\partial x} + \frac{\partial v}{\partial y}\right) + k\left(\frac{\partial^2 T}{\partial x^2} + \frac{\partial^2 T}{\partial y^2}\right) + \phi \tag{4.44}$$

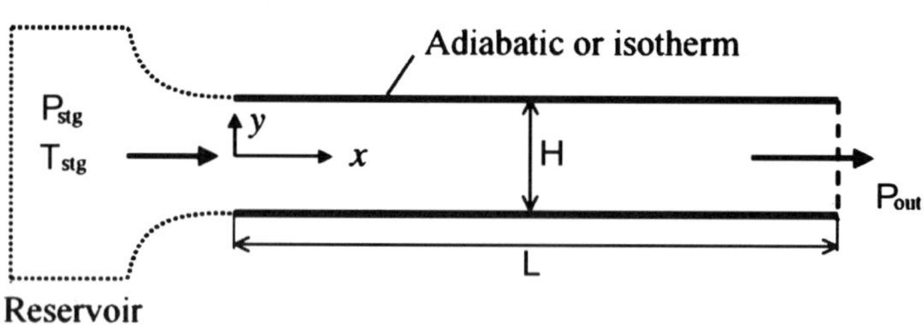

Fig. 4.19 A schematic diagram of parallel-plate micro-channels. Reprinted from Asako and Toriyama (2005) with permission

where

$$\phi = 2\mu\left(\left(\frac{\partial u}{\partial x}\right)^2 + \left(\frac{\partial v}{\partial y}\right)^2\right) - \frac{2\mu}{3}\left(\frac{\partial u}{\partial x} + \frac{\partial v}{\partial y}\right)^2 + \mu\left(\frac{\partial u}{\partial x} + \frac{\partial v}{\partial y}\right)^2. \quad (4.45)$$

The equation of the state for the ideal gas is expressed by

$$h = \frac{\gamma}{\gamma - 1}\frac{P}{\rho} = \frac{\gamma R}{\gamma - 1}T \quad (4.46)$$

where $\gamma = C_p/C_v$, R is the gas constant.

If the range of the channel height is limited to be above 10 μm, then the no-slip boundary condition can be adopted. Furthermore, with the assumptions of uniform inlet velocity, pressure, density, and specified pressure P_{out} at the outlet, the boundary conditions can be expressed as follows:

$$\begin{aligned} &\text{On the walls } (y = \pm 0.5H): && u = v = 0 \\ &\text{At the inlet } (x = 0): && u = u_{in}, v = 0, P = P_{in}, \rho = \rho_{in} \\ &\text{At the outlet } (x = \ell): && P = P_{out}. \end{aligned} \quad (4.47)$$

The thermal boundary conditions on the channel walls are

$$\frac{\partial T}{\partial y} = 0 \text{ for the adiabatic channel} \quad (4.48)$$

$$T = T_w \text{ for the isothermal channel.} \quad (4.49)$$

Adiabatic wall temperature

The computations were performed for air of $R = 287\,\mathrm{J/kg\,K}$, $\gamma = 1.4$, $\mu = 1.862 \times 10^{-5}\,\mathrm{Pa\,s}$, and $k = 0.0261\,\mathrm{W/m\,K}$ to obtain the adiabatic wall temperature. The adiabatic wall temperature is a wall temperature of the channel with the adiabatic walls. The channel height ranged from 10 to 100 μm and length was fixed at 30 mm. The stagnation temperature was kept at $T_{stg} = 300\,\mathrm{K}$. The stagnation pressure P_{stg} varied between 1.3×10^5 and 2.5×10^6 Pa. The outlet pressure was maintained at atmospheric condition $P_{out} = 10^5$ Pa.

Since the kinetic energy is related to Ma^2, the adiabatic wall temperature might be reduced by a function of Ma^2 for the cases where the viscous heat dissipation is negligibly small. Then, the values of T_w/T_{stg} for all channels are plotted as a function of Ma^2 in Fig. 4.20.

Channels with isothermal walls

The computations were also performed for the channels with isothermal walls of $T_w = 305$, 310, and 350 K. Air was assumed for the working fluid and the identical thermophysical properties mentioned earlier were used. The channel height ranged from 10 to 100 μm and the ratio of the channel length to its height was either 100 or 200. The stagnation temperature was kept at $T_{stg} = 300\,\mathrm{K}$. The stagnation pressure

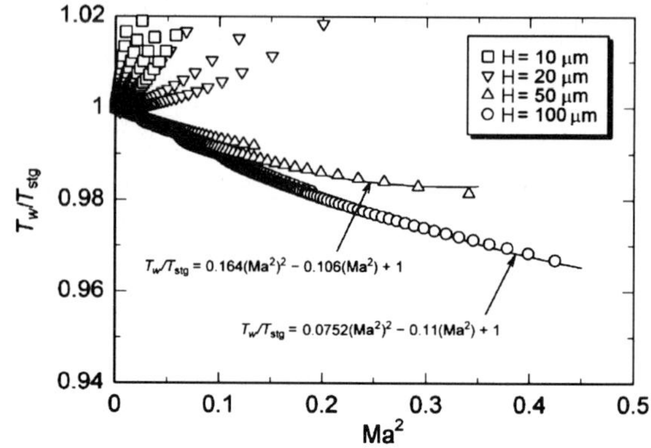

Fig. 4.20 T_w/T_{stg} as a function of Ma^2. Reprinted from Asako and Toriyama (2005) with permission

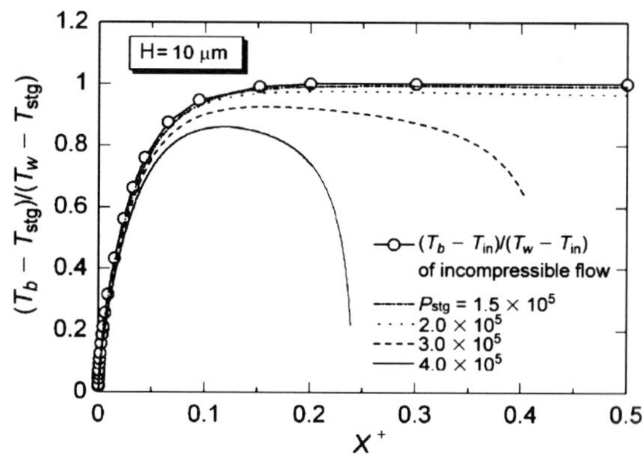

Fig. 4.21 Bulk temperature for $H = 10\,\mu m$ and $T_w = 350$ K. Reprinted from Asako and Toriyama (2005) with permission

P_{stg} was varied between 1.2×10^5 and 4×10^5 Pa. The outlet pressure was maintained at atmospheric condition $P_{out} = 10^5$ Pa.

The values of $(T_b - T_{stg})/(T_w - T_{stg})$ for the channel of $H = 10\,\mu m$ and $T_w = 350$ K are plotted as a function of X^+ in Fig. 4.21 with the bulk temperature for the incompressible flow. The value of the bulk temperature for the incompressible flow is normalized as $(T_b - T_{in})/(T_w - T_{in})$. As seen in the figure, the bulk temperature of the gaseous flow increases along the channel downstream and then levels off. It decreases when approaching the outlet due to conversion of the thermal energy into the kinetic energy. The same trend can be seen for the channels of $h = 20$, 50, and 100 µm. Therefore, the bulk temperature of the gas flow in the micro-channel cannot be estimated from the correlation for the incompressible flow.

4.7 Electro-Osmotic Heat Transfer in a Micro-Channel

Electro-osmosis generated flows are interesting for micro-electronics, biomedical diagnostic techniques, and a number of other applications. Important results related to heat transfer in such flows were obtained recently by Maynes and Webb

(2003) and Horiuchi and Dutta (2004). Below we follow the first of these works and consider some results corresponding to fully developed electro-osmotic heat transfer in circular micro-tubes. At the assumption of no pressure-driven contribution to the velocity field, constant electrical and thermal conductivities and constant wall ζ-potential, the momentum and energy equations that describe flow in circular micro-tubes are

$$\mu\frac{1}{r}\frac{\mathrm{d}}{\mathrm{d}r}\left(r\frac{\mathrm{d}u}{\mathrm{d}r}\right)+\frac{\varepsilon}{r}\frac{\mathrm{d}}{\mathrm{d}r}\left(r\frac{\mathrm{d}\psi}{\mathrm{d}r}\right)\frac{\mathrm{d}\Phi}{\mathrm{d}x}=0 \tag{4.50}$$

$$\frac{\partial^2 T}{\partial x^2}+\frac{1}{r}\frac{\partial}{\partial r}\left(r\frac{\partial T}{\partial r}\right)=\frac{u}{\alpha}\frac{\partial T}{\partial x}-\frac{s}{k} \tag{4.51}$$

where u and T are the velocity and temperature, Φ is the applied potential field, s is the volumetric energy generation ($s=i_e^2\sigma$, i_e is the conduction current density, σ is the liquid electrical resistivity), ε is the fluid dielectric constant, ψ is the excess charge distribution, k is the thermal conductivity, α is the thermal diffusivity, and μ is the viscosity.

For low wall potentials the Debye–Huckel linearization holds and the excess charge distribution is

$$\psi=\zeta\frac{I_0(r/\lambda)}{I_0(r_0/\lambda)} \tag{4.52}$$

where ζ is the zeta potential, λ is the Debye length, r_0 is the micro-channel radius, and I_0 is the modified Bessel function of the first kind of order zero.

Integration of Eqs. (4.50) and (4.51), with correlation (4.52), yields the following expression for velocity and temperature distribution, at $q_{\mathrm{w}}=\text{const.}$:

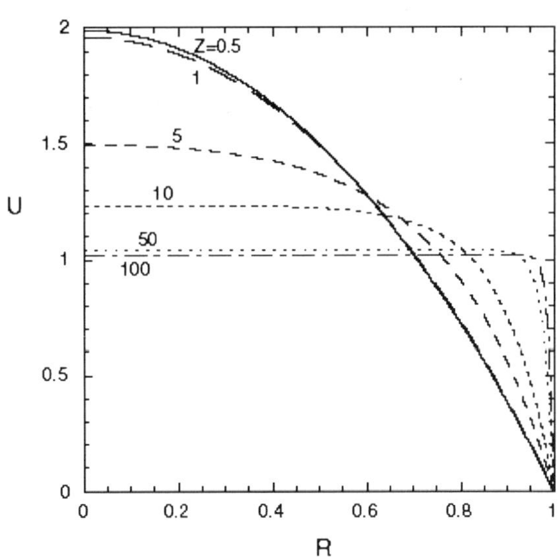

Fig. 4.22 Normalized electro-osmotically driven velocity profiles as a function of z for circular tube. Reprinted from Maynes and Webb (2003) with permission

$$\frac{u}{u_{\text{max}}} = \left[1 - \frac{I_0(ZR)}{I_0(Z)}\right] \tag{4.53}$$

$$\overline{U}_{\text{max}} = \frac{U}{u_{\text{max}}} = \left[1 - \frac{2I_1(Z)}{ZI_0(Z)}\right] \tag{4.54}$$

$$\theta(R) = \left\{\frac{2}{\overline{U}_{\text{max}}} + S\left(\frac{1}{\overline{U}_{\text{max}}} - 1\right)\right\}\left[\frac{R^2-1}{4} + \frac{1}{8\overline{U}_{\text{max}}}\right] + \frac{(2+S)}{Z^2\overline{U}_{\text{max}}}\left[1 - \frac{I_0(ZR)}{I_0(Z)}\right] - \frac{(2+S)}{\overline{U}^2_{\text{max}}}C_1(Z) + \frac{S}{\overline{U}_{\text{max}}}C_2(Z) \tag{4.55}$$

where $R = r/r_0$, $Z = r_0/\lambda$, λ is the Debye length, $\overline{U}_{\text{max}}$ is the maximum possible electro-osmotic velocity for a given applied potential field, $\theta - (T - Tm)(q_{\text{w}}r_0/k)$ is the dimensionless temperature, T_{m} is the mixed mean temperature, q_{w} is the wall heat flux, $S = sr_0/q_{\text{w}}$ is the dimensionless energy generation, $\overline{u}$ is the average velocity,

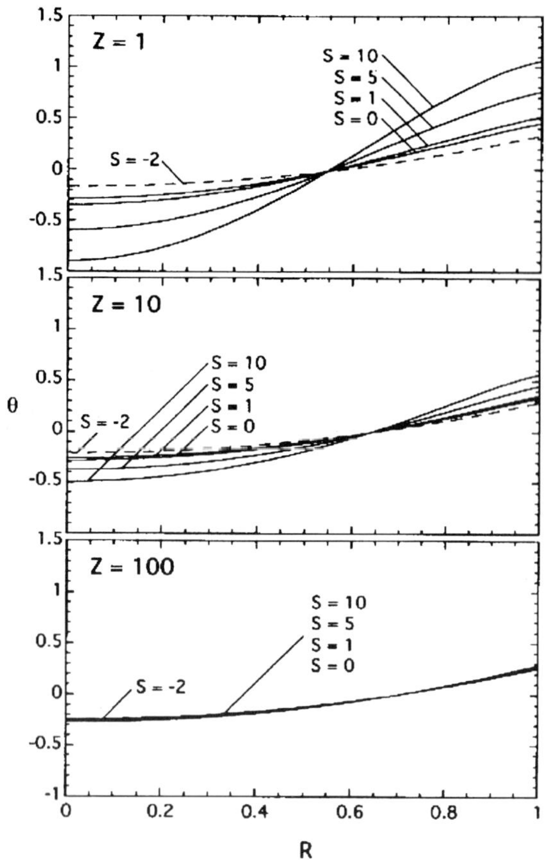

Fig. 4.23 Normalized temperature profile for the circular tube as a function of R and Z. Reprinted from Maynes and Webb (2003) with permission

$$C_1(Z) = \frac{1}{Z^2}\left[3 - \frac{2I_1(Z)}{I_0(Z)} - \frac{I_1^2(Z)}{I_0^2(Z)} - \frac{4I_1(Z)}{zI_0(Z)}\right], \quad C_2(Z) = \frac{1}{Z^2}\left[1 - \frac{2I_1(Z)}{ZI_0(Z)}\right],$$

and I_1 is the modified Bessel function of the first kind of order one.

The velocity and temperature distributions in a cross-section of a circular micro-tube are plotted in Figs. 4.22 and 4.23. It is seen that the velocity profile is determined by a single parameter, Z, whereas the temperature profile depends on two dimensionless groups, Z and S.

The shape of the velocity profile changes from parabolic corresponding to Poiseuille flow ($Z < 1$) to a uniform, slug-like one which is typical for flow with concentrated momentum source near the wall ($Z \gg 1$) and thin boundary layer. As can be seen in Fig. 4.23, the temperature distribution is independent from parameter s at large enough values of the ratio of micro-channel radius to the Debye length. Taking into account that the Debye length in pure water is about 1 μm (Hunter 1981), it is possible to estimate the micro-channel diameter, at which the electro-osmotic effect on heat transfer is negligible, as about 200 μm.

A general case of heat transfer under the conditions of combined action of electro-osmotic forces and imposed pressure gradient was considered by Chakraborty (2006). The analysis showed that in this case the Nusselt number depends not only on parameters z and S, but also on an additional dimensionless group, which is a measure of the relative significance of the pressure gradient and osmotic forces.

4.8 Closing Remarks

Heat transfer in micro-channels occurs under superposition of hydrodynamic and thermal effects, determining the main characteristics of this process. Experimental study of the heat transfer in micro-channels is problematic because of their small size, which makes a direct diagnostics of temperature field in the fluid and the wall difficult. Certain information on mechanisms of this phenomenon can be obtained by analysis of the experimental data, in particular, by comparison of measurements with predictions that are based on several models of heat transfer in circular, rectangular and trapezoidal micro-channels. This approach makes it possible to estimate the applicability of the conventional theory, and the correctness of several hypotheses related to the mechanism of heat transfer. It is possible to reveal the effects of the Reynolds number, axial conduction, energy dissipation, heat losses to the environment, etc., on the heat transfer.

Theoretical models used for this purpose can be subdivided into two groups depending on the degree of accuracy of their assumptions. The first of these groups includes the simplest one-dimensional models assuming uniform heat flux, constant heat transfer coefficient, etc. The comparison of these models with experiments shows significant discrepancy between the measurements and the theoretical predictions. Using some "new effects," which are not accounted for by conventional Navier–Stokes and energy equations, the simple models may explain experimental results. The second group is based on numerical solution of full Navier–Stokes and energy equations, which account for the real geometry of the micro-channel, axial conduction in the fluid and wall, energy dissipation, non-adiabatic thermal boundary condition at the inlet and outlet of the heat sink, dependence of physical properties of fluid on temperature, etc. These models demonstrate a fairly good agreement with available experimental data. As a rule, the numerical calculations using simple models were performed for the following hydraulic boundary conditions:

1. A uniform velocity profile was set at the channel inlet.
2. The flow was assumed to be fully developed in the test section.
3. All the fluid properties were constant.

The thermal boundary conditions were set as follows:

1. Constant heat flux at the walls
2. Adiabatic conditions at the inlet and outlet

These boundary conditions are not in agreement with experiments for which the "new effects" were assumed. As a result, some researchers concluded that conventional Navier–Stokes and energy equations are not valid, and that only "new effects" can explain the experimental data. The numerical solutions based on the Navier–Stokes and energy equations with the proper boundary conditions demonstrate a fairly good agreement with available experimental data. The results can be generalized as follows:

1. The effect of energy dissipation on heat transfer in micro-channels is negligible under typical flow conditions.
2. Axial conduction in the fluid and wall affects significantly the heat transfer in micro-channels. In laminar flow, two heat transfer regimes may be considered. The first takes place when $\mathrm{Re} > 150$ and the axial conduction number $M <$ 0.01, or estimated using the relation $k_\mathrm{w}\left(d_\mathrm{out}^2 - d_0^2\right)/k_\mathrm{f} d_0 L.01\mathrm{Re}\,\mathrm{Pr}$. Under these conditions the heat transferred through the solid substrate may be neglected and adiabatic boundary conditions may be imposed at the inlet and outlet manifolds to solve a conjugate three-dimensional heat transfer problem. The second regime occurs at $\mathrm{Re} < 150$, $M > 0.01$. In this case, the heat transferred through the solid substrate should be taken into account.
3. The following considerations must be taken into account in the evaluation of any experimental results:

   a. The experimental results based on the measurements of the fluid temperature only at the inlet and the outlet manifolds of the heat sink may lead to incorrect values of the Nusselt number.
   b. Since it is difficult to measure the local heat flux at the inner channel wall, the definition of the heat transfer coefficient is very important and will strongly influence its value.

4. Accurate estimation of the heat transferred through the solid substrate in experiments should be obtained from energy balance that includes electric power (simulating the electronic components on the top or bottom wall of the heat

sink), convective heat transfer to the fluid (based on fluid mass flow rate and bulk fluid temperature measured at the inlet and outlet manifolds), and heat losses.

5. The heat transfer coefficient calculated numerically using an exact model with regard to the heat transferred through the solid substrate represents the correct variation of the Nusselt number with respect to the Reynolds number.
6. The thermal entry length should be considered by comparison between experimental and numerical results.

Summary

1. A variety of studies can be found in the literature for the solution of the convection heat transfer problem in micro-channels. Some of the analytical methods are very powerful, computationally very fast, and provide highly accurate results. Usually, their application is shown only for those channels and thermal boundary conditions for which solutions already exist, such as circular tube and parallel plates for constant heat flux or constant temperature thermal boundary conditions. The majority of experimental investigations are carried out under other thermal boundary conditions (e.g., experiments in rectangular and trapezoidal channels were conducted with heating only the bottom and/or the top of the channel). These experiments should be compared to solutions obtained for a given channel geometry at the same thermal boundary conditions. Results obtained in devices that are built up from a number of parallel micro-channels should account for heat flux and temperature distribution not only due to heat conduction in the streamwise direction but also conduction across the experimental set-up, and new computational models should be elaborated to compare the measurements with theory.
2. A number of physical parameters or dimensionless groups formulated to describe the same phenomenon have different interpretations. For example, in some studies heat flux is defined as based on a planform area of heat sink top surface, and in other studies as based on a channel heated inside area. The Nusselt number is defined in the literature as based on the difference between wall and fluid temperature. It may be for developed or developing flows, and may present a peripheral local or average value. Additionally this temperature difference may be defined differently. For example, the averaged Nu in micro-channels may be based on the difference $(T_{\mathrm{w,av}} - T_{\mathrm{f,av}})$ where $T_{\mathrm{w,av}}$ is the average temperature of the channel bottom, or averaged peripheral temperature, or may be based on the logarithmic temperature difference, etc. For the most part the temperature $T_{\mathrm{f,av}}$ was defined as $(T_{f,\mathrm{in}} + T_{f,\mathrm{out}})/2$. Such an assumption of linear varying the fluid temperature along a micro-channel may lead to the conclusion which is not physically sound, that the Nusselt number depends on the Reynolds number under laminar flow. An effort to standardize the definitions of physical parameters and dimensional groups for flow through micro-channels should be made.

3. Only a small number of solutions for the laminar forced convection problem and experimental investigations are available in the literature with some variations in the associated thermophysical properties. To the authors' knowledge, for example, no experimental study is available to clarify the effect of the Prandtl number on the heat transfer in micro-channels with different duct geometries.
4. Depending on the particular design of inlet and outlet manifolds, the difference between the flow rates into some parallel micro-channels may be up to 20%. Idealizing the flow rate as uniform can result in significant error in prediction of the temperature distribution of a heated electronic device.
5. For channels above $d_h = 1$ mm the surface roughness generally does not affect the Nusselt number as long as the height of the relative average surface roughness is less than 1%. For small diameters tubes ($d_h < 0.6$ mm) the relative average roughness $k_s/d_h > 0.003$ increases heat transfer up to 25–30% in the range of Re = 1,000–2,000. This effect is more pronounced at higher values of relative surface roughness and Reynolds numbers.

References

Adams TM, Abdel-Khalik SI, Jeter SM, Qureshi ZH (1998) An experimental investigation of single-phase forced convection in micro-channels. Int J Heat Mass Transfer 41:851–857

Asako Y, Toriyama H (2005) Heat transfer characteristics of gaseous flows in micro-channels. Microscale Thermophys Eng 9:15–31

Baehr HD, Stephan K (1998) Heat and mass transfer. Springer, Berlin Heidelberg New York

Bailey DK, Ameel T, Warrington RO, Savoie TI (1995) Single phase forced convection heat transfer in micro-geometries: a review. In: Proceedings the 13th of Intersociety Energy Conversion Engineering Conference, San Diego, 20–25 August 1978. American Society of Mechanical Engineers, New York, pp 301–310

Bastanjian SA, Merzhanov AG, Xudiaev SI (1965) On hydrodynamic thermal explosion. Sov Phys Docl 163:133–136

Celata GP, Como M, Zummo G (2004) Thermal-hydraulic characteristics of single-phase flow in capillary pipes. Exp Thermal Fluid Sci 28:87–95

Celata GP, Cumo M, Marconi V, McPhail SJ, Zummo Z (2005) Micro-tube heat transfer scaling effects: an experimental validation. In: Proceedings of ECI International Conference on Heat Transfer and Fluid Flow in Microchannels, Caste/Vecchio Pascoli, Italy, 25–30 September 2005

Celata GP, Cumo M, Marconi V, McPhail SJ, Zummo Z (2006) Micro-tube liquid single phase heat transfer in laminar flow. Int. J. Heat Mass Transfer 49:3538–3546

Chakraborty S (2006) Analytical solutions of Nusselt number for thermally fully developed flow in microtubes under a combined action of electroosmotic forces and imposed gradients. Int J Heat Mass Transfer 49:810–813

Choi SB, Barron R, Warrington RQ (1991) Fluid flow and heat transfer in micro-tubes. In: Choi D et al. (eds) Micro-mechanical sensors, actuators and systems. ASME DSC 32:121–128

Davis EJ, Gill WN (1970) The effect of axial conduction in the wall on heat transfer with laminar flow. Int J Heat Mass Transfer 23:459–470

Dittus FW, Boelter LMK (1930) Heat transfer in automobile radiators of tubular type. University of California, Berkley. Publ Eng 2(13):443–461

Eckert E, Weise W (1941) Die Temperatur unbeheizter Körper in einem Gasstrom hoher Geschwindigkeit. Forsch Ing Wes 12:40–50

Gad-el-Hak M (1999) The fluid mechanics of micro-devices. The Freeman Scholar Lecture. J Fluid Eng 121:5–33
Gad-el-Hak M (2003) Comments on critical view on new results in micro-fluid mechanics. Int J Heat Mass Transfer 46:3941–3945
Gamart G, Favre-Marinet M, Asendrych D (2005) Conduction and entrance effects on laminar liquid flow and heat transfer in rectangular micro-channels. Int J Heat Mass Transfer 48:2943–2954
Gao P, Le Person S, Favre-Marinet M (2002) Scale effects on hydrodynamics and heat transfer in two-dimensional mini and micro-channels. Int J Thermal Sci 41:1017–1027
Garimella SV, Sobhan CB (2003) Transport in micro-channels – a critical review. Ann Rev Heat Transfer 13:1–50
Gnielinski V (1976) New equations for heat and mass transfer in turbulent pipe and channel flow. Int Chem Eng 16:359–368
Gruntfest J, Young JP, Johnson NL (1964) Temperatures generated by the flow of liquids in pipers. J Appl Phys 35:18–23
Gua Z-Y, Li Z-X (2003) Size effect on micro-scale single-phase flow and heat transfer. Int J Heat Mass Transfer 46:149–159
Harms TM, Kazmierzak MJ, Gerner FM (1999) Developing convective heat transfer in deep rectangular micro-channels. Int J Heat Fluid Flow 20:149–157
Hassan I, Phuttavong P, Abdelgawad M (2004) Micro-channel heat sinks: an overview of the state of the art. Microscale Thermophys Eng 8:183–204
Hehnecke DK (1968) Heat transfer by Hagen–Poiseuille flow in the thermal development region with axial conduction. Wazme Stoffubertz 1:177–184
Herwig H (2000) Flow and heat transfer in micro systems. Is everything different or just smaller. ZAMM 82:579–586
Herwig H, Hausner O (2003) Critical view on new results in micro-fluid mechanics: an example. Int J Heat Mass Transfer 46:935–937
Hetsroni G, Gurevich M, Mosyak A, Rozenblit R (2003) Surface temperature measurement of a heated capillary tube by means of an infrared technique. Meas Sci Technol 14:807–814
Hetsroni G, Gurevich M, Mosyak A, Rozenblit R (2004) Drag reduction and heat transfer of surfactants flowing in a capillary tube. Int J Heat Mass Transfer 47: 3797–3809
Hetsroni G, Mosyak A, Pogrebnyak E, Yarin LP (2005) Fluid flow in micro-channels Int J Heat Mass Transfer 48:1982–1998
Hetsroni G, Mosyak A, Segal Z (2001) Nonuniform temperature distribution in electronic devices cooled by flow in parallel micro-channels. IEEE Trans Comp Packag Technol 24(1):16–23
Ho CM, Tai Y-C (1998) Micro-electronic mechanic systems (MEMS) and fluid flows. Ann Rev Fluid Mech 30:5–33
Horiuchi K, Dutta P (2004) Joule heating effects in electroosmotically driven microchannel flows. Int J Heat Mass Transfer 47:3085–3095
Hunter RJ (1981) Zeta potential in colloid science: principles and applications. Academic, New York
Incropera FP, De Witt DP (1996) Fundamentals of heat and mass transfer, 4th edn. Wiley, New York
Judy J, Maynes D, Webb BW (2002) Characterization of friction pressure drop for liquid flows through micro-channels. Int J Heat Mass Transfer 45:3477–3489
Kandlikar S, Grande W (2002) Evolution of micro-channel flow passages – thermohydraulic performance and fabrication technology. In: Proceedings of IMEECE, ASME International Mechanical Engineering Congress and Exposition, New Orleans, November 2002, pp 1–13
Kandlikar SG, Joshi S, Tian S (2003) Effect of surface roughness on heat transfer and fluid flow characteristics at low Reynolds numbers in small diameter tubes. Heat Transfer Eng 24(3):4–16
Kays WM, Crawford ME (1993) Convective heat and mass transfer. McGraw-Hill, New York
Klein D, Hetsroni G, Mosyak A (2005) Heat transfer characteristics of water and APG surfactant solution in a micro-channel heat sink. Int J Multiphase Flow 31:393–415

Koo J, Kleinstreuer C (2004) Viscous dissipation effects in micro-tubes and micro-channels. Int J Heat Mass Transfer 47:3159–3169

Kostic M (1994) On turbulent drag and heat transfer reduction phenomena and laminar heat transfer enhancement in non-circular duct flow of certain non-Newtonian fluid. Int J Heat Mass Transfer 37:133–147

Kroeker CJ, Soliman HM, Ormiston SJ (2004) Three-dimensional thermal analysis of heat sinks with circular cooling micro-channels. Int J Heat Mass Transfer 47:4733–4744

Lee PS, Garimella SV, Liu D (2005) Investigation of heat transfer in rectangular micro-channels. Int J Heat Mass Transfer 48:1688–1704

Lelea D (2005) Some considerations on frictional losses evaluation of a water flow in micro-tubes. Int Comm Heat Mass Transfer 32:964–973

Lelea D, Nishio S, Takano K (2004) The experimental research on micro-tube heat transfer and fluid flow of distilled water. Int J Heat Mass Transfer 47:2817–2830

Li J, Peterson GP, Cheng P (2004) Three-dimensional analysis of heat transfer in a micro-heat sink with single phase flow. Int J Heat Mass Transfer 47:4215–4231

Lin TY, Yang CY (2007) An experimental investigation by method of fluid crystal thermography. Int. J. Heat Mass Transfer 50(23-24):4736-4742

Ma HB, Peterson GP (1997) Laminar friction factor in micro-scale ducts of irregular cross section. Microscale Thermophys Eng 1:253–265

Mala GM, Li D (1999) Flow characteristics of water in micro-tubes. Int J Heat Fluid Flow 20:142–148

Mala GM, Li D, Dale JD (1997a) Heat transfer and fluid flow in micro-channels. Int J Heat Mass Transfer 40:3079–3088

Mala GM, Li D, Werner C (1997b) Flow characteristics of water through a micro-channel between two parallel plates with electro kinetic effects. Int J Heat Fluid Flow 18:491–496

Male van P, Croon de MHJM, Tiggelaar RM, Derg van den A, Schouten JC (2004) Heat and mass transfer in a square micro-channel with asymmetric heating. Int J Heat Mass Transfer 47:87–99

Maranzana G, Perry I, Maillet D (2004) Mini- and micro-channels: influence of axial conduction in the walls. Int J Heat Mass Transfer 47:3993–4004

Maynes D, Webb BW (2003) Full developed electro-osmotic heat transfer in microchannels. Int J Heat Mass Transfer 46:1359–1369

Morini GL (2004) Single-phase convective heat transfer in micro-channels: overview of experimental results. Int J Thermal Sci 43:631–651

Morini GL (2005) Viscous heating in liquid flows in micro-channels. Int J Heat Mass Transfer 48:3637–3647

Nguyen NT, Bochnia D, Kiehnscherrf R, Dozel W (1996) Investigation of forced convection in micro-fluid systems. Sens Actuators A 55:49–55

Nguyen TV (1992) Laminar heat transfer for thermal developing flow in ducts. Int J Heat Mass Transfer 35:1733–1741

Owhaib W, Palm B (2004) Experimental investigation of single-phase convective heat transfer in circular micro-channels. Exp Thermal Fluid Sci 28:105–110

Peng XF, Peterson GP (1995) The effect of thermofluid and geometric parameters on convection of liquid through rectangular micro-channels. Int J Heat Mass Transfer 38:755–758

Peng XF, Peterson GP (1996) Convective heat transfer and flow friction for water flow in micro-channel structures. Int J Heat Mass Transfer 39:2599–2608

Peng XF, Wang BX, Peterson GP, Ma NB (1995) Experimental investigation of heat transfer in flat plates with rectangular micro-channels. Int J Heat Mass Transfer 38:127–137

Petukhov BS, Kurgano V, Gladuntsov A (1973) Heat transfer in turbulent pipe flow of gases with variable properties. Heat Transfer Sov Res 5:109–116

Petukhov BS (1967) Heat transfer and drag of laminar flow of liquid in pipes. Energy, Moscow

Qu W, Mala GM, Li D (2000) Heat transfer for water flow in trapezoidal silicon micro-channels. Int J Heat Mass Transfer 43:3925–3936

Qu W, Mudawar I (2002) Analysis of three-dimensional heat transfer in micro-channel heat sinks. Int J Heat Mass Transfer 45:3973–3985

Qu W, Mudawar I (2002) Experimental and numerical study of pressure drop and heat transfer in a single-phase micro-channel heat sink. Int J Heat Mass Transfer 45:2549–2565

Qu W, Mudawar I (2004) Measurement and correlation of critical heat flux in two-phase micro-channel heat sinks. Int J Heat Mass Transfer 47:2045–2059

Ren L, Qu W, Li D (2001) Interfacial electro kinetic effects on liquid flow in micro-channels. Int J Heat Mass Transfer 44:3125–3134

Reynaud S, Debray F, Frans J-P, Maitre T (2005) Hydrodynamics and heat transfer in two-dimensional mini-channels. Int J Heat Mass Transfer 48:3197–3211

Schlichting H (2000) Boundary layer theory, 8th edn. Springer, Berlin Heidelberg New York

Shah RK, London AL (1978) Laminar flow forced convection in ducts. Academic, New York

Sobhan CB, Garimella SV (2001) A comparative analysis of studies on heat transfer and fluid flow in micro-channels. Microscale Thermophys Eng 5:293–311

Tiselj I, Hetsroni G, Mavko B, Mosyak A, Pogrebhyak E, Segal Z (2004) Effect of axial conduction on the heat transfer in micro-channels. Int J Heat Mass Transfer 47:2551–2565

Toh KC, Chen XY, Chai JC (2002) Numerical computation of fluid flow and heat transfer in micro-channels. Int J Heat Mass Transfer 45:5133–5141

Tso CP, Mahulikar SP (1998) The use of the Brinkman number for single phase forced convective heat transfer in micro-channels. Int J Heat Mass Transfer 41:1759–1769

Tso CP, Mahulikar SP (1999) The role of the Brinkman number in analyzing flow transitions in micro-channels. Int J Heat Mass Transfer 42:1813–1833

Tso CP, Mahulikar SP (2000) Experimental verification of the role of the Brinkman number in micro-channels using local parameters. Int J Heat Mass Transfer 43:1837–1849

Tunc G, Bayazitoglu Y (2001) Heat transfer in micro-tubes with viscous dissipation. Int J Heat Mass Transfer 44:2395–2403

Turner SE, Sun H, Faghri M, Gregory OJ (1999) Local pressure measurement of gaseous flow through micro-channels. ASME HTD 364(3):71–80

Turner SE, Sun H, Faghri M, Gregory OJ (2000) Effect of surface roughness on gaseous flow through micro-channels. ASME HTD 366(2):291–298

Wang BX, Peng XF (1994) Experimental investigation on liquid forced-convection heat transfer through micro-channels. Int J Heat Mass Transfer 37(1):73–82

Warrier GR, Dhir VK, Momoda LA (2002) Heat transfer and pressure drop in narrow rectangular channels. Exp Thermal Fluid Sci 26:53–64

Weigand B, Lauffer D (2004) The extended Graetz problem with piecewise constant wall temperature for pipe and channel flows. Int J Heat Mass Transfer 47:5303–5312

Wu HY, Cheng P (2003) An experimental study of convective heat transfer in silicon micro-channels with different surface conditions. Int J Heat Mass Transfer 46:2547–2556

Wu PY, Little WA (1984) Measuring of the heat transfer characteristics of gas flow in fine channel heat exchangers for micro-miniature refrigerators. Cryogenics 24:415–420

Yang C, Li D (1998) Analysis of electro kinetic effects on the liquid flow in micro-channels. Coll Surf A Physicochem Eng Aspects 143:339–353

Yang CY, Lin TY (2007) Heat transfer characteristics of water flow in micro-tubes. Exp. Thermal and Fluid Science 32:432-439

Yoo JY (2006) Recent studies of fluid flow and heat transfer in thermal micro-devices. Nanoscale Microscale Thermophys Eng 10:67–81

Zel'dovich YaB, Barenblatt GI, Librovich VB, Maxhviladse GM (1985) Mathematical theory of combustion and explosion. Plenum, New York

Zhao CY, Lu TJ (2002) Analysis of micro-channels for electronic cooling. Int J Heat Mass Transfer 45:4857–4869

Nomenclature

A	Cross-section
A_d	Ratio of convective heat transfer to dissipation one
a	Relative surface roughness
c_p	Specific heat
C	Concentration
d	Inner diameter
d_h	Hydraulic diameter
h	Heat transfer coefficient, enthalpy
H	Micro-channel height
I_0	Bessel function
i_e	Current density
k	Thermal conductivity
k_s	Average height of roughness
L	Length
M	Conductive to convective heat flux ratio
m	Flow rate
Ma	Mach number
N	Power
P	Pressure
q	Heat flux
Re_{sh}	Shear Reynolds number, based on the shear viscosity
r	Radius
r_0	Tube radius, inner hydraulic radius
R	Gas constant, $R = r/r_0$ dimensionless distance from tube axis
s	Volumetric energy generation
T	Temperature
$\overline{T}$	Average fluid temperature
U	Average velocity
U_m	Normalized local velocity
u	Streamwise velocity
u_{max}	Maximum possible electro-osmotic velocity
v	Spanwise velocity
W	Channel width
W_b	Width of the channel bottom
W_c	Spacing
W_t	Width of the channel top
x	Longitudinal coordinate
$\mathrm{Br} = \dfrac{\mu U^2}{k(T_w - T_f)}$	Brinkman number
$\mathrm{Ec} = \dfrac{U^2}{2c_p \Delta \mathrm{T}}$	Eckert number

$\mathrm{Kn} = \dfrac{\overline{\lambda}}{d_\mathrm{h}}$	Knudsen number
$\mathrm{Na} = -\dfrac{4\beta\mu U^2}{k}$	Nahme number
$\mathrm{Nu} = \dfrac{h d_\mathrm{h}}{k_\mathrm{f}}$	Nusselt number
Nu_0	Nusselt number corresponds to negligible viscous dissipation
$\mathrm{Pe} = \dfrac{U d_\mathrm{h}}{\alpha}$	Peclet number
$\mathrm{Pr} = \dfrac{\nu}{\alpha}$	Prandtl number
$\mathrm{Re} = \dfrac{U d_\mathrm{h}}{\nu}$	Reynolds number
$X^+ = \dfrac{x}{d_\mathrm{h}\mathrm{Pe}}$	Dimensionless longitudinal coordinate

Greek symbols

α	Thermal diffusivity
β	Temperature sensitivity
$\Delta T = T_\mathrm{w} - T_\mathrm{f}$	Temperature difference
δ	Thickness of tube
γ	Aspect ratio, relation of $c_\mathrm{p}/c_\mathrm{v}$
ε	Fluid dielectric constant
ζ	Wall zeta potential
θ	Dimensionless temperature
λ	Friction factor, Debye length
$\overline{\lambda}$	Mean free path
μ	Dynamic viscosity
ν	Kinematic viscosity
Π	Bejan number
ρ	Density
σ	Liquid electrical resistivity
Φ	Applied potential field
ψ	Excess charge distribution

Subscripts

ad	Adiabatic
av	Average

ax	Axial
b	Bulk
cond	Conduction
conv	Convection
d	Developed
f	Fluid
h	Heated
in	Inner, inlet
inv	Inversion
out	Outer, outlet
sh	Shear
sur	Surfactant
stg	Stagnation
th	Theoretical
therm	Thermal
tot	Total
v	Volume
w	Wall
wat	Water
1	Fluid
2	Wall
2.1	Ratio parameter of wall to fluid

Chapter 5
Gas–Liquid Flow

In Chap. 5 the available data related to flow and heat transfer of a gas–liquid mixture in single and parallel channels of different size and shape are presented. These data concern flow regimes, void fraction, pressure drop and heat transfer. The effects of different parameters on flow patterns and hydrodynamic and thermal characteristics of gas–liquid flow are discussed.

Understanding the differences in two-phase flow characteristics between conventional size channels and micro-channels is also important for designing mini- or micro-heat exchangers, since the flow characteristics will affect the phase change heat transfer.

5.1 Two-Phase Flow Characteristics

Gas–liquid flows occur widely in both nature and industrial applications, including energy production (e.g., oil transportation, steam generators, cooling systems) and chemical engineering (e.g., bubble columns, reactors, aeration systems). Two-phase flows in micro-channels have attracted attention because of its wide applicability to such advanced fields as MEMS, electronic cooling, medical and genetic engineering, bioengineering, etc. At present, the knowledge of flow and heat transfer in micro-scale flow passages of a size less than 100 μm is thus strongly demanded. Specifically, fundamental knowledge of two-phase flow characteristics in small flow passages, such as the flow pattern, void fraction, pressure drop, and heat transfer coefficient, is crucial for engineering design purposes as well as for evaluation of practical performance. Papers by Ghiaasiaan and Abdel-Khalik (2001), Serizawa et al. (2002), Kawahara et al. (2002), Garimella and Sobhan (2003), Celata (2004), and Cheng and Wu (2006) extensively reviewed the literature on two-phase flow in micro-channels.

However, our current knowledge is still limited and in reality only a small number of literature sources are available. One of the questions is whether the two-phase flow patterns in small size channels are different from those encountered in "ordinary" size channels. In particular, in large tubes, as well as in tubes of a few millimeters in diameter, two-phase flow patterns are dominated in general by gravity with minor surface tension effects. On the other hand, in micro-channels with the diameter on the order of a few microns to a few hundred microns, two-phase flow is believed to be influenced mainly by surface tension, viscosity and inertia forces. Entrance effects, effects of surface roughness and wettability should also be taken into account. There is an important resemblance between two-phase flow in micro-channels and the flow in large channels at micro-gravity. In both system types the surface tension, inertia, and viscosity are important, while buoyancy is suppressed. Consequently, two-phase dimensionless parameters that have previously been developed for micro-gravity might be useful for micro-channels.

For adiabatic, steady-state, and developed gas–liquid two-phase flow in a smooth pipe, assuming immiscible and incompressible phases, the essential variables are ρ_L, ρ_G, μ_L, μ_G, σ, d_h, g, θ, U_{LS}, and U_{GS}, where subscripts L and G represent liquid and gas (or vapor), respectively, ρ is the density, μ is the viscosity, σ is the surface tension, d_h is the channel hydraulic diameter, θ is the channel angle of inclination with respect to the gravity force, or the contact angle, g is the acceleration due to gravity, and U_{LS} and U_{GS} are the liquid and gas superficial velocities, respectively. The independent dimensionless parameters can be chosen as $\Delta\rho/\rho_L$ (where $\Delta\rho = \rho_L - \rho_G$), and

$$\mathrm{Eo} = \frac{\Delta\rho g d_h^2}{\sigma} \tag{5.1}$$

$$\mathrm{We}_{LS} = \frac{U_{LS}^2 d_h \rho_L}{\sigma} \tag{5.2}$$

$$\mathrm{We}_{GS} = \frac{U_{GS}^2 d_h \rho_G}{\sigma} \tag{5.3}$$

$$\mathrm{Re}_{LS} = U_{LS} d_h/\nu_L \tag{5.4}$$

$$\mathrm{Re}_{GS} = U_{GS} d_h/\nu_G \tag{5.5}$$

where ν is the kinematic viscosity, and Eo, We, and Re are Eotvos, Weber and Reynolds numbers, respectively. Other relevant and widely used dimensionless parameters include the Bond number, $\mathrm{Bn} = d_h/\sqrt{\sigma/(g\Delta\rho)}$, Capillary number, $\mathrm{Ca} = \mu_L U_{LS}/\sigma$, phase Froude numbers, $\mathrm{Fr}_{GS} = U_{GS}/\sqrt{g d_h}$ and $\mathrm{Fr}_{LS} = U_{LS}/\sqrt{g d_h}$. In a class of micro-channels of interest for a wide range of applications, $\mathrm{Eo} < 1$, at least one of the Weber numbers is on the order of 1–10^2, and $\mathrm{Re}_{LS} \geq 1$. Thus, the surface tension dominates over buoyancy while inertia is important. Similar conditions apply to the two-phase flow at micro-gravity as well, indicating that important similarities between two-phase flow processes in the two-system categories should be expected.

Zhao and Rezkallah (1993), Rezkallah (1996), and more recently Lowe and Rezkallah (1999) developed two-phase flow transition models for micro-gravity channel flows based on liquid and gas Weber numbers. Zhao and Rezkallah (1993) suggested $\mathrm{We}_{GS} \approx 1$ as the upper boundary for the surface tension-dominated zone, and $\mathrm{We}_{GS} \approx 20$ as the lower boundary for the inertia-dominated zone.

Although many theories, models and correlations have been developed for two-phase flow in relatively large diameter tubes on the order of 10 mm, their applicability to channels of $d_h < 10$ mm also needs to be clarified. With an order-of-magnitude reduction in the hydraulic diameter of the flow channel from 10 to 1 mm, significant differences have been reported by Kawaji (1999) in the two-phase flow pattern map, void fraction and pressure drop. In narrow channels, the gravitational effect is diminished so that the channel orientation no longer has a significant effect on the two-phase flow map. A small channel height results in the disappearance of the stratified flow pattern, and some of the flow pattern transition boundaries are also shifted. We discuss the flow characteristics in the channels of $d_h \geq 1$ mm, and identify the changes that occur due to a reduction in the channel diameter from about 1 mm to 20 µm. Table 5.1 is a summary of the recent relevant experimental investigations considered extensively below.

Table 5.1 Experimental investigations

Authors	Number of channels and inclination	Channel shape	Hydraulic diameter, mm	Superficial gas velocity, U_{GS}, m/s	Superficial liquid velocity, U_{LS}, m/s	Object of study: Flow pattern	Flow regime map	Void fraction	Pressure drop	Heat transfer
Hetsroni et al. (1998a)	Single horizontal	Circular	49.2	0.02–0.3	0.62–0.87					+
Hetsroni et al. (1998b)	Single inclination 2°, 5°	Circular	49.2	0.02–0.39	0.40–1.4					+
Triplett et al. (1999a)	Single	Circular Semi-triang.	1.1, 1.45 1.09, 1.49	0.02–80	0.02–8	+				
Triplett et al. (1999b)	Single	Circular Semi-triang.	1.1, 1.45 1.1 1.45	0.02–80	0.02–8			+	+	
Zhao and Bi (2001a)	Single	Triang.	2.886 1.443 0.866	0.1–100	0.08–6.0 0.1–10	+	+			
Zhao and Bi (2001b)	Single	Triang.	2.886 1.443 0.866	0.1–100	0.08–6.0 0.1–3			+	+	
Kawahara et al. (2002)	Single	Circular	0.100	0.1–60	0.02–4	+	+	+	+	

Table 5.1 (continued)

Authors	Number of channels and inclination	Channel shape	Hydraulic diameter, mm	Superficial gas velocity, U_{GS}, m/s	Superficial liquid velocity, U_{LS}, m/s	Object of study: Flow pattern	Flow regime map	Void fraction	Pressure drop	Heat transfer
Serizawa et al. (2002)	Single	Circular	0.020 0.025 0.100	0.0012– 295.3	0.003– 17.52	+	+	+		
Hetsroni et al. (2003a)	21 Parallel channels	Triang.	0.130	0.63– 3.1	0.13	+				
Hetsroni et al. (2003b)	Single horizontal inclination 8°	Circular	49.2 25.0	20.0 24.0– 55.0	0.005– 0.03 0.016– 0.17	+				+
Qu et al. (2004)	Single	Rectang.	0.406 × 2.032	0.08– 81.92	0.04– 10.24	+	+			+
Ghajar et al. (2004)	Single horizontal inclination 2°, 5°, 7°	Circular	25.4	Re_{GS}= 560– 48000	Re_{LS}= 820– 26000	+				+
Hetsroni et al. (see this book)	21 Parallel channels	Triang.	0.130	Re_{GS}= 4.7– 270	Re_{LS}= 4.0–60					+
Zimmerman et al. (2006)	Single horizontal	Circular	25.0	24.2– 41.5	0.02– 0.09	+				+
Ide et al. (2006)	Single	Circular Rectang. Circular	0.5– 6.0 1 × 1 1 × 5 1 × 9.9 0.5, 0.05	0.5– 30	0.1–0.5	+		+	+	

5.2 Flow Patterns in a Single Conventional Size Channel

Flow patterns in conventional size channels deviate significantly from those in micro-channels. Slug and annular flow constitute dominant flow patterns in conventional size channels, and transition to annular flow occurs at much lower values of superficial gas velocity than in micro-channels. There are also appreciable discrepancies in both flow patterns and transition boundaries among various conventional size/micro-channel studies due to different channel shape, entrance effects, etc. It is difficult to generate a universal flow pattern map for adiabatic conventional size/micro-channel flows because of the need to maintain simultaneously values of a large number of dimensionless parameters.

5.2.1 Circular Channels

The data of Triplett et al. (1999a) were obtained with air and water at near-atmospheric pressure and room temperature.

The flow regimes in the test sections were identified visually with the aid of a strobe and a digital camera. The camera was always targeted at the test section center. No systematic attempt was made to assess and eliminate the test section entrance effects on the flow regimes. However, the distance between the point pictured by the camera and the test section inlet was well over 100 channel diameters everywhere. Thus, although the possibility exists that the reported flow regimes are influenced by the test section entrance conditions, this influence may not be significant.

Figure 5.1 displays representative photographs of the identified flow patterns in the 1.1 mm inner diameter test section. Five distinct flow patterns can be identified. Bubbly flow (Fig. 5.1a,b) was characterized by distinct and distorted (non-spherical) bubbles, generally considerably smaller in diameter than the channel diameter. With increasing U_{GS} (which leads to increasing void fraction) the bubbles crowded near the channel top and eventually led to the development of the slug flow (Fig. 5.1c,d), characterized by elongated cylindrical bubbles. This flow pattern has been referred to by others as slug (Suo and Griffith 1964), plug (Damianides and Westwater 1988), and bubble-train. Parameter changes leading to higher void fraction (e.g., increasing U_{GS} and/or decreasing U_{LS}) lead to longer bubbles and shorter liquid slugs. The bubbles, however, appear to effectively occupy most of the channel cross-section, with the liquid film at the bottom only slightly thicker than the liquid film at the top.

At high liquid superficial velocities, U_{LS}, by increasing the mixture volumetric flux ($U_{LS}+U_{GS}$), churn flow was established. Two processes were assumed to characterize churn flow. In some cases the elongated bubbles in the slug flow pattern became unstable near their trailing ends, leading to their disruption (Fig. 5.1e). In others, flooding-type churning waves periodically disrupted an otherwise apparently wavy-annular flow (Fig. 5.1f). The churn flow pattern defined here thus includes the aerated slug flow pattern, as the flow pattern represented by Fig. 5.1e is sometimes referred to.

At relatively low liquid superficial velocities, increasing the mixture volumetric flux led to longer bubbles and shorter liquid slugs, eventually leading to the merging of elongated bubbles, and the development of the slug-annular flow pattern, repre-

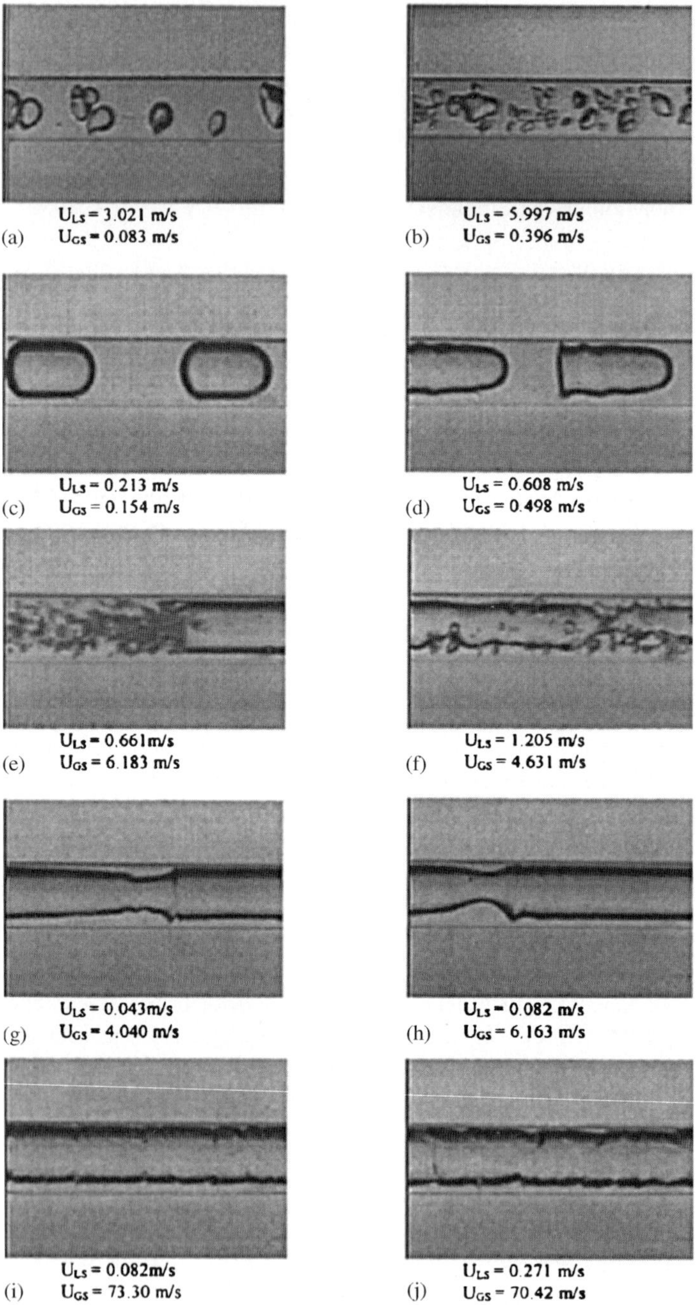

Fig. 5.1a–j Representative photographs of flow patterns in the 1.097 mm diameter circular test section. Test section: (**a**), (**b**) bubbly; (**c**), (**d**) slug; (**e**), (**f**) churn; (**g**), (**h**) slug-annular; (**i**), (**j**) annular. Reprinted from Triplett et al. (1999a) with permission

sented in Fig. 5.1g,h. In this flow pattern, long segments of the channel support an essentially wavy-annular flow, and are interrupted by large-amplitude solitary waves which do not grow sufficiently to block the flow path. With further increasing of U_{GS}, these large-amplitude solitary waves disappear and an annular flow pattern, represented by Fig. 5.1i,j, is obtained.

Comparison among the micro-channel two-phase flow regime data is complicated, among others, due to the inconsistent terminology used by various authors for some flow patterns (Triplett et al. 1999a; Ghiaasiaan and Abdel-Khalik 2004; Akbar et al. 2003). The flow pattern distinction is made between the slug flow regime, when it is defined as the pattern dominated by elongated, large bubbles (e.g., in Damianides and Westwater 1988), and the flow pattern that represents transition from plug to annular flow, and is characterized by large waves superimposed on otherwise separated phases that intermittently block the channel (e.g., Yang and Shieh 2001). Also, the flow pattern referred to as churn flow by Triplett et al. (1999a) in fact included two different regimes according to other investigators: churn flow, characterized by unstable and aerated long bubbles similar to the pseudo-slug regime as defined by Suo and Griffith (1964), and frothy slug defined by Zhao and Rezkallah (1993). The latter flow pattern, furthermore, appears to have been identified as dispersed flow by some of the other investigators (Damianides and Westwater 1988; Barajas and Panton 1993; Coleman and Garimella 1999; Yang and Shieh 2001).

5.2.2 *Triangular and Rectangular Channels*

The experiments of co-current upward air–water two-phase flow patterns in vertical triangular micro-channels were carried out by Zhao and Bi (2001a). Three equilateral triangular channels 270 mm in length, having side lengths of 5.0, 2.5, and 1.5 mm, corresponding to hydraulic diameters of 2.886, 1.443, and 0.866 mm, were tested.

Flow patterns in the triangular channels of $d_h = 2.886$ and 1.443 mm

Dispersed bubbly flow

Dispersed bubbly flow (DB) is usually characterized by the presence of discrete gas bubbles in the continuous liquid phase. As indicated in Fig. 5.2, for the channel of $d_h = 2.886$ mm, dispersed bubbles appeared at a low gas superficial velocity but a very high liquid superficial velocity. It is known that for large circular tubes dispersed bubbles usually take a sphere-like shape. For the triangular channel of $d_h = 2.886$ mm, however, it is observed from Fig. 5.2 that the discrete bubbles in the liquid phase were of irregular shapes. The deformation of the gas bubbles was caused by rather high liquid velocities in the channel.

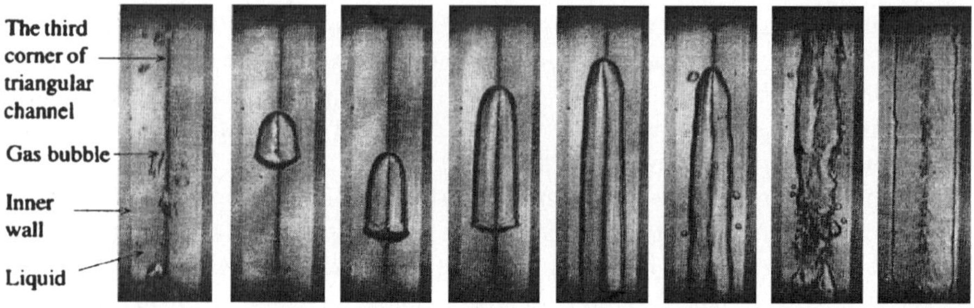

U_{GS} (m/s)=	0.1	0.1	0.2	0.3	0.9	2.2	6.4	72
U_{LS}(m/s)=	4.5	1	1	1	1	1	1	1
flow pattern:	DB	S	S	S	S	S	C	A

Fig. 5.2 Representative flow patterns in the triangular channel of $d_h = 2.886$ mm. Reprinted from Zhao and Bi (2001a) with permission

Slug flow

Slug flow (S) was observed at low gas superficial velocities in the entire range of the liquid superficial velocities. It is seen from Fig. 5.2 that gas slugs existed in the continuous liquid phase, spanned most of the channel cross-section, and were separated by liquid plugs in the axial direction. The gas slug had a semi-spherically shaped nose, a smooth body, and a flattened tail at low gas superficial velocities. As the gas superficial velocity increased, the slugs were deformed. It is also evident from Fig. 5.2 that the gas slug length increased as the gas superficial velocity increased.

Churn flow

As shown in Fig. 5.2, churn flow (C) appeared at moderate gas superficial velocities and the entire range of liquid superficial velocities. Flow was extremely chaotic and the gas–liquid interface was rather irregular. The gas phase and liquid phase had no distinct shapes.

Annular flow

Annular flow (A) existed at high gas superficial velocities and at the entire range of liquid superficial velocities. In annular flow, liquid film formed at the side wall with part of the liquid remaining in the three corners of the channel, while the continuous gas core flowed concurrently with the liquid phase.

Flow patterns in the triangular channel of $d_h = 0.866$ mm

The images of the representative flow patterns, together with the measured simultaneous pressure fluctuations, for the equilateral triangular channel having a side length of 1.5 mm (corresponding to 0.866 mm in hydraulic diameter) are displayed in Fig. 5.3. Generally, as the size of the channel was reduced to $D_h = 0.866$ mm, two peculiar phenomena were found: (1) dispersed bubbly flow, which existed in the channels of $d_h = 2.886$ mm (as presented in Fig. 5.2) and $d_h = 1.443$ mm, was never observed in the channel of $d_h = 0.866$ mm; (2) as shown in Fig. 5.3a, a new flow pattern, referred to as the capillary bubbly flow here, was found for the channel of $d_h = 0.866$ mm. This capillary bubbly flow has the following characteristics. First, as seen from Fig. 5.3a, the gas bubbles are more regularly distributed in the liquid phase, represented by a train of bubbles, essentially ellipsoidal in shape, spanning almost the entire cross-section with their centers located along the center line of the channel. This bubbly behavior is rather different from that for conventional large-sized channels including the two larger triangular channels considered in the same work, where the sizes of bubbles are usually much smaller than the channel size and the bubbles are more randomly distributed in the continuous liquid phase. Secondly, the capillary bubbly flow in the triangular channel of $d_h = 0.866$ mm occurred at low gas flow rates and moderate liquid flow rates.

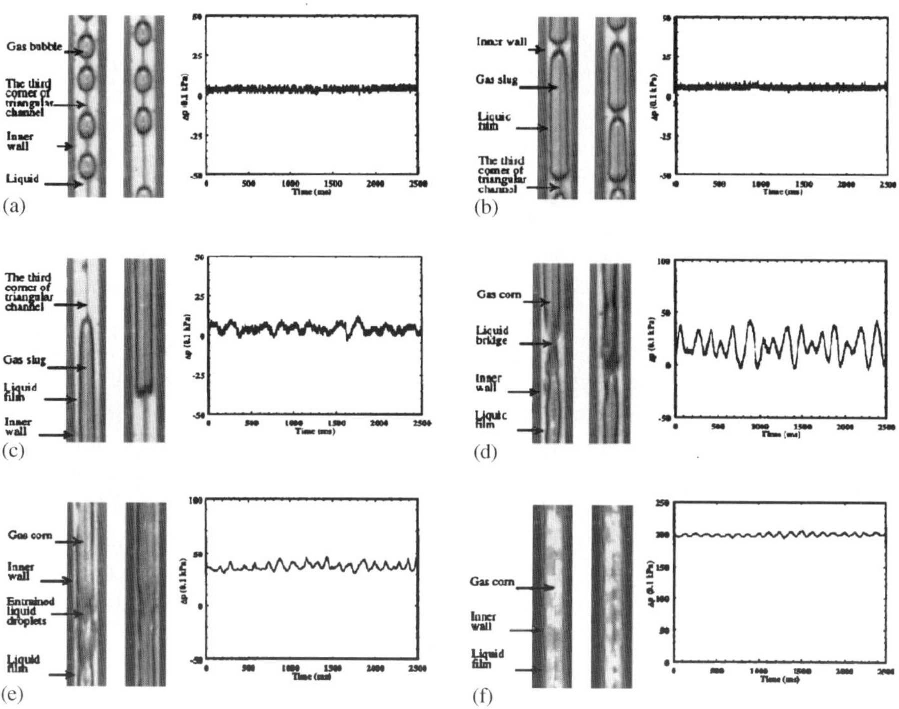

Fig. 5.3a–f Flow patterns and pressure fluctuations in the equilateral triangular channel of $d_h = 0.866$ mm: (**a**) capillary bubble flow ($U_{LS} = 0.1$ m/s, $U_{GS} = 0.2$ m/s); (**b**) slug flow ($U_{LS} = 0.1$ m/s, $U_{GS} = 0.4$ m/s); (**c**) slug flow ($U_{LS} = 0.1$ m/s, $U_{GS} = 2$ m/s); (**d**) churn flow ($U_{LS} = 0.1$ m/s, $U_{GS} = 5.5$ m/s); (**e**) annular flow ($U_{LS} = 0.1$ m/s, $U_{GS} = 20$ m/s); (**f**) annular flow ($U_{LS} = 0.1$ m/s, $U_{GS} = 85$ m/s). Reprinted from Zhao and Bi (2001a) with permission

With an increase in the air velocity, bubble disturbances, collisions, and coalescence became stronger, and gas slugs were formed. Figure 5.3b,c represent typical slug flows in the triangular channel of $d_h = 0.866\,\text{mm}$. It is seen from this figure that both the nose and the tail of the distinct slugs took the shape of a semi-sphere at low mixture velocities. The slugs approached the side walls of the channel although the channel corners were always resided by liquid because of surface tension. As compared with the flow in large-sized channels, the gas slugs were substantially elongated. The measured pressure traces, presented in the right side of Fig. 5.3b, show that the flow fluctuated at a lower frequency than in the capillary bubbly flow shown in Fig. 5.3a. For a higher air velocity, as shown in Fig. 5.3c, the slugs in the channel became irregular in shape and were further elongated. It is also observed from the right side of Fig. 5.3c that the pressure oscillatory amplitude became significantly higher.

As more air was added to the channel, the slug flow became unstable, the slug bubble broke down, and eventually the churn flow occurred in the channel. As shown in Fig. 5.3d, the most significant feature of flow characteristics in the churn flow is that the pressure oscillated at a relatively high amplitude, since the gas plug and liquid bridge flowed through the test section alternatively.

Figure 5.3e shows the situation when the air velocity was increased to $U_{GS} = 20\,\text{m/s}$. It is seen from this figure that the liquid bridges in churn flow disappeared and a liquid film formed at the side walls of the channel with a continuous gas core, in which a certain amount of liquid droplets existed. The pressure fluctuations in this case became relatively weaker in comparison with the case of the churn flow. The flow pattern displayed in Fig. 5.3f indicates that as the air velocity became high enough, such as $U_{GS} = 85\,\text{m/s}$, the liquid droplets entrained in the gas core disappeared and the flow became a pure annular flow. It is also observed from Fig. 5.3f that the flow fluctuation in this flow regime became weaker than that for the case shown in Fig. 5.3e, where $U_{GS} = 20\,\text{m/s}$.

The experimental results show that the typical flow patterns encountered in the conventional, large-sized vertical circular tubes, such as dispersed bubbly flow, slug flow, churn flow, and annular flow were also found in the two larger triangular channels. However, dispersed bubbly flow was not found in the smallest triangular channel. Instead, a new type of flow pattern, referred to as the capillary bubbly flow, has been identified. Unlike randomly discrete bubbles in continuous liquid phase in large channels, in the capillary bubbly flow regime, the gas bubbles were more regularly distributed in the liquid phase, represented by a single train of bubbles with ellipsoidal shape that flowed upwards along the channel centerline.

In the study by Qu et al. (2004), experiments were conducted with adiabatic nitrogen-water two-phase flow in a rectangular micro-channel. The bubbly, stratified and churn flow patterns commonly encountered in macro-channels were never observed in the study. No water droplets were observed in the nitrogen bubble, nor were any nitrogen bubbles present in the water slugs.

Annular flow pattern is characterized by a thin water film, which flows along the channel wall with the nitrogen comprising the central core. Unlike annular macro-channel flow, the interface between the two phases is fairly smooth, and no droplets are entrained in the nitrogen core.

From the visual studies on the flow patterns for circular, trapezoidal and rectangular channels it may be concluded that as the tube diameter decreases, transitions between flow regimes occur at different combinations of superficial gas and liquid velocities.

5.3 Flow Patterns in a Single Micro-Channel

Knowledge of dominant two-phase flow patterns in micro-channels is a key factor in developing accurate and physically sound predictive tools for heat sink design. Unfortunately, interfacial interactions between the vapor and liquid phases during flow boiling in a micro-channel are often far too complex to permit accurate measurement or quantitative assessment of flow patterns.

5.3.1 Experimental Observations

Serizawa et al. (2002) studied experimentally, through visualization, the two-phase flow patterns in air–water two-phase flows in round tubes. The test section for air–water experiments consisted of a transparent silica or quartz capillary tube with circular cross-section positioned horizontally. The two-phase flow was realized through a mixer with different designs, as shown in Figs. 5.4 and 5.5. The air was injected into the mixer co-axially while water was introduced peripherally.

Air–water flow in 25 μm tube

Figure 5.6 shows typical two-phase flow patterns observed in an air–water flow in a 25 μm silica tube at nearly atmospheric pressure.

Bubble flow

Dispersed bubbles are observed (Fig. 5.6a) when the gas flow rate is very small such as $U_{GS} = 0.0083\,\mathrm{m/s}$. Two kinds of bubbles are observed: one type is finely dispersed with a size smaller than the tube diameter, and the other type has a length of near to or a little larger than the tube diameter with spherical cap and tail. The distance between two consecutive bubbles may be longer than ten times the tube diameter. This flow pattern is also considered as a dispersed bubbly flow. Often in air–water flow two kinds of bubbles appear together as pairs of bubbles in which the small-sized bubbles follow the larger ones.

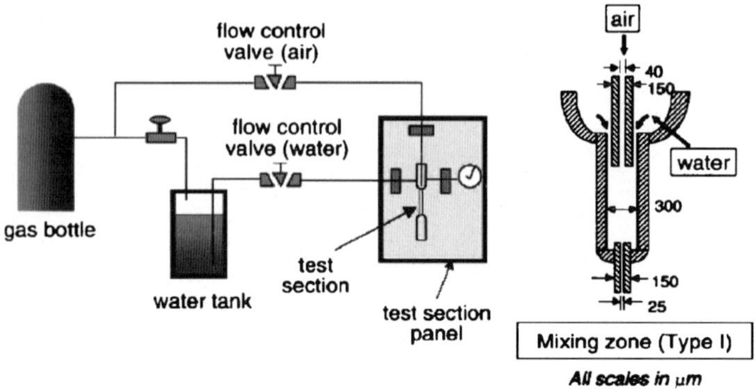

Fig. 5.4 Schematic diagram of the experimental facility for air–water flows. Reprinted from Serizawa et al. (2002) with permission

Fig. 5.5 Mixing chamber. Reprinted from Serizawa et al. (2002) with permission

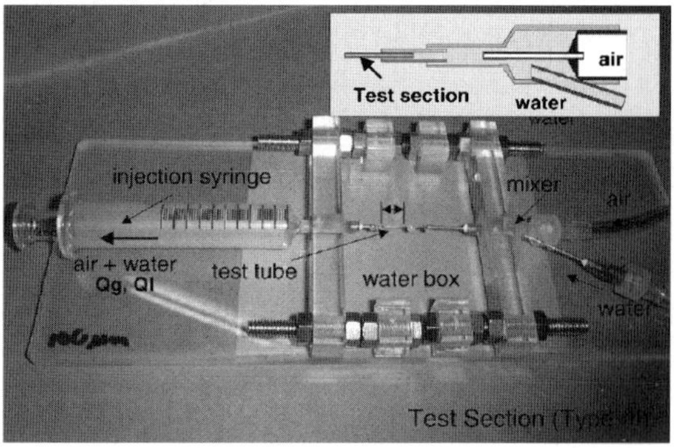

Slug flow

From the experimental observation it is quite clear that the occurrence of the slug flow is rather an entrance phenomenon than one induced from the tube. Slug flow occurs if the speed of long gas bubbles is not high enough to overcome the strong surface tension force of the liquid bridge between them (Fig. 5.6b).

Liquid ring flow

Figure 5.6c shows the typical liquid ring flow structure where the liquid film on the wall is symmetrically distributed. From experimental observation it is evident that the slug flow turns into the liquid ring flow when the gas velocity is high. It is interesting to note that Rezkallah (1998) observed under micro-gravity conditions frothy slug-annular flow, which is similar to the present liquid ring flow but contains small liquid droplets in the gas phase.

Fig. 5.6a–d Two-phase flow patterns. Reprinted from Serizawa et al. (2002) with permission

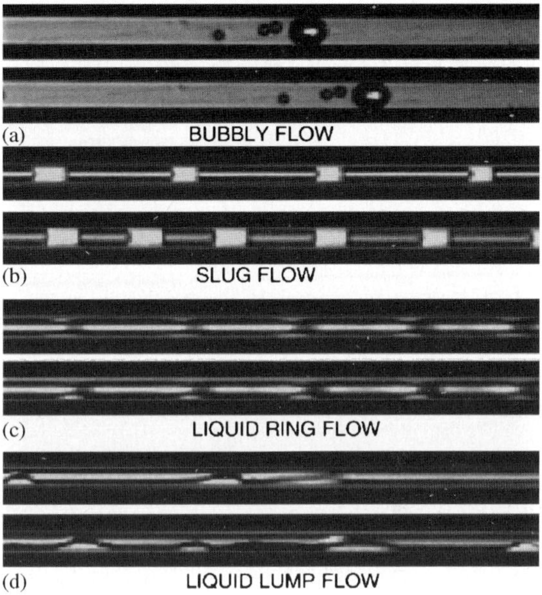

Liquid lump flow

Further increase in the gas flow rate in liquid ring flow leads to a liquid lump flow, of which the high-speed core gas entrains the liquid phase and liquid lumps slide on the wall, as shown in Fig. 5.6d. The shape of the liquid lump is very similar to that of the wavy flow in a horizontal large tube. The liquid lump shifts from side to side. When a liquid lump contacts the tube wall, the strong surface tension force will prevent it from spreading into the liquid film.

Air–water two-phase flow in a 100 μm tube

As demonstrated in Fig. 5.7, the result indicates that two-phase flow patterns observed in a 100 μm quartz tube are almost similar to those observed in a 25 μm silica capillary tube with several exceptions. One such exceptions is that in slug flow encountered at low velocities, small liquid droplets in a gas slug stick to the tube wall (Fig. 5.8). This fact is evidence that no liquid film exists between the gas slug and the tube wall.

5.3.2 Effect of Surface Wettability and Dryout

In order to examine how two-phase flow patterns in small channels are sensitive to surface contamination, Serizawa et al. (2002) observed visually the flow patterns

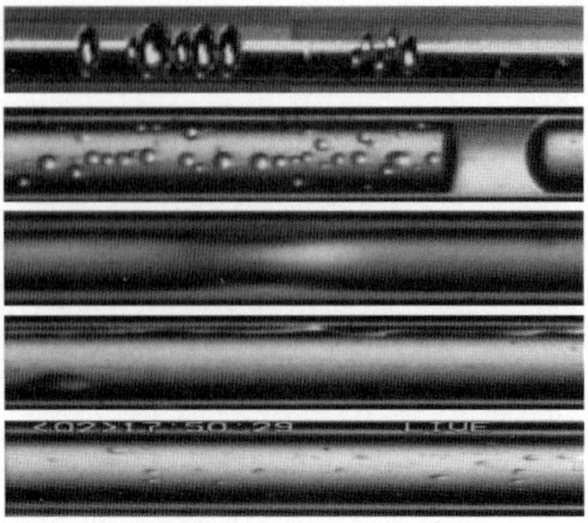

Fig. 5.7 Two-phase flow patterns in a 100 μm quartz tube. Reprinted from Serizawa et al. (2002) with permission

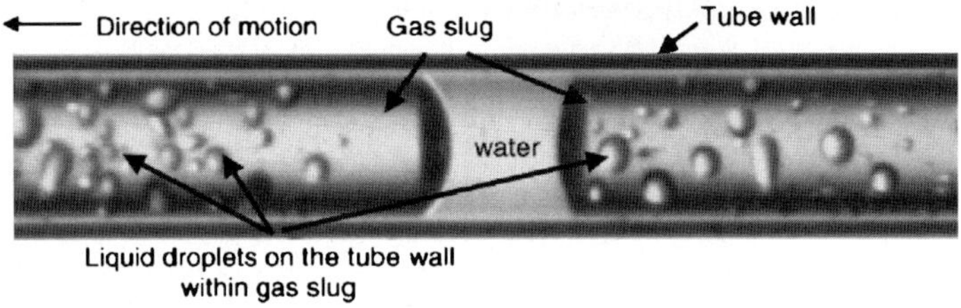

Fig. 5.8 Liquid droplets sticking onto the tube wall in a gas slug. Reprinted from Serizawa et al. (2002) with permission

formation using a 100 μm i.d. quartz tube test section, carefully cleaned by ultrasonic vibration in pools of high-purity distilled water, ethanol and dilute hydrochloride acid solution.

Figure 5.9 shows various interesting aspects of two-phase flow patterns obtained in this observation. It should be noticed from these pictures that a variety of two-phase flow patterns were encountered in a clean micro-channel. The authors noticed that in a very clean tube, many small individual bubbles flow in a discrete way in the tube without coalescence in bubbly flow. The most interesting thing is the special flow pattern given in Fig. 5.9d, where several bubbles with various shapes are connected in a series by the gas stems located at the tube center line. The liquid ring flow is also clearly seen in Fig. 5.9e.

One of the key questions is whether the inner wall of the test tube is wet or dry during the passage of a gas slug in slug flow pattern. Serizawa et al. (2002) checked this problem using a high-speed video and a high-precision laser confocal displacement meter, with an accuracy of 0.4 μm in thickness. Figure 5.10 shows a moving boundary between wet and dry areas on the tube wall during a passage of gas slugs. A detailed inspection of the signals from the laser confocal displacement meter suggests the existence of a dry area underneath a gas slug as well.

The effect of surface contamination and the wettability between the tube wall and the fluids were also studied experimentally. It has been shown that a stable annular flow and gas slug formation with a stable thin liquid film formed between the tube wall and gas slugs, which appeared at high velocities under carefully treated, clean

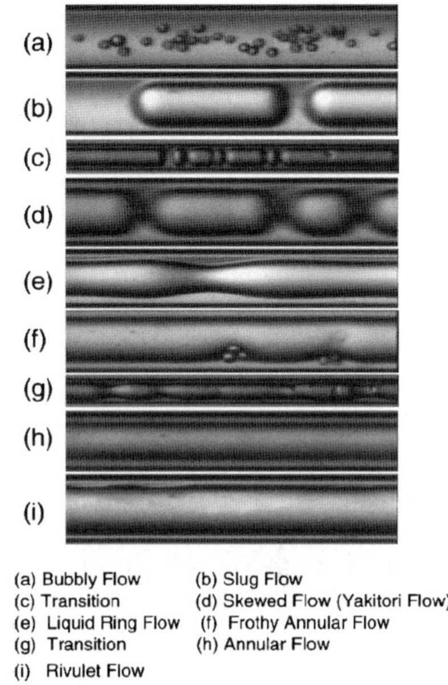

Fig. 5.9a–i Air–water two-phase flow patterns in a 100 μm i.d. clean quartz tube treated with ultrasonic vibration in distilled water, in ethanol and in dilute hydrochloride acid solution. Reprinted from Serizawa et al. (2002) with permission

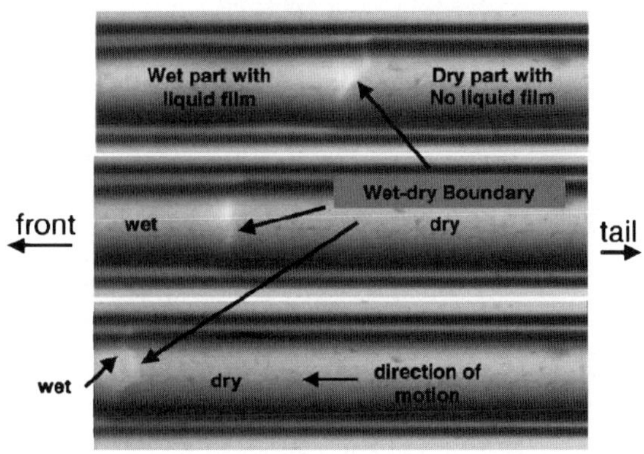

Fig. 5.10 Wet–dry boundary at the tube wall. Reprinted from Serizawa et al. (2002) with permission

surface conditions. However, at lower velocities, dry and wet areas exist between gas slug and the tube wall.

5.3.3 Probability of Appearance of Different Flow Patterns

The objectives of the study by Kawahara et al. (2002) were to experimentally investigate the probability of appearance of different flow patterns in a circular microchannel. The test section was a circular transparent channel made of fused silica with an internal diameter of 100 μm and length of 64.5 mm, providing an L/d ratio of 645.

At a given flow condition, different flow patterns were observed which can be classified into five distinct patterns depending on the interfacial configuration: liquid alone (or liquid slug), gas core with a smooth thin liquid film, gas core with a smooth thick liquid film, gas core with a ring-shaped liquid film, and gas core with a deformed interface.

For all flow conditions tested in that study, a bubbly flow pattern with bubbles much smaller than the channel diameter (100 μm) was never observed. While liquid-only flows (or liquid slugs) containing small spherical bubbles were not observed, small droplets were observed inside gas core flows. Furthermore, no stratified flow occurred in the micro-channel as reported in previous studies of two-phase flow patterns in channels with a diameter close to 1 mm (Damianides and Westwater 1988; Fukano and Kariyasaki 1993; Triplett et al. 1999a; Zhao and Bi 2001a).

In developing a two-phase flow pattern map it became clear that new flow patterns need to be defined to fully describe the flow characteristics, due to the simultaneous occurrence of different flow patterns in the channel under any given flow condition as shown in Fig. 5.11. Thus, the time fractions of different flow patterns were obtained for each two-phase flow condition, as typically shown in Figs. 5.11 and 5.12 for low and high liquid flow rates, respectively. In the case of a low liquid flow rate (Fig. 5.11), three flow patterns alternately appeared at the observation section of the micro-channel: (A) liquid alone, (B) gas core with a smooth thin liquid film, and (C) gas core with a ring-shaped liquid film. At high liquid flow rates (Fig. 5.12), two additional flow patterns were observed in addition to the three above: (D) gas core with a smooth thick liquid film, and (E) a serpentine-like gas core with a deformed liquid film. The number of images containing each flow pat-

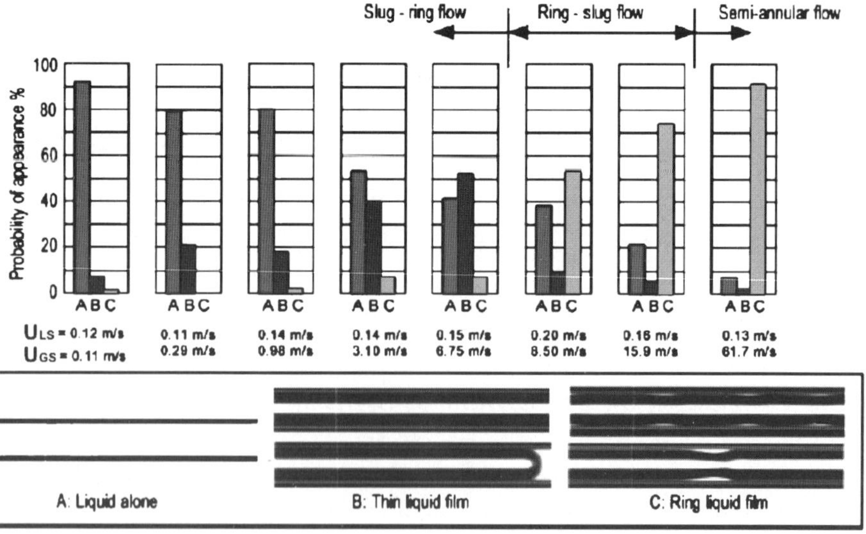

Fig. 5.11 Probability of appearance of different two-phase flow patterns at low liquid flow rates. Reprinted from Kawahara et al. (2002) with permission

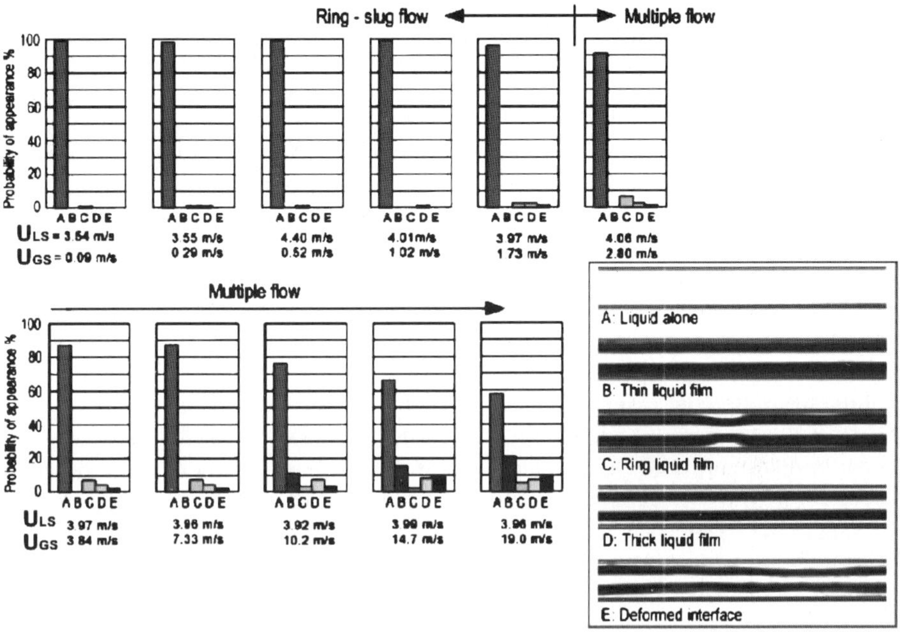

Fig. 5.12 Probability of appearance of different two-phase flow patterns at high liquid flow rates. Reprinted from Kawahara et al. (2002) with permission

tern was then counted and the probability of appearance was computed for a given flow condition. Based on these probabilities and the time-averaged void fraction for each flow condition, four flow regimes were defined as follows:

1. Slug-ring flow is the flow in which the probability of B is larger than that of C and the time-averaged void fraction is less than 0.8.
2. Ring-slug flow is the flow in which the probability of C is greater than that of B and the time-averaged void fraction is less than 0.8.
3. Semi-annular flow is the flow in which the flow alternates mostly between A and C. The time-averaged void fraction is greater than 0.8.
4. Multiple flow contains all five flow patterns, A–E, and the time-averaged void fraction is less than 0.8.

5.4 Flow Patterns in Parallel Channels

Two-phase flow in parallel pipes, fed from a common manifold, displays interesting phenomena, as two phases may split unevenly when entering the parallel piping. Ozawa et al. (1979, 1989) performed experimental studies on two-phase flow systems in parallel pipes of 3.1 mm diameter. They simulated the flow in boiling channels by injection of air and water into the pipes.

Hetsroni et al. (2003a) investigated two-phase air–water flow in 21 triangular parallel micro-channels of $d_h = 130\,\mu m$. The experimental test facility and flow loop, for liquid flowing through micro-channels is shown in Fig. 5.13.

At different flow conditions, different flow patterns were observed which can now be classified depending on the interfacial configuration according to Lowe and Rezkallah (1999): liquid alone (or single-phase flow), bubbly flow, slug flow, and annular flow (gas core with a thin liquid film, gas core with a thick liquid film). In the parallel channels having common inlet and outlet collectors, non-uniform distribution of the working fluid occurs. Tshuva et al. (1999) investigated the splitting of adiabatic two-phase flow in a system of two parallel pipes with a common feed and common exit. The results showed that the flow was non-symmetric. Hetsroni et al. (2003a) observed simultaneous different flow patterns, in different parallel micro-channels. This point is illustrated in Fig. 5.14 for superficial liquid velocity $U_{LS} = 0.13\,m/s$ and superficial gas velocities $U_{GS} = 0.63$, 1.7, and $3.1\,m/s$, respectively. In this figure four parallel channels of the test module can be seen, the flow is from left to right. The field of view is 2.4 mm in the streamwise direction and 2.2 mm in the spanwise direction.

The "liquid alone" pattern showed no entrained bubbles or gas–liquid interface in the field of view. The capillary bubbly flow, in the upper part of Fig. 5.14a, is characterized by the appearance of distinct non-spherical bubbles, generally smaller in the streamwise direction than at the base of the triangular channel. This flow pattern was also observed by Triplett et al. (1999a) in the 1.097 mm diameter circular tube, and by Zhao and Bi (2001a) in the triangular channel of hydraulic diameter of 0.866 mm. This flow, referred to by Zhao and Bi (2001a) as "capillary bubbly

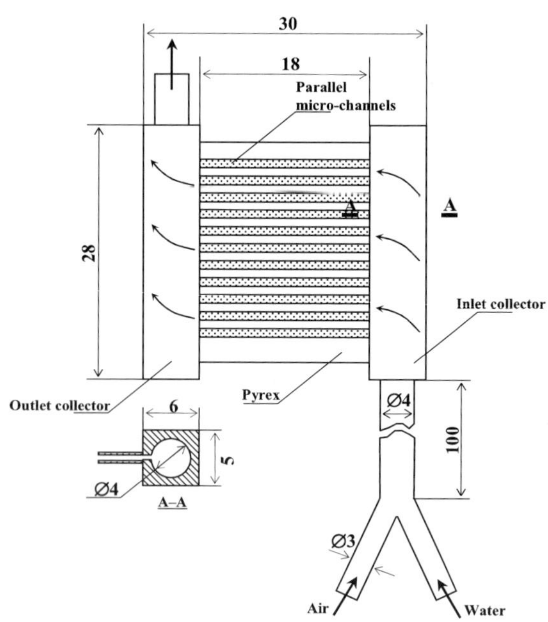

Fig. 5.13 Scheme of the test section

flow," has the following characteristics (as seen from Fig. 5.14a): the gas bubbles are regularly distributed in the liquid phase, represented by a train of bubbles, essentially ellipsoidal in shape, spanning almost the entire cross-section with their centers located along the center line of the channel.

The gas core with a thick liquid film is also shown in Fig. 5.14a, the second channel from the top. It is seen from this figure that a liquid film formed at the side walls of the channel with a continuous gas core in which a certain amount of liquid droplets exist. The flow with elongated cylindrical bubbles may be referred to as "slug" flow (Fig. 5.14b, the third channel from the top).

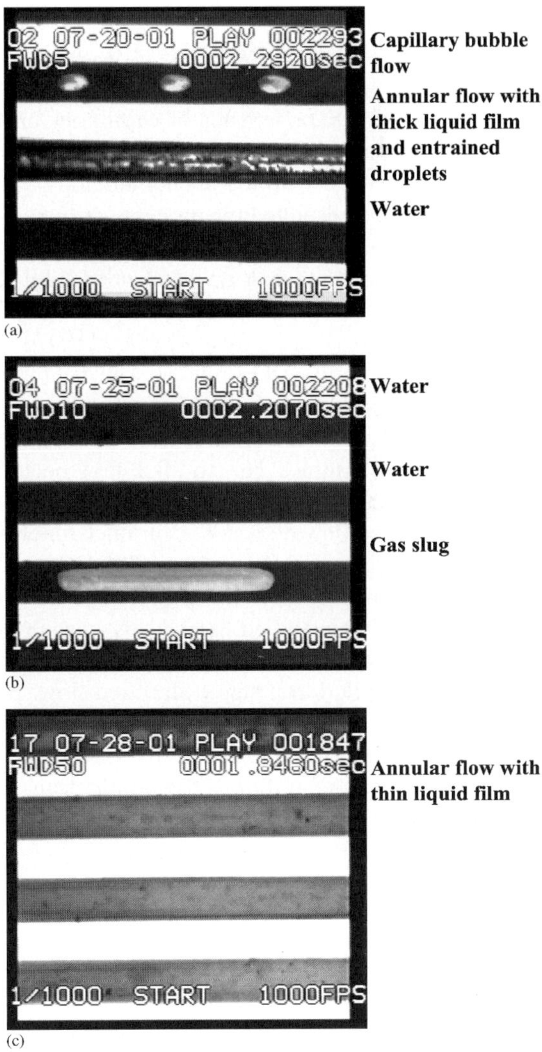

Fig. 5.14a–c Air–water flow in different parallel microchannels. $U_{LS} = 0.13$ m/s. Type B plenum (**a**) $U_{GS} = 0.63$ m/s; (**b**) $U_{GS} = 1.7$ m/s; (**c**) $U_{GS} = 3.1$ m/s. Reprinted from Hetsroni et al. (2003a) with permission

With increasing superficial gas velocity the gas core with a thin liquid film was observed. The flow pattern, displayed in Fig. 5.14c (the second, third and fourth channels from the top), indicates that as the air velocity increased, the liquid droplets entrained in the gas core disappeared such that the flow became annular.

In air–water flow the different flow patterns occur simultaneously in different micro-channels. Although the gas core may occupy almost the entire cross-section of the triangular channel, making the side walls partially dry, the liquid phase always remained continuous due to the fact that the liquid was drawn into the triangular corners by surface tension.

5.5 Flow Regime Maps

Two-phase flow characteristics of capillaries are known to be significantly different from the characteristics of larger channels, and consequently the existing vast literature associated with the phenomenology of change-of-phase heat transfer and two-phase flow hydrodynamic processes generally do not apply to capillaries.

Suo and Griffith (1964) performed experiments in horizontal channels with 0.5 and 0.7 mm diameters, and could identify slug, slug-bubbly, and annular flow patterns. For transition from slug to slug-bubbly, they suggest: $\mathrm{ReWe} = 2.8 \times 10^5$, where $\mathrm{Re} = \rho_\mathrm{L} d U_\mathrm{B}/2\mu_\mathrm{L}$, $\mathrm{We} = d\rho_\mathrm{L} U_\mathrm{B}^2/2\sigma$, $U_\mathrm{B} = 1.2(U_\mathrm{LS} + U_\mathrm{GS})$. Re and We represent Reynolds and Weber numbers, respectively, d is the channel diameter, U_LS and U_GS are liquid and gas superficial velocities, respectively, ρ_L and μ_L are the liquid density and viscosity, respectively, σ is the surface tension, and U_B is the velocity of the gas bubbles. No stratified flow occurred in the experiments of Suo and Griffith (1964), and subsequent studies have confirmed that, with gas and water-like liquids, stratified flow does not occur in channels with diameters smaller than about 1 mm (Damianides and Westwater 1988; Fukano and Kariyasaki 1993).

Transition from stratified to slug flow patterns is an important flow regime transition and has been studied extensively in the past. A widely used semi-analytical model, due to Taitel and Dukler (1976), is based on a Kelvin–Helmholtz instability criterion applied to a stable stratified flow. Barnea et al. (1983) argued that in micro-channels the predominance of surface tension on gravitational force, and not the Kelvin–Helmholtz type instability, is responsible for flow regime transition from stratified to slug. Accordingly, the regime transition was shown to occur when the liquid depth, H_L found from the solution of one-dimensional steady-state stratified momentum equations satisfied the following condition:

$$d - H_\mathrm{L} \leq \frac{\pi}{4}\left[\frac{\sigma}{\rho_\mathrm{G}\left(1-\frac{\pi}{4}\right)}\right]^{1/2}. \tag{5.6}$$

Furthermore, when d is smaller than the right-hand side of the above equation, the flow transition criterion is, according to Barnea et al. (1983):

$$H_L \geq \left(1 - \frac{\pi}{4}\right) d \,. \tag{5.7}$$

The criterion of Barnea et al. (1983) predicts well flow patterns of air and water in channels several millimeters in diameter. For smaller channels, however, the criteria of Taitel and Dukler (1976) and Barnea et al. (1983) are not suitable.

Fukano and Kariyasaki (1993) reported bubbly, intermittent, and annular regimes, and noted poor agreement with the flow regime map of Mandhane et al. (1974). Furthermore, among the semi-analytical flow regime transition models of Taitel and Dukler (1976) and Taitel et al. (1980), only the following relation for transition to dispersed bubbly flow agreed with their data:

$$U_{LS} + U_{GS} = 4.0 \left(\frac{d^{0.429} (\sigma/\rho_L)^{0.089}}{\nu_L^{0.072}} \left[\frac{g(\rho_L - \rho_G)}{\rho_L} \right]^{0.446} \right) \tag{5.8}$$

where ν_L represents the liquid kinematic viscosity.

5.5.1 Circular Channels

A comparison between the flow patterns observed by Triplett et al. (1999a) and the flow regime map of Mandhane et al. (1974), which is representative of flow patterns in larger channels, show poor agreement, confirming previously reported observations (Fukano and Kariyasaki 1993). The experimental flow patterns obtained in that study are compared with the correlation of Suo and Griffith (1964), and the mechanistic model of Taitel et al. (1980) for the flow pattern transition line leading to a dispersed bubbly flow pattern (Eq. 5.8). The expression suggested by Suo and Griffith (1964) significantly disagrees with the data of Taitel et al. (1980). The model of Taitel et al. (1980) satisfactorily predicts the bubbly-slug transition line for all four test sections, and in particular for the circular test sections.

The flow patterns for the 1.1 mm diameter circular test section observed by Triplett et al. (1999a) are compared with the experimental flow pattern transition lines of Damianides and Westwater (1988), representing their circular, 1 mm inner diameter test section, in Fig. 5.15. The flow pattern names displayed in the figure represent the notation of Damianides and Westwater (1988). The latter authors appear to have identified churn flow as a dispersed flow pattern. The two data sets are in relative agreement with respect to slug and slug-annular (referred to as plug and slug, respectively, by Damianides and Westwater 1988) flow patterns and the flow conditions leading to annular flow.

The experimental data obtained by Triplett et al. (1999a) are also in satisfactory overall agreement with similar experimental data of Fukano and Kariyasaki (1993), when inconsistencies associated with the identified flow patterns are removed. These experimental data, however, have all been obtained with air and water. Similar experiments using other fluids are recommended in order to examine the effects of their properties on flow patterns.

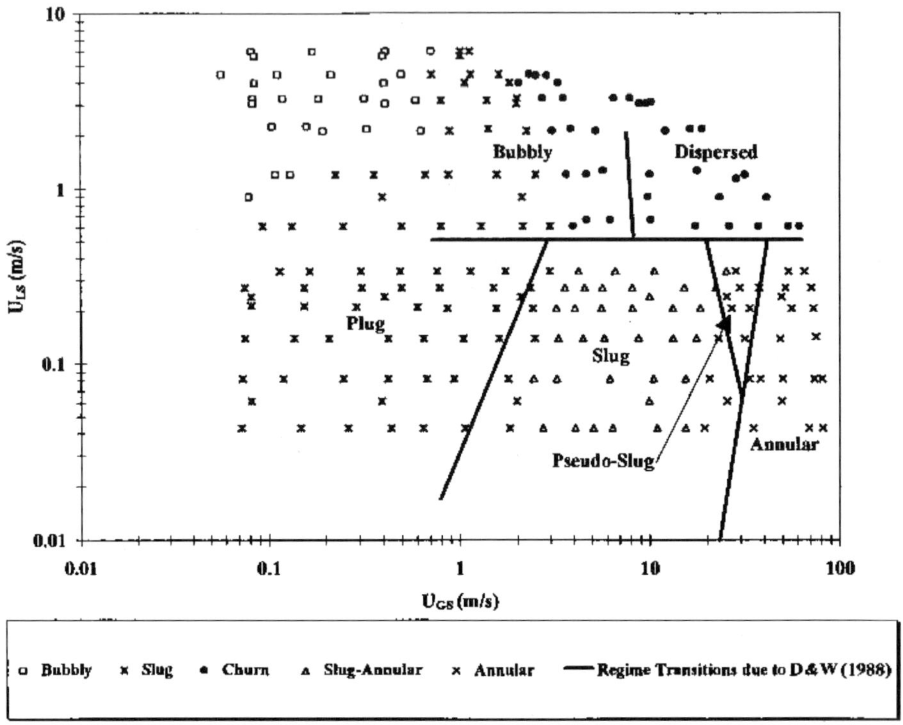

Fig. 5.15 Comparison between the experimental flow patterns obtained by Triplett et al. (1999a) and the experimental flow regime transition lines of Damianides and Westwater (1988) representing their 1 mm diameter circular test section. Reprinted from Triplett et al. (1999a) with permission

5.5.2 Triangular and Rectangular Channels

Zhao and Bi (2001a) presented the flow regime maps for the three tested triangular channels. The results are shown in Fig. 5.16. In these figures, various flow patterns such as dispersed bubbly flow, capillary bubbly flow, slug flow, churn flow, and annular flow are denoted, respectively, by the symbols D, B, S, C, and A. The flow regime maps, using gas and liquid superficial velocities as coordinates, were obtained for liquid velocities in the range 0.08–10 m/s, and air velocities in the range 0.1–100 m/s.

The flow regime maps shown in Fig. 5.16a,b indicate that typical flow patterns encountered in the conventional, large-sized vertical circular tubes, such as bubbly flow, slug flow, churn flow and annular flow, were also observed in the channels having larger hydraulic diameters ($d_h = 2.886$ and 1.443 mm).

Figure 5.16c indicates that as the channel size was reduced to $d_h = 0.866$ mm, the dispersed bubbly flow pattern vanished from the flow regime map. Figure 5.16a–c indicates that the slug-churn flow transition line shifted to the right, as the channel size was reduced. Similar trends were also found in small circular tubes by the

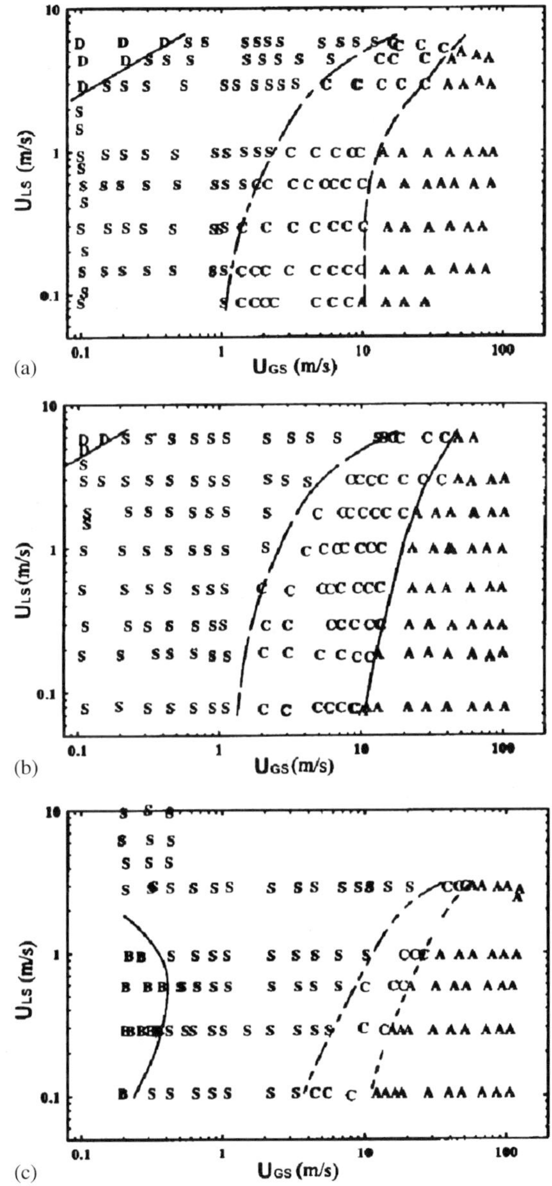

Fig. 5.16a–c Flow regime maps for upward flow in conventional size triangular channels: (**a**) side length is 5 mm ($d_h = 2.886$ mm); (**b**) side length is 2.5 mm ($d_h = 1.443$ mm); (**c**) side length is 1.5 mm ($d_h = 0.886$ mm). Reprinted from Zhao and Bi (2001a) with permission

previous investigators (for example, Barnea et al. 1983; Galbiati and Andreini 1992; Mishima and Hibiki 1996; Ide et al. 1997). However, it should be pointed out that for large circular tubes the slug-churn flow transition line shifted to the left, as the tube diameter was decreased (Taitel et al. 1980; Mishima and Ishii 1984). It has also been reported that for large circular tubes, the transition from churn to annular flow was not affected by tube diameters (Taitel et al. 1980) or the churn-annular transition line shifted to the left as the tube diameter was decreased (Mishima and Ishii 1984). Apparently, all these findings by previous investigators for large circular tubes are in conflict with the results presented by Zhao and Bi (2001a).

The comparisons between previous theories and the experimental data obtained by Zhao and Bi (2001a) suggest that both the model by Taitel et al. (1980) and the model by Mishima and Ishii (1984) significantly deviate from experimental data of the flow regime transitions in triangular channels. The discrepancies might be attributed to the fact that previous phenomenological models were developed for moderate and large circular channels, in which the effect of surface tension is insignificant. For triangular channels, however, the meniscus effect induced by the sharp corners may play an important role in the flow regime transitions, especially at high gas void fractions.

Qu et al. (2004) presented adiabatic channel two-phase flow pattern maps. Figure 5.17 depicts these flow patterns and the flow pattern transition boundaries in the flow pattern map utilizing the superficial velocities of nitrogen and water, respectively, as coordinates. At low values of U_{LS}, the flow is slug or bubbly/slug for low U_{GS}, but predominantly annular for high U_{GS}. For high U_{LS}, the flow transition from liquid/slug to liquid/annular occurs with increasing U_{GS}.

Mandhane et al. (1974) developed one of the most popular flow pattern maps for horizontal adiabatic two-phase flow in macro-channels. Figure 5.18a shows significant discrepancy between their map and the data of Qu et al. (2004). Most notably, certain flow patterns in their map (stratified, wavy, dispersed) were never observed by Qu et al. (2004). Furthermore, the conventional size channel data show an order of magnitude shift in U_{GS} values corresponding to the transition to annular flow. These unique features of adiabatic two-phase conventional size channel flow can be attributed to several factors, including strong effects of surface tension and negligible gravitational effects. In addition, low flow rates in conventional size channels

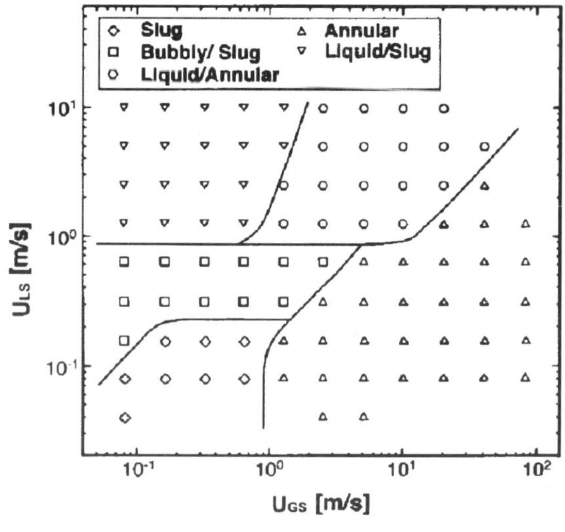

Fig. 5.17 Adiabatic micro-channel two-phase flow pattern map. Reprinted from Qu et al. (2004) with permission

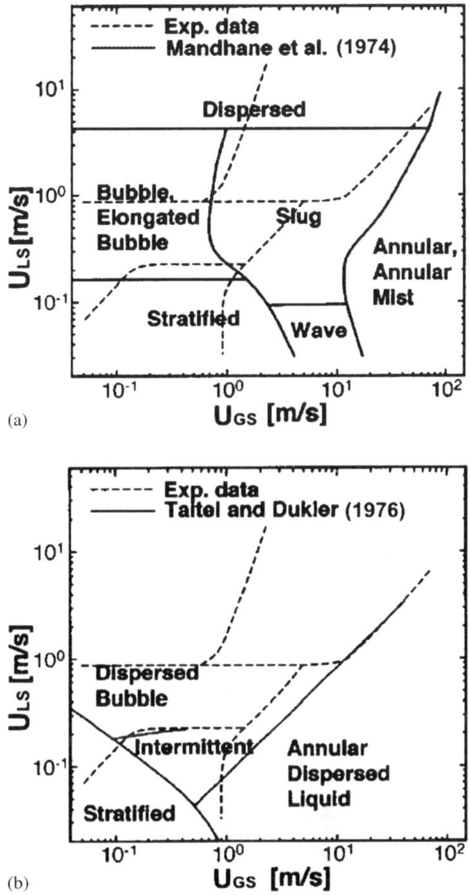

Fig. 5.18a,b Comparison of flow pattern map of Qu et al. (2004) with predictions of (**a**) Mandhane et al. (1974), (**b**) Taitel and Dukler (1976). Reprinted from Qu et al. (2004) with permission

often produce laminar liquid and gas flows, a combination not commonly encountered in macro-channels. Figure 5.18b shows a significant departure of the Taitel and Dukler models (1976) from the data of Qu et al. (2004), especially at low U_{GS}. However, predictions of the transition to annular flow are reasonably accurate.

5.6 Flow Regime Maps in Micro-Channels

There has been relatively little work done on the development of two-phase flow regime maps for micro-channels. The general trends of how the transition lines are shifted as the diameter is decreased are unclear. Figure 5.19 shows a flow pattern map obtained for air–water two-phase flow in a 20 μm i.d. silica tube by Serizawa et al. (2002) at nearly atmospheric pressure.

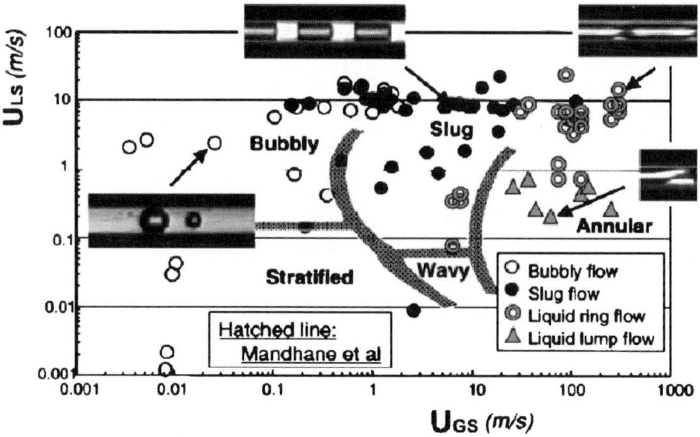

Fig. 5.19 Two-phase flow pattern transition. Reprinted from Serizawa et al. (2002) with permission

Figure 5.20 shows the overall two-phase flow regime map developed for micro-channels by Kawahara et al. (2002). The lines represent the boundaries at which flow regime transition occurred among the four flow patterns, i.e., slug-ring, ring-slug, multiple and semi-annular flows.

This flow regime map can be compared with other maps developed previously for air–water two-phase flow in small-diameter horizontal and vertical channels. Figure 5.21a–d shows comparisons with the results of Damianides and Westwater (1988), Fukano and Kariyasaki (1993), Triplett et al. (1999a) and Zhao and Bi (2001a), respectively. The solid lines represent the flow regime transition boundaries observed in the ~1 mm diameter channels and the flow regime names in parentheses are those given by the respective authors.

Damianides and Westwater (1988) used a 1 mm inner diameter, circular horizontal channel, and identified five flow regimes, i.e., bubbly, plug, slug, pseudo-slug,

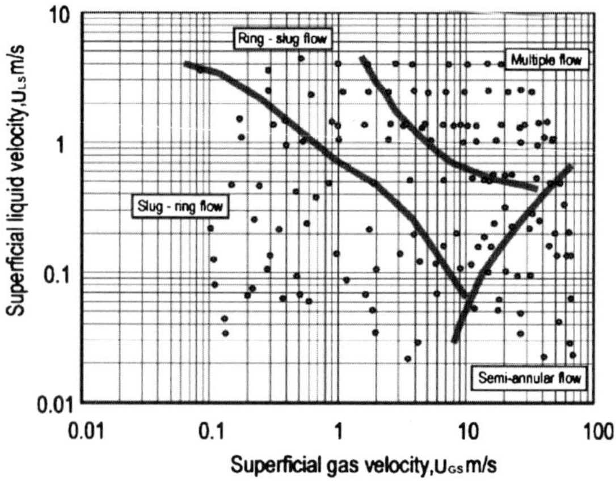

Fig. 5.20 Two-phase flow regime map for a 100 μm micro-channel. Reprinted from Kawahara et al. (2002) with permission

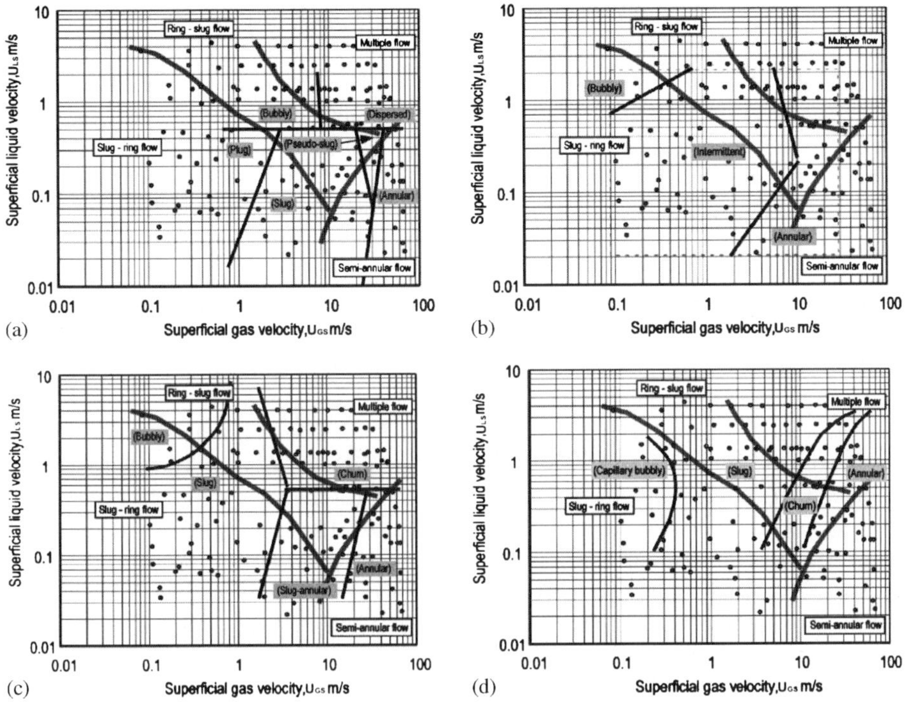

Fig. 5.21a–d Comparison of two-phase flow regime maps presented by Kawahara et al. (2003) with: (**a**) Damianides and Westwater (1988), (**b**) Fukano and Kariyasaki (1993), (**c**) Triplett et al. (1999a) and (**d**) Zhao and Bi (2001a). Reprinted from Kawahara et al. (2002) with permission

dispersed and annular flows. Fukano and Kariyasaki (1993) identified three flow regimes, namely, bubbly, intermittent and annular flows, in a horizontal capillary tube having a 1 mm inner diameter. Triplett et al. (1999a) distinguished five flow regimes in a horizontal circular channel having a 1.1 mm inner diameter: bubbly, slug, churn, slug-annular and annular flows. Most recently, Zhao and Bi (2001a) observed "capillary bubbly," slug, churn and annular flows in a vertical equilateral triangular channel with a hydraulic diameter of 0.87 mm.

There are similarities and differences between the flow regime map presented by Kawahara et al. (2002) and other flow regime maps obtained in ~1 mm diameter channels. While intermittent flow patterns such as plug and slug flow, and annular (or semi-annular) flow patterns have been observed in both the micro-channel of $d = 100\,\mu m$ and $d \sim 1$ mm diameter channels, bubbly and churn flow patterns are absent in the micro-channel results. Differences in the locations of the transition boundaries are also found, but some variations already exist among the flow regime maps for ~1 mm diameter channels, so this aspect will not be pursued in detail. Instead, we will focus on the significant differences concerning the bubbly flow and multiple flow patterns. The multiple flow pattern has been observed by Kawahara et al. (2002) in a micro-channel at high liquid and gas flow rates, with alternating occurrences of liquid alone and gas core flows with different liquid film geometries. In ~1 mm diameter channels, both Zhao and Bi (2001a) and Triplett et al. (1999a) observed a churn flow under the similar flow conditions. The churn flow involves a highly agitated gas–liquid interface, indicative of the high turbulence level in the two-phase mixture. Although slight interface deformation was observed in the serpentine-like gas core, the degree of interface deformation seen in the multiple flow pattern in the present micro-channel is incomparably smaller than that obtained in churn flow.

The bubbly flow pattern always appears in larger channels typically at high liquid and low gas flow rates, due to turbulence in the liquid breaking up the gas phase into small bubbles. However, bubbly flow was not observed at all in the 100 μm micro-channel within the present flow conditions, even at high superficial liquid velocities up to 4 m/s. The lowest superficial gas velocity covered in the present work was ~0.1 m/s, and it may not be considered as sufficiently low for micro-channels. Serizawa and Feng (2001) showed images of very small bubbles and a chain of short bubbles in micro-channels under certain conditions. Also, in flow boiling situations, small bubbles nucleating in a heated micro-channel may be swept up by high-speed liquid flow to form a bubbly flow pattern. However, when conducting experiments of steady two-phase flow at low superficial gas velocities ($U_{GS} \sim 0.1$ m/s) the measurement of such extremely small volumes of gas continuously injected into the micro-channel is somewhat difficult. The gas subsequently expands within the micro-channel due to the large pressure drop, especially at high liquid flow rates.

The absence of the bubbly and churn flow patterns may be mainly attributed to the laminar nature of the liquid flow in the micro-channel, even at the highest superficial liquid velocity tested. For example, the liquid Reynolds number based on $U_{LS} = 4$ m/s is only about 400 for water flowing through a 100 μm micro-channel. Even with the injection of gas flow, not enough turbulence can be induced and sustained in the liquid to break up the gas phase into small bubbles with diameters less than the channel diameter. Thus, from a liquid turbulence point of view, the bubbly flow regime is not expected to occur in micro-channels except at much higher superficial liquid velocities, possibly over 20 m/s, and sufficiently low gas velocities. This would mean that the frictional pressure drop would be so large that the bubbly flow regime could only occur over very short distances in the micro-channels.

5.7 Void Fraction

5.7.1 Void Fraction Definition and Correlations

There are two different methods to experimentally define void fraction. The first is based on analyzing the images of gas–liquid interfaces recorded in the observation window of the channel (Triplett et al. 1999b; Serizawa et al. 2002; Kawahara et al. 2002). For example, in experiments by Kawahara et al. (2002) at low liquid flow rates, most of the recorded images showed either liquid flowing alone or a gas core flow with a smooth thin liquid film or ring-shaped liquid film. For the latter flow pattern, the void fraction was assumed to be equal to unity. By counting the number of images containing each flow type, the time-averaged void fraction was determined.

The second method to calculate of void fraction is based on gas velocity measurement. A high-speed motion analyzer was employed by Zhao and Bi (2001b) to visualize bubble behavior in the test sections.

For slug flow in large-sized channels, the liquid bridges usually contain a number of small isolated gas bubbles, but for slug flow through miniature triangular channels, Zhao and Bi (2001b) found that, in most cases, almost no isolated gas bubbles existed in the liquid bridges between the relatively long gas slugs. This fact suggests that the slug bubble velocity can be regarded as the gas velocity in the slug flow regime. According to the drift-flux model given by Wallis (1969)

$$u_g = U_{GS}/\alpha = C_0 j + V_b \tag{5.9}$$

where α is the void fraction, $j = U_{GS} + U_{LS}$, C_0 the distribution parameter, and V_b is the so-called drift velocity (Benjamin 1968). For slug flow in a large circular tube, it has been shown by Ishii (1977) that

$$C_0 = 1.2 - 0.2\sqrt{\rho_G/\rho_L} \tag{5.10}$$

$$V_b = 0.35\sqrt{\Delta\rho g d/\rho_L} \tag{5.11}$$

where d is the tube diameter, g the acceleration due to gravity, and ρ is the fluid density, with the subscripts G and L denoting the quantities of gas and liquid.

Zhao and Bi (2001b) concluded that the drift-flux model with zero drift velocity and $C_0 = 1.2 - 0.2\sqrt{\rho_G/\rho_L}$ agrees with the measured gas velocities for the three tested miniature channels.

Homogeneous flow may be described using volumetric gas quality.

For homogeneous flow the void fraction is:

$$\alpha = \frac{\frac{x}{\rho_G}}{\frac{x}{\rho_G} + \frac{(1-x)}{\rho_L}} \tag{5.12}$$

where x is the volumetric quality.

Empirical void fraction correlations, relating to the Lockhart–Martinelli factor X and from there to volumetric quality x, have been provided by Lockhart and Martinelli (1949), Baroczy (1963), Wallis (1969), and have been discussed by Butterworth (1975) and Chen and Spedding (1983). Butterworth (1975) showed that Lockhart and Martinelli's correlation (1949) for void fraction, as well as several other void fraction correlations, can be represented in the following generic form:

$$\frac{1-\alpha}{\alpha} = A\left(\frac{1-x}{x}\right)^p \left(\frac{\rho_G}{\rho_L}\right)^q \left(\frac{\mu_L}{\mu_G}\right)^r \tag{5.13}$$

where $A = 0.28$, $p = 0.64$, $q = 0.36$ and $r = 0.07$ for Lockhart and Martinelli (1949), and $A = 1$, $p = 0.74$, $q = 0.65$ and $r = 0.13$ for a correlation due to Baroczy (1963).

5.7.2 Experiments in Conventional Size Channels

Void fraction and two-phase frictional pressure drop in conventional size channels were experimentally investigated by Triplett et al. (1999b). Measured void fractions were compared with several correlations, and shown in Figs. 5.22a,b and 5.23. These figures indicate that the homogeneous flow model provides a satisfactory prediction of the channel void fractions at low U_{GS} values, corresponding to bubbly and slug two-phase flow patterns. For churn and annular flow patterns, where relatively significant interface slip is likely due to the separation of the liquid and gas phases, the homogeneous flow model evidently overpredicts the experimental data by large margins. The correlations of Butterworth (1975) also predict well in bubbly and slug flow patterns.

It was shown by Zhao and Bi (2001b) that the void fractions for slug flow in the triangular channels of $d_h = 0.87–2.9$ mm can be obtained from the measured

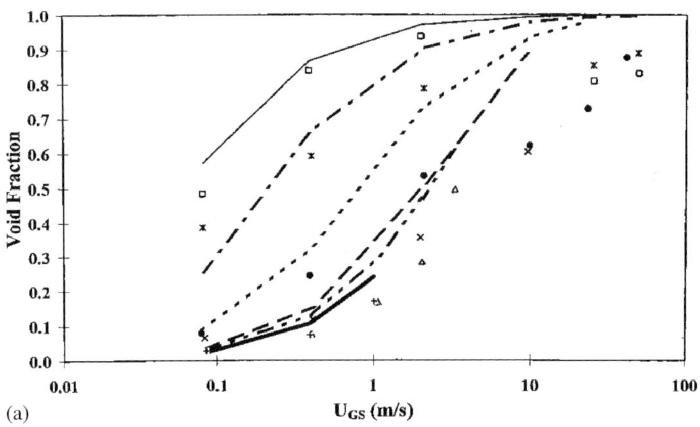

(a)

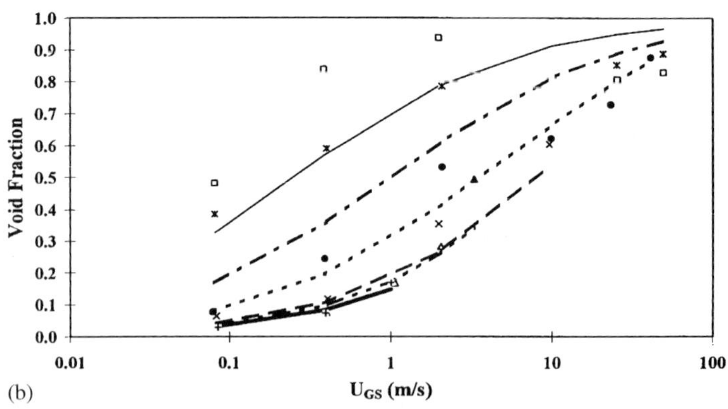

(b)

Fig. 5.22a,b Comparison of measured void fractions by Triplett et al. (1999b) for circular test section with predictions of various correlations: (**a**) homogeneous flow model; (**b**) Lockhart–Martinelli–Butterworth (Butterworth 1975). Reprinted from Triplett et al. (1999b) with permission

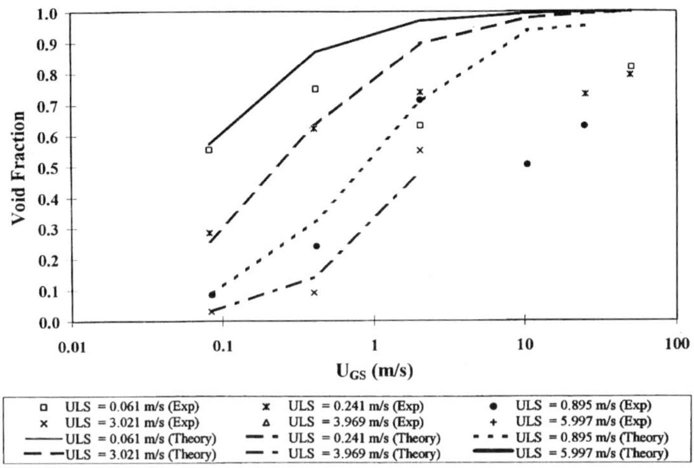

Fig. 5.23 Comparison of measured void fractions by Triplett et al. (1999b) for a trapezoid test section with predictions of various correlations for the homogeneous flow model. Reprinted from Triplett et al. (1999b) with permission

superficial gas and liquid velocities. Figure 5.24 presents the void fractions for the tested channels in terms of the volumetric quality defined as

$$\beta = \frac{U_{GS}}{U_{GS} + U_{LS}} \quad \text{and} \quad \alpha = 0.838\beta\,. \tag{5.14}$$

Note that Eq. (5.14) is very close to $\alpha = 0.833\beta$ for large circular tubes given by Armand (1946). Equation (5.14) is compared with the experimental data in Fig. 5.24. It is evident from Fig. 5.24 that the experimental data for the three tested channels can be best approximated by Eq. (5.14), 95% of the data falling within the deviation of 10% when $\beta < 0.8$. Equation (5.14) may be used to obtain the pressure drop of two-phase flow through the triangular channels.

5.7.3 Experiments in Micro-Channels

Micro-channels on the order of a few microns to a few hundred microns

An experimental investigation has been carried out by Kawahara et al. (2002). Water and nitrogen gas were injected at superficial velocities of $U_{GS} = 0.1-60\,\mathrm{m/s}$ for gas, and $U_{LS} = 0.02-4\,\mathrm{m/s}$ for liquid.

In Fig. 5.25 the void fraction α is plotted versus a homogeneous void fraction β with different symbols used for different ranges of liquid superficial velocity U_{LS}. The void fraction can be correlated with the homogeneous void fraction:

$$\alpha = \frac{0.03\beta^{0.5}}{1 - 0.97\beta^{0.5}}\,. \tag{5.15}$$

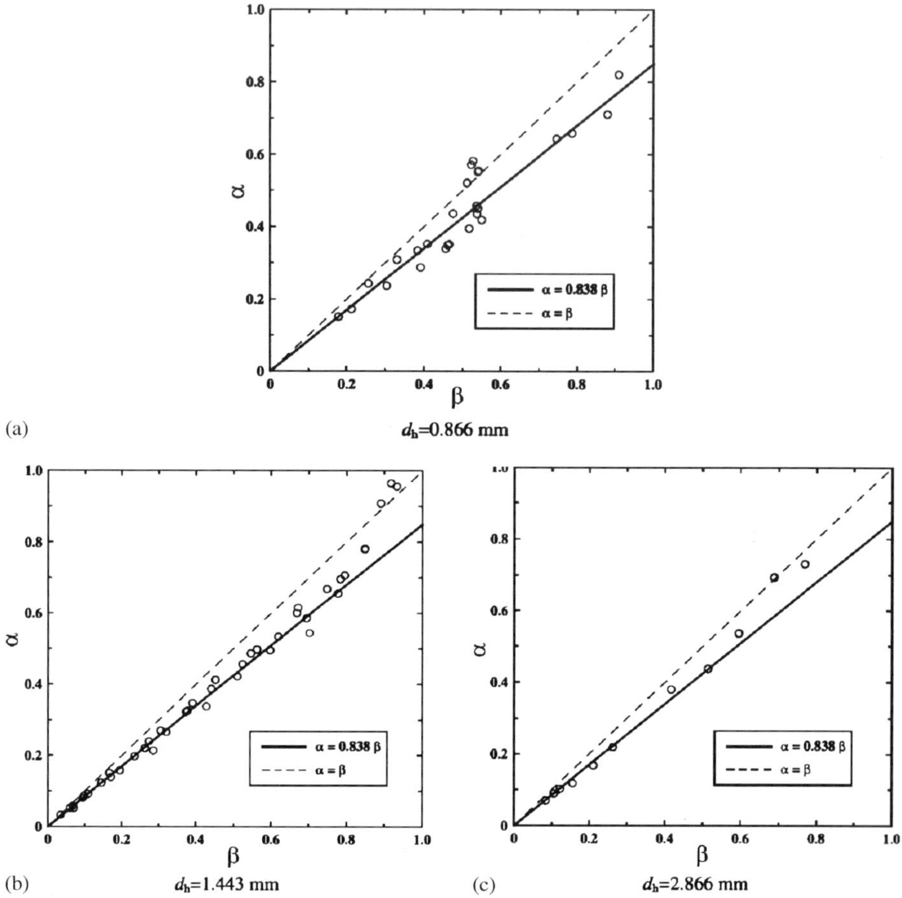

Fig. 5.24a–c Void fraction vs. mixture volumetric quality β. Reprinted from Zhao and Bi (2001b) with permission

The time-averaged void fraction remained low even for relatively high homogeneous void fractions, $\beta < 0.8$, but increased rapidly for $0.8 < \beta < 1$. The data and best-fit curve are highly non-linear indicating strong deviations from linear relations for a homogeneous flow (i.e., $\alpha = \beta$, shown by a dotted line in Fig. 5.25). For the latter, Ali et al. (1993) have reported that the void fraction in narrow channels with $d_{\mathrm{h}} \sim 1$ mm can be approximately given by an Armand-type (1946) correlation, $\alpha = 0.8\beta$, which is shown by a dashed line in Fig. 5.25.

On the other hand, in the study by Serizawa et al. (2002) the cross-sectional averaged void fraction was correlated with the Armand (1946) correlation as shown in Fig. 5.26. This trend does not contradict the data reported for conventional size channels, but it is different from results obtained by Kawahara et al. (2002). Disagreement between results of void fraction in micro-channels obtained by different investigators was shown by Ide et al. (2006) and will be discussed in the next section.

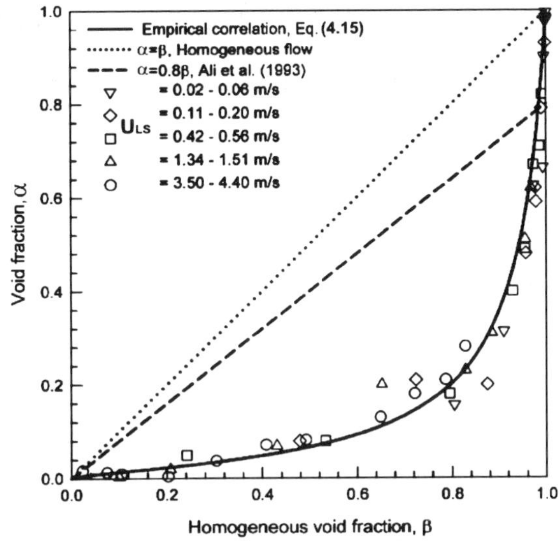

Fig. 5.25 Relation between measured and homogeneous void fractions. Reprinted from Kawahara et al. (2002) with permission

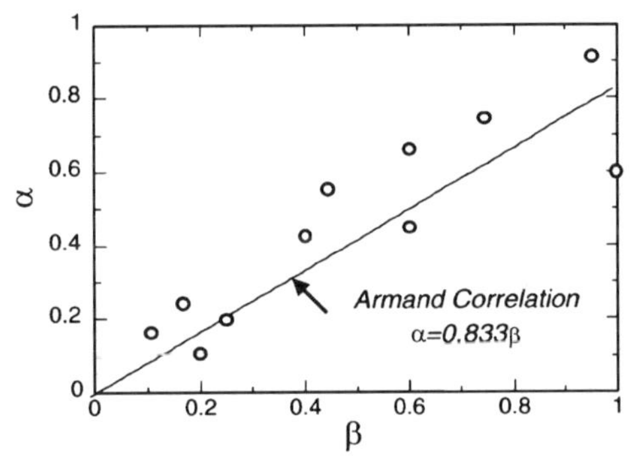

Fig. 5.26 Cross-sectional average void fraction in air–water two-phase flow in a 20 μm i.d. silica tube. Reprinted from Serizawa et al. (2002) with permission

5.8 Pressure Drop

5.8.1 Frictional Pressure Drop Correlations

Homogeneous model

The homogeneous mixture model is the simplest method for calculating the frictional two-phase pressure drop, and has been found by Ungar and Cornwell (1992) to agree reasonably well with their experimental data representing the flow of two-phase ammonia in channels with $d = 1.46–3.15$ mm.

The two-phase frictional pressure gradient is obtained from:

$$\left(-\frac{\mathrm{d}P}{\mathrm{d}z}\right)_{\mathrm{TP}} = \Phi_{\mathrm{L}}^2 \left(-\frac{\mathrm{d}P}{\mathrm{d}z}\right)_{\mathrm{L}} \tag{5.16}$$

where $(-\mathrm{d}P/\mathrm{d}z)_\mathrm{L}$ represents the frictional pressure gradient when all fluid is assumed to be liquid.

In the homogeneous flow model, ρ_H is the homogenous mixture density defined by,

$$\frac{1}{\rho_\mathrm{H}} = \frac{x}{\rho_\mathrm{G}} + \frac{1-x}{\rho_\mathrm{L}} \tag{5.17}$$

where x is the volumetric quality.

Several models have been proposed to evaluate the two-phase mixture viscosity, and the model selected may affect the predicted two-phase frictional pressure drop:

Owens (1961): $$\mu_\mathrm{H} = \mu_\mathrm{L} \tag{5.18}$$

McAdams (1954): $$\mu_\mathrm{H} = \left(\frac{x}{\mu_\mathrm{G}} + \frac{1-x}{\mu_\mathrm{L}}\right)^{-1} \tag{5.19}$$

Cicchitti et al. (1960): $$\mu_\mathrm{H} = x\mu_\mathrm{G} + (1-x)\mu_\mathrm{L} \tag{5.20}$$

Dukler et al. (1964): $$\mu_\mathrm{H} = \beta\mu_\mathrm{G} + (1-\beta)\mu_\mathrm{L} \tag{5.21}$$

Beattie and Whalley (1982): $$\mu_\mathrm{H} = \mu_\mathrm{L}(1-\beta)(1+2.5\beta) + \mu_\mathrm{G}\beta \tag{5.22}$$

Lin et al. (1991): $$\mu_\mathrm{H} = \frac{\mu_\mathrm{L}\mu_\mathrm{G}}{\mu_\mathrm{G} + x^{1.4}(\mu_\mathrm{L} - \mu_\mathrm{G})} \tag{5.23}$$

where subscripts H, G, and L denote homogeneous, gas and liquid, respectively.

The Lockhart and Martinelli model

The Lockhart and Martinelli (1949) correlation also uses a two-phase friction multiplier, defined by Eq. (5.16). The friction multiplier has been correlated in terms of the Lockhart–Martinelli parameter, X, given by

$$X^2 = \frac{(\Delta P/\Delta z)_\mathrm{L}}{(\Delta P/\Delta z)_\mathrm{G}} \tag{5.24}$$

where $(\Delta P/\Delta z)_\mathrm{G}$ is the frictional pressure drop when the gas is assumed to flow alone in the channel. A widely used correlation to calculate the friction multiplier is that proposed by Chisholm (1983)

$$\Phi_\mathrm{L}^2 = 1 + \frac{C}{X} + \frac{1}{X^2} \tag{5.25}$$

where the coefficient, C, is a constant ranging in value from 5 to 20 depending on whether the liquid and gas flows are laminar or turbulent.

The C-value in Eq. (5.25) is known to decrease as the channel hydraulic diameter is reduced (Kawaji 1999). Mishima and Hibiki (1996) have proposed a correlation for the C-value for small diameter channels as follows:

$$C = 21(1 - \mathrm{e}^{-0.319 d_\mathrm{h}}) . \tag{5.26}$$

Equation (5.26) was developed using the available air–water two-phase flow data obtained in circular and rectangular channels with $d_h = 1-4$ mm.

5.8.2 Experiments in Conventional Size Channels

Triplett et al. (1999b) used the value of two-phase liquid viscosity, suggested by McAdams (1954) and compared experimental results with the homogeneous model. The predicted pressure drops, obtained with a homogeneous pressure drop model, normalized everywhere with corresponding experimentally measured pressure drops, are displayed in Fig. 5.27 for circular channels, where the channel two-phase flow patterns are also specified. Overall, the homogeneous pressure drop model, as expected, predicts well the experimental data in bubbly and slug flow patterns and at high Re_L (where $Re_L = \rho_L U_{LS} d/\mu_L$), where the homogeneous flow assumption is well applicable. Relatively significant deviations are mostly associated with slug-annular and annular flow patterns, and slug flow at very low Re_L.

Figure 5.28 displays the predicted pressure drops, everywhere normalized with experimentally measured pressure drops for a trapezoidal channel. The model predictions in this figure were obtained by using the homogeneous flow (no-slip) assumption, and using the Friedel correlation (1979) for two-phase wall friction. Once again model predictions are in satisfactory agreement with data for bubbly and slug flow patterns at high Re_L. They, however, deviate from experimental data rather significantly at low Re_L values. Compared with the homogeneous two-phase frictional pressure drop model, the Friedel correlation (1979) evidently provides less accu-

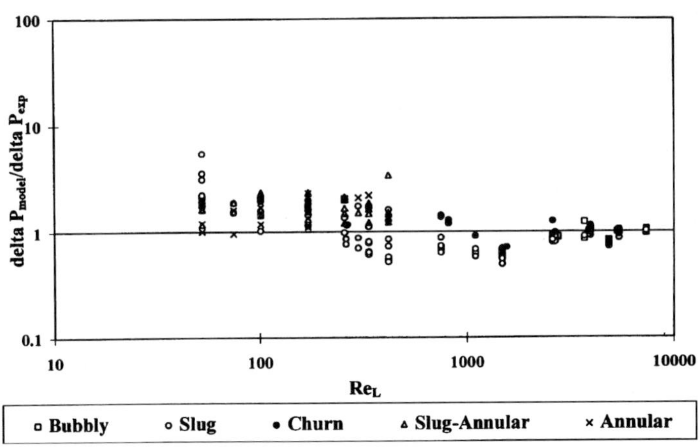

Fig. 5.27 Model-predicted pressure drops normalized with experimentally measured pressure drops for a circular test section (Triplett et al. 1999b). Model predictions represent the homogeneous wall friction model. Reprinted from Triplett et al. (1999b) with permission

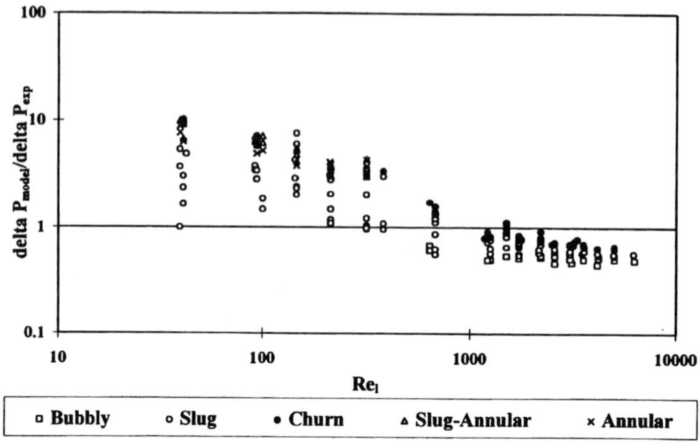

Fig. 5.28 Model-predicted pressure drops normalized with experimentally measured pressure drops for a trapezoid test section (Triplett et al. 1999b). Model predictions represent Friedel correlation (1979). Reprinted from Triplett et al. (1999b) with permission

rate predictions for the experimental data. At low $\mathrm{Re_L}$ homogeneous models did not agree well with experimental results.

Zhao and Bi (2001b) measured pressure drop in triangular conventional size channels ($d_\mathrm{h} = 0.866-2.866\,\mathrm{mm}$). The variations of the measured two-phase frictional multiplier Φ_L^2 with the Martinelli parameter X for the three miniature triangular channels used in experiments are displayed, respectively, in Fig. 5.29a–c. In Fig. 5.29 also shown are the curves predicted by Eq. (5.25) for $C = 5$ and $C = 20$. It is evident from Fig. 5.29 that the experimental data are reasonably predicted by the Lockhart–Martinelli correlation, reflected by the fact that all the data largely fall between the curves for $C = 5$ and $C = 20$, except for the case at very low superficial liquid velocities.

5.8.3 Experiments in Micro-Channels

The two-phase pressure drop was measured by Kawahara et al. (2002) in a circular tube of $d = 100\,\mu\mathrm{m}$. In Fig. 5.30, the data are compared with the homogeneous flow model predictions using the different viscosity models. It is clear that the agreement between the experimental data and homogeneous flow model is generally poor, with reasonably good predictions (within $\pm 20\%$) obtained only with the model from Dukler et al. (1964) for the mixture viscosity.

Figure 5.31 shows a comparison of the two-phase friction multiplier data with the values predicted by Eq. (5.25) with $C = 5$, for both phases being laminar, and with $C = 0.66$ given by Mishima and Hibiki's (1996) correlation. It is clear that the data correlate well using a Lockhart–Martinelli parameter, but the predictions of

Fig. 5.29a–c Two-phase frictional multiplier Φ_L^2 vs. Lockhart–Martinelli parameter X (Lockhart and Martinelli 1949). Reprinted from Zhao and Bi (2001b) with permission

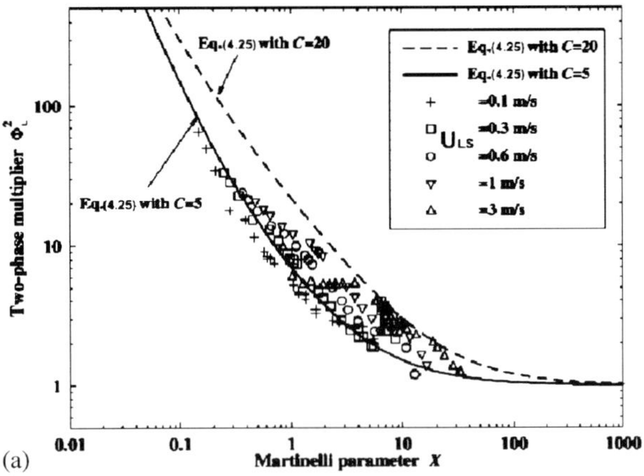

(a) d_h=0.866 mm

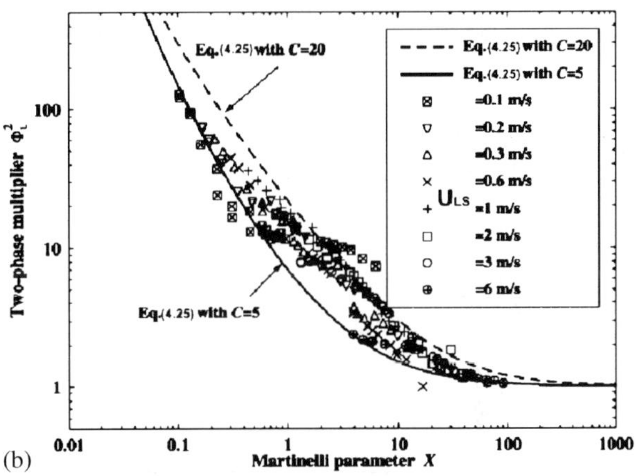

(b) d_h=1.443 mm

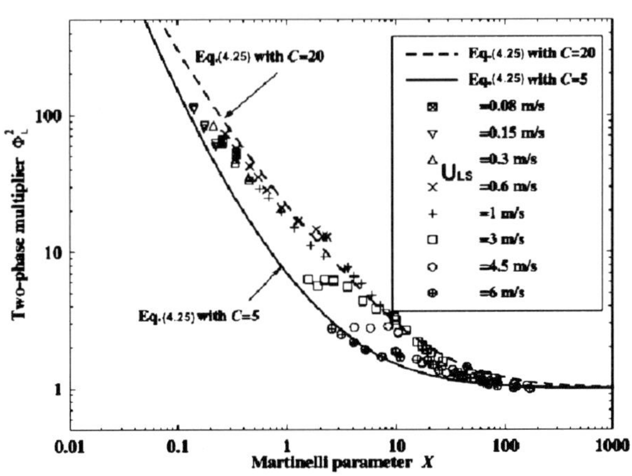

(c) d_h=2.886 mm

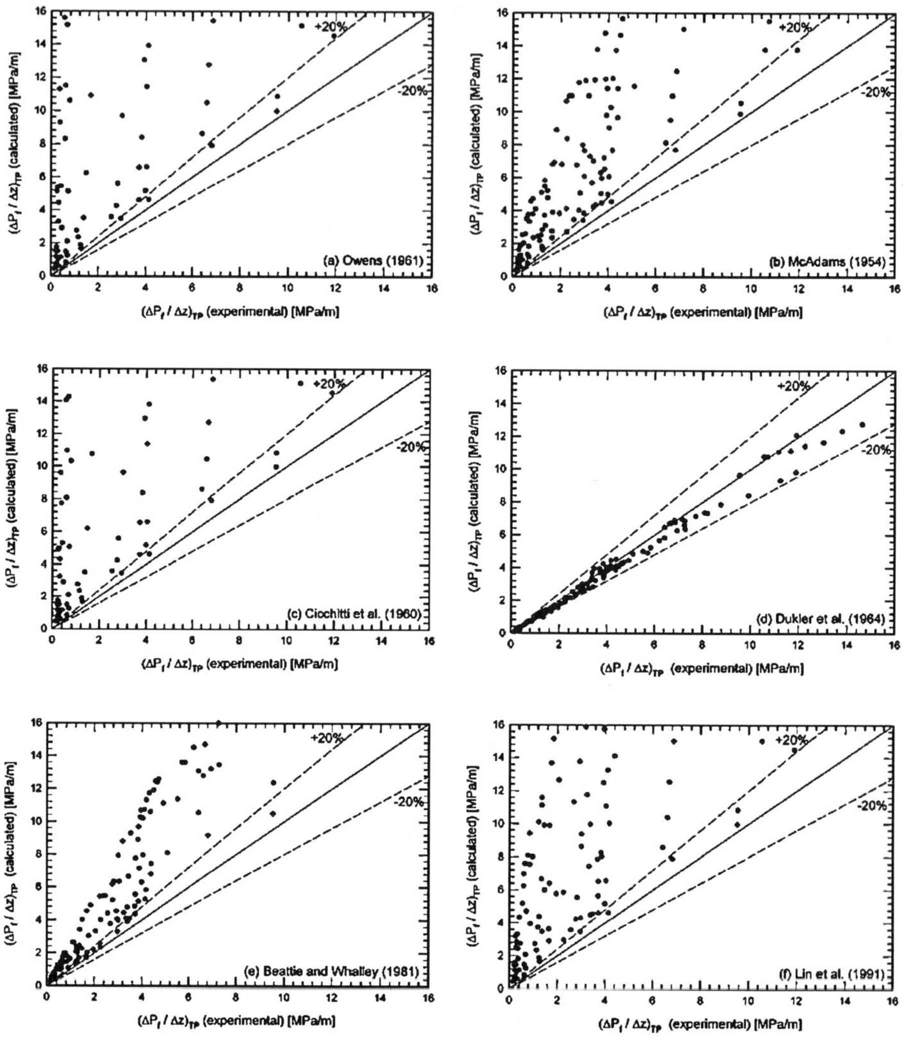

Fig. 5.30 Comparison of the two-phase frictional pressure gradient between micro-channel data and homogeneous flow model predictions using different viscosity formulations. Reprinted from Kawahara et al. (2002) with permission

Eq. (5.25) with $C = 5$ are well over the present data. On the other hand, the friction multiplier values calculated with $C = 0.66$ are slightly higher than the experimental data, while a slightly smaller value of $C = 0.24$ was found by Kawahara et al. (2002) to best fit the results.

Finally, a comparison of the two-phase frictional pressure gradient data with the predictions of the Lockhart–Martinelli correlation using different C-values is shown in Fig. 5.32, including $C = 5$, $C = 0.66$, C calculated from the Lee and Lee model (2001), and $C = 0.24$. The conventional value of $C = 5$ again significantly over-

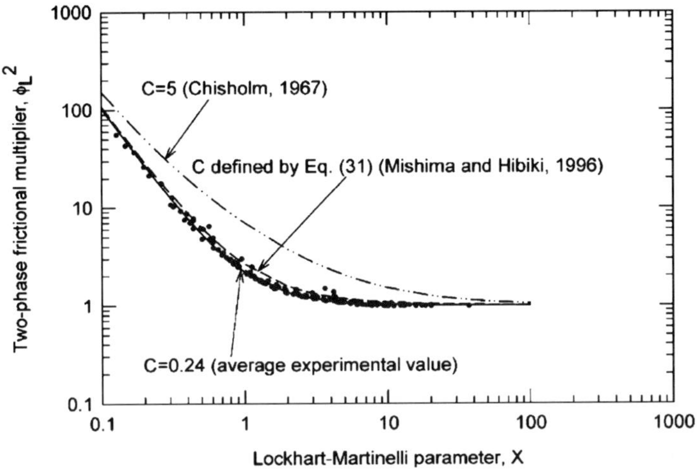

Fig. 5.31 Variation of two-phase friction multiplier data with Lockhart–Martinelli parameter. Reprinted from Kawahara et al. (2002) with permission

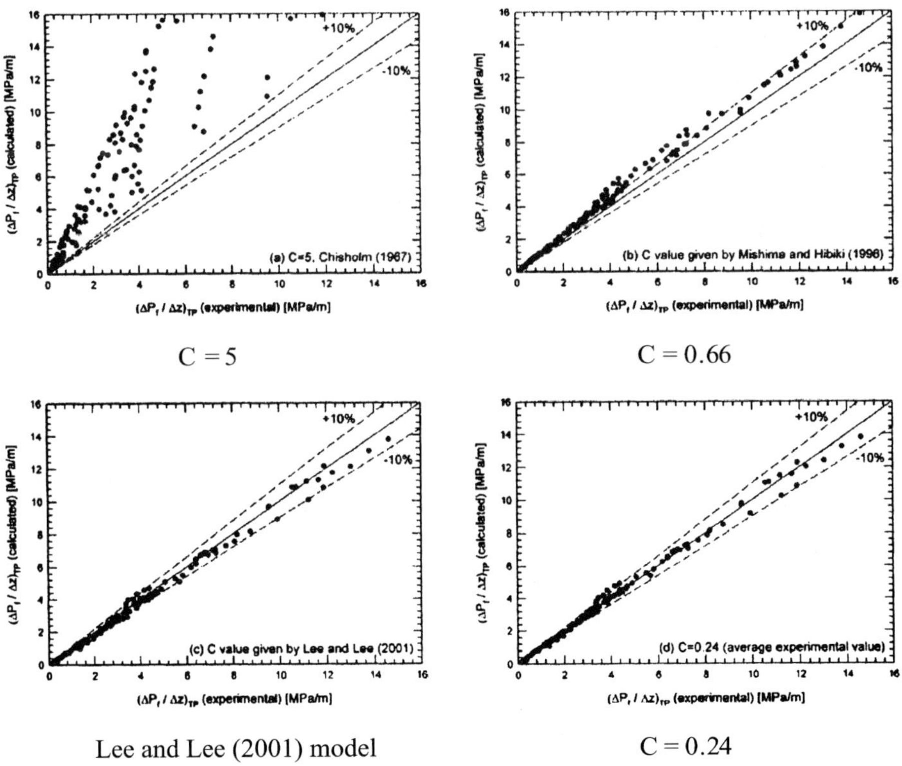

Fig. 5.32 Predictions of two-phase friction pressure gradient data by a Lockhart–Martinelli correlation with different *C*-values. Reprinted from Kawahara et al. (2002) with permission

predicted the present data, while Mishima and Hibiki's (1996) correlation generally overpredicted the data by about 10%. On the other hand, good agreement (within ±10%) was obtained with the use of the C-value given by the Lee and Lee (2001) model and the present value of $C = 0.24$.

5.9 Heat Transfer

Most of heat transfer correlations are based on data obtained in flow boiling from relatively large diameter conduits and the predictions from these correlations show considerable variability. Effects of superficial liquid and gas velocity on heat transfer in gas–liquid flow and its connection to flow characteristics were studied by Hetsroni et al. (1998a,b, 2003b), Zimmerman et al. (2006), Kim et al. (1999), and Ghajar et al. (2004). However these investigation were carried out for tubes of $D = 25-42$ mm. These data, as well as results presented by Bao et al. (2000) in tubes of $D = 1.95$ mm and results obtained by Hetsroni et al. (2001), Mosyak and Hetsroni (1999) are discussed in the next sections to clarify how gas and liquid velocities affect heat transfer. Effects of the channel size and inclination are considered.

5.9.1 *Effect of Superficial Liquid Velocity*

Conventional size channels. Inclined tubes

Experiments in annular flow were performed by Hetsroni et al. (2003b) to study the flow regimes and heat transfer in air–water flow in 8° inclined tubes of inner diameter 49.2 mm and 25 mm.

Flow in a 49.2 mm pipe. Visual observations

The purpose of these experiments was to characterize different flow details under conditions when the superficial gas velocity is constant and the superficial liquid velocity increases. The upward flow regimes are presented in Fig. 5.33. Figure 5.33a shows the stratified flow pattern at $U_{GS} = 20\,\mathrm{m/s}$ and $U_{LS} = 0.005\,\mathrm{m/s}$. In the region of pure stratified flow the liquid layer is drawn upward by the gas via the interfacial shear stress. No droplets could be observed at the interface. Such a regime was also observed by Taitel and Dukler (1976), and Spedding et al. (1998).

Droplets appeared on the surface of the pipe (Fig. 5.33b) after increasing the water flow rate up to $U_{LS} = 0.007\,\mathrm{m/s}$. Spedding et al. (1998) referred to this regime as "film plus droplet pattern." When the water flow rate increased and superficial liquid velocity was $U_{LS} = 0.03\,\mathrm{m/s}$ (Fig. 5.33c) droplets began to roll back into the liquid film. Kokal and Stanislav (1989) identified such a regime as "annular plus roll wave flow pattern." The experimental facility used in the present study allowed us to achieve values of superficial gas velocities up to 20 m/s in the 49.2 mm pipe.

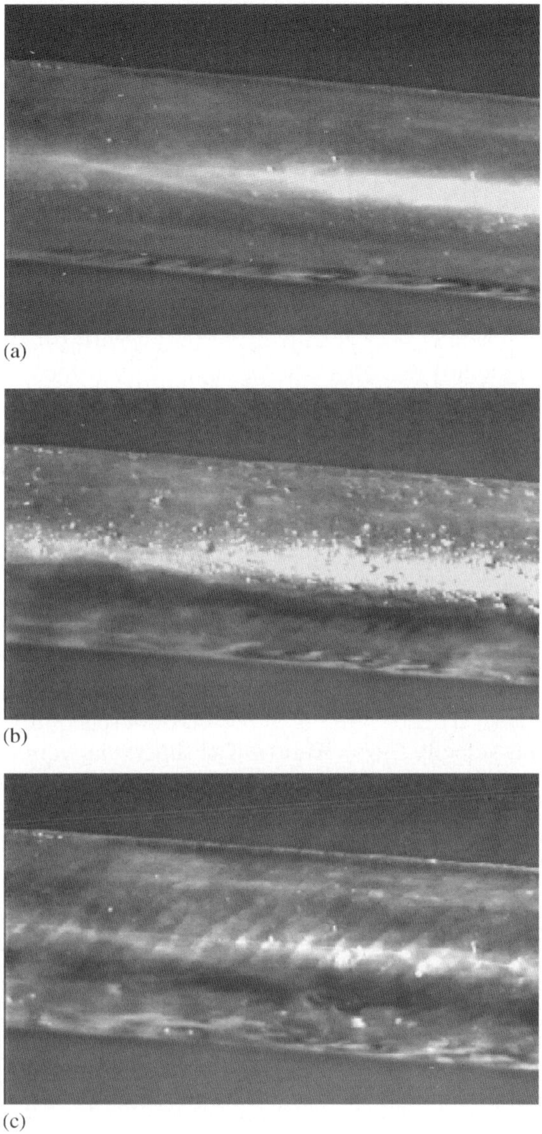

Fig. 5.33 Flow patterns in the inclined pipe 49.2 mm, $U_{GS} = 20\ \mathrm{m/s}$ (**a**) $U_{LS} = 0.005\ \mathrm{m/s}$; (**b**) $U_{LS} = 0.007\ \mathrm{m/s}$; (**c**) $U_{LS} = 0.03\ \mathrm{m/s}$. Reprinted from Hetsroni et al. (2003b) with permission

Therefore, to study the flow regimes at higher superficial gas velocities the pipe diameter was decreased.

Flow in a 25 mm pipe. Visual observations

The second part of the experiments was carried out in a pipe of inner diameter 25 mm. The experiments were carried out in the range of the $U_{GS} = 24{-}55\ \mathrm{m/s}$ and $U_{LS} = 0.0016{-}0.17\ \mathrm{m/s}$.

Experimental investigations in the 25 mm pipe reveal significant differences from the results obtained in the 49.2 mm pipe. Figure 5.34 shows an instantaneous side view of air–water flow into the tube, obtained by a high-speed video camera. The flow moves from right to left. The distance in the flow direction is 280 mm. The picture was captured at the following values of the superficial velocities $U_{GS} = 36\,\mathrm{m/s}$ and $U_{LS} = 0.045\,\mathrm{m/s}$. The light regions on the left and right side of the image are air–water clusters, the smooth light bands reflect the disturbance wave flow at the lower part of the tube and liquid film at the upper part of the tube, respectively.

It was observed that as some new waves were created, others disappeared. This process occurred either by faster waves catching up with slower waves and engulfing them or occasionally by waves appearing to die spontaneously (an event, which apparently occurs when the amplitude of the disturbance wave got too large). In Fig. 5.34 one can distinguish the regions of disturbance wave without air–water clusters and air–water clusters that occupied significant cross-section area. This flow pattern is difficult to identify since a problem existed in distinguishing visually between highly aerated slugs and wavy annular flow. This difficulty was overcome by the study of these patterns separately. Qualitative behavior of clusters showed that within a certain distance they propagated with some change in shape. Using high-speed video, it was possible to show separately the sequence of the frames with wave regimes and sequence of frames with air–water clusters.

Detailed typical instantaneous images of disturbance wave regions without air–water clusters are shown in Fig. 5.35a–d. The clusters, also observed at these regions, are shown in Fig. 5.35e–h. Both sets of images were obtained at a superficial gas velocity $U_{GS} = 36\,\mathrm{m/s}$. The superficial liquid velocity varied from $U_{LS} = 0.016$ (Fig. 5.35a,e) to $U_{LS} = 0.17$ (Fig. 5.35d,h). The flow moves from right to left, the distance in the streamwise direction, shown in those figures, is about two tube diameters. Figure 5.35a ($U_{LS} = 0.016\,\mathrm{m/s}$) shows open annular flow with disturbance waves. Figure 5.35e shows rather small air–water clusters that at some time were observed instead of the flow pattern shown in Fig. 5.35a. It was determined from visual observations that in this case the drops on the upper part of the pipe almost did not move. When the liquid flow rate increased (Fig. 5.35b,f) ($U_{LS} = 0.027\,\mathrm{m/s}$) the drops on the upper part of the pipe began to move. Velocity of the displacement of droplets in the flow direction is an order of magnitude less than the superficial liquid velocity. Further increase in the liquid flow rate (Fig. 5.35c,g) ($U_{LS} = 0.045\,\mathrm{m/s}$) leads to closed annular flow with liquid film on the upper part of the tube, in addition to air–water clusters. Figure 5.35d,h ($U_{LS} = 0.17\,\mathrm{m/s}$) shows the closed annular flow with huge air–water clusters, that often block the tube cross-section. Visual observations showed that under flow conditions, of the last two flow regimes, dryout

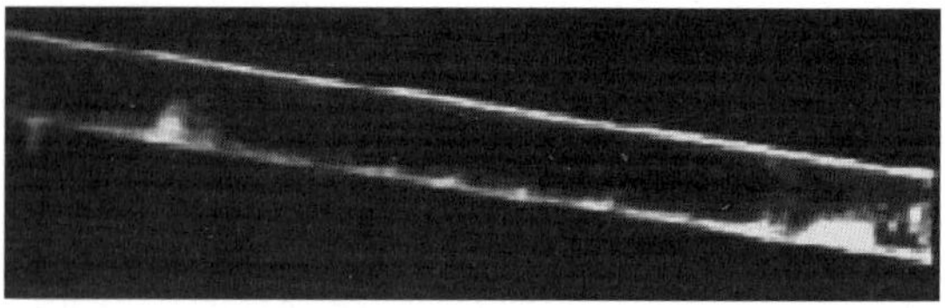

Fig. 5.34 Annular flow in the inclined pipe 25 mm with air–water clusters. Reprinted from Hetsroni et al. (2003b) with permission

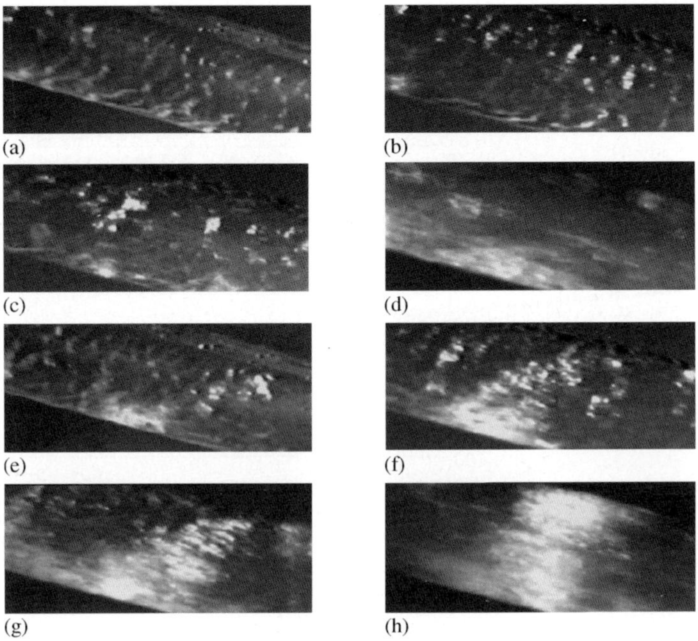

Fig. 5.35a–h Flow regimes in the pipe of 25 mm at $U_{GS} = 36$ m/s: (**a**) $U_{LS} = 0.016$ m/s, disturbance waves with motionless droplets; (**b**) $U_{LS} = 0.027$ m/s, disturbance waves with moving droplets; (**c**) $U_{LS} = 0.045$ m/s, disturbance waves and liquid film on the upper tube part; (**d**) $U_{LS} = 0.17$ m/s, disturbance air–water waves and liquid film on the upper tube part; (**e**) $U_{LS} = 0.016$ m/s, small air–water clusters; (**f**) $U_{LS} = 0.027$ m/s, air water clusters; (**g**) $U_{LS} = 0.045$ m/s, huge air–water clusters; (**h**) $U_{LS} = 0.17$ m/s, huge air–water clusters that block the tube cross-section. Reprinted from Hetsroni et al. (2003b) with permission

was not observed. Parameters of air–water clusters were obtained from histograms of velocity and amplitude distribution. Figure 5.35e–h shows typical images that were used to describe characteristics of clusters in statistical terms for air–water regimes. When the clusters blocked the entire pipe cross-section, they touched the top wall only momentarily. It is of interest to note that the picture of air–water clusters, for the 25 mm pipe, resembles the "pseudo-slugs" described by Lin and Hanratty (1987).

Heat transfer in the 49.2 mm pipe

The local heat transfer coefficients on the surface of the pipe may not be uniform, though the surface is heated by uniform heat flux. This irregularity is due to the distribution of the air and liquid phase in the pipe. The temperature distribution along the pipe perimeter shows a maximum at the top and a minimum at the bottom of the pipe. In Fig. 5.36a–c, the heat transfer coefficients h_θ are plotted versus angle θ. These results were compared to simultaneous visual observations of the flow pat-

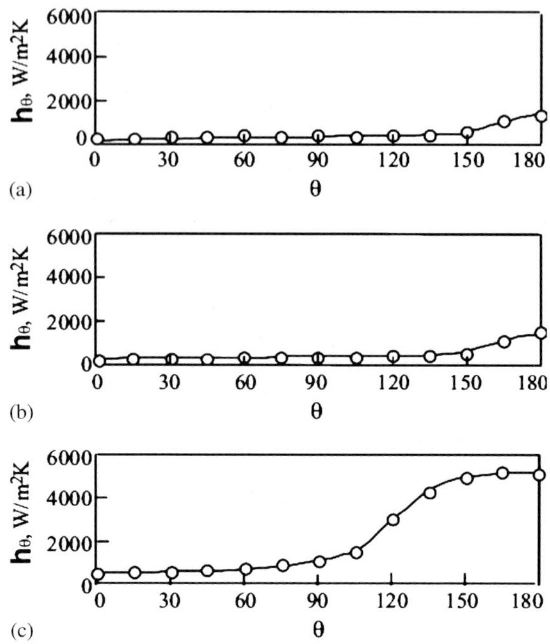

Fig. 5.36a–c Pipe 49.2 mm. $U_{GS} = 20\,m/s$. Local heat transfer coefficients: (**a**) $U_{LS} = 0.005\,m/s$; (**b**) $U_{LS} = 0.007\,m/s$; (**c**) $U_{LS} = 0.03\,m/s$. Reprinted from Hetsroni et al. (2003b) with permission

terns. The distributions of h_θ plotted in Fig. 5.36a–c correspond to conditions under which the dryout of the upper part takes place.

Heat transfer in the 25 mm pipe

Figure 5.37a–d illustrates a typical temperature distribution in the range of the angle $0 \leq \theta < 180°$ (where $\theta = 0°$ is at the top of the tube). The heat flux was $q = 8{,}000\,W/m^2$, the superficial gas velocity was $U_{GS} = 36\,m/s$. The superficial liquid velocities were 0.016, 0.027, 0.045 and 0.099 m/s, respectively. The flow moves from the right to the left. The color shades are indicative of the wall temperature. Comparison to simultaneous visual observations shows that the distribution of heat transfer coefficient at $U_{LS} = 0.0016\,m/s$ corresponds to dryout on the upper part of the pipe.

Experiments in annular and slug flow were carried out also by Ghajar et al. (2004). The test section was a 25.4 mm stainless steel pipe with a length-to-diameter ratio of 100. The authors showed that heat transfer coefficient increases with increase in liquid superficial velocity not only in annular, but also in slug flow regimes.

Horizontal pipes. Annular and slug flow

For conventional size pipes the flow regimes depend on orientation. Two-phase air–water flow and heat transfer in a 25 mm internal diameter horizontal pipe were investigated experimentally by Zimmerman et al. (2006). Figure 5.38 shows the flow

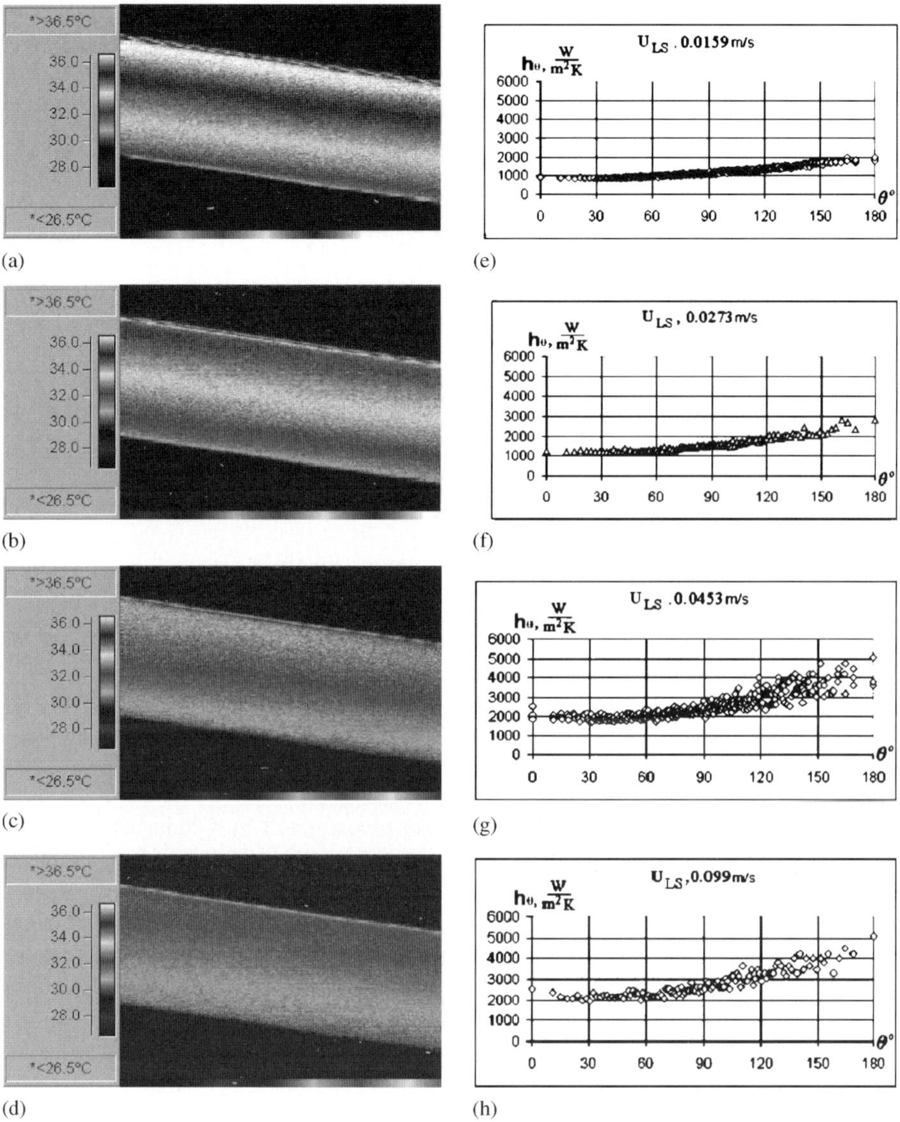

Fig. 5.37a–h Thermal patterns on the heated wall and local heat transfer coefficients, $U_{GS} = 36\ \mathrm{m/s}$, $q = 8{,}000\ \mathrm{W/m^2}$. Thermal patterns: **(a)** $U_{LS} = 0.016\ \mathrm{m/s}$; **(b)** $U_{LS} = 0.027\ \mathrm{m/s}$; **(c)** $U_{LS} = 0.045\ \mathrm{m/s}$; **(d)** $U_{LS} = 0.099\ \mathrm{m/s}$. Local heat transfer coefficient: **(e)** $U_{LS} = 0.016\ \mathrm{m/s}$; **(f)** $U_{LS} = 0.027\ \mathrm{m/s}$; **(g)** $U_{LS} = 0.045\ \mathrm{m/s}$; **(h)** $U_{LS} = 0.099\ \mathrm{m/s}$. Reprinted from Hetsroni et al. (2003b) with permission

regime map reported by Lin and Hanratty (1987) for air and water flowing in a horizontal pipe of inner diameter 25.4 mm and experimental conditions of the study by Zimmerman et al. (2006).

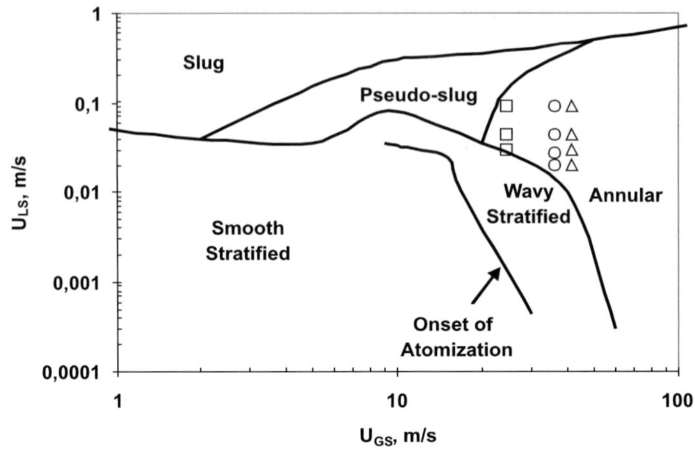

Fig. 5.38 Flow regime map. Reprinted from Zimmerman et al. (2006) with permission

In Fig. 5.39a–d the local heat transfer coefficients derived in the horizontal tube are compared to those obtained in the 8° upward inclined pipe and presented by Hetsroni et al. (2006). The results show a clear improvement of the heat transfer coefficient with the pipe inclination. Taitel and Dukler (1976) showed that the flow regimes are very sensitive to the pipe inclination angle. In the flow regime maps presented in their work, the transition from stratified to annular flow in the inclined tube occurs for a smaller air superficial velocity than for the case of the horizontal tube.

Flow patterns and heat transfer were also investigated by Ghajar et al. (2004) in slug and annular flow. The different flow regimes depicted in Fig. 5.40 illustrate parameters in their experiments in the tube of $d = 25.4$ mm.

The results presented in Fig. 5.41 clearly show that two-phase mean heat transfer coefficients are strongly influenced by the liquid superficial Reynolds number (Re_{LS}). As shown in Fig. 5.41, the heat transfer coefficient increases proportionally as Re_{LS} increases.

Micro-channels

To the best of our knowledge there is a paucity of data in the literature on gas–liquid (without boiling) heat transfer in micro-channels. Hetsroni et al. (2003a) studied this matter in a test section, that contains 21 parallel triangular micro-channels of $d_h = 130\,\mu$m. A scheme of the test section used in that work is shown in Fig. 5.13 (see Sect. 5.4).

As shown in Fig. 5.42 an increase in superficial liquid velocity involves an increase in heat transfer (Nu_L). This effect falls off with increasing superficial gas velocity in the range $Re_{GS} = 4.7–270$.

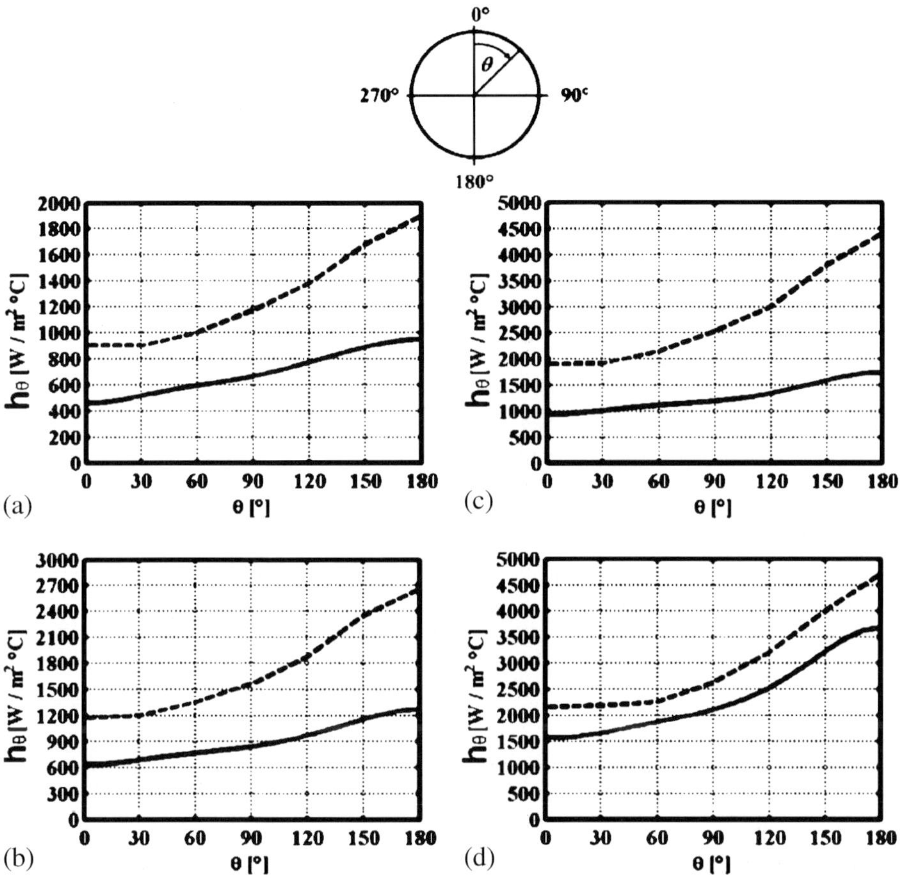

Fig. 5.39 Local heat transfer comparison to 8° upward inclination pipe. $U_{GS} = 36.3\,m/s$. The *solid line* (———) represents the horizontal; the *dashed line* (– – –) represents 8° upwards. (**a**) $U_{LS} = 0.020\,m/s$; (**b**) $U_{LS} = 0.028\,m/s$; (**c**) $U_{LS} = 0.045\,m/s$; (**d**) $U_{LS} = 090\,m/s$. Reprinted from Zimmerman et al. (2006) with permission

5.9.2 *Effect of Superficial Gas Velocity*

Conventional size channels

The experiments were performed by Mosyak and Hetsroni (1999) in a horizontal and upward inclined tube of $d_h = 49.2\,mm$ at $\beta = 5°$. For horizontal flow, the superficial velocities of the liquid phase were $U_{LS} = 0.62-1.40\,m/s$ ($Fr_{LS} = 0.9-2.0$), the superficial velocities of the air phase were $U_{GS} = 0.02-0.30\,m/s$ ($Fr_{GS} = 0.03-0.43$). For flow into the inclined tube, the parameters were $U_{LS} = 0.40-1.40\,m/s$ ($Fr_{LS} = 0.59-2.0$), $U_{GS} = 0.02-0.39\,m/s$ ($Fr_{GS} = 0.03-0.57$). The Froude numbers are defined as $Fr_L = U_{LS}/\sqrt{gd}$, $Fr_G = U_{GS}/\sqrt{gd}$, where d is the inner pipe diameter, and g is the acceleration due to gravity. The experiments

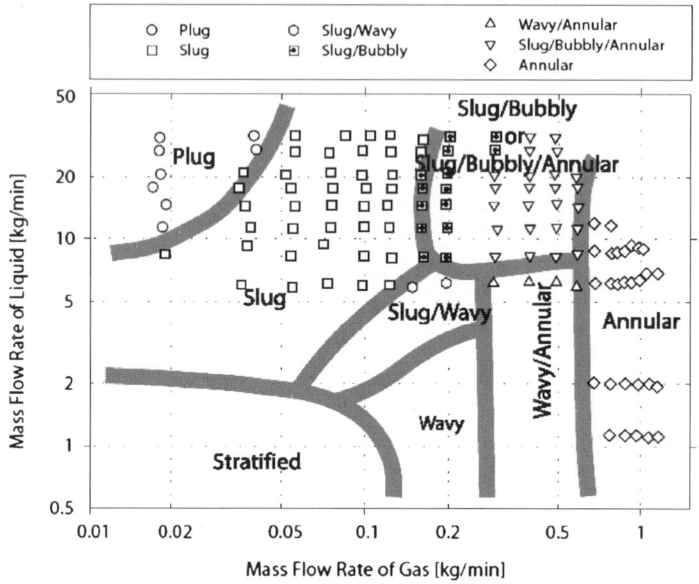

Fig. 5.40 The flow pattern map for horizontal flow. Reprinted from Ghajar et al. (2004) with permission

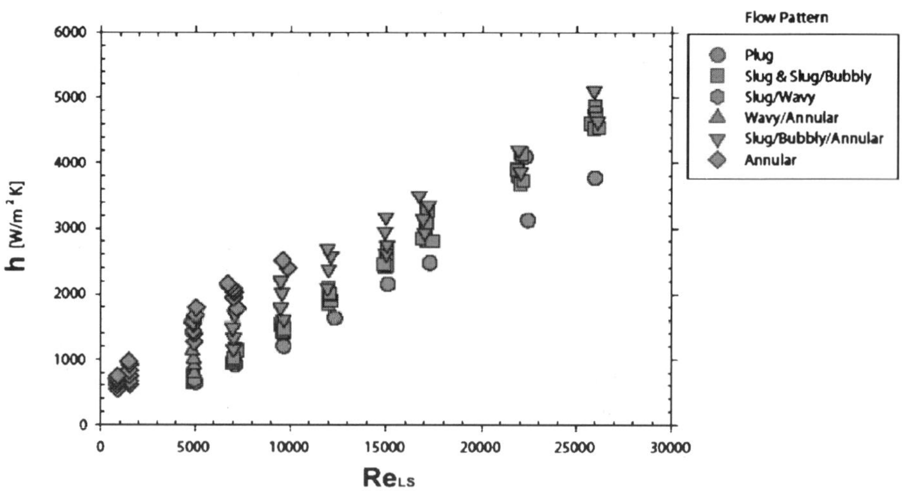

Fig. 5.41 Variation of heat transfer coefficient in horizontal flow vs. Re_{LS}. Reprinted from Ghajar et al. (2004) with permission

for the air–water mixture in a horizontal pipe were carried out at the intermittent flow regime. The data for an inclination of 5° also correspond to intermittent regime with elongated bubbles (Taitel and Dukler 1976). In Fig. 5.43 the ratio of the time-averaged heat transfer coefficient at the top of the tube ($\theta = 0$) to that for single-

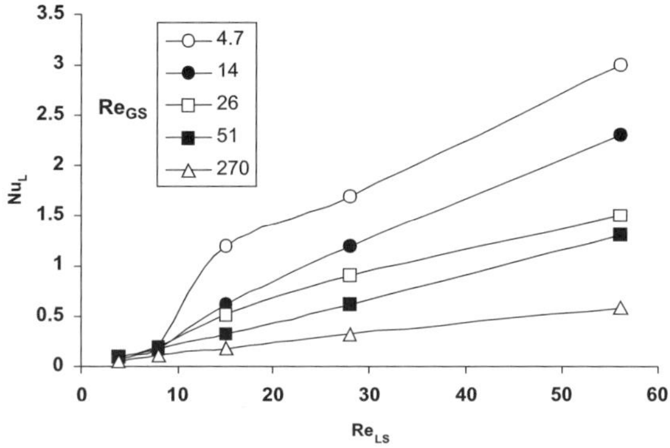

Fig. 5.42 Effect of superficial liquid velocity on heat transfer in parallel triangular micro-channels of $d_h = 130\ \mu m$

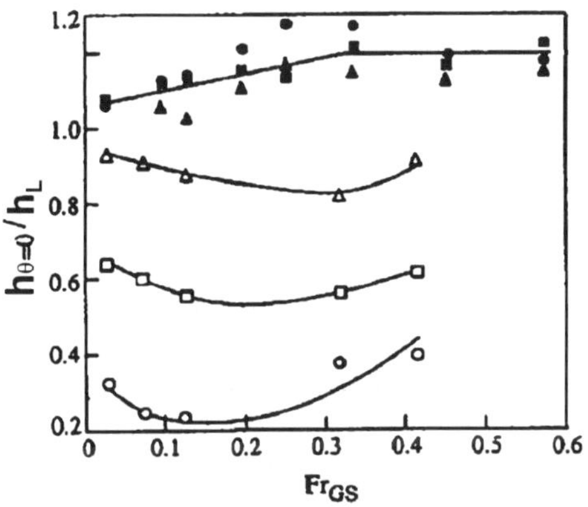

Fig. 5.43 Relative heat transfer coefficients at the top of horizontal and inclined tubes. Horizontal: *circles* (○) indicate $Fr_{LS} = 0.9$, *squares* (□) indicate $Fr_{LS} = 1.3$, *triangles* (△) indicate $Fr_{LS} = 2.0$. Inclined at $\beta = 5°$: *filled circles* (•) indicate $Fr_{LS} = 0.59$, *filled squares* (■) indicate $Fr_{LS} = 1.2$, *filled triangles* (▲) indicate $Fr_{LS} = 2.0$. Reprinted from Mosyak and Hetsroni (1999) with permission

phase liquid, is plotted versus the gas Froude number. The open symbols denote the data for the horizontal tube, the closed symbols denote the data for the inclined tube.

On the top of the horizontal tube ($d_h = 49.2$ mm, $\theta = 0°$) the heat transfer coefficient in two-phase flow is less than that for single-phase water flow.

Ghajar et al. (2004) studied heat transfer of two-phase flow in a horizontal tube of $d_h = 25.4$ mm and reported that the effect of superficial gas velocity on heat transfer depended on flow pattern and showed its own distinguished trend (Fig. 5.44).

Authors also demonstrated the effect of inclination by varying the inclination angle of the pipe, going from the horizontal position to each 2°, 5°, and 7° upward inclined positions. Figure 5.45 shows the heat transfer results for these cases. The figure shows that not only do Re_{LS}, Re_{GS}, and flow pattern but also inclination affect the two-phase heat transfer. The results clearly show that a slight change in the inclination angle has a significant effect on the two-phase heat transfer, especially

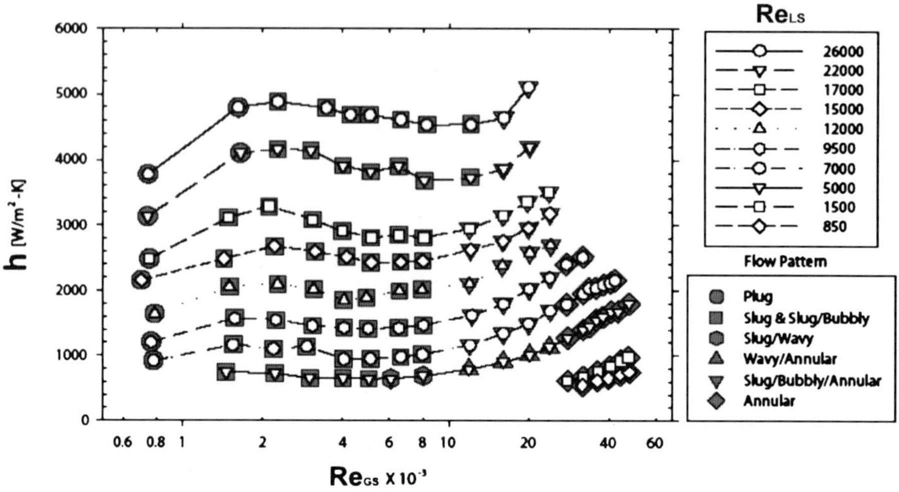

Fig. 5.44 Variation of heat transfer coefficient in horizontal flow vs. Re_{GS}. Reprinted from Ghajar et al. (2004) with permission

in the middle range of Re_{GS}. It is seen from Fig. 5.45 that at some range of Re_{GS} the heat transfer coefficient decreases with increase in Re_{GS}.

On the other hand Bao et al. (2000) reported that the measured heat transfer coefficients for the air–water system are always higher than would be expected for the corresponding single-phase liquid flow, so that the addition of air can be considered to have an enhancing effect. This paper reports an experimental study of non-boiling air–water flows in a narrow horizontal tube (diameter 1.95 mm). Results are presented for pressure drop characteristics and for local heat transfer coefficients over a wide range of gas superficial velocity (0.1–50 m/s), liquid superficial velocity (0.08–0.5 m/s) and wall heat flux (3–58 kW/m^2).

Figure 5.46 shows the experimental heat transfer coefficients for horizontal flows with: (a) a liquid mass flux of 78.6 kg/m^2 s and a heat flux of 20 kW/m^2, and (b) a liquid mass flux of 290 kg/m^2 s and heat flux of 33 kW/m^2 plotted against the gas superficial velocity and the gas phase Reynolds number. The heat transfer coefficient increases with increasing liquid mass flux and increasing gas superficial velocity or Reynolds number. The data show a sharp increase in the heat transfer coefficient at a gas velocity of about 3 m/s ($Re_{GS} \approx 1{,}600$) for the case of $G_{LS} = 78.6$ kg/m^2 s and a gas superficial velocity of 5.5 m/s for the case of $G_{LS} = 290$ kg/m^2 s. This is most likely caused by a change in flow regime.

Figure 5.47 shows a plot of the ratio of the experimental heat transfer coefficient obtained by Bao et al. (2000) divided by the predicted values of Chen (1966) and Gungor and Winterton (1986) for heat transfer to saturated flow boiling in tubes versus liquid Reynolds number. It can be seen that both methods provide reasonable predictions for $Re_{LS} > 500$, but that both overpredict the heat transfer coefficient at lower values of Re_{LS}. For comparison it was assumed that the boiling term of these correlations is zero.

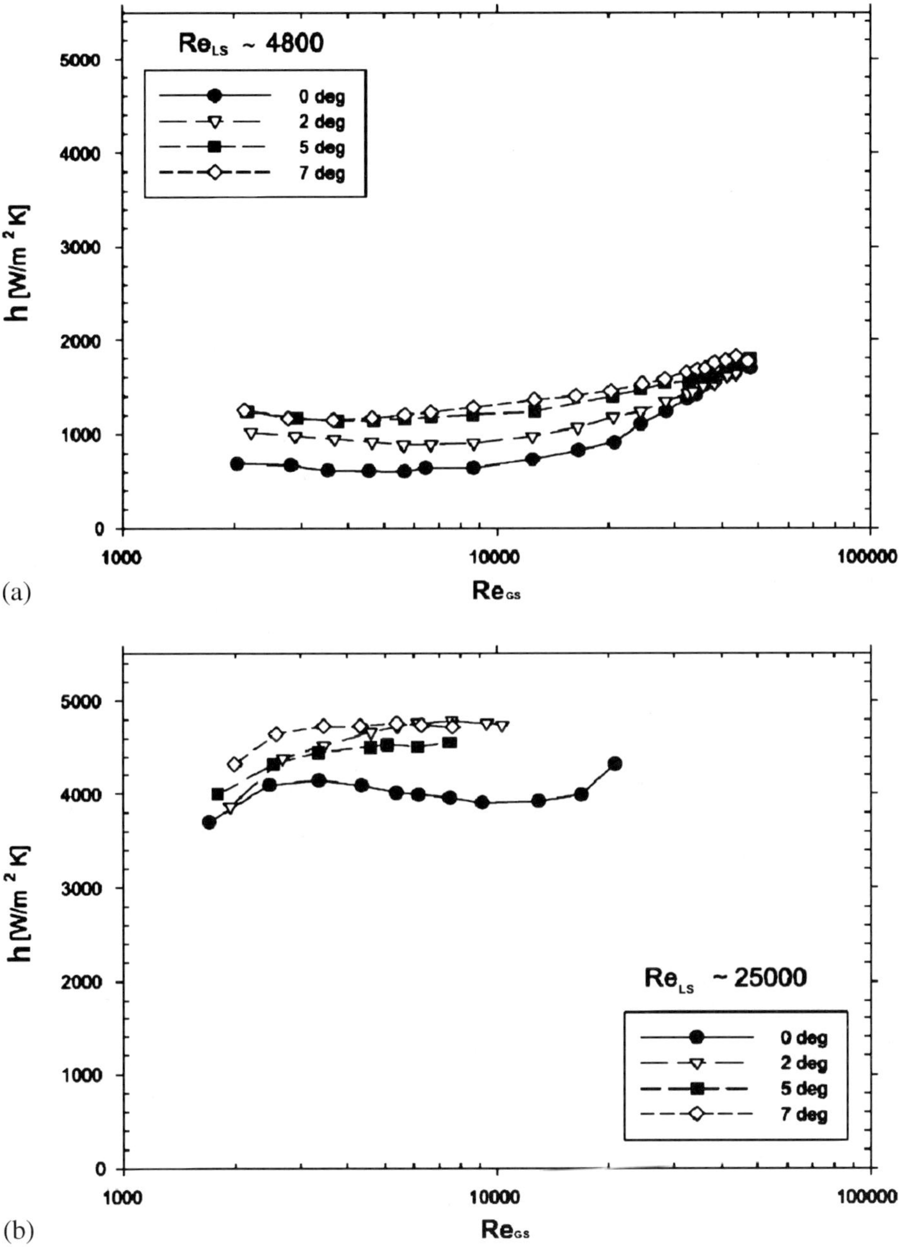

Fig. 5.45 Inclination effects on heat transfer. Reprinted from Ghajar et al. (2004) with permission

From the results discussed above one may conclude that due to the complex nature of two-phase gas–liquid flow there is a large variation in experimental results and heat transfer correlations presented by different investigators. For channels of

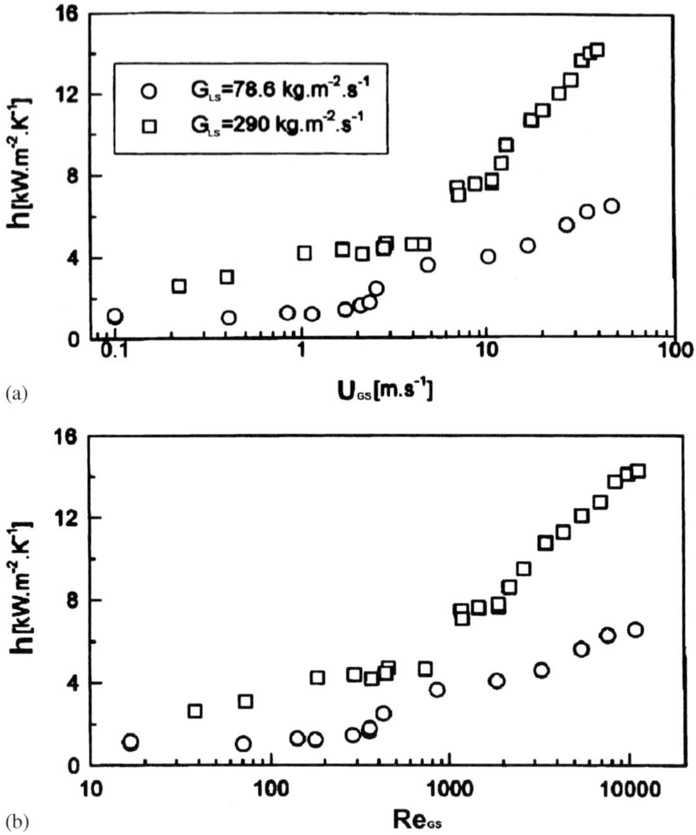

Fig. 5.46 A plot of the experimentally determined heat transfer coefficient as a function of the superficial gas velocity and the gas Reynolds number. The liquid mass fluxes are 78.6 and 290 kg/m^2s, the heat fluxes are 20 and 33 kW/m^2 and the pressure ranges from 140 to 200 kPa. Reprinted from Bao et al. (2000) with permission

$d_h = 1-100$ mm the prediction of Chen (1966) and Gungor and Winterton (1986) (assuming that the boiling term is zero) may be used to calculate the heat transfer coefficient in the range of $Re_{LS} = 500-1{,}500$ and $Re_{GS} = 100-10{,}000$.

Micro-channels

Figure 5.48a–e illustrates the effect of Re_{GS} on heat transfer depending on Re_{LS}. In the range of $Re_{LS} = 4-56$ heat transfer decreases with increasing Re_{GS}.

The Nusselt number may be calculated as:

$$Nu_L = 0.044 Re_{LS}^{0.96} Re_{GS}^{-0.18} \text{ for } Re_{GS} = 4.7-270, Re_{LS} = 4.0-8.0 \quad (5.27)$$

$$Nu_L = 0.13 Re_{LS}^{0.96} Re_{GS}^{-0.40} \text{ for } Re_{GS} = 4.7-270, Re_{LS} = 8.0-56 \quad (5.28)$$

where $Nu_L = hd_h/k_L$, $Re_{LS} = U_{LS}d_h/\nu_L$, $Re_{GS} = U_{GS}d_h/\nu_G$, h is the heat transfer

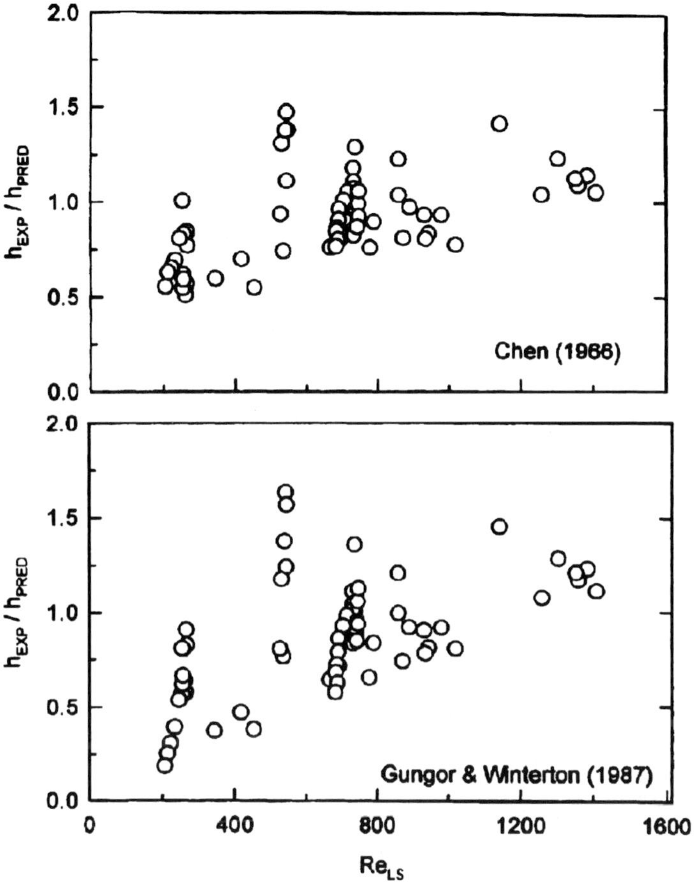

Fig. 5.47 A plot of the ratio of the experimentally determined heat transfer coefficients divided by the predicted values from the Chen (1966) and Gungor and Winterton (1986) correlations as a function of the liquid phase Reynolds number. Reprinted from Bao et al. (2000) with permission

coefficient, k_L is the liquid thermal conductivity, U_{LS} and U_{GS} are the liquid and gas superficial velocities, respectively, and ν_L and ν_G are the liquid and gas kinematic viscosity, respectively (see Fig. 5.49 for comparison of experimental Nusselt numbers to predictions by Eqs. (5.27), (5.28)).

5.9.3 Heat Transfer in Micro-Channels and Dryout

The quasi-steady-state analysis approach to the dryout problem

The important reason for the quasi-steady-state approach arises from the difficulty in obtaining a solution to the transient convection problem for two-phase situations.

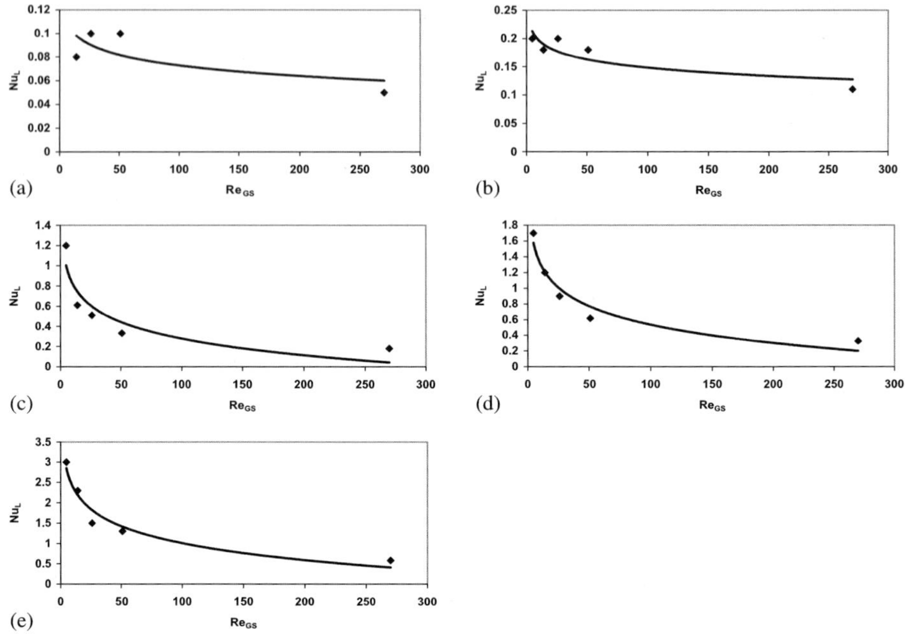

Fig. 5.48 Effect of superficial gas velocity on heat transfer in parallel triangular micro-channels of $d_{\mathrm{h}} = 130\,\mu\mathrm{m}$: (**a**) $\mathrm{Re}_{\mathrm{LS}} = 4$, (**b**) $\mathrm{Re}_{\mathrm{LS}} = 8$, (**c**) $\mathrm{Re}_{\mathrm{LS}} = 15$, (**d**) $\mathrm{Re}_{\mathrm{LS}} = 28$, (**e**) $\mathrm{Re}_{\mathrm{LS}} = 56$

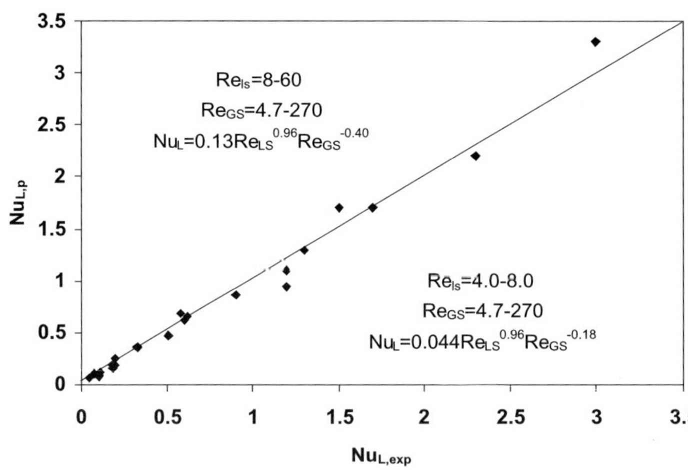

Fig. 5.49 Comparison of experimental Nusselt numbers to predictions by Eqs. (5.27) and (5.28)

We will discuss the question in terms of the interface between the two-phase liquid (liquid–gas) and the wall. For this, consider the transient local heat flux as

$$q = q(\tau, T_{\mathrm{w}}, T_{\mathrm{f}}, P, G, x, \ldots) \tag{5.29}$$

where q is the total heat flux from the wall, τ is the time, T_w is the wall temperature, T_f is the fluid temperature, P is the pressure, G is the total mass flux, and x is the void fraction.

When $G = x = 0$, P and T_f are constant, the total derivative of (5.29) with respect to time can be written as:

$$\frac{dq}{d\tau} = \left(\frac{\partial q}{\partial T_w}\right)\left(\frac{dT_w}{d\tau}\right). \tag{5.30}$$

This equation has been discussed by Nelson and Pasamehmetoglu (1992) relative to the application of the quasi-steady-state model for the convection problem.

In general, the term dryout indicates a local continuous contact of the gas phase with the surface. This term contains no statement as to the average (either spatial or temporal) value of the surface temperature at which dryout may occur. The term used in the present study is defined with respect to the averaged viewpoint of the process time.

It is worthwhile to develop some understanding of certain aspects of the more general topic of two-phase modeling and the assumptions which are involved. The discussion will be limited to the definition of the heat transfer coefficient, because it is the averaging value for which the dryout must be specified. Since the microscopic details of this process are rarely needed for engineering problems, the macroscopic nature of the process is usually much more important. The averaging operation effectively eliminates information on local instantaneous fluctuation. Eulerian averaging is of the group that is of most interest, because it involves averaging over time and space, which are the independent variables. The time-averaged value of the heat transfer coefficient may be calculated at a fixed point located at the angle θ, from the top of the pipe:

$$h_\theta = \frac{1}{\tau}\int_0^\tau h(\tau)\,d\tau \tag{5.31}$$

and the time and space mean value, of the heat transfer coefficient within the angle θ and $\theta + \Delta\theta$, when $\Delta\theta \to 0$ is

$$h_\theta = \frac{1}{\tau\Delta\theta}\int_0^\tau\int_\theta^{\theta+\Delta\theta} h(\tau,\theta)\,d\tau\,d\theta\,. \tag{5.32}$$

It may be used, the relation of the time-averaged heat transfer coefficients on the top and bottom, as a criterion for determination of dryout. It was assumed that the relation $h_\theta/h_\ell < 1$ indicates dryout, i.e., the surface superheat $\overline{T}_w - \overline{T}_f$ is greater than that, when the surface contacts single-phase water only (h_ℓ is the heat transfer at the bottom of the channel). This method can be applied to connect dryout with hydraulic conditions, if the value of h_θ may be associated with intermittent flow parameters.

5.10 Comparison of Gas–Liquid Two-Phase Flow Characteristics Between Conventional Size Channels and Micro-Channels

Ide et al. (2006) addressed the differences in gas–liquid two-phase flow characteristics that occur in conventional size channels and micro-channels by examining the two-phase flow pattern, interfacial wave, void fraction and friction pressure drop data obtained in circular and rectangular channels with a hydraulic diameter ranging from 50 μm to 6.0 mm.

The experimental data obtained in conventional size channels and micro-channels with diameters between 100 μm and 6.0 mm are examined to further elucidate and understand the differences in two-phase flow characteristics between the micro-channels and conventional size channels. Since two separate sets of experiments have been conducted using air and water in acrylic channels with diameters between 500 μm and 6.0 mm, and nitrogen gas–water in fused silica channels with diameters between 50 and 500 μm, the authors refer to the former channels as conventional size channels, and the latter channels as micro-channels for convenience. Two different inlet sections were covered in micro-channel experiments, a gradually reducing section and a T-junction.

The void fraction data obtained in micro-channels and conventional size channels showed significant differences depending on the channel cross-section and inlet geometry. For the micro-channel with a diameter of 100 μm, the effects of the inlet geometry and gas–liquid mixing method on the void fraction were seen to be quite strong, while the conventional size channels have shown a much smaller effect of inlet geometry on the void fraction.

The void fraction data obtained with a T-junction inlet showed a linear relationship between the void fraction and volumetric quality, in agreement with the homogeneous flow model predictions. The flow patterns observed for this inlet geometry included plug and slug flows with short gas plugs separated by short liquid slugs. Both phases likely moved with the same velocity, giving rise to the good agreement between the void fraction data with the homogeneous flow model. On the contrary, the void fraction data from the reducing section inlet experiments by Kawahara et al. (2002) showed a non-linear void fraction-to-volumetric quality relationship.

This trend implies a strong departure from homogeneous flow and indicates a large slip between the liquid and gas phases. Kawahara et al. (2002) observed mainly long gas core flows surrounded by thin, thick and wavy liquid films, which likely moved with a much slower velocity due to a large viscous effect near the channel wall. They observed few plug flows with short gas plugs for the same micro-channel connected to a reducer inlet. From these data, it is clear that the void fraction in a micro-channel strongly depends on the inlet geometry and gas–liquid mixing method due to the significantly different flow patterns that occur in the micro-channel.

In contrast, the conventional size channel void fraction data conform to the Armand correlation (1946). To our knowledge, no unusually low void fraction data have been reported for conventional size channels, so the two-phase flow in conventional size channels can be considered to be relatively independent of the gas–liquid mixing method.

Thus, similar void fraction data can be obtained in micro-channels and conventional size channels, but the micro-channel void fraction can be sensitive to the inlet geometry and deviate significantly from the Armand correlation.

For a micro-channel connected to a 100 μm T-junction the Lockhart–Martinelli model correlated well with the data, however, different C-values were needed to correlate well with all the data for the conventional size channels. In contrast, when the 100 μm micro-channel was connected to a reducing inlet section, the data could be fit by a single value of $C = 0.24$, and no mass velocity effect could be observed. When the T-junction diameter was increased to 500 μm, the best-fit C-value for the 100 μm micro-channel again dropped to a value of 0.24. Thus, as in the void fraction data, the friction pressure drop data in micro-channels and conventional size channels are similar, but for micro-channels, significantly different data can be obtained depending on the inlet geometry.

In contrast to conventional size channels, the flow regimes and heat transfer coefficients in micro-channels are not sensitive to the channels inclination. In the range of $\mathrm{Re}_{\mathrm{LS}} = 4-56$ and $\mathrm{Re}_{\mathrm{GS}} = 4.7-270$ an increase in $\mathrm{Re}_{\mathrm{GS}}$ leads to a decrease in the heat transfer coefficient (Hetsroni et al.) as opposed to results reported for mini-channels.

Summary

1. The flow pattern observed in two-phase gas–liquid flow may take many different configurations with respect to the distribution gas–liquid interface. The designation of the flow pattern should be based on the flow configuration that has basically the same character, pertaining to the distribution of the interfaces and the mechanisms dominating pressure drop and heat and mass transfer. It depends largely on the individual interpretation of the different research. Standardization is essential so that data from different groups and laboratories can be correctly interpreted and compared.
2. In large tubes, as well as in tubes of a few millimeters in diameter, two-phase flow patterns are dominated in general by gravity with minor surface tension effects. In micro-channels with the diameter on the order of a few microns to a few hundred microns, two-phase flow is influenced mainly by surface tension, viscosity and inertia forces. The stratified flow patterns commonly encountered in single macro-channels were not observed in single micro-channels.
3. The micro-channels are sensitive not only to inlet geometry but also to the method of gas–liquid injection.
4. Experimental data on flow patterns and the transition boundaries are usually mapped on a two-dimensional plot. Two basic types of coordinates may be used for the flow regime maps – one that uses dimensional coordinates such as superficial velocities, and another that uses some kind of dimensionless group. Di-

mensionless analysis is the straightforward approach to select the proper dimensionless coordinates. The flow pattern transition boundaries should be a function of all these variables. Maps based on dimensionless coordinates are more general, for example, they should be independent of fluid properties and channel size. Unfortunately, determining the correct dimensionless coordinates for flow pattern is not at all trivial. Moreover, there is no guarantee that two dimensionless coordinates are sufficient in the majority of cases. The maps obtained for a given geometry and channel size are usually based on superficial liquid and gas velocities. The usual experimental approach is to apply these variables to all transitions in all channels, assuming that entrance effects, channel geometry and size, roughness, solid–liquid–gas contact wetting angle, etc., have no influence on flow pattern. The chance of this approach being correct, outside the specific range of operating conditions that has been used to correlate the data, is very slim.

5. Two-phase flow in parallel micro-channels, feeding from a common manifold shows that different flow patterns occur simultaneously in different micro-channels. The probability of appearance of different flow patterns should be taken into account for developing flow pattern maps.
6. Similar void fraction data can be obtained in micro-channels and conventional size channels, but the micro-channel void fraction can be sensitive to the inlet geometry and deviate significantly from the homogeneous flow model.
7. The Lockhart–Martinelli model can correlate the data obtained from pressure drop measurements in gas–liquid flow in channels with hydraulic diameter of 0.100–1.67 mm. The friction multiplier is $\Phi_L = 1 + C/X + 1/X^2$.
   Different C-values are needed to correlate all the data. For example, when the 100 μm micro-channel was connected to a reducing inlet section, the data could be fit by a single value of $C = 0.24$.
8. Only a few experimental investigations deal with heat transfer of gas–liquid flow in the conventional size channels. There is a significant discrepancy between experimental results on heat transfer presented for channels of $d_h = 1-100$ mm. No data is available in the literature on gas–liquid heat transfer in micro-channels, except for the results on the study of heat transfer in the test section that contains 21 parallel triangular micro-channel of $d_h = 130$ μm reported in the present chapter. In the range of superficial velocities $U_{LS} = 0.015-0.244$ m/s, $U_{GS} = 0.50-28.6$ m/s the heat transfer coefficient increases with increasing liquid velocity and decreases with increasing air velocity.

References

Akbar MK, Plummer DA, Ghiaasiaan SM (2003) On gas–liquid two-phase flow regimes in micro-channels. Int J Multiphase Flow 29:855–865

Ali MI, Sadatomi M, Kawaji M (1993) Two-phase flow in narrow channels between two flat plates. Can J Chem Eng 71:657–666

Armand AA (1946) The resistance during the movement of a two-phase system in horizontal pipes. Izv Vses Teplotekh Inst 1:16–23 (AERE-Lib/Trans 828)

Bao ZY, Fletcher DF, Haynes BS (2000) An experimental study of gas–liquid flow in a narrow conduit. Int J Heat Mass Transfer 43:2313–2324

Barajas AM, Panton RL (1993) The effect of contact angle on two-phase flow in capillary tubes. Int J Multiphase Flow 19:337–346

Barnea D, Luniski Y, Taitel Y (1983) Flow pattern in horizontal and vertical two phase flow in small diameter pipes. Can J Chem Eng 61:617–620

Baroczy CJ (1963) Correlation of liquid fraction in two-phase flow with application to liquid metals, NAA-SR-8171. Butterworth, London

Beattie DRH, Whalley PB (1982) A simple two-phase flow frictional pressure drop calculation method. Int J Multiphase Flow 8:83–87

Benjamin TB (1968) Gravity currents and related phenomena. J. Fluid Mechanics 31(2):209–248

Butterworth D (1975) A comparison of some void-fraction relationships for co-current gas–liquid flow. Int J Multiphase Flow 1:845–850

Celata GP (ed) (2004) Heat transfer and fluid flow in micro-channels. Bergel, New York

Chen JC (1966) Correlation for boiling heat transfer to saturated fluids in convective flow. Ind Eng Chem Process Des Dev 5:322–329

Chen JJJ, Spedding PL (1983) An analysis of holdup in horizontal two-phase gas–liquid flow. Int J Multiphase Flow 9:147–159

Cheng P, Wu WY (2006) Mesoscale and microscale phase change heat transfer. In: Greene G, Cho Y, Hartnett J, Bar-Cohen A (eds) Advances in heat transfer 39. Academic, New York

Chisholm D (1983) Two-phase flow in pipelines and heat exchangers. Pitman, Bath, England

Cicchitti A, Lombardi C, Silvestri M, Solddaini G, Zavalluilli R (1960) Two-phase cooling experiments: Pressure drop, heat transfer and burnout measurement. Energia Nucl 7(6):407–425

Coleman JW, Garimella SV (1999) Characteristics of two-phase patterns in small diameter round and rectangular tubes. Int J Heat Mass Transfer 42:2869–2881

Damianides CA, Westwater JW (1988) Two-phase flow patterns in a compact heat exchanger and in small tubes. In: Proceedings of the 2nd UK National Conference On Heat Transfer, Glasgow, 14–16 September 1988. Mechanical Engineering, London, pp 1257–1268

Dukler AE, Wicks IM, Cleveland RG (1964) Pressure drop and hold-up in two-phase flow. AIChE J 10(1):38–51

Friedel L (1979) Improved friction pressure drop correlations for horizontal and vertical two-phase pipe flow. In: 3rd International European Two-Phase Group Meeting, Ispra, Italy, 1979, vol 18, issue 7

Fukano T, Kariyasaki A (1993) Characteristics of gas–liquid two-phase flow in a capillary. Nucl Eng Des 141:59–68

Galbiati L, Andreini P (1992) Flow patterns transition for vertical downward two-phase flow in capillary tubes. Inlet mixing effects. Int Comm Heat Mass Transfer 19:791–799

Garimella S, Sobhan C (2003) Transport in microchannels – a critical review. Ann Rev Heat Transfer 13:1–50

Ghajar AJ, Kim J-Y, Malhotra K, Trimble S (2004) Systematic heat transfer measurements for air–water two-phase flow in horizontal and slightly upward inclined pipe. In: Proceedings of the 10th Brazilian Congress of Thermal Sciences and Engineering – ABCM, Rio de Janeiro, 29 Nov–3 Dec 2004

Ghiaasiaan SM, Abdel-Khalik SI (2001) Two-phase flow in micro-channels. Adv Heat Transfer 34:145–254

Gungor KE, Winterton RHS (1986) A general correlation for flow boiling in tubes and annuli. Int J Heat Mass Transfer 29:351–358

Hetsroni G, Gurevich M, Mosyak A, Rozenblit R (2001) Dryout in inclined gas–liquid pipe-lines. Trans IChemE 79(A):376–382

Hetsroni G, Hu BG, Yi JH, Mosyak A, Yarin LP, Ziskind G (1998a) Heat transfer in intermittent air–water flows: Part I. Int J Multiphase Flow 24(2):165–188

Hetsroni G, Hu BG, Yi JH, Mosyak A, Yarin LP, Ziskind G (1998b) Heat transfer in intermittent air–water flows: Part II. Int J Multiphase Flow 24(2):189–212

Hetsroni G, Mewes D, Enke C, Gurevich M, Mosyak A, Rozenblit R (2003b) Heat transfer of two-phase flow in inclined tubes. Int J Multiphase Flow 29:173–194

Hetsroni G, Mosyak A, Segal Z, Pogrebnyak E (2003a) Two-phase flow patterns in parallel micro-channels. Int J Multiphase Flow 29:341–360

Ide H, Kawahara A, Kawaji M (2006) Comparison of gas–liquid two-phase flow characteristics between mini-channels and micro-channels. In: Proceedings of 13th International Heat Transfer Conference, Sydney Convention and Exibition Centre, Sydney, Australia, 13–18 August 2006, MPH-51

Ishii (1977) One-dimensional drift-flux model and constitutive equations for relative motion between phases in various two-phase regimes. ANL Report ANL-77-47

Ide H, Matsumura H, Tanaka Y, Fukano T (1997) Flow patterns and frictional pressure drop in gas–liquid two-phase flow in vertical capillary channels with rectangular cross section, Trans JSME Ser B 63:452–160

Kariyasaki A, Fukano T, Ousaka A, Kagawa M (1991) Characteristics of time-varying void fraction in isothermal air–water co-current flow in horizontal capillary tube Trans JSME 57(544):4036–4043

Kawahara A, Chung PM, Kawaji M (2002) Investigation of two-phase flow pattern, void fraction and pressure drop in a micro-channel. Int J Multiphase Flow 28:1411–1435

Kawaji M (1999) Fluid mechanics aspects of two-phase flow: Flow in other geometries. In: Kandlikar SG, Shoji M, Dhir VK (eds) Handbook of phase change: boiling and condensation. Taylor and Francis, Washington, DC, pp 205–259

Kim D, Ghajar AJ, Dougherty RL, Ryali VK (1999) Comparison of 20 two-phase heat transfer correlations with seven sets of experimental data, including flow pattern and tube inclination effects. J Heat Transfer Eng 29(1):15–40

Kokal SL, Stanislav JF (1989) An experimental study of two- phase flow in slightly inclined pipes. 1 Flow patterns, 2 Liquid hold-up and pressure drop. Chem Eng Sci 44:655–679, 681–693

Lee HJ, Lee SY (2001) Pressure drop correlations for two-phase flow within horizontal rectangular channels with small height. Int J Multiphase Flow 27:783–796

Lin PY, Hanratty TJ (1987) The effect of pipe diameter on flow patterns for air–water flow in horizontal pipes. Int J Multiphase Flow 13:549–563

Lin S, Kwok CCK, Li RY, Chen ZH, Chen ZY (1991) Local frictional pressure drop during vaporization for R-12 through capillary tubes. Int J Multiphase Flow 17:95–102

Lockhart RW, Martinelli RC (1949) Proposed correlation of data for isothermal two-phase two-component flow in pipes. Chem Eng Prog 45:39–18

Lowe DC, Rezkallah KS (1999) Flow regime identification in microgravity two-phase flow using void fraction signals. Int J Multiphase Flow 25:433–457

Mandhane JM, Gregory GA, Aziz K (1974) A flow pattern map for gas–liquid flow in horizontal pipes. Int J Multiphase Flow 1:537–553

McAdams WH (1954) Heat transmission, 3rd edn. McGraw-Hill, New York

Mishima K, Hibiki T (1996) Some characteristics of air–water two-phase flow in small diameter vertical tubes. Int J Multiphase Flow 22:703–712

Mishima K, Ishii M (1984) Flow regime transition criteria for upward two-phase flow in vertical tubes. Int J Heat Mass Transfer 27:723–737

Mosyak A, Hetsroni G (1999) Analysis of dryout in horizontal and inclined tubes. Int J Multiphase Flow 25:1521–1543

Nelson RA, Pasamehmetoglu KO (1992) Quenching phenomena. In: Hewitt GF, Delhaye JM, Zuber N (eds) Post-dryout Heat transfer. CRC, Boca Raton, pp 39–184

Owens WL (1961) Two-phase pressure gradient. In: ASME International Developments in Heat Transfer, Part II. ASME, New York

Ozawa M, Akagawa K, Sakaguchi T (1989) Flow instabilities in parallel-channel flow systems of gas–liquid two-phase mixtures. Int J Multiphase Flow 15:639–657

Ozawa M, Akagawa K, Sakaguchi T, Tsukahara T, Fujii T (1979) Oscillatory flow instabilities in air–water two-phase flow systems. Report. Pressure drop oscillation. Bull JSME 22:1763–1770
Qu W, Yoon S-M, Mudawar I (2004) Two-phase flow and heat transfer in rectangular micro-channels. J Electron Packag 126:288–300
Rezkallah KS (1998) Heat transfer and flow characteristics of liquid–gas flows at reduced gravity. Trends Chem Eng 4:161–170
Rezkallah KS (1996) Weber number based flow-pattern maps for liquid–gas flows at microgravity. Int J Multiphase Flow 22:1265–1270
Sadatomi Y, Sato Y, Saruwatari S (1982) Two-phase flow in vertical noncircular channels. Int J Multiphase Flow 8:641–655
Serizawa A, Feng Z, Kawara Z (2002) Two-phase flow in micro-channels. Exp Thermal Fluid Sci 26:703–714
Serizawa A, Feng Z (2001) Two-phase flow in micro-channels. In: Proceedings of the 4th International Conference on Multiphase Flow, New Orleans, 27 May–1 June 2001
Spedding PL, Watterson JK, Raghunathan SR, Ferguson MEG (1998) Two-phase co-current flow in inclined pipe. Int J Heat Mass Transfer 41:4205–4228
Suo M, Griffith P (1964) Two-phase flow in capillary tubes. J Basic Eng 86:576–582
Taitel Y, Dukler AE (1976) A model for predicting flow regime transitions in horizontal and near horizontal gas–liquid flow. AIChE J 22:47–55
Taitel Y, Barnea D, Dukler AE (1980) Modeling flow pattern transitions for steady upward gas–liquid flow in vertical tubes. AIChE J 26:345–354
Triplett KA, Ghiaasiaan SM, Adbel-Khalik SI, Sadowski DL (1999a) Gas–liquid two-phase flow in microchannels. Part I: two-phase flow patterns. Int J Multiphase Flow 25:377–394
Triplett KA, Ghiaasiaan SM, Abdel-Khalik SI, LeMouel A, McCord BN (1999b) Gas–liquid two-phase flow in microchannels. Part II: void fraction and pressure drop. Int J Multiphase Flow 25:395–410
Tshuva M, Barnea D, Taitel Y (1999) Two-phase flow in inclined parallel pipes. Int J Multiphase Flow 25:1491–1503
Ungar EK, Cornwell JD (1992) Two-phase pressure drop of ammonia in small diameter horizontal tubes. In: AIAA 17th Aerospace Ground Testing Conference, Nashville, 6–8 July 1992
Wallis GB (1969) One dimensional two-phase flow. McGraw-Hill, New York
Yang CY, Shieh CC (2001) Flow pattern of air–water and two-phase R-134a in small circular tubes. Int J Multiphase Flow 27:1163–1177
Zhao L, Rezkallah KS (1993), Gas–liquid flow patterns at microgravity conditions. Int J Multiphase Flow 19:751–763
Zhao TS, Bi QC (2001a) Co-current air–water two-phase flow patterns in vertical triangular microchannels. Int J Multiphase Flow 27:765–782
Zhao TS, Bi QC (2001b) Pressure drop characteristics of gas–liquid two-phase flow in vertical miniature triangular channels. Int J Heat Mass Transfer 44:2523–2534
Zimmerman R, Gurevich M, Mosyak A, Rozenblit R, Hetsroni G (2006) Heat transfer to air–water annular flow in a horizontal pipe. Int J Multiphase Flow 32:1–19

Nomenclature

C	Constant used in Eq. (5.25)
C_0	Distribution parameter
d	Diameter
g	Acceleration due to gravity
G	Mass flux
h	Heat transfer coefficient

H	Height, depth
$j = U_{GS} + U_{LS}$	Sum of the superficial gas and liquid velocities
K	Thermal conductivity
q	Heat flux
P	Pressure
T	Temperature
U	Velocity
u_g	Slug bubble velocity
V_b	Drift velocity
X	Lockhart–Martinelli parameter
x	Volumetric quality, void fraction
z	Streamwise coordinate
$\mathrm{Eo} = \frac{\Delta\rho g d_h^2}{\sigma}$	Eotvos number
$\mathrm{We}_{LS} = \frac{U_{LS}^2 d_h \rho_L}{\sigma}$	Weber number based on superficial liquid velocity
$\mathrm{We}_{GS} = \frac{U_{GS}^2 d_h \rho_G}{\sigma}$	Weber number based on superficial gas velocity
$\mathrm{Re}_{LS} = \frac{U_{LS} d_h}{\nu_L}$	Reynolds number based on superficial liquids velocity
$\mathrm{Re}_{GS} = \frac{U_{GS} d_h}{\nu_G}$	Reynolds number based on superficial gas velocity
$\mathrm{Bn} = d_h \sqrt{\frac{g(\rho_L - \rho_G)}{\sigma}}$	Bond number
$\mathrm{Ca} = \frac{\mu_L U_{LS}}{\sigma}$	Capillary number
$\mathrm{Fr}_{LS} = \frac{U_{LS}}{\sqrt{g d_h}}$	Froude number based on superficial liquids velocity
$\mathrm{Fr}_{GS} = \frac{U_{GS}}{\sqrt{g d_h}}$	Froude number based on superficial gas velocity
$\mathrm{Nu}_L \frac{h_L d_h}{k_L}$	Nusselt number based on liquid parameters

Greek symbols

α	Void fraction
β	Homogeneous (volumetric) void fraction
$\Delta\rho = \rho_L - \rho_G$	Difference between liquid and gas density
θ	Angle, contact angle
ρ	Density
μ	Dynamic viscosity

σ	Surface tension
τ	Time
ν	Kinematic viscosity
Φ	Two-phase multiplier

Subscripts

B	Gas bubbles
b	Bottom
f	Fluid
G	Gas
GS	Gas superficial
exp	Experimental
H	Homogeneous
h	Hydraulic
L	Liquid
LS	Liquid superficial
model	Model prediction
pred	Predicted
TP	Two-phase
w	Wall
θ	Angle

Chapter 6
Boiling in Micro-Channels

The subject of Chap. 6 is boiling in micro-channels. Several aspects of boiling are also considered for conventional size channels and comparison with micro-channels was carried out. Significant differences of ONB in micro-channels have been discussed compared to conventional channels. Effect of dissolved gases on boiling in water and surfactant solution was revealed. Attention was paid on pressure drop and heat transfer, critical heat flux and instabilities during flow boiling in micro-channels.

6.1 Onset of Nucleate Boiling in Conventional Size Channels

In this section the general approach used to describe the parameters corresponding to the onset of nucleate boiling (ONB) in a single channel are considered, and models and experiments are compared. The mode of ONB in parallel micro-channels is presented in Sect. 6.2. We will also look at the effect of dissolved gases on ONB during flow boiling of water and surfactant solutions in micro-channels. Section 6.3 deals with the dynamics of vapor bubble in a single micro-channel and in parallel micro-channels. Pressure drop in two-phase flow boiling in parallel micro-channels is considered in Sect. 6.4. In this section we look at a new correlation incorporating the effects of both channel size and coolant mass velocity, which shows better accuracy than prior correlations. In Sect. 6.4 heat transfer correlations in two-phase flow boiling in parallel micro-channels are examined and correlations based on the Martinelli parameter are considered. Parameters that affect the explosive boiling are discussed in Sect. 6.5. Existing critical heat flux (CHF) correlations for flow boiling of water with available databases taken from small-diameter channels are also evaluated.

6.1.1 Models for Prediction of Incipient Boiling Heat Flux and Wall Superheat

The onset of bubble nucleation usually requires that the temperature on the heated surface exceeds the saturation temperature of the liquid corresponding to a given pressure. There has been a great deal of analysis on bubble nucleation and a number of semi-empirical models of boiling incipience have been proposed during the past few decades (Carey 1992; Dhir 1998). Two general approaches were used to describe the parameters corresponding to the onset of nucleate boiling. The first is based on analysis of the behavior of a single bubble on a rough surface where the temperature of the surrounding liquid exceeds the saturation temperature. The second approach considers vapor bubbles generated in liquid (homogeneous nucleation), on a surface or at a corner (heterogeneous nucleation) based on the classical kinetics of nucleation.

The first approach developed by Hsu (1962) is widely used to determine ONB in conventional size channels and in micro-channels (Sato and Matsumura 1964; Davis and Anderson 1966; Celata et al. 1997; Qu and Mudawar 2002; Ghiaasiaan and Chedester 2002; Li and Cheng 2004; Liu et al. 2005). These models consider the behavior of a single bubble by solving the one-dimensional heat conduction equation with constant wall temperature as a boundary condition. The temperature distribution inside the surrounding liquid is the same as in the undisturbed near-wall flow, and the temperature of the embryo tip corresponds to the saturation temperature in the bubble $T_{S,b}$. The vapor temperature in the bubble can be determined from the Young–Laplace equation and the Clausius–Clapeyron equation (assuming a spherical bubble):

$$T_{S,b} - T_S = \frac{2\sigma T_S}{h_{LG}\rho_G r_b} \tag{6.1}$$

where σ is the surface tension, h_{LG} is the latent heat of vaporization, ρ_G is the vapor density, and r_b is the radius of bubble embryo assumed to be proportional to the radius of the cavity r_c.

The temperature distribution in the undisturbed fluid over the heated surface is described by the equation

$$\frac{\partial\theta}{\partial t} = \alpha\frac{\partial^2\theta}{\partial y^2} \tag{6.2}$$

where $\theta = T - T_\infty$, T_∞ is the temperature of the surrounding fluid, α is the thermal diffusivity.

Integrating Eq. (6.2) with the initial and boundary conditions

$$\begin{aligned} &t = 0, \quad 0 \le y \le \delta, \quad \theta = 0 \\ &t > 0, \quad y = 0, \quad \theta = \theta_w \end{aligned} \tag{6.3}$$

leads to the following expression for the liquid temperature

$$\frac{\theta}{\theta_w} = \eta + \frac{2}{\pi}\sum_{n=1}^{\infty}\frac{\cos\pi n}{n}\exp(-n^2\pi^2\mathrm{Fo})\sin\pi n\eta \tag{6.4}$$

where $\theta_w = T_w - T_\infty$, $\eta = y/\delta$, $Fo = 2\alpha t/\delta^2$ is the Fourier number, δ is the thickness of thermal boundary layer, t is the time, y is the distance from the wall.

From Eqs. (6.1) and (6.4) one can find the limiting values of the cavity radius of active sizes on the heated surface

$$r_{crit} = \frac{\delta}{4}\left(1 - \frac{\theta_S}{\theta_w} \pm \sqrt{\left(1 - \frac{\theta_S}{\theta_w}\right)^2 - \frac{12.8\sigma T_S}{\rho_G h_{LG}\delta\theta_w}}\right) \qquad (6.5)$$

where plus and minus signs on the right-hand side of Eq. (6.5) correspond to maximum and minimum active cavity, respectively.

It should be stressed that such models are valid only when the medium is continuous. Measurements by Ye et al. (2004) showed that bubble nucleation in water may be considered as continuous when $r_c \geq 10$ nm.

Thermodynamic and mechanical equilibrium on a curved vapor–liquid interface requires a certain degree of superheat in order to maintain a given curvature. Characteristics of homogeneous and heterogeneous nucleation can be estimated in the frame of classical theory of kinetics of nucleation (Volmer and Weber 1926; Farkas 1927; Becker and Doring 1935; Zel'dovich 1943). The vapor temperature in the bubble $T_{S,b}$ can be computed from equations (Bankoff and Haute 1957; Cole 1974; Blander and Katz 1975; Li and Cheng 2004) for homogeneous nucleation in superheated liquids

$$J_{hom} = N_0\left(\frac{kT_{S,b}}{h}\right)\exp\left(-\frac{16\pi\sigma^3}{3kT_{S,b}(P_G - P_L)}\right) \qquad (6.6)$$

for heterogeneous nucleation on a surface

$$J_{het} = N_0^{2/3}\psi\left(\frac{kT_{S,b}}{h}\right)\exp\left(-\frac{16\pi\sigma^3\omega}{3kT_{S,b}(P_G - P_L)^2}\right) \qquad (6.7)$$

where J is the bubble nucleation density, N_0 is the molecule number per unit volume, k is the Boltzmann constant, P_G is the pressure in a bubble, P_L is the pressure in liquid, and $h = kT_{S,b}/(2\sigma/\pi mB)^{0.5}$, where m is the mass of molecule, $B \approx 2/3$. The term ω is the geometric correction factor for the minimum work required to form the critical nucleus (for heterogeneous nucleation on a smooth surface with no cavities, $\omega = (1/2)(1+\cos\theta)$; $\psi = (1/4)(1+\cos\theta)^2(2-\cos\theta)$, θ is the contact angle).

Using the Clausius–Clapeyron equation and Eqs. (6.6) and (6.7) one can obtain the saturation temperature in the bubble $T_{S,b}$ (Li and Cheng 2004):

$$T_{S,b} - T_S = \frac{T_S}{h_{LG}\rho_G}\sqrt{\frac{16\pi\sigma^3\omega}{3kT_{S,b}\ln(N_0 kT_{S,b}\psi/J\cdot h)}}. \qquad (6.8)$$

6.1.2 Comparison Between Models and Experiments

There are a number of experiments directed at studying the incipience of nucleate boiling in heated channels of hydraulic diameter $d_h = 0.04$–20 mm. The overall

Table 6.1 Onset of nucleate boiling. Characteristics of experiment

Author	Channel geometry	Hydraulic diameter (mm)	Mass flux (kg/m^{2}s)	Fluid	Heat flux (kW/m^2)	Pressure (MPa)
Bergles and Rohsenow (1964)	Cylindrical tube	2.387	(2.1–19.4) $\times 10^3$	Distilled water	1,890–18,900	0.261
Sato and Matsumara (1964)	Cylindrical tube	–	–	Water	100–750	0.1
Unal (1975)	Cylindrical tube	4–20	132–2,818	Water; refrigerant-22	20–1,920	0.1–15.8
Hino and Ueda (1985)	Cylindrical tube	7.0	158–1,600	Fluorocarbon R-113	11.78–46.90	0.147
Kennedy et al. (2000)	Cylindrical tube	1.17, 1.45	800–4,500	De-ionized and de-gassed water	0–4,000	0.344–1.034
Hapke et al. (2000)	Cylindrical tube	1.5	100–500	De-ionized water	50–200	0.1
Hetsroni et al. (see this book)	Cylindrical tube	1.07	49–146	Distilled water	62–162	0.1
Stoddard et al. (2002)	Annular channel	0.724–1.0	85–1,428	Fully de-gassed water	124–1,000	0.344–1.034
Su et al. (2001)	Annular channel	1.0, 1.5	45–180	Pure water	40–210	0.2–3.5
Qu and Mudawar (2002)	Rectangular channel	0.231$\times$0.713	130–1,440	De-ionized water	200–2,000	0.12
Liu et al. (2005)	Rectangular channel	0.275$\times$0.636	309–883	De-ionized water	100–730	0.1
Lee et al. (2004)	Trapezoidal channel	0.413	170–899	De-ionized water	1.49–500	0.161

characteristics of experiments are presented in Table 6.1. Measurements of hydrodynamic and thermal parameters corresponding to onset of nucleate boiling (ONB) were performed in circular, annular, rectangular, and trapezoidal channels, and water, refrigerant-22, and fluorocarbon R-113 were used as working fluid. ONB was recorded by various ways: analysis of the pressure drop-mass flux characteristic curve; relation between the heat flux and the wall superheat at the start point of the sharp drop in wall temperature ($\Delta T_{S,ONB} = T_{W,ONB} - T_S$, where $T_{W,ONB}$ and T_S is the temperature of the heated wall, corresponding to onset of nucleate boiling and saturation temperature, respectively); visual detection of bubble growth and departure from the heated wall. Experiments were carried out in the range of mass flux $G = 45-2{,}100\,\mathrm{kg/m^2s}$, heat flux $q = 1.49-18{,}900\,\mathrm{kW/m^2}$, and pressure $P = 0.1-15.8\,\mathrm{MPa}$.

Wall superheat

Bergles and Rohsenow (1964) studied the wall superheat that corresponded to the nucleate boiling of distilled water in a stainless steel tube of inner diameter $d = 2.387\,\mathrm{mm}$. The measurements of $\Delta T_{S,ONB} = T_{W,ONB} - T_S$ corresponding to the ONB were performed at different values of inlet temperature, T_{in}, in the range of inlet velocity $U_{in} = 3.74-19.42\,\mathrm{m/s}$, and Reynolds number corresponding to inlet parameters $\mathrm{Re} = 9 \times 10^3 - 2 \times 10^5$. Wall temperature measurements were taken at the heated length to the tube diameter ratio of $L/d = 29$ and 48. Experiments were restricted to low heat fluxes. This usually limits subcooled-boiling tests to the region of low wall superheat.

During the subcooled nucleate flow boiling of a liquid in a channel the bulk temperature of the liquid at ONB, T_B, is less than the saturation temperature, and at a given value of heat flux the difference $\Delta T_{sub,ONB} = T_S - T_B$ depends on L/d. The experimental parameters are presented in Table 6.2.

Table 6.2 shows that the ratio of $q_{ONB}/\rho U_{in} c_p$ changes relatively weakly in the range of wide variation of heat flux and inlet flow velocity. The wall superheat significantly depends on heat flux. The data of Bergles and Rohsenow (1964) are shown in Fig. 6.1. This dependence is close to $\Delta T_{S,ONB} \sim q_{ONB}^{0.5}$. The results presented by

Table 6.2 Parameters of experiments by Bergles and Rohsenow (1964)

Inlet velocity	Inlet temperature		Subcooling		Heat flux	Wall superheat	Parameter	
	$L_{ONB}/d = 29$	$L_{ONB}/d = 48$	$L_{ONB}/d = 29$	$L_{ONB}/d = 48$			$L/d = 29$	$L/d = 48$
U_{in} [m/s]	T_{in} [°C]		$\Delta T_{sub,ONB}$ [K]		q_{ONB} [MW/m^2]	$\Delta T_{S,ONB}$ [K]	$q_{ONB}/\rho_{ONB} U_{in} c_p$ [K]	
19.42	8	40	108.3	66.6	9.974	27	0.126	0.126
9.54	13	52	103.3	50	5.67	19.4	0.149	0.146
2.145	22	47	92.2	56.6	3.71	12.2	0.144	0.144

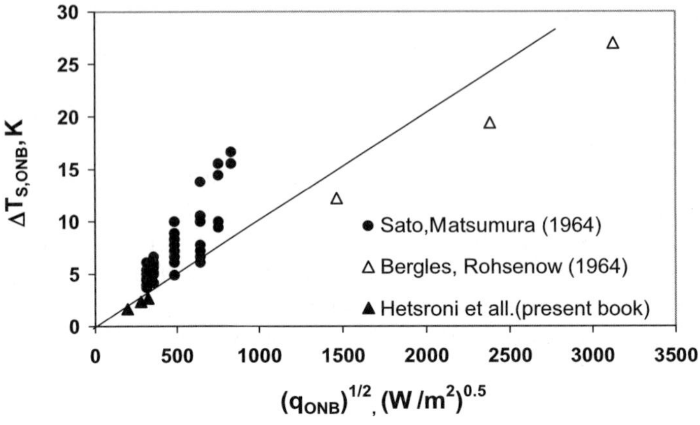

Fig. 6.1 Dependence of wall superheat on heat flux

Sato and Matsumura (1964) and the results of flow boiling in the 1.07 mm tube with average roughness of $k_s = 0.29\,\mu$m, obtained in the range of Reynolds number Re = 50–150 by Hetsroni et al. are also shown in Fig. 6.1. The values of $\Delta T_{S,ONB}$ presented by Sato and Matsumura (1964) are higher than those obtained at the same heat flux by Bergles and Rohsenow (1964), and Hetsroni et al. According to studies by Hsu (1962), Sato and Matsumura (1964), Davis and Anderson (1966), Kandlikar et al. (1997), and Liu et al. (2005) one can conclude that $\Delta T_{S,ONB} \sim q_{ONB}^{0.5}$

$$q_{ONB} = M \frac{k_L h_{LG} \rho_G}{\sigma T_S} \Delta T_{S,ONB}^2 \tag{6.9}$$

where M is constant in a wide range of $q_{ONB} = 10^3 - 10^6\,\mathrm{W/m^2}$, and k_L, ρ_G, σ are the thermal conductivity of liquid, the vapor density, and the surface tension, respectively. For example M is $1/12$ in Hsu (1962), $1/8$ in Sato and Matsumura (1964), $1/8(1+\cos\theta)$ in Davis and Anderson (1966), where θ is contact angle, and $1/9.2$ (Kandlikar et al. 1997).

Experiments by Bergles and Rohsenow (1964) corresponded to boiling incipience in relatively short channels when subcooling of the working fluid at the ONB point was about $\Delta T_{sub,ONB} = 50-100$ K, where $\Delta T_{sub,ONB} = T_S - T_{B,ONB}$ and $T_{B,ONB}$ is the bulk (mean mass) fluid temperature at the channel cross-section, where ONB occurs. Unal (1975) studied the incipience of boiling at relatively high values of L_{ONB}/d_h. The experimental determination of the ONB in this study has been carried out in such a way: The number of bubbles appearing on the developed films have been counted for different inlet temperatures at a constant heat flux q, pressure P, and mass flux G. It was observed that not all the bubbles were attached to the wall of the test section, but some were in the bulk of the flow. In these experiments the value of $\Delta T_{sub,ONB}$ did not exceed 4.3 K.

Heat transfer characteristics during flow boiling of water in the $d = 1.5$ mm tube were studied by Hapke et al. (2000). The measured roughness was on the order of magnitude of 5 μm. The axial distribution of the external wall temperature was measured using the thermographic method. The temperature reached a maximum at the initial point after the single-phase liquid heating. A wall superheat occurred at this point. The local temperature varied near the initial point when the boiling started. Amplitudes of 2 K with a frequency of approximately 2 Hz were achieved. The results measured were compared with those presented by Sato and Matsumura (1964) and by Bergles and Rohsenow (1964). A satisfactory agreement existed only for heat fluxes of about $q = 50\,\mathrm{W/m^2}$. However, the wall has to be superheated to a relatively great extent to initiate the nucleate boiling in the experiments by Hapke et al. (2000) compared to those reported by Sato and Matsumura (1964) and by Bergles and Rohsenow (1964). A mass flux dependence of the wall superheat was reported by Hapke et al. (2000). Figure 6.2 shows the dependence of ΔT_S on the parameter q_{ONB}/G.

Qu and Mudawar (2002) performed experiments to measure the incipient boiling heat flux q_{ONB}, in a heat sink containing 21 rectangular micro-channels 231 μm wide and 713 μm deep. Tests were performed using de-ionized water with inlet liquid velocities of 0.13–1.44 m/s, inlet temperatures of 30, 60, and 90 °C, and outlet pressure of 0.12 MPa. Using a microscope, boiling incipience was identified when the first bubbles were detected growing at, and departing from the micro-channel wall. The authors conclude that bubble behavior at incipient boiling in micro-channels is quite different from that in large channels. At incipient boiling, a small number of nucleation sites were observed close to the exit of several micro-channels. The detachment size was comparable to that of the micro-channel cross-section for lower velocities and decreased progressively with increasing velocity. Results of these experiments are presented in Fig. 6.3. Figure 6.3 shows that the wall superheat is directly proportional to the heat flux, $\Delta T_{S,ONB} \sim q_{ONB}$.

Sato and Matsumura (1964), and Bergles and Rohsenow (1964) have proposed the equation for the incipient boiling condition in the case that surface cavities of all sizes are available for nucleation. Hino and Ueda (1985) studied incipient boiling of fluorocarbon R-113 in a stainless steel tube of $d = 7$ mm at mass velocity $G = 158-1{,}600\,\mathrm{kg/m^2s}$, and inlet subcooling $\Delta T_{sub,in} = T_S - T_{in}$ ranged from 10–

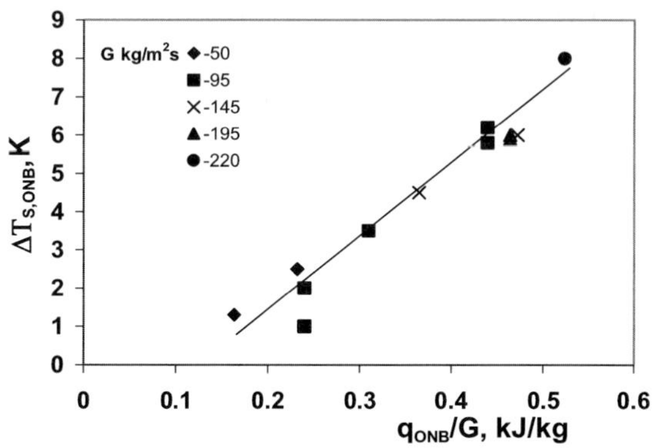

Fig. 6.2 Dependence of wall superheat $\Delta T_{S,ONB} = T_{W,ONB} - T_S$ at ONB point on the parameter q_{ONB}/G observed by Hapke et al. (2000) in the $d = 1.5$ mm tube

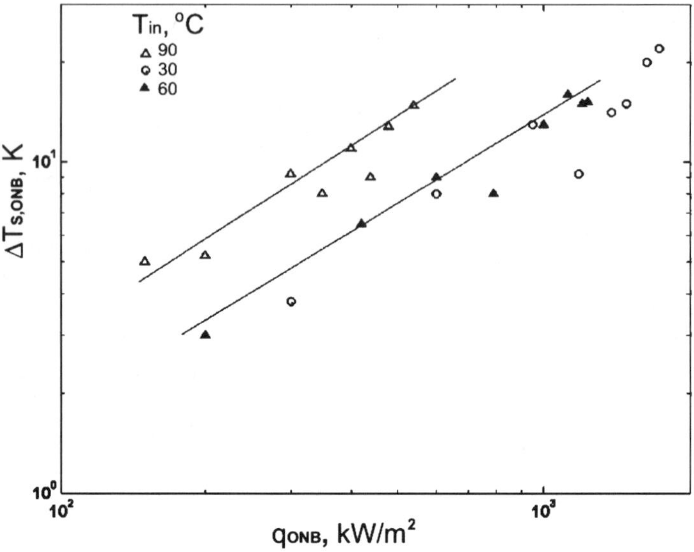

Fig. 6.3 Dependence of wall superheat on heat flux. Experiments performed by Qu and Mudawar (2002) in rectangular parallel micro-channels 231 μm wide and 713 μm deep

30 K. In the case where the upper limit of available cavity sizes was restricted to radius r^*_{max} the incipient boiling condition was expressed as follows by Hino and Ueda (1985):

$$q_{ONB} = \frac{k_L}{r^*_{max}} \Delta T_{ONB} - \frac{2\sigma k_L T_S}{h_{LG}\rho_G (r^*_{max})^2}. \tag{6.10}$$

Figure 6.4 shows the relation between the heat flux and the wall superheat at the ONB position obtained by Hino and Ueda (1975) in the range of the largest cavity radius $r^*_{max} = 0.22{-}0.34\,\mu m$. Experimental points show that the wall superheat at the ONB position was practically independent of the mass flux and the inlet subcooling. The lines shown in this figure represent the values of Sato and Matsumura (1964), and Bergles and Rohsenow (1964). The wall superheats reported by Hino and Ueda (1975) were much greater than those predicted by Eq. (6.9).

The data of ONB in trapezoidal micro-channels of $d_h = 41.3\,\mu m$ presented by Lee et al. (2004) are shown in Fig. 6.5. Figure 6.5 illustrates a comparison of the data of the results reported by Lee et al. (2004) and prediction of Eq. (6.10) with various different values of r^*_{max}. For the experimental data points in Fig. 6.5, the saturation temperature is corresponding to the local pressure at each of the ONB locations. The local pressure is estimated by assuming a linear pressure distribution in the channel between the inlet and exit ones. The system pressure may vary from case to case. For Fig. 6.5 an average system pressure of 161.7 kPa over various different cases of this study was employed. As for the wall temperature, it is assumed that the channel wall temperature is uniform as the channel is relatively short and the wall material, silicon, has relatively good thermal conductivity. The figure indi-

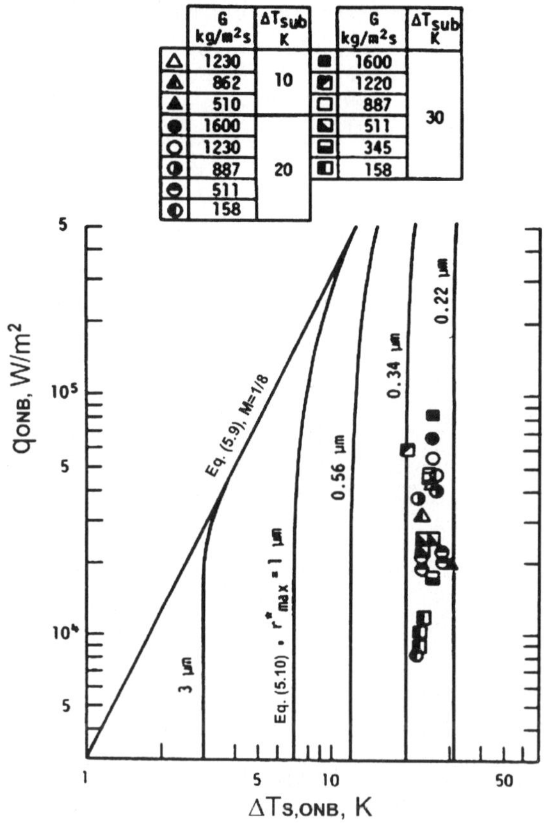

Fig. 6.4 Relation between heat flux and wall superheat at the position of incipient boiling. Reprinted from Hino and Ueda (1975) with permission

cates that most of the cavity sizes ranged from 1.5 to 4 μm. This is consistent with the maximum roughness on the side wall of the channels used in experiments. Experiments showed that for a given value of r^*_{max}, wall superheat, $\Delta T_{S,ONB}$, does not depend on mass flux.

Effect of pressure

The wall superheat in Eq. (6.9) depends on saturation temperature T_S and consequently on pressure P_S:

$$\frac{\sigma T_S}{k_L \rho_G} = \psi(P_S)\,. \tag{6.11}$$

The analysis shows that Eq. (6.11) can be presented for water in the pressure range $P_S = 1-20$ bar as follows

$$\left(\frac{\sigma T_S}{k_L \rho_G}\right)^{1/2} = 7.33 P_S^{-0.5}\,. \tag{6.12}$$

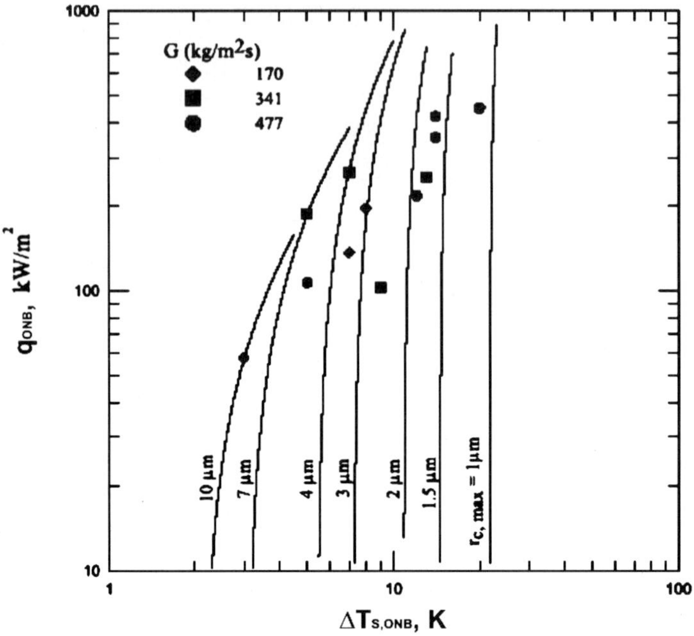

Fig. 6.5 Relationship between heat flux and wall superheat in micro-channels of $d_h = 41.3\,\mu$m. Reprinted from Lee et al. (2004) with permission

Using Eq. (6.9) and Eq. (6.12) the wall superheat is given by

$$\Delta T_{S,ONB} = M P_S^{\alpha} (q_{ONB})^{\beta} \tag{6.13}$$

where M is constant.

Analysis of experimental data presented by Sato and Matsumura (1964), Bergles and Rohsenow (1964), Hetsroni et al. as well as theoretical analysis by Hsu (1962), Davis and Anderson (1966), Kandlikar et al. (1997) shows that the values of exponents α and β are close to $\alpha = -0.50$, $\beta = 0.50$ (Table 6.3).

Table 6.3 Exponents in Eq. (6.13)

Author	Exponents α, β		Source
	α	β	
Hsu (1962)	−0.50	0.50	Theoretical analysis
Sato, Matsumura (1964)	−0.50	0.50	α Theoretical analysis β Empirical correlation
Bergles, Rohsenow (1964)	−0.51	0.49	Empirical correlation
Davis, Anderson (1966)	−0.50	0.50	Theoretical analysis
Kandlikar (1997)	−0.50	0.50	Theoretical analysis
Hetsroni et al.	–	0.50	Empirical correlation

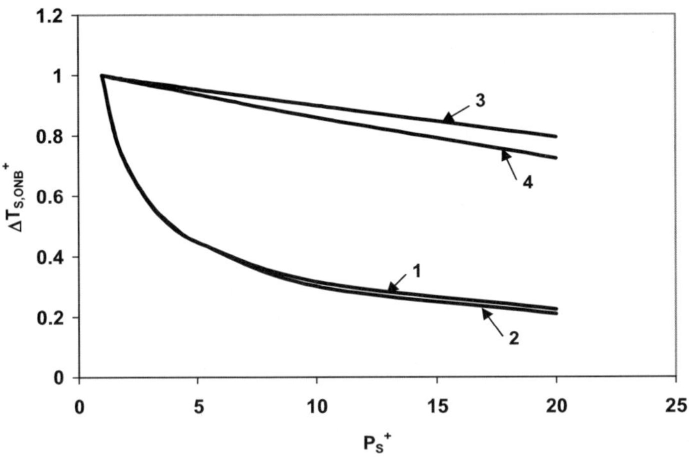

Fig. 6.6 Dependence of dimensionless wall superheat $\Delta T^+_{\mathrm{S,ONB}} = \Delta T_{\mathrm{ONB}}/\Delta T^*_{\mathrm{ONB}}$ on dimensionless pressure $P^+_{\mathrm{S}} = P_{\mathrm{S}}/P^*_{\mathrm{S}}$. *1* Hsu (1962), *2* Bergles and Rohsenow (1964), *3* Thom et al. (1965), *4* Jens and Lottes (1951)

On the other hand, Jens and Lottes (1951) and Thom et al. (1965) presented empirical correlations that showed another dependence of wall superheat on pressure. Dependence of the dimensionless wall superheat $\Delta T^+_{\mathrm{S,ONB}} = \Delta T_{\mathrm{ONB}}/\Delta T^*_{\mathrm{ONB}}$ on dimensionless pressure $P^+_{\mathrm{S}} = P_{\mathrm{S}}/P^*_{\mathrm{S}}$ is presented in Fig. 6.6 where $\Delta T^*_{\mathrm{ONB}}$ and P^*_{S} corresponds to wall superheat and pressure of 1 bar. In Fig. 6.6 the theoretical prediction based on the Hsu (1962) model, and empirical correlation of Bergles and Rohsenow (1964) are shown.

The Bergles and Rohsenow correlation (1964) is

$$\Delta T_{\mathrm{S,ONB}} = 0.555\left(\frac{q_{\mathrm{ONB}}}{1{,}082P^{1.156}}\right)^{P^{0.0234}/2.16} \tag{6.14}$$

where P is in bar, $\Delta T_{\mathrm{S,ONB}}$ in K, and q_{ONB} in W/m^2.

The empirical correlation by Jens and Lottes (1951) is

$$\Delta T_{\mathrm{S,ONB}} = 25(q_{\mathrm{ONB}})^{0.25}\exp\left(-\frac{P}{62}\right) \tag{6.15}$$

where P is in bar, $\Delta T_{\mathrm{S,ONB}}$ in K, and q_{ONB} in W/m^2.

The empirical correlation by Thom et al. (1965) is

$$\Delta T_{\mathrm{S,ONB}} = 22.65(q_{\mathrm{ONB}})^{0.5}\exp\left(-\frac{P}{87}\right) \tag{6.16}$$

where P is in bar, $\Delta T_{\mathrm{S,ONB}}$ in K, and q_{ONB} in W/m^2.

It is seen that the empirical correlation by Bergles and Rohsenow (1964) agrees fairly well with prediction of theoretical analysis based on the Hsu (1962) model. Empirical correlations by Jens and Lottes (1951) and by Thom et al. (1965) show that wall superheat weakly depends on pressure.

Comparison of wall superheat predicted by classical kinetics of nucleation to experimental results

Using the properties of water Li and Cheng (2004) computed from the classical kinetics of nucleation the homogeneous nucleation temperature and the critical nucleation radius r_{cr}. The values are: $T_{S,B} = 303.7\,°C$ and $r_{crit} = 3.5$ nm. However, the nucleation temperatures of water in heat transfer experiments in micro-channels carried out by Qu and Mudawar (2002), and Hetsroni et al. (2002b, 2003, 2005) were considerably less that the homogeneous nucleation temperature of $T_{S,B} = 303.7\,°C$. The nucleation temperature of a liquid may be considerably decreased because of the following effects: dissolved gas in liquid, existence of corners in a micro-channel, surface roughness.

Liquid subcooling at ONB point

The data of liquid subcooling at ONB point presented in Table 6.4 indicate that the value of $\Delta T_{sub,ONB} = T_S - T_{B,ONB}$ changes in the wide range of $\Delta T_{sub,ONB} =$

Table 6.4 Liquid subcooling at ONB point

Author and method of ONB detection	Heat flux q_{ONB} MW/m^2	Inlet flow velocity U_{in} m/s	Pressure P MPa	Relative heated length L_{ONB}/d	Fluid subcooling $\Delta T_{sub,ONB}$ K
Unal (1975)	0.38	2.121	13.9	1250	4.3
High-speed photographic technique	0.45	2.121	15.8	1250	4.1
Bergles and	9.774	19.2	0.261	29	108.3
Rohsenow	9.774	19.2	0.261	48	66.6
(1964)	6.67	9.54	0.261	29	103.3
Dependence	5.67	9.54	0.261	48	50.0
of wall	2.145	3.74	0.261	29	92.2
superheat on heat flux	2.145	3.74	0.261	48	56.6
Kennedy et al.	1.5–4.0	1.0–4.0	1.034	137	5–12
(2000)	1.5–3.0	1.3–2.6	1.034	110	13–19
Dependence	1.5–3.0	3.5	0.69	137	4.0
of pressure	1.5–3.0	1.3	0.69	110	4.0
drop on mass	1.5–3.0	2.0	0.344	137	5.0
flux	1.5–3.0	4.3	0.344	110	5.0

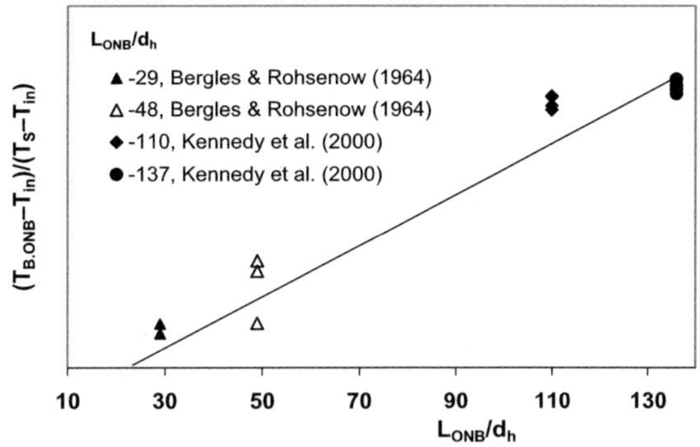

Fig. 6.7 Relationship of $(T_{B,ONB} - T_{in})/(T_S - T_{in})$ on L_{ONB}/d_h

4.0–108.3 K, where $T_{B,ONB}$ is the average mass liquid temperature at the cross-section where ONB occurs.

Figure 6.7 shows the relationship of $(T_{B,ONB} - T_{in})/(T_S - T_{in})$ on L_{ONB}/d_h. The results obtained by different investigators indicate that in the range of $q_{ONB} = 1.5–9.8\,MW/m^2$ the value of $(T_{B,ONB} - T_{in})/(T_S - T_{in})$ depends only on the parameter L_{ONB}/d_h.

For qualitative analysis of the conditions at which the ONB phenomenon was studied experimentally in conventional size channels, as well as in micro-channels the parameter $D = \Delta T_{sub,ONB}/T_S$ may be used. Depending on the value of D the channels can be subdivided into two groups: (1) $D < 1$, and (2) $D \ll 1$. When the value of D was in the range 0.125–0.25, as in experiments by Bergles and Rohsenow (1964), the onset of nucleate boiling occurred at values of the bulk temperature, $T_{B,ONB}$, significantly less than the saturation temperature. It should be noted that at low values of $T_{B,ONB}$ the Prandtl number of liquid is large enough. As a result the thin thermal layer formed over the channel wall. When $D \ll 1$, the onset of nucleate boiling occurred at values of the bulk temperature, $T_{B,ONB}$, that were close to the saturation temperature. Experiments by Unal (1975), Kennedy et al. (2000), and by Liu et al. (2005) were carried out under such conditions. For example, in experiments by Kennedy et al. (2000) the value of D was in the range 0.011–0.043. Figure 6.8 illustrates the fluid temperature distribution depending on the relation of $D = \Delta T_{sub,ONB}/T_S$. The dark area corresponds to the region where the local fluid temperature T exceeds the saturation temperature T_S. In this region the maximum probability of the bubble embryo formation takes place.

6.1.3 Effect of Inlet Velocity on Wall Superheat

The point of ONB located at the channel exit corresponds to minimum heat flux at which the boiling incipience is possible for the channel of the given length. Kennedy et al. (2000) presented the measurements that were carried out at different values of the inlet average velocity U_{in}, whereas the inlet temperature T_{in}, was constant.

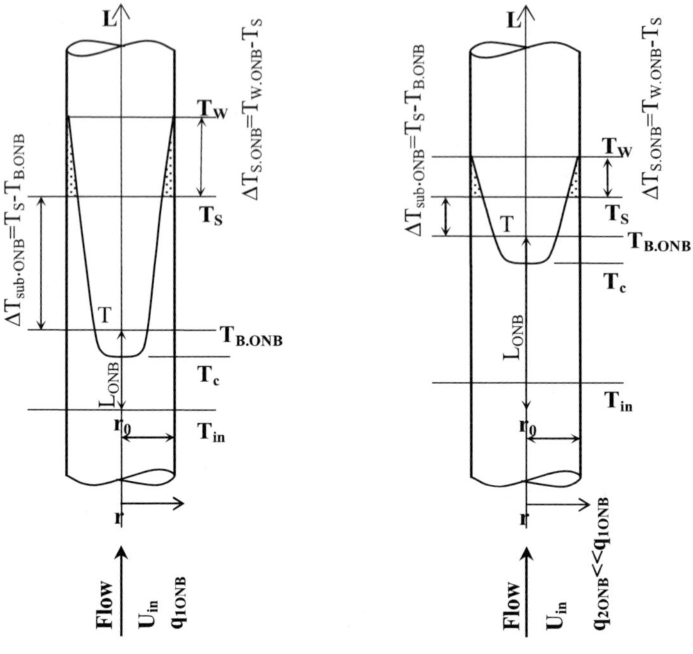

$T_{B,ONB} < T_S$

$$D = \frac{\Delta T_{sub.ONB}}{T_S} < 1$$

$T_{B,ONB} \approx T_S$

$$D = \frac{\Delta T_{sub.ONB}}{T_S} << 1$$

Fig. 6.8 Fluid temperature distribution depending on the relation of $D = \Delta T_{\mathrm{sub,ONB}}/T_{\mathrm{S}}$

Fig. 6.9 Dependence of the boiling Stanton number at ONB point. Experiments by Kennedy et al. (2000)

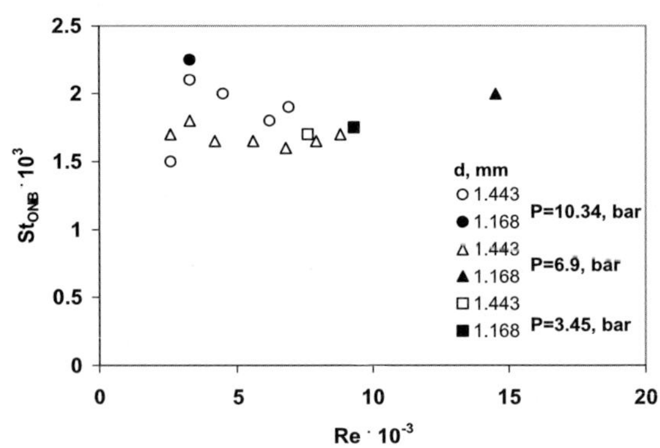

Figure 6.9 illustrates the dependence of the boiling Stanton number at ONB point, $\mathrm{St_{ONB}} = q_{\mathrm{ONB}}/[\rho_{\mathrm{L}} U_{\mathrm{in}} c_{\mathrm{p}} (T_{\mathrm{S}} - T_{\mathrm{in}})]$, on the Reynolds number, $\mathrm{Re} = U_{\mathrm{in}} d_{\mathrm{h}}/\nu$. One can see that the value of the St_{ONB} does not change significantly in the range of Re $\sim 0.4 \times 10^4 – 1.6 \times 10^4$, i.e., the value of q_{ONB} depends on the inlet average velocity when its temperature is constant.

To estimate the value of the bulk liquid temperature at ONB in conventional size channels, as well as in micro-channels the energy and continuity equations should be considered.

Energy and continuity equations for cylindrical channels are:

$$\rho u c_{\mathrm{p}} \frac{\partial T}{\partial x} + \rho v c_{\mathrm{p}} \frac{\partial T}{\partial y} = \frac{k}{y} \frac{\partial}{\partial y}\left(y \frac{\partial T}{\partial y} \right) + k \frac{\partial^2 T}{\partial x^2} \tag{6.17}$$

$$\frac{\partial(\rho u)}{\partial x} + \frac{1}{y} \frac{\partial(\rho u)}{\partial y} = 0 \tag{6.18}$$

where c_{p} and k are specific heat and thermal conductivity of the liquid, respectively, and u, v are streamwise and spanwise components of the velocity.

The boundary conditions for Eqs. (6.17) and (6.18) read as

$$\begin{aligned} &x = 0\,, \quad u = U_{\mathrm{in}}\,, \quad \rho = \rho_{\mathrm{in}}\,, \quad T = T_{\mathrm{in}} \\ &x > 0 \quad \begin{cases} y = 0\,, & \frac{\partial T}{\partial y} = 0 \\ y = r_0\,, & -\lambda \frac{\partial T}{\partial y} = q \end{cases} \end{aligned} \tag{6.19}$$

where r_0 is the tube radius, q is the heat flux on the wall.

Restricting our consideration analysis of the developed flow we rewrite Eqs. (6.17) and (6.18) in the form

$$\rho u c_{\mathrm{p}} \frac{\partial (T - T_{\mathrm{in}})}{\partial x} = \frac{k}{y} \frac{\partial}{\partial y}\left(y \frac{\partial T - T_{\mathrm{in}}}{\partial y} \right) + k \frac{\partial^2 (T - T_{\mathrm{in}})}{\partial x^2} \tag{6.20}$$

$$\frac{\partial \rho u}{\partial x} = 0\,. \tag{6.21}$$

Multiplying Eq. (6.21) on $c_{\mathrm{p}}(T - T_{\mathrm{in}})$ and summing this equation with Eq. (6.20) we obtain

$$\frac{\partial \rho u c_{\mathrm{p}} (T - T_{\mathrm{in}})}{\partial x} = \frac{k}{y} \frac{\partial}{\partial y}\left(y \frac{\partial (T - T_{\mathrm{in}})}{\partial y} \right) + k \frac{\partial^2 (T - T_{\mathrm{in}})}{\partial x^2}\,. \tag{6.22}$$

Integration of Eq. (6.22) gives

$$\frac{\partial}{\partial x} \int\limits_0^{r_0} \rho u c_{\mathrm{p}} (T - T_{\mathrm{in}}) y\,\mathrm{d}y = k y \left. \frac{\partial (T - T_{\mathrm{in}})}{\partial y} \right|_0^{r_0} + k \frac{\partial^2}{\partial x^2} \int\limits_0^{r_0} (T - T_{\mathrm{in}}) y\,\mathrm{d}y\,. \tag{6.23}$$

Omitting in Eq. (6.23) the last term that allows for axial heat transfer, Eq. (6.23) may be expressed:

$$\frac{\partial}{\partial x} \int\limits_0^{r_0} \rho u c_{\mathrm{p}} (T - T_{\mathrm{in}}) y\,\mathrm{d}y = q r_0\,. \tag{6.24}$$

Integration of Eq. (6.24) from $x = 0$ to $x = L_{\mathrm{ONB}}$ gives

$$\frac{\partial}{\partial x} \int\limits_0^{r_0} \rho u_{\mathrm{ONB}} c_{\mathrm{p}} (T_{\mathrm{ONB}} - T_{\mathrm{in}}) y\,\mathrm{d}y = q_{\mathrm{ONB}} r_0 L_{\mathrm{ONB}}\,. \tag{6.25}$$

Introducing the mean mass temperature gives

$$T_{\mathrm{B,ONB}} = \frac{1}{2\rho U_{\mathrm{in}} c_{\mathrm{p}} r_0^2} \int_0^{r_0} \rho u c_{\mathrm{p}} T_{\mathrm{ONB}} y \, \mathrm{d}y \,. \tag{6.26}$$

Equation (6.26) may be presented in the form

$$\frac{q_{\mathrm{ONB}}}{\rho U_{\mathrm{in}} c_{\mathrm{p}} (T_{\mathrm{B.ONB}} - T_{\mathrm{in}})} \frac{2L_{\mathrm{ONB}}}{r_0} = 1 \,. \tag{6.27}$$

Equation (6.27) was obtained for circular channels. Rectangular channels with four or three conductive walls are shown in Fig. 6.10.

The thermal balance equation for rectangular channels with four or three heated walls shown in Fig. 6.10 are:

$$\iint_{\mathrm{S}} \rho u c_{\mathrm{p}} (T_{\mathrm{B,ONB}} - T_{\mathrm{in}}) \, \mathrm{d}y \, \mathrm{d}z = q_{\mathrm{ONB}} L_{\mathrm{ONB}} \cdot 4(a+b) \tag{6.28}$$

and

$$\iint_{\mathrm{S}} \rho u c_{\mathrm{p}} (T_{\mathrm{B,ONB}} - T_{\mathrm{in}}) \, \mathrm{d}y \, \mathrm{d}z = q_{\mathrm{ONB}} L_{\mathrm{ONB}} \cdot 2(2a+b) \tag{6.29}$$

where S is the area of channels cross-section:

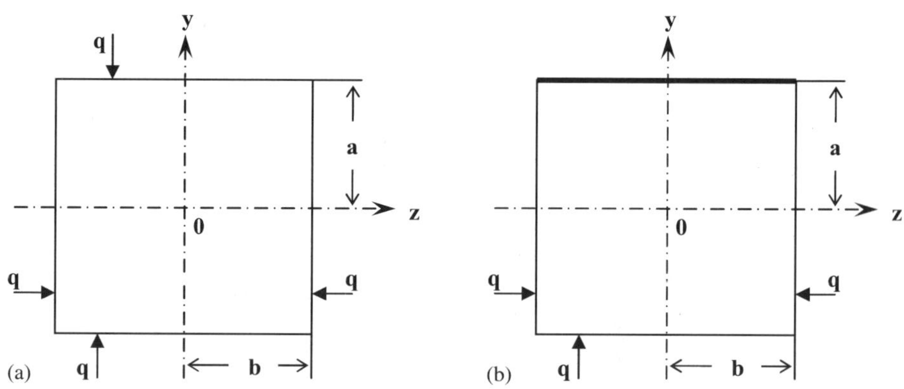

Fig. 6.10 Rectangular channels: (**a**) four heated walls, (**b**) three heated walls

In the terms of mean mass temperature Eqs. (6.28) and (6.29) take the form

$$\frac{q_{\mathrm{ONB}}}{\rho U_{\mathrm{in}} c_{\mathrm{p}} (T_{\mathrm{B,ONB}} - T_{\mathrm{in}})} \frac{L_{\mathrm{ONB}}(a+b)}{ab} = 1 \tag{6.30}$$

and

$$\frac{q_{\mathrm{ONB}}}{\rho U_{\mathrm{in}} c_{\mathrm{p}} (T_{\mathrm{B,ONB}} - T_{\mathrm{in}})} \frac{L_{\mathrm{ONB}}(a+b)}{2ab} = 1 \,. \tag{6.31}$$

If the value of $T_{B,ONB} \approx T_S$, ($D \ll 1$) the integral characteristics for the circular channels, the rectangular channels with four and three heated walls, respectively, are:

$$\frac{q_{ONB}}{\rho U_{in} c_p (T_S - T_{in})} \frac{2L_{ONB}}{r_0} = I_{c,ONB} \tag{6.32}$$

$$\frac{q_{ONB}}{\rho U_{in} c_p (T_S - T_{in})} \frac{L_{ONB}(a+b)}{ab} = I_{4,ONB} \tag{6.33}$$

$$\frac{q_{ONB}}{\rho U_{in} c_p (T_S - T_{in})} \frac{L_{ONB}(2a+b)}{2ab} = I_{3,ONB} \,. \tag{6.34}$$

The bulk temperature $T_{B,ONB}$ is close to saturation temperature T_S, when the values calculated using Eqs. (6.32), (6.33) and (6.34) do not differ significantly from unity. In Fig. 6.11 the experimental results reported by Kennedy et al. (2000) are presented as the dependence of the value $I_{c,ONB}$ (Eq. 6.32) on the Peclet number. The data may be described by the single line of $I_{c,ONB} = 0.96$. In this case the bulk temperature $T_{B,ONB}$, at ONB point should not differ significantly from T_S. Experimental results given in Table 6.4 support this statement.

The onset of nucleate boiling in the flow of water through a micro-channel heat sink was investigated by Liu et al. (2005). The rectangular micro-channels of $d_h = 384\,\mu\text{m}$ and length of 25.4 mm were used. Holes were drilled into the bottom of the copper block to house eight cartridge heaters that can provide heating through three walls. Onset of nucleate boiling was identified with a high-speed imaging system. To complement the incipient heat flux results identified from the visualization approach, the micro-channel wall temperatures and pressure drop along the micro-channels were analyzed. Experimental parameters obtained in this study are presented in Fig. 6.12 as the dependence of the value $I_{3,ONB}$ on the Peclet number. In this case the parameter $I_{3,ONB} \sim 0.93$, and T_b should also not differ significantly from T_S. Unfortunately, the experimental data of T_b were not reported by Liu et al. (2005).

Parameters $I_{c,ONB}$, $I_{3,ONB}$, and $I_{4,ONB}$ change in the range of $0 < T < 1$. They account for a specific temperature field in heated micro-channels and are criteria for the relative micro-channel length. Note, if $D < 1$ the value of parameter I is significantly less than unity. The paper by Celata et al. (1997) reports the results of experimental research of the onset of subcooled water boiling in the circular

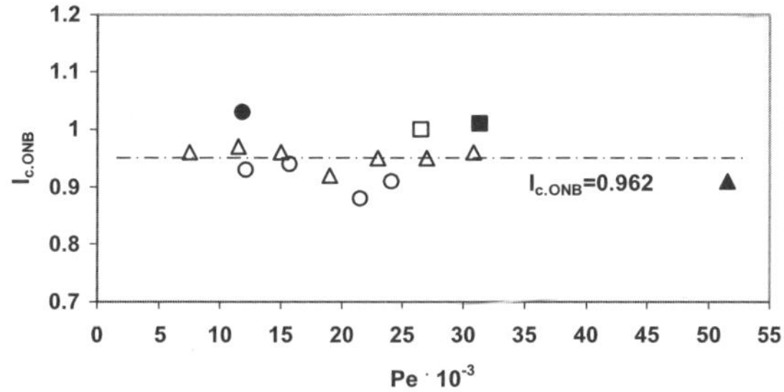

Fig. 6.11 Dependence of the value $I_{c,ONB}$ on the Peclet number (the legend is given in Fig. 6.9). Experiments by Kennedy et al. (2000)

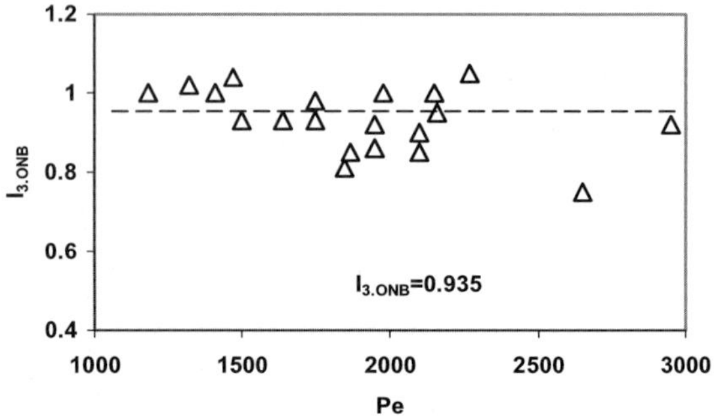

Fig. 6.12 Dependence of the value $I_{3,ONB}$ on the Peclet number. Experiments by Liu et al. (2005)

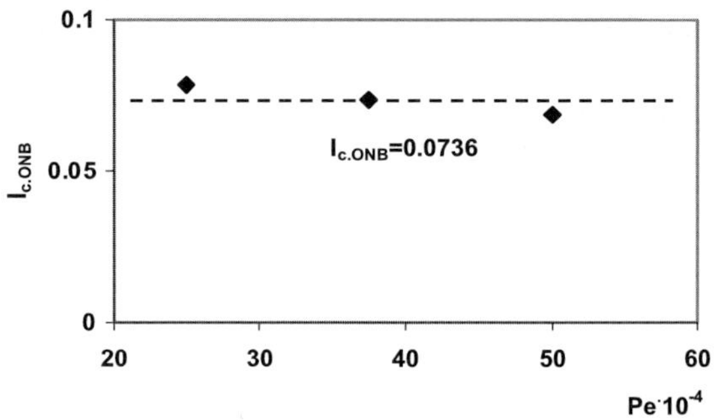

Fig. 6.13 Value of I_c as a function of the Peclet number. Experiments by Celata et al. (1997)

channel with $d = 8$ mm, and $L = 100$ mm. Test conditions were as follows: $U_{in} = 5-10$ m/s, $P_{out} = 1.0-2.5$ MPa, $T_{in} = 30-60\,°$C, inlet subcooling $\Delta T_{sub,in} = T_S - T_{in}$ from 120 to 194 K, $q_{ONB} = 3-11\,\mathrm{MW/m^2}$. Figure 6.13 shows the value of $I_{c,ONB}$ as a function of the Peclet number. The authors did not report liquid subcooling at ONB. In this study the parameter was $I_{c,ONB} \sim 0.074$. According to an estimation based on such a small value of $I_{c,ONB}$ one can conclude that onset of nucleate boiling in experiments by Celata et al. (1997) occurred at high liquid subcooling.

The wall superheat that corresponds to bubble formation in liquid flow can be estimated using an approach that is not connected to the mechanism of bubble formation. Such tentative estimation makes it possible to consider only the low level of wall superheat. According to Kays and Krawford (1993) the temperature distribution in turbulent flow and Pr ~ 1 is

$$\frac{T_W - T}{T_W - T_c} = \left(\frac{r_0 - r}{r_0}\right)^{1/7} \tag{6.35}$$

where T is the current fluid temperature, and T_c is the liquid temperature at the tube axis.

From the analytical solution based on the thermal boundary approach:

$$T_w - T_c = \frac{q}{\rho c_p U}\frac{1}{\sqrt{f/2}}\left(5\,\mathrm{Pr} + 5\ln(5\,\mathrm{Pr}+1) + 2.5\ln\frac{\mathrm{Pr}\sqrt{f/2}}{60}\right) \tag{6.36}$$

where $f = 2\tau_w/(\rho U^2)$, τ_w is the wall shear stress.

Dependence of wall superheat on inlet fluid velocity

For $D \ll 1$, $T_{B,ONB} \approx T_S$ and

$$\Delta T_{S,ONB} = T_{W,ONB} - T_S \,. \tag{6.37}$$

Under these conditions the wall superheat depends weakly on the Reynolds number (Kennedy et al. 2000).

The same conclusion is evident from results obtained by Hino and Ueda (1975) and presented above in Fig. 6.4. The conclusion that ΔT_S is almost unaffected by inlet flow velocity as at $D \ll 1$ as at $D < 1$ was established from experiments carried out in the channels of diameters about $d = 1-10$ mm. What has been commonly observed at incipient boiling for subcooled flow in channels of this size is that small bubbles nucleate, grow and collapse while still attached to the wall, as a thin bubble layer formed along the channel wall.

6.1.4 Effect of Inlet Parameters on Incipient Boiling Heat Flux

Effect of inlet velocity

Qualitative analysis of the Eqs. (6.32), (6.33), and (6.34) makes it possible to illustrate the salient features of the dependence q_{ONB} on U_{in}. They could be approximated by the following lines shown in Fig. 6.14a. The solid line corresponds to the case of $T_{in} \ll T_S$ ($D < 1$), the dotted line corresponds to the case of $T_{in} \approx T_S$ ($D \ll 1$). Experimental results reported by Liu et al. (2005) (see Fig. 6.14b) agree qualitatively with analytical prediction. The same tendency was also observed by Qu and Mudawar (2002).

Effect of inlet temperature on q_{ONB}

From Eqs. (6.32), (6.33) and (6.34) one can conclude that incipient boiling heat flux q_{ONB} depends on the inlet temperature T_{in}. Such a behavior is shown in Fig. 6.15a. Note that at the same value of T_{in} the value of q_{ONB} increases with increasing inlet velocity. Experimental data by Liu et al. (2005) shown in Fig. 6.15b agree qualitatively with analysis.

6.1.5 Incipience of Boiling in Surfactant Solutions

Under some conditions boiling incipience in surfactant solutions may be quite different from that in Newtonian fluids. Hetsroni et al. (2007) presented results for

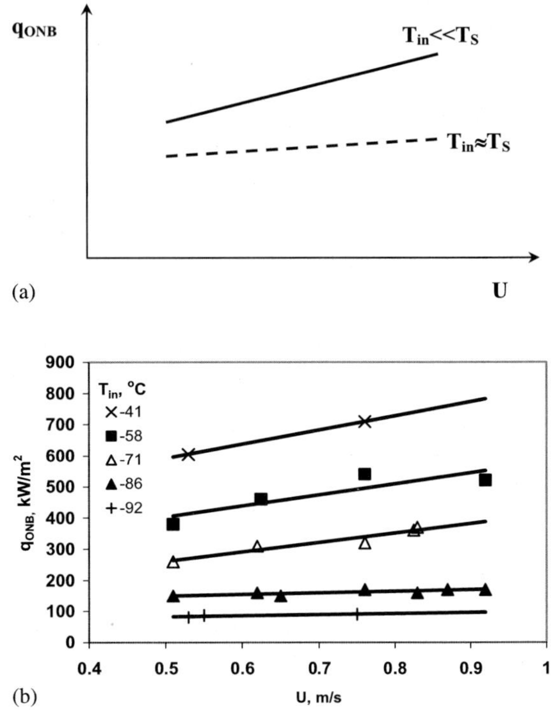

Fig. 6.14 Variation of incipient boiling heat flux with inlet flow velocity: (**a**) analytical predictions, (**b**) experimental data by Liu et al. (2005)

natural convection boiling in narrow horizontal annular channels of the gap size 0.45–2.2 mm for Alkyl (8-16) degraded solutions, i.e., solutions that were used after 6 to 10 runs.

For degraded Alkyl (8-16) solutions boiling occurred at wall superheat higher than that observed in fresh solutions and water. Incipience of boiling in both water and fresh surfactant solutions was accompanied by formation of small bubbles on the heated surface. However, a significant difference in the behavior of boiling patterns was observed. The formation of big vapor clusters took place before boiling incipience in degraded Alkyl (8-16) solutions in the range of concentrations $C = 10{-}600$ ppm (weight part per million). This process is shown in Fig. 6.16a–c. The burst of such a cluster is shown in Fig. 6.16d. The cluster formation was accompanied by high wall superheat ($T_W - T_S$) in heat flux controlled experiments, where T_W is the temperature measured on the heated wall, and T_S is the saturation temperature measured in the vessel. It should be stressed that these clusters were not gas (air) bubbles. The desorption of the dissolved gases formed bubbles of gas and a limited amount of bubbles containing gas–water vapor mixture. As a result, boiling incipience occurred at a heated wall temperature below that of saturation temperature. In the present study such a phenomenon was not observed. We also measured fluid temperature T_f in the annular space between the heated tube and the inner wall of glass tube by a thermocouple. This temperature exceeded over 4–12 K the saturation temperature depending on solution concentration. Finally, the collapse

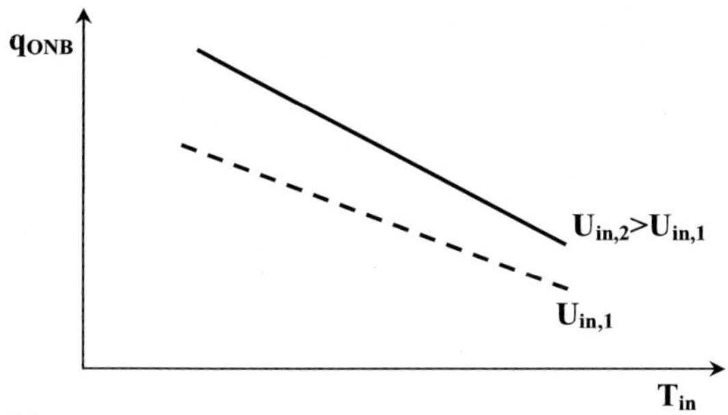

(a)

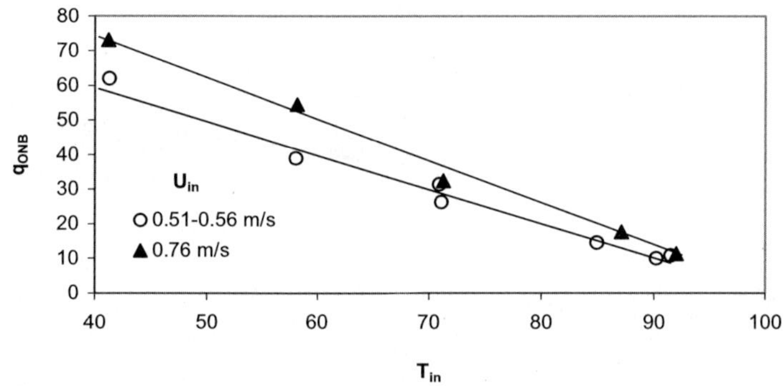

(b)

Fig. 6.15 Variation of incipient boiling heat flux with inlet flow temperature: (**a**) analytical predictions, (**b**) experimental data by Liu et al. (2005)

of the cluster led to a reduction in wall superheat and the saturated boiling regime occurred. For water boiling we did not observe the bubble coalescence at very small scales. For pool boiling of surfactant solutions bubble coalescence was observed. There were clusters of small bubbles, which rose from the cavity. These bubbles were adjacent to each other and the cluster neck was not observed. The bursting of vapor clusters before boiling incipience of degraded cationic surfactant Habon G solution was also observed by Hetsroni et al. (2002b).

Data were taken for both increasing and decreasing heat fluxes. The total mass of the liquid in the test facility remained constant, thus no fresh liquid was introduced to "top off" the system. For water boiling in the gap sizes of 0.45, 1.2, 2.2, and 3.7 mm, the Bond numbers, $\mathrm{Bn} = \delta(\sigma/g(\rho_L - \rho_G))^{-0.5}$, were 0.185, 0.493, 0.9 and 1.52, respectively, where δ is the gap size, σ is the surface tension, g is the acceleration due to gravity, and ρ_L and ρ_G are the liquid and the vapor densities. Boiling of surfactant solutions was investigated in a gap size of 0.45 and 2.2 mm in the range of Bond numbers $\mathrm{Bn} = 0.26-1.26$.

The results obtained at $\mathrm{Bn} = 1.26$ are presented in Fig. 6.17, for different concentrations of surfactant solutions. The onset of boiling corresponds to the curve ABCD for the runs with increasing heat flux. It follows the curve DCA for decreas-

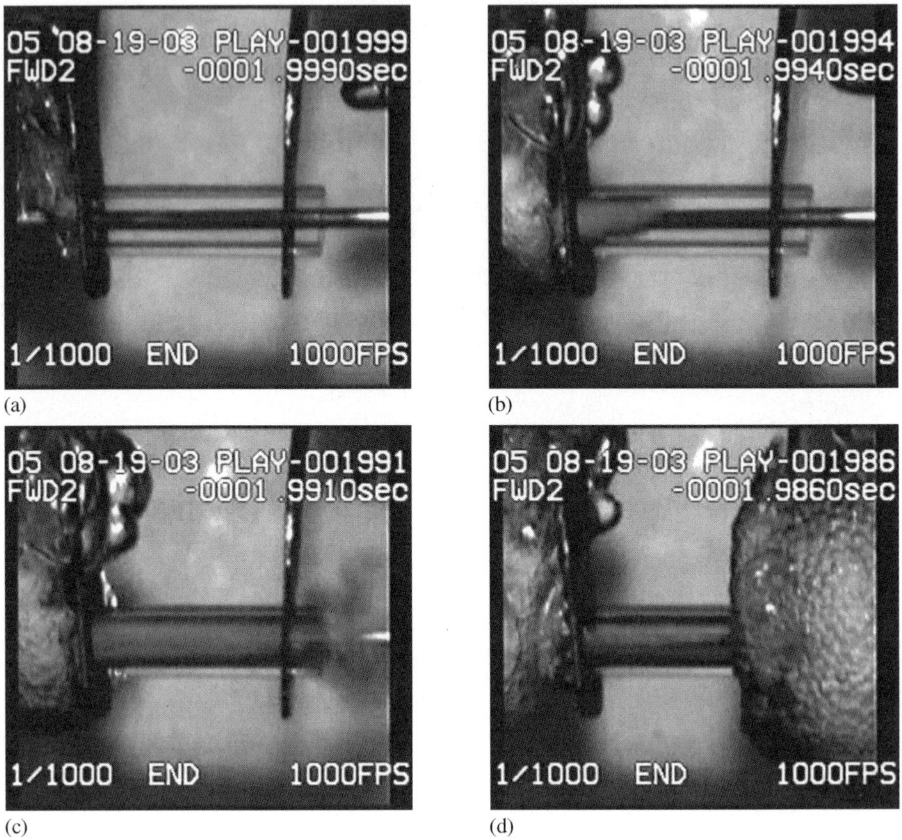

Fig. 6.16 Boiling incipience in degraded solutions. Reprinted from Hetsroni et al. (2007) with permission

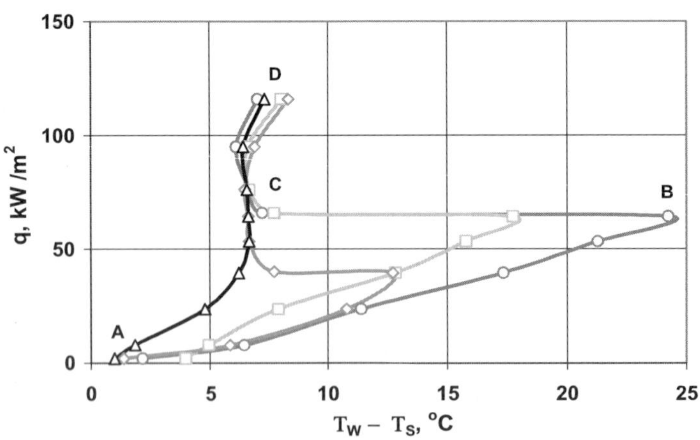

Fig. 6.17 Boiling hysteresis in degraded Alkyl (8-16) solutions. (○) $C = 300$ ppm, (□) $C = 100$ ppm, (◇) $C = 25$ ppm, (△) $C = 10$ ppm. Reprinted from Hetsroni et al. (2007) with permission

ing heat flux. The measurements were repeated several times and the same phenomena were observed. Point B stands for the condition at which the fluid starts to boil when the heat flux is increasing (the typical process is shown in Fig. 6.16a–d). Zhang and Manglik (2005) concluded that hysteresis occurred due to high wettability, which takes place at very high concentrations, $C > C_{CMC}$, where C_{CMC} is the critical micelle concentration. It should be stressed that in the present study hysteresis was observed in restricted boiling of degraded solutions as for pre-CMC solutions ($C < 300$ ppm) as for post-CMC solutions. It is speculated that molecules of degraded surfactant are more amenable to formation of a surfactant monolayer, which renders the interface less flexible and results in the dampening of interfacial motion. For Alkyl (8-16) hysteresis occurs only in degraded solutions.

6.2 Onset of Nucleate Boiling in Parallel Micro-Channels

6.2.1 Physical Model of the Explosive Boiling

Flow and boiling in parallel micro-channels occur under conditions of significant hydrodynamic and thermal heterogeneity due to finite lateral size of the heat sink, as well as interaction between the steams in individual channels connected by inlet and outlet collectors. Some data on such flows are presented by Ozawa et al. (1989), Peles (1999), and Peles et al. (2001). The temperature field of the heat sinks involving 13 to 26 micro-channels with hydraulic diameters of 0.10–0.22 mm under conditions of low quality flow boiling was studied by Hetsroni et al. (2002a, 2002b, 2003). The results showed that the wall temperature of the channels located at the central part of the heat sink was higher than that of the channels located at the peripheral part. The successive images obtained using high-speed video visualization showed that flow patterns in a given channel were quasi-periodic. The onset of nucleate boiling was established when the first bubbles were observed. This process is accompanied by pressure drop and temperature fluctuations, so that the boiling incipience is a local phenomenon. As discussed previously, the vapor accumulated in the inlet plenum. That leads to the increase of inlet temperature to a value close to T_S. According to analysis presented in the previous section, $T_{B,ONB}$ is close to the saturation temperature T_S, and wall superheat ΔT_S, since the onset of nucleation boiling is much lower than that calculated from the classical kinetics of nucleation for homogeneous nucleation in superheated fluid. For example, the wall superheat was about 4 K for flow boiling in the 26 micro-channels of $d_h = 0.103$ mm at $q = 220\,\mathrm{kW/m^2}$ and $U_{in} = 0.14$ m/s (Hetsroni et al. 2003).

Visual observation in the studies by Hetsroni et al. (2002a,b, 2003), Qu and Mudawar (2002) proved bubble behavior at incipient boiling in micro-channels ($d_h < 1$ mm) and concluded that it was quite different from that in larger channels. After nucleation, bubbles first grew to detachment size before departing into the liquid flow. The detachment size was comparable to that of the micro-channel cross-section for lower velocities and decreased progressively with increasing velocity. The detached bubbles moved to the downstream plenum where they collapsed. These visual observations proved that boiling incipience in the micro-channel heat sink was accompanied by both bubble growth and departure.

A physical model of ONB for the explosive boiling and dryout, was suggested. In order to understand why dryout occurred even at a low value of vapor quality x, it is important to keep in mind that the liquid film does not cover the entire heated surface of the micro-channel, and two-phase flow is characterized by an unsteady cyclic behavior. The following assumptions are made in the development of the model:

1. The bubble nucleation occurs at the location where the wall temperature exceeds the saturation temperature.
2. The heat flux is uniform and constant along the inner wall of the micro-channel.
3. All energy entering the fluid is used to vaporize the liquid. The temperatures of the liquid and vapor remain at saturation temperature.
4. After the bubble venting the liquid remains attached to the wall as droplets or clusters of droplets. It evaporates during the period of the cycle.
5. The thermal inertia of the wall is negligible, i.e., we assumed no phase shift between temperature of the channel wall and the heater.

Figure 6.18 shows a representation of the explosive boiling process. The bubble is assumed to nucleate at the ONB point (Fig. 6.18a). Then the bubble quickly grows to the channel size and an elongated bubble is formed. During this process some amount of liquid remains in front of the bubble (Fig. 6.18b). Figure 6.18c shows bubble venting. During this process the bubble expands not only in the upstream but also in the downstream direction. An interesting result is that a single bubble cyclically growing and collapsing away from the outlet manifold is capable of inducing a mean unidirectional fluid flow. This trend was also reported by Ory et al. (2000).

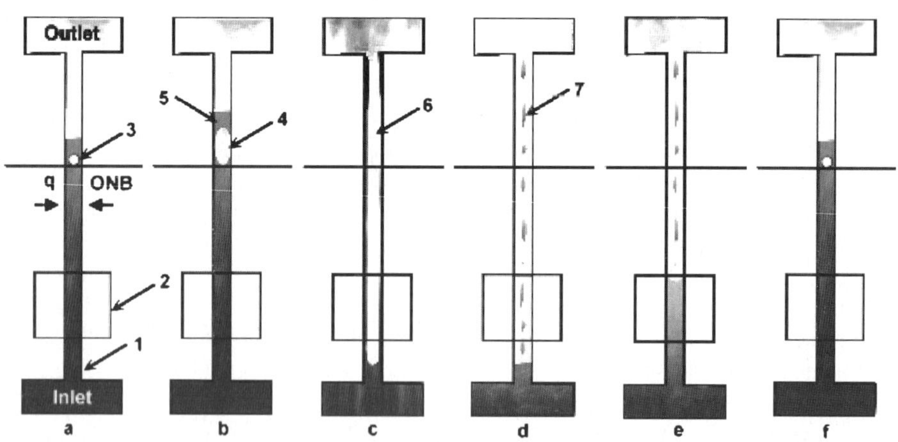

Fig. 6.18 Scheme of explosive boiling: *1* micro-channel, *2* main area of visual observation, *3* ONB point, *4* elongated cylindrical bubble, *5* liquid in front of the bubble, *6* vapor, *7* liquid droplets and clusters. Reprinted from Hetsroni et al. (2005) with permission

Figure 6.18d shows the appearance of liquid droplets or clusters of liquid droplets on the wall after the bubble venting. The pressure in the micro-channel decreases and water starts to move into it from the inlet manifold (Fig. 6.18e). Figure 6.18f shows the start of a new cycle.

6.2.2 Effect of Dissolved Gases on ONB During Flow Boiling of Water and Surfactant Solutions in Micro-Channels

Desorption of the dissolved gases formed bubbles of gas and a limited amount of bubbles containing gas–water vapor mixture. As a result, boiling incipience occurred at a channel wall temperature below the saturation temperature. Steinke and Kandlikar (2004a) studied flow boiling in six parallel micro-channels, each having hydraulic diameter of 0.207 mm. During the flow boiling studies with water in these micro-channels, nucleation was observed at a surface temperature of $T_W = 90.5\,°C$ for the dissolved oxygen content of 8.0 parts per million (ppm) at a pressure of $P = 1$ bar.

Comparison between water flow and surfactant solution was investigated by Klein et al. (2005). The experimental facility was designed and constructed as illustrated schematically in Fig. 6.19.

The test module consisted of inlet and outlet manifolds that were jointed to the test chip (Fig. 6.20). The tested chip with heater is shown in Fig. 6.21. It was made from a square shape 15 × 15 mm and 0.5 mm thick silicon wafer, which was later bonded to a 0.53 mm thick Pyrex cover. On one side of the silicon wafer 26 micro-channels were etched, with triangular shaped cross-sections, with a base of 0.21 mm

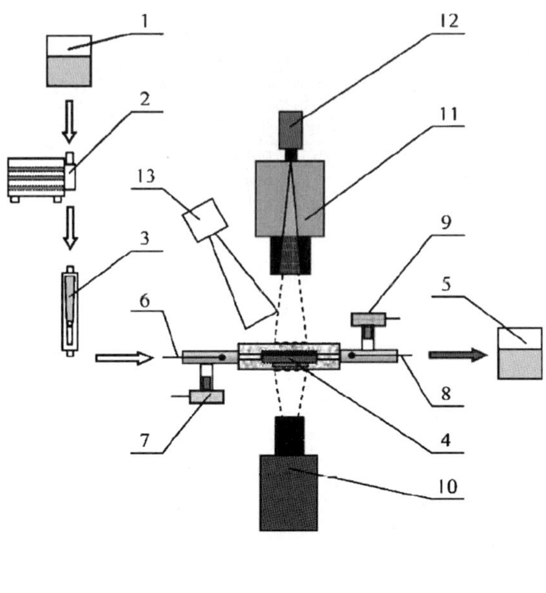

Fig. 6.19 Schematic view of the experimental facility. *1* Inlet tank, *2* mini-gear pump, *3* rotameter, *4* test module, *5* exit tank, *6* inlet thermocouple, *7* inlet pressure gauge, *8* outlet thermocouple, *9* outlet pressure gauge, *10* high-speed IR camera, *11* microscope, *12* high-speed CCD camera, *13* external light source. Reprinted from Klein et al. (2005) with permission

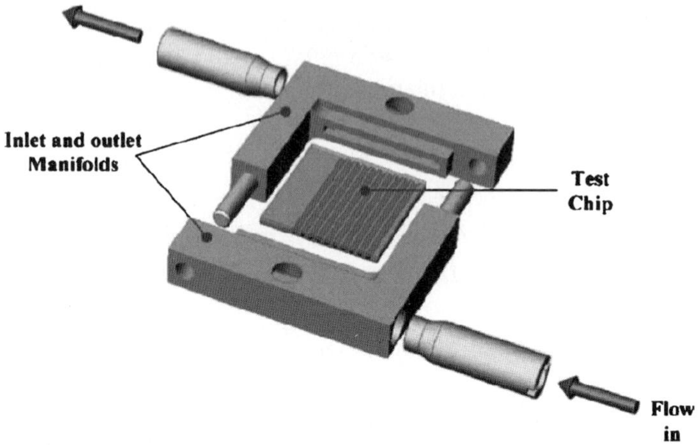

Fig. 6.20 Test module. Reprinted from Klein et al. (2005) with permission

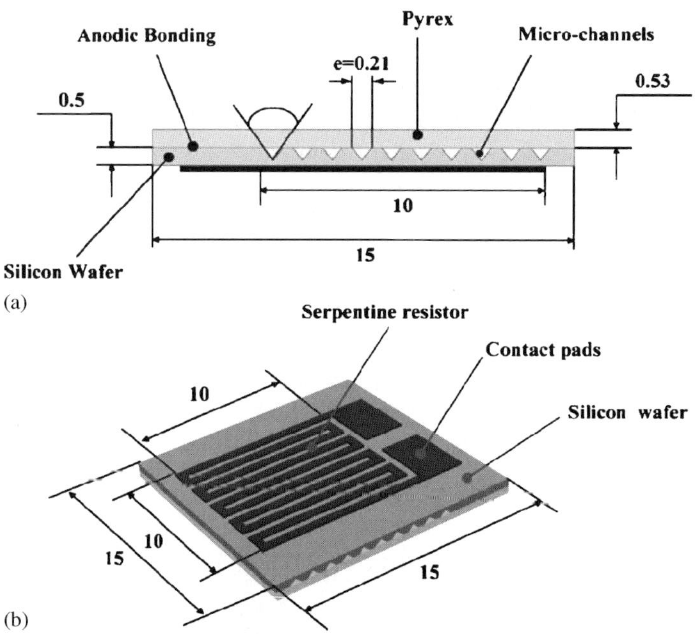

Fig. 6.21 Test chip with heater: (**a**) cross-section, (**b**) heater. Reprinted from Klein et al. (2005) with permission

and a base angle of 54.7°. Using a microscopic lens, IR measurements can be taken up to 800 Hz with a 30 μm spatial resolution. The surfactant used was of the Alkyl polyglucosides (APG) type.

Figure 6.22 shows the effect of APG additives on the dynamic and the static surface tension for different mass concentrations, measured at 75 and 95 °C. The

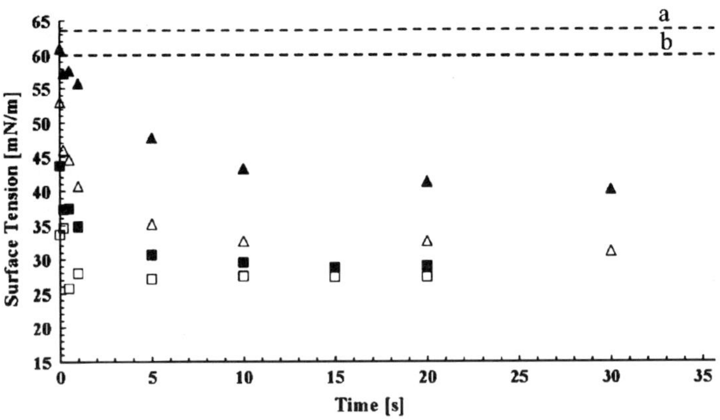

Fig. 6.22 Surface tension of the APG solutions. Concentration and solution temperature (°C): $C=100$ ppm, *filled triangles* (▲) represent 75 °C, *empty triangles* (△) represent 95 °C; $C=300$ ppm, *filled squares* (■) represent 75 °C, *empty squares* (□) represent 95 °C. Reprinted from Klein et al. (2005) with permission

dashed lines a,b represent the surface tension value for pure water at 75 and 95 °C. Solid points represent the APG data at 75 °C and the hollow points represent the APG data at 95 °C. Note that an increase in concentration decreases surface tension down to a value of 31 mN/m, compared to 59.9 mN/m for pure water.

The temperatures on the heater $T_{\mathrm{W,ONB}}$, and heat fluxes q_{ONB} corresponding to onset of nucleate boiling in water and surfactant solution that contain dissolved gases are presented in Table 6.5.

As can be seen in Table 6.5, ONB in APG solution of concentration $C=100$ ppm took place at significantly higher surface temperatures. It should be noted that the ONB in surfactant solutions may not be solely associated with static surface tension Sher and Hetsroni (2002). Other parameters such as heat flux, mass flux, kind of surfactant, surface materials, surface treatments, surface roughness, dynamic surface tension and contact angle need to be considered as well.

Table 6.5 Onset of nucleate boiling in fluids that contain dissolved gases. Reprinted from Klein et al. (2005) with permission

De-ionized water			APG-100 ppm, surfactant solution		
Mass flux (kg/m^2 s)	q_{ONB} (W/cm^2)	$T_{\mathrm{W,ONB}}$ (°C)	Mass flux (kg/m^2 s)	q_{ONB} (W/cm^2)	$T_{\mathrm{W,ONB}}$ (°C)
37.9	5.2	81.6	39.4	7.7	107.3
57.7	8.2	91.6	57.2	9.3	101.4
84.9	9.9	81.6	83.3	15.9	116.6
116.2	16.2	96.6	117.6	23.2	120.3
172.3	21.0	91.6	171.2	32.3	121.1

6.2.3 Effect of Roughness

On roughened surface boiling enhancement occurs due to the increased number of active cavities that promote bubble nucleation and provide more and larger sites for bubble growth. However, during boiling the effect of surface roughness may diminish generally due to phenomenon known as "aging". Cavities formed by roughening the surface were often not stable vapor traps and reentrant cavities may be used as a vapor trap for conventional size channels. Effect of reentrant cavities on boiling in micro-channels was studied by Kosar et al. (2005) and by Kuo and Peles (2007). Intensified nucleation activity was observed in the micro-channels with reentrant cavities accompanied by significantly better uniformity of nucleation site distribution. Significant reductions in the wall superheat required to initiate boiling were measured in structured surface micro-channels compared to the plain-wall micro-channel.

6.3 Dynamics of Vapor Bubble

6.3.1 The State of the Art of the Problem

The bubble dynamics was a subject of numerous theoretical investigations starting from pioneering work by Rayleigh (1917). The results of the study of bubble growth in superheated liquid, as well as their response to different disturbances are collected in a number of monographs and surveys devoted to mechanisms of two-phase media and the theory of heat transfer and boiling (Kutateladze 1963; Nigmatulin 1991; Carey 1992; Dhir 1998; Nakoryakov et al. 2000).

There are many criteria for bubble shape or bubble detachment suggested for the description of bubble growth dynamics. The distinct feature of bubble dynamics is that contact angle deviates from the static value due to the fast growth. The well-known Fritz equation (1936) predicts bubble departure diameter for contact angle of 48°. This agrees well with the experimental data at atmospheric pressure. Hsu and Graham (1961) assumed a contact angle of 53.1°. Han and Griffith (1965) presented a theory and experimental results of bubble growth of pool boiling at low heat flux. Bubble departure was considered by the authors and it was found that the Fritz relation works as the non-equilibrium bubble contact angle. Most of the previous research on single bubble growth has been performed using a constant wall heat flux created by heating a metallic block beneath the bubble (Staniszewski 1959; Han and Griffith 1965; Cole and Shulman 1966; van Stralen 1966). The effect of pressure for contact angle of 48° was described by Cole and Shulman (1966).

Analytical analyses for the growth of a single bubble have been performed for simple geometrical shapes, using a simplified heat transfer model. Plesset and Zwick (1954) solved the problem by considering the heat transfer through the bubble interface in a uniformly superheated fluid. The bubble growth equation was obtained from the conduction through the thermal boundary layer around the bubble. The interface cooling effect was also reported by Zuber (1961) and Mikic et al. (1970).

Zeng et al. (1993) proposed that the dominant forces leading to bubble detachment could be the unsteady growth force and buoyancy force. In order to derive an accurate detachment criterion from a force balance, all forces should be accurately known. If a mechanism is not known precisely, then approximate expressions, one or two fitted parameters and comparison with experiments might offer a solution. Such fitting procedures have indeed been applied (Klausner et al. 1993; Mei et al. 1995a; Helden et al.1995).

Mei et al. (1995b) developed a numerical model for vapor bubble growth in saturated pool boiling and demonstrated that the bubble growth rate is reduced due to the resulting temperature gradients beneath the nucleation site. Thorncroft et al. (1998) carried out experimental investigations of bubble growth and detachment in vertical up flow and down flow boiling. They observed that the growth data fit a power law ranging from about $t^{0.33}$ to $t^{0.5}$, where t is the time. The models of bubble growth and correlations for the bubble radius are discussed by Thorncroft et al. (2001).

Robinson and Judd (2001) considered four regions of bubble growth: surface tension controlled region, transition domain, inertial controlled growth and heat transfer controlled growth. A theory has been developed that is able to accommodate both spatial and temporal variations in the temperature and velocity fields in the liquid surrounding the bubble as it grows. Overall agreement between the theory and experimental data is very good. Bubble growth and variation of bubble lifetime and size with flux, subcooling, heat flux and pressure were examined by Prodanovic et al. (2002).

Nucleate pool boiling experiments with constant wall temperature were performed by Lee et al. (2003) using R-11 and R-113 for saturated boiling conditions. The geometry of the bubble was obtained from the images. The bubble growth rate was proportional to $t^{0.2}$, which was slower than the growth rate proposed in previous studies. Detachment of a vapor bubble from a plane, solid wall has been studied theoretically by Geld (2004). The vapor–liquid interface shape was approximated by a truncated sphere. The forces related to gravity and surface energy densities were found to be major contributors to departure time.

Boiling incipience and vapor bubble growth dynamics in aqueous surfactant solutions were studied using high-speed photography by Wu and Yang (1992). The bubble growth period was observed to increase slightly, while the waiting period and the time interval between two consecutive bubbles were reduced drastically. The experimental results of bubble dynamics for pure water and 100 ppm SDS solution at relatively low heat flux of $23\,\mathrm{kW/m^2}$ were presented by Yang and Maa (2003). It was shown that the departure diameter decreases considerably with addition of surfactant. Bubble growth in saturated pool boiling in water and surfactant solution was studied by Hetsroni et al. (2006a). It was shown that at relatively low heat fluxes on the wall ($q \leq 10\,\mathrm{kW/m^2}$), shape, lifetime and the volume of bubble growth in surfactant solution did not differ significantly with those in water. The time behavior of contact angle of bubble growth in surfactant solution is qualitatively similar to that of water.

At high enough heat flux of $q \geq 50\,\mathrm{kW/m^2}$, the boiling in surfactant solution, when compared to pure water, was observed to be more vigorous. Surfactant promotes activation of nucleation sites. The bubbles appeared in clusters. The lifetime of each bubble in the cluster is shorter than that of a single water bubble. The detachment diameter increases with increasing heat flux, whereas analysis of bubble growth in surfactant solution reveals the opposite effect: the detachment diameter of the bubble decreases with increasing heat flux.

These studies consider the dynamics of a single bubble that grows in infinity space, which is filled by superheated liquid. Under these conditions the bubble expansion depends on inertia forces or on intensity of heat transfer. In the case when inertia forces are dominant the bubble radius grows linearly in time (Carey 1992):

$$r(t) = \left(\frac{2}{3} \frac{T_\infty - T_\mathrm{S}}{T_\mathrm{S}} \frac{h_\mathrm{LG}\rho_\mathrm{G}}{\rho_\mathrm{L}} \right)^{1/2} t \tag{6.38}$$

where T_∞ and T_S are the ambient and saturated temperature, h_LG is the latent heat of evaporation, and ρ_G and ρ_L are the vapor and liquid density, respectively.

When the heat transfer is dominant, bubble radius is directly proportional to the square root of time

$$r(t) = 2C\sqrt{\alpha t} \tag{6.39}$$

where $C = \mathrm{Ja}\sqrt{3/\pi}$, $\mathrm{Ja} = (T_\infty - T_\mathrm{S})\rho_\mathrm{L} c_\mathrm{pL}/\rho_\mathrm{G} h_\mathrm{LG}$ is the Jacob number, and α is the liquid thermal diffusivity.

The bubble dynamics in a confined space, in particular in micro-channels, is quite different from that in infinity still fluid. In micro-channels the bubble evolution depends on a number of different factors such as existence of solid walls restricting bubble expansion in the transversal direction, a large gradient of the velocity and temperature field, etc. Some of these problems were discussed by Kandlikar (2002), Dhir (1998), and Peng et al. (1997). A detailed experimental study of bubble dynamics in a single and two parallel micro-channels was performed by Lee et al. (2004) and Li et al. (2004).

6.3.2 Dimensional Analysis

From the physical point of view it is possible to suggest that the rate of bubble growth in micro-channel is determined by the following parameters:

1. Physical properties of the fluid and its vapor: ρ_G, ρ_L, μ_L, k_L, c_pL, σ, h_LG
2. Average inlet flow velocity: U
3. Micro-channel characteristic size: d_*
4. Initial diameter of the bubble: d_0 (6.40)
5. Wall superheat: $\Delta T_\mathrm{S} = T_\mathrm{w} - T_\mathrm{S}$
6. Heat flux on the wall: q
7. Acceleration due to the gravity: g
8. Time: t

In accordance with (6.40) one can present the functional equation for rate of bubble growth as follows

$$\frac{\mathrm{d}r}{\mathrm{d}t} = f(q, h_{\mathrm{LG}}, \rho_{\mathrm{L}}, \rho_{\mathrm{G}}, \mu_{\mathrm{L}}, k_{\mathrm{L}}, c_{\mathrm{pL}}, d_0, d_*, \sigma, \Delta T_{\mathrm{S}}, g, t, U) \tag{6.41}$$

where $\mathrm{d}r/\mathrm{d}t$ is the rate of bubble growth, t is the time.

Using dimensions of length L, mass M, time τ, temperature T, and energy J, one can obtain dimensions of parameters on the right-hand side of Eq. (6.41):

$$\begin{gathered} q\,[J\tau^{-1}L^{-2}],\ h_{\mathrm{LG}}\,[JM^{-1}],\ \rho_{\mathrm{L}}\,[ML^{-3}],\ \rho_{\mathrm{G}}\,[ML^{-3}],\ \mu_{\mathrm{L}}\,[ML^{-1}T^{-1}], \\ k_{\mathrm{L}}\,[J\tau^{-1}L^{-1}T^{-1}],\ c_{\mathrm{pL}}\,[JM^{-1}T^{-1}],\ U\,[L\tau^{-1}],\ d_0\,[L],\ d_*\,[L],\ \sigma\,[M\tau^{-2}], \\ T_{\mathrm{S}}\,[T],\ g\,[L\tau^{-2}],\ t\,[\tau]. \end{gathered} \tag{6.42}$$

Among the dimensional variables of the problem, five parameters have independent dimensions and Eq. (6.41) may be written in dimensionless form. Choosing parameters ρ_{L}, c_{pL}, U, ΔT_{S}, d_* and taking into account π-theorem (Sedov 1993), Eq. (6.41) can be represented as:

$$\Pi = \varphi(\Pi_1, \Pi_2, \ldots \Pi_9) \tag{6.43}$$

where $\Pi = \frac{\dot{r}}{U}$, $\Pi_1 = \frac{q}{\rho_{\mathrm{L}} U c_{\mathrm{pL}} \Delta T_{\mathrm{S}}}$, $\Pi_2 = \frac{h_{\mathrm{LG}}}{c_{\mathrm{pL}} \Delta T_{\mathrm{S}}}$, $\Pi_3 = \frac{\rho_{\mathrm{G}}}{\rho_{\mathrm{L}}}$, $\Pi_4 = \frac{\mu_{\mathrm{L}}}{\rho U d_*}$, $\Pi_5 = \frac{k_{\mathrm{L}}}{\rho_{\mathrm{L}} c_{\mathrm{pL}} U d_*}$, $\Pi_6 = \frac{d_0}{d_*}$, $\Pi_7 = \frac{\sigma}{\rho U^2 d_*}$, $\Pi_8 = \frac{g d_*}{U^2}$, $\Pi_9 = \frac{\tau U}{d_*}$, $\dot{r} = \frac{\mathrm{d}r}{\mathrm{d}t}$.

Analysis of experimental data presented by Lee et al. (2004) revealed that parameters Π_3–Π_9 did not change significantly. Parameter Π_2 depends mainly on the wall superheat, $\Delta T_{\mathrm{S}} = T_{\mathrm{W}} - T_{\mathrm{S}}$. It should be noted that the value of wall temperature T_{W} in experimental investigations was obtained from the measurements, whereas the value of saturation temperature T_{S} was calculated using the dependence of T_{S} on saturation pressure P_{S}. Assuming linear pressure distribution along the micro-channel Lee et al. (2004) presented two approaches to calculate T_{S}. The first of them is based on the local pressure at the nucleate boiling incipience. The second one is based on the average value of the pressure in the micro-channel $P_{\mathrm{S}} = (P_{\mathrm{in}} + P_{\mathrm{out}})/2$. Depending on the method of calculation of saturation temperature, the results differ. We assumed that $\Pi_2 \approx$ const. Thus, one can consider Π_1 as the dimensionless parameter that determines the rate of bubble growth in the linear regime:

$$\frac{\mathrm{d}r}{\mathrm{d}t} = \varphi(\Pi_1)\,. \tag{6.44}$$

6.3.3 Experimental Data

Single micro-channel

Data by Lee et al. (2004) and Li et al. (2004) contain the results related to bubble dynamics in a single micro-channel and two parallel ones. The experimental setup used by Lee et al. (2004) consisted of the test section, a syringe pump, a heating module and a flow visualization system. The test section was a 5 mm wide silicon strip, etched with the micro-channel on its top surface. The top surface dimension was 5 × 20 mm. Figure 6.23 displays the top and cross-section views of the test section. The top and bottom widths of the trapezoid channel are 102.8 and 59.12 μm, respectively, and the channel depth is 30.1 μm resulting in a hydraulic diameter of 41.3 μm. The channel length is 28 mm leading to a length-to-diameter ratio of 678. Only the central 20 mm was heated. The rms roughness of the bottom surface was 206 nm, while it was 82 nm for the side wall.

The results are presented in Table 6.6 (the data related to flow regimes at which the wall temperature T_W was less than saturated, are excluded from this table).

According to Eq. (6.44) such a behavior may be analyzed using parameter Π_1. In the range of $\Pi_1 = 0.00791–0.0260$ linear behavior of the bubble radius was observed, when $\Pi_1 > 0.0260$ exponential bubble growth took place (Fig. 6.24).

Table 6.6 also demonstrates extraordinarily high bubble growth rates of 94.63, 72.8 and 95.3 μm/ms. For these three cases, the growth rates are two orders of magnitude higher than the other cases. The authors noted that it is unclear why the bubble growth rate for such cases is much higher than the other cases.

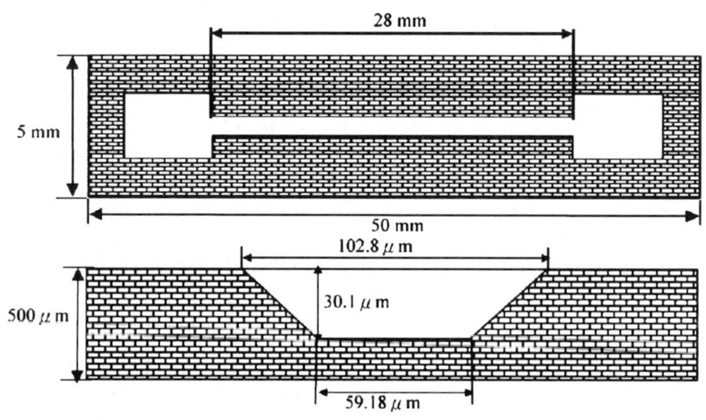

Fig. 6.23 The top and cross-section views of the test section with a trapezoid micro-channel (length shown not in scale). Reprinted from Lee et al. (2004) with permission

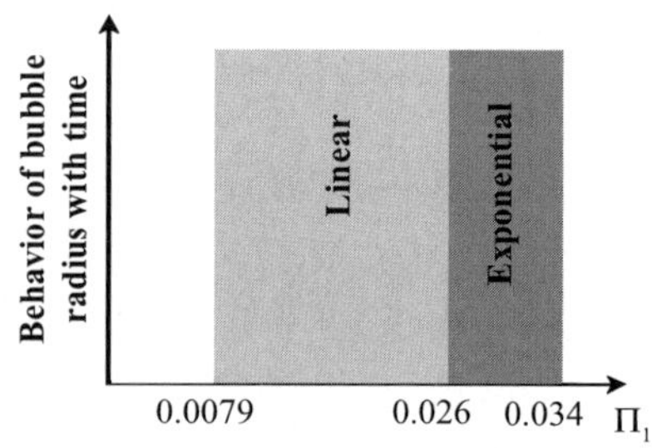

Fig. 6.24 Behavior of bubble radius with time. The values of Π_1 were calculated from experimental results presented by Lee et al. (2004)

Table 6.6 Behavior of bubble radius with time. Data by Lee et al. (2004)

Number	Heat flux q [W/m^2]	Mass flux G [kg/m^2 s]	Bubble growth rate $\mathrm{d}r/\mathrm{d}t$ [μm/ms]	Parameter $\Pi_1 = \frac{q}{\rho_L U c_{pL} \Delta T_S}$	Behavior of bubble radius with time
1	103,000	341	94.63	0.00791	Linear
2	189,000	341	0.24	0.0260	
3	264,000	341	0.32	0.0259	
4	255,000	341	0.22	0.0135	
5	107,000	477	7.09	0.0106	
6	218,000	477	72.8	0.0019	
7	353,000	477	0.54	0.0124	
8	415,000	477	0.10	0.0145	
9	449,000	477	0.10	0.0110	
10	57,600	170	2.62	0.0267	Exponential
11	137,000	170	95.3	0.0272	
12	196,000	170	4.91	0.0339	

Parallel micro-channels

The bubble dynamics under conditions corresponding to flow in two parallel trapezoidal micro-channels with hydraulic diameter 47.7 μm was studied by Li et al. (2004). The bubbles in two parallel micro-channels generally grow similarly to that in a single micro-channel. The authors reported on the presence of two-phase flow instability.

Hetsroni et al. (2003) studied bubble growth in 26 parallel triangular micro-channels of $d_h = 103$ μm and length of 10 mm. When incipient boiling superheat is achieved, the vapor bubble nucleates and grows rapidly. The bubble is initially a small sphere and then the axisymmetric growth in a narrow channel takes place. Temporal variation of bubble size is shown in Fig. 6.25. Figure 6.25a,b shows the variation of bubble size in the streamwise direction L_p, and in the spanwise direction L_n, respectively. From these figures one can conclude that at time $t = 0.375$ s the maximum length of the bubble in the streamwise direction L_p is about eight times larger than that in the spanwise direction L_n.

The observed ratio $f = L_p/L_n$, is quite different from that reported for subcooled flow boiling of water in tubes of 17–22 mm inner diameter. Prodanovic et al. (2002) reported that this ratio was typically around 0.8 for experiments at 1.05–3 bar. The situation considered in experiments carried out by Hetsroni et al. (2003) is however different as the bubbles undergo a significant volume change and the flow is unstable. Ory et al. (2000) studied numerically the growth and collapse of a bubble in a narrow tube filled with a viscous fluid. The situation considered in that study is also quite different from experiments by Hetsroni et al. (2003) as, in that case, heat was added to the system impulsively, rather than continuously as we do here.

Visual observation of bubble growth in parallel triangular micro-channels showed that the majority of the first bubbles were observed on the channel bottom wall, though a few bubbles did appear on the side walls. After nucleation, bubbles first grew to detachment size before departing into the liquid flow. The detached bubbles moved to the downstream plenum.

Bubble velocity

Figure 6.26 shows the velocity of displacement of the bubble tail in the streamwise direction U_b (m/s) versus bubble lifetime t, at fixed conditions as described

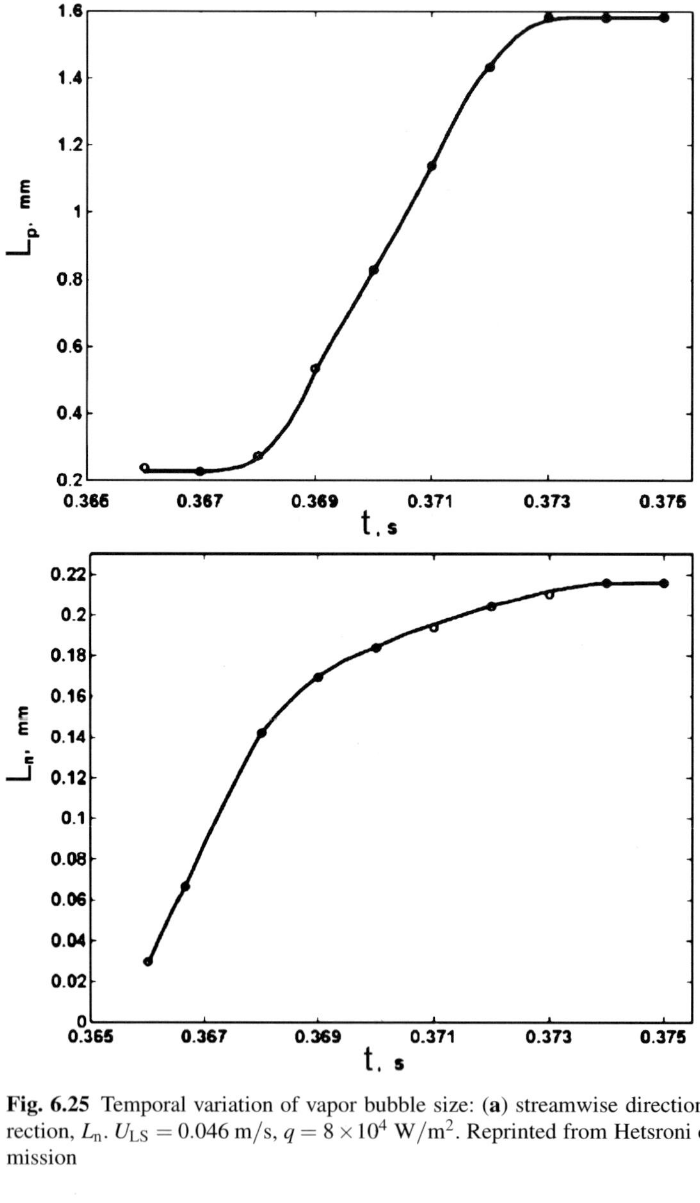

Fig. 6.25 Temporal variation of vapor bubble size: (**a**) streamwise direction, L_p (**b**) spanwise direction, L_n. $U_{LS} = 0.046$ m/s, $q = 8 \times 10^4$ W/m^2. Reprinted from Hetsroni et al. (2003) with permission

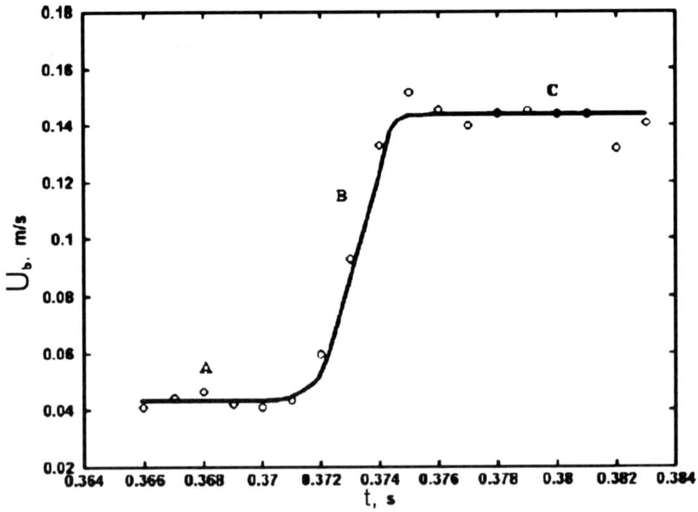

Fig. 6.26 Velocity of bubble displacement. $U_{\mathrm{LS}} = 0.046\,\mathrm{m/s}$, $q = 80\,\mathrm{kW/m^2}$. Reprinted from Hetsroni et al. (2003) with permission

above, i.e., $q = 8 \times 10^4\,\mathrm{W/m^2}$, $U_{\mathrm{LS}} = 0.046\,\mathrm{m/s}$. One may conclude that the velocity of bubble displacement varies, depending on the given range of lifetime. In region A the boiling process during the time of about 0.005 s from the appearance of the first bubble, ONB, the bubble velocity is equal to the superficial liquid velocity. It should be noted that the term ONB, known as the onset of nucleate boiling, was "borrowed" from the terminology of subcooled flow boiling in larger tubes. Region B is characterized by a sharp increase in the bubble velocity. One may conclude that the bubble is accelerated in the streamwise flow direction. Figure 6.26 shows that in this region the bubble velocity increases about threefold, during a time interval of about 0.003 s. After the time when U_{b} reaches maximum value it remains constant, as shown in Fig. 6.26, region C.

Such a behavior agrees with results reported by Agostini et a. (2008). It was found that the elongated bubble velocity increased with increasing bubble length until a plateau was reached. An analytical model has been proposed that is able to predict this trend.

The collision of elongated bubbles has been studied by Revellin et al. (2008) along adiabatic glass micro-channels of 509 and 709 μm internal diameters for refrigerant R-134a. A model for the collision of elongated bubbles in micro-channels was proposed to predict the bubble length distribution at the exit of the micro-evaporator.

In micro-channels bubbles cause a significant volume change (relative to the channel size). As a result, pressure fluctuations were observed. The temporal behavior of the pressure drop is shown in Fig. 6.27. The data were obtained at $q = 220\,\mathrm{kW/m^2}$ and $U_{\mathrm{LS}} = 0.14\,\mathrm{m/s}$. Such a behavior is a result of vapor formation in each micro-channel.

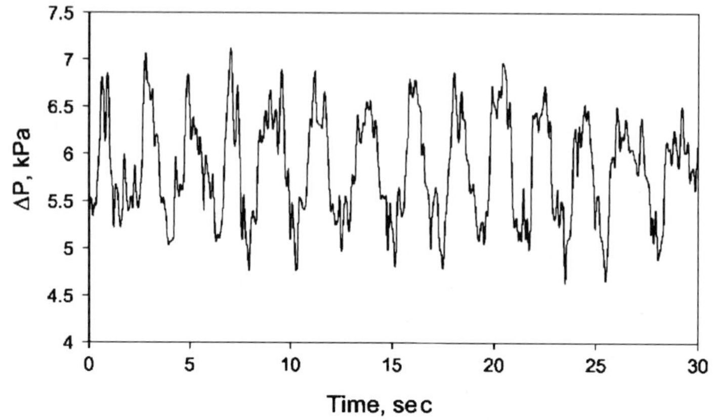

Fig. 6.27 Pressure drop fluctuations. $U_{LS} = 0.14\,\text{m/s}$, $q = 220\,\text{kW/m}^2$. Reprinted from Hetsroni et al. (2003) with permission

The pressure spike introduces a disruption in the flow. Depending on the local conditions, the excess pressure inside the bubble may overcome the inertia of the incoming liquid and the pressure in the inlet manifold, and cause a reverse flow of varying intensity depending on the local conditions. There are two ways to reduce the flow instabilities: reduce the local liquid superheat at the ONB and introduce a pressure drop element at the entrance of each channel, Kandlikar (2006). Kakac and Bon (2008) reported that density-wave oscillations were observed also in conventional size channels. Introduction of additional pressure drop at the inlet (small diameter orifices were employed for this purpose) stabilized the system.

A simultaneous visualization and measurement study has been carried out by Wang et al. (2008) to investigate effects of inlet/outlet configurations on flow boiling instabilities in parallel micro-channels having a length of 30 mm and a hydraulic diameter of 186 μm. It was found that nearly steady flow boiling existed in the parallel micro-channels through an inlet restriction.

6.4 Pressure Drop and Heat Transfer

6.4.1 Pressure Drop in Two-Phase Flow Boiling

The measurements show that the pressure drop in circular and rectangular micro-channels depend strongly on the mass and heat fluxes (Tran et al. 2000; Yu et al. 2002; Shuai et al. 2003).

The flow patterns (expansion of the bubbly, slug and annular regions of flow) affect the local pressure drop, as well as the pressure oscillations in micro-channels (Kandlikar et al. 2001; Wu and Cheng 2003a,b, 2004; Qu and Mudawar 2003; Hetsroni et al. 2005; Lee and Mudawar 2005a).

The influence of inlet conditions on stability of flow boiling in micro-channels was analyzed by Brutin and Tadrist (2004). The set-up with rectangular micro-channel $500 \times 4{,}000\,\mu m$ was used to study flow boiling at two kinds of upstream conditions, which corresponded to constant liquid velocity at channel entrance (confinement condition) and constant velocity at the syringe outlet. The flow characteristics corresponding to steady and unsteady regimes were studied and the Reynolds number that subdivided these states was found.

Hwan and Kim (2006) investigated the pressure drop in circular stainless steel tubes with inner diameter of 244, 430, and 792 μm. These data show that mass flux strongly affects two-phase pressure drop in micro-channels of different diameters.

A detailed analysis of several correlations for calculating pressure drop in boiling flow was carried out by Qu and Mudawar (2003a). They tested two groups of correlations, the first of them corresponds to boiling flow in conventional channels (Collier and Thome 1994; Lockhart and Martinelli 1949), and the second ones to boiling in conventional size channels and micro-channels (Mishima and Hibiki 1996; Tran et al. 2000; Lee and Lee 2001a; Yu et al. 2002; Qu and Mudawar 2003a).

In the study by Qu and Mudawar (2003a) experiments were performed to measure pressure drop in two-phase micro-channel heat sink containing 21 parallel $231 \times 713\,\mu m$ micro-channels. The pressure drop in the micro-channel two-phase region ΔP_{tp} was expressed as the sum of acceleration and friction components:

$$\Delta P_{tp} = \Delta P_{tp,a} + \Delta P_{tp,f} \tag{6.45}$$

where $\Delta P_{tp,a}$ is the acceleration component of two-phase pressure drop, and $\Delta P_{tp,f}$ is the friction component of two-phase pressure drop.

The predictive capability of the proposed correlation for all operating conditions of the study by Qu and Mudawar (2003a) was illustrated. The mean absolute error (MAE) of each correlation

$$\mathrm{MAE} = \frac{1}{N}\sum\frac{\left|\Delta P_{pred} - \Delta P_{exp}\right|}{\Delta P_{exp}}100\% \tag{6.46}$$

was used to discuss the results, where N is the number of experimental data. The comparison showed that the MAE for the first group of correlations, denoted as Eqs. (1) to (6) in the study of Qu and Mudawar (2003a), was in the range of $\mathrm{MAE} = 29-378\%$. The correlations of the second group, Eqs. (7) to (11), are presented in Table 6.7.

In Table 6.7, C is the Martinelli–Chisholm constant, f is the friction factor, f_f is the friction factor based on local liquid flow rate, f_{fo} is the friction factor based on total flow rate as a liquid, G is the mass velocity in the micro-channel, L is the length of micro-channel, P is the pressure, ΔP is the pressure drop, $P_{tp,a}$ is the acceleration component of two-phase pressure drop, $\Delta P_{tp,f}$ is the frictional component of two-phase pressure drop, v is the specific volume, x_e is the thermodynamic equilibrium quality, X_{vt} is the Martinelli parameter based on laminar liquid-turbulent vapor flow, X_{vv} is the Martinelli parameter based on laminar liquid-laminar vapor flow, α is the void fraction, μ is the viscosity, ρ is the density, Φ_f^2 is the two-phase frictional multiplier based on local liquid flow rate, Φ_{fo}^2 is the two-phase frictional multiplier based on total flow considered as a liquid, Re_{g} is the Reynolds number based on local vapor flow, $\mathrm{Re}_{\mathrm{fo}}$ is the Reynolds number based on total flow as a liquid; the subscripts are f for liquid, frictional, g for vapor, in for test module inlet, out for test module outlet, pred for predicted, and tp for two-phase.

It should be noted that for convenience in Table 6.7 we used the same nomenclature as in original paper by Qu and Mudawar (2003a).

Figures 6.28 and 6.29 show the comparison of pressure drop data with predictions of conventional size/micro-channels correlations ((7)–(11), Table 6.7).

A new approach was developed by Lee and Mudawar (2005a) to improve the accuracy of pressure drop prediction in two-phase micro-channels. Since the bubbly and churn flow patterns are rarely detected in high-flux micro-channel flow, the separated flow model was deemed more appropriate than the homogeneous.

It was assumed that the added complexity of two-phase flow in a micro-channel is the result of interactions between liquid inertia, the liquid viscous force, and surface tension. Two key measures of these interactions are the Reynolds and Weber numbers based on liquid properties:

$$\mathrm{Re}_{\mathrm{L}} = \frac{U d_{\mathrm{h}}}{\nu_{\mathrm{L}}} \tag{6.47}$$

and

$$\mathrm{We}_{\mathrm{L}} = \frac{G^2 d_{\mathrm{h}}}{\sigma \rho_{\mathrm{L}}} \tag{6.48}$$

The two-phase pressure drop multiplier

$$\Phi_{\mathrm{L}}^2 = 1 + \frac{C}{X} + \frac{1}{X^2} \tag{6.49}$$

is modified with a new dimensionless parameter defined as

$$C = C_1 \mathrm{Re}_{\mathrm{L}}^{C_2} \mathrm{We}_{\mathrm{L}}^{C_3} \tag{6.50}$$

where C_1, C_2, C_3 are parameters presented in Eqs. (6.51–6.52).

To enhance the predictive capability of the new correlation, both the present R-134a data and prior micro-channel water data of Qu and Mudawar (2003a) were examined. Large differences between the thermophysical properties of the two coolants were deemed highly effective at broadening the application range of the new correlation. Another key difference between the two data sets is both the liquid and vapor flows are laminar for the water data, while low viscosity rendered the vapor flow turbulent for R-134a. Typical micro-channel operating conditions rarely produced turbulent liquid flow. Therefore, two separate correlations were derived for C, based on the flow states of the liquid and vapor,

$$C_{\mathrm{vv}} = 2.16 \mathrm{Re}_{\mathrm{L}}^{0.047} \mathrm{We}_{\mathrm{L}}^{0.60} \text{ (laminar liquid–laminar vapor)} \tag{6.51}$$

$$C_{\mathrm{vt}} = 1.45 \mathrm{Re}_{\mathrm{L}}^{0.25} \mathrm{We}_{\mathrm{L}}^{0.23} \text{ (laminar liquid–turbulent vapor)} \tag{6.52}$$

Table 6.7 Pressure drop correlations. Reprinted from Qu and Mudawar (2003a) with permission

Author	Component of ΔP	Definitions	Max average error
Misima, Hibiki (1996) (7)	$\Delta P_{\mathrm{tp,f}} = \frac{L_{\mathrm{tp}}}{x_{\mathrm{e,out}}} \int_0^{x_{\mathrm{e,out}}} \frac{2 f_{\mathrm{f}} G^2 (1-x_{\mathrm{e}})^2 v_{\mathrm{f}}}{d_{\mathrm{h}}} \Phi_{\mathrm{f}}^2 \, \mathrm{d}x_{\mathrm{e}}$ $\Delta P_{\mathrm{tp,a}} = G^2 v_{\mathrm{f}} \left[\frac{x_{\mathrm{e,out}}^2}{\alpha_{\mathrm{out}}} \left(\frac{v_{\mathrm{g}}}{v_{\mathrm{f}}} \right) + \frac{(1-x_{\mathrm{e,out}})^2}{1-\alpha_{\mathrm{out}}} - 1 \right]$	$\Phi_{\mathrm{f}}^2 = 1 + \frac{C}{X_{\mathrm{vv}}} + \frac{1}{X_{\mathrm{vv}}^2} X_{\mathrm{vv}} = \left(\frac{\mu_{\mathrm{f}}}{\mu_{\mathrm{g}}} \right)^{0.5} \left(\frac{1-x_{\mathrm{e}}}{x_{\mathrm{e}}} \right)^{0.5} \left(\frac{v_{\mathrm{f}}}{v_{\mathrm{g}}} \right)^{0.5}$ $\alpha_{\mathrm{out}} = \left[1 + \left(\frac{1-x_{\mathrm{e,out}}}{x_{\mathrm{e,out}}} \right) \left(\frac{v_{\mathrm{f}}}{v_{\mathrm{g}}} \right)^{2/3} \right]^{-1}$ $C = 21 \left[1 - \exp(-0.319 \cdot 10^3 d_{\mathrm{h}}) \right]$	13.9
Tran et al. (2000) (8)	$\Delta P_{\mathrm{tp,f}} \frac{2 f_{f0} G^2 v_{\mathrm{f}}}{d_{\mathrm{h}}} \frac{L_{\mathrm{tp}}}{x_{\mathrm{e,out}}} \int_0^{x_{\mathrm{e,out}}} \Phi_{\mathrm{fo}}^2 \, \mathrm{d}x_{\mathrm{e}}$ $\Delta P_{\mathrm{tp,a}} = G^2 v_{\mathrm{f}} \left[\frac{x_{\mathrm{e,out}}^2}{\alpha_{\mathrm{out}}} \left(\frac{v_{\mathrm{g}}}{v_{\mathrm{f}}} \right) + \frac{(1-x_{\mathrm{e,out}})^2}{1-\alpha_{\mathrm{out}}} - 1 \right]$	$\Phi_{f0}^2 = 1 + (4.3\Gamma^2 - 1) \cdot \left[N_{\mathrm{conf}} x_{\mathrm{e}}^{0.875} (1-x_{\mathrm{e}})^{0.875} + x_{\mathrm{e}}^{1.75} \right]$ $\Gamma = \left(\frac{v_{\mathrm{g}}}{v_{\mathrm{f}}} \right)^{0.5} \left(\frac{\mu_{\mathrm{g}}}{\mu_{\mathrm{f}}} \right)^{0.5}$ $\alpha_{\mathrm{out}} = \left[1 + \left(\frac{1-x_{\mathrm{e,out}}}{x_{\mathrm{e,out}}} \right) \left(\frac{v_{\mathrm{f}}}{v_{\mathrm{g}}} \right)^{2/3} \right]^{-1}$ $N_{\mathrm{conf}} = \left[\frac{\sigma}{g(\rho_{\mathrm{f}} - \rho_{\mathrm{g}})} \right]^{0.5} / d_{\mathrm{h}}$	828.3
Lee, Lee (2001a) (9)	$\Delta P_{\mathrm{tp,f}} = \frac{L_{\mathrm{tp}}}{x_{\mathrm{e,out}}} \int_0^{x_{\mathrm{e,out}}} \frac{2 f_{\mathrm{f}} G^2 (1-x_{\mathrm{e}})^2 v_{\mathrm{f}}}{d_{\mathrm{h}}} \Phi_{\mathrm{f}}^2 \, \mathrm{d}x_{\mathrm{e}}$ $\Delta P_{\mathrm{tp,a}} = G^2 v_{\mathrm{f}} \left[\frac{x_{\mathrm{e,out}}^2}{\alpha_{\mathrm{out}}} \left(\frac{v_{\mathrm{g}}}{v_{\mathrm{f}}} \right) + \frac{(1-x_{\mathrm{e,out}})^2}{1-\alpha_{\mathrm{out}}} - 1 \right]$	$\Phi_{\mathrm{f}}^2 = 1 + \frac{C}{X_{\mathrm{vt}}} + \frac{1}{X_{\mathrm{vt}}^2}, X_{\mathrm{vt}} = \left(\frac{f_{\mathrm{f}} \mathrm{Re}_{\mathrm{g}}^{0.25}}{0.079} \right)^{0.5} \left(\frac{1-x_{\mathrm{e}}}{x_{\mathrm{e}}} \right) \left(\frac{v_{\mathrm{f}}}{v_{\mathrm{g}}} \right)^{0.5}$ $\alpha_{\mathrm{out}} = \left[1 + \left(\frac{1-x_{\mathrm{e,out}}}{x_{\mathrm{e,out}}} \right) \left(\frac{v_{\mathrm{f}}}{v_{\mathrm{g}}} \right)^{2/3} \right]^{-1}$ $C = 6.185 \cdot 10^{-2} \mathrm{Re}_{f0}^{0.726}$	19.1
Yu et al. (2002) (10)	$\Delta P_{\mathrm{tp,f}} = \frac{L_{\mathrm{tp}}}{x_{\mathrm{e,out}}} \int_0^{x_{\mathrm{e,out}}} \frac{2 f_{\mathrm{f}} G^2 (1-x_{\mathrm{e}})^2 v_{\mathrm{f}}}{d_{\mathrm{h}}} \Phi_{\mathrm{f}}^2 \, \mathrm{d}x_{\mathrm{e}}$ $\Delta P_{\mathrm{tp,a}} = G^2 v_{\mathrm{f}} \left[\frac{x_{\mathrm{e,out}}^2}{\alpha_{\mathrm{out}}} \left(\frac{v_{\mathrm{g}}}{v_{\mathrm{f}}} \right) + \frac{(1-x_{\mathrm{e,out}})^2}{1-\alpha_{\mathrm{out}}} - 1 \right]$	$\Phi_{\mathrm{f}}^2 = \frac{1}{X_{\mathrm{vt}}^{1.9}} X_{\mathrm{vt}} = \left(\frac{f_{\mathrm{f}} \mathrm{Re}_{\mathrm{g}}^{0.25}}{0.046} \right)^{0.5} \left(\frac{1-x_{\mathrm{e}}}{x_{\mathrm{e}}} \right) \left(\frac{v_{\mathrm{f}}}{v_{\mathrm{g}}} \right)^{0.5}$ $\alpha_{\mathrm{out}} = \left[1 + \left(\frac{1-x_{\mathrm{e,out}}}{x_{\mathrm{e,out}}} \right) \left(\frac{v_{\mathrm{f}}}{v_{\mathrm{g}}} \right)^{2/3} \right]^{-1}$	48.0
Qu, Mudawar (2003a) (11)	$\Delta P_{\mathrm{tp,f}} = \frac{L_{\mathrm{tp}}}{x_{\mathrm{e,out}}} \int_0^{x_{\mathrm{e,out}}} \frac{2 f_{\mathrm{f}} G^2 (1-x_{\mathrm{e}})^2 v_{\mathrm{f}}}{d_{\mathrm{h}}} \Phi_{\mathrm{f}}^2 \, \mathrm{d}x_{\mathrm{e}}$ $\Delta P_{\mathrm{tp,a}} = G^2 v_{\mathrm{f}} \left[\frac{x_{\mathrm{e,out}}^2}{\alpha_{\mathrm{out}}} \left(\frac{v_{\mathrm{g}}}{v_{\mathrm{f}}} \right) + \frac{(1-x_{\mathrm{e,out}})^2}{1-\alpha_{\mathrm{out}}} - 1 \right]$	$\Phi_{\mathrm{f}}^2 = 1 + \frac{C}{X_{\mathrm{vv}}} + \frac{1}{X_{\mathrm{vv}}^2}, X_{\mathrm{vv}} = \left(\frac{\mu_{\mathrm{f}}}{\mu_{\mathrm{g}}} \right)^{0.5} \left(\frac{1-x_{\mathrm{e}}}{x_{\mathrm{e}}} \right)^{0.5} \left(\frac{v_{\mathrm{f}}}{v_{\mathrm{g}}} \right)^{0.5}$ $\alpha_{\mathrm{out}} = \left[1 + \left(\frac{1-x_{\mathrm{e,out}}}{x_{\mathrm{e,out}}} \right) \left(\frac{v_{\mathrm{f}}}{v_{\mathrm{g}}} \right)^{2/3} \right]^{-1}$ $C = 21 \left[1 - \exp(-0.319 \cdot 10^3 d_{\mathrm{h}}) \right] \cdot (0.00418G + 0.0613)$	12.4

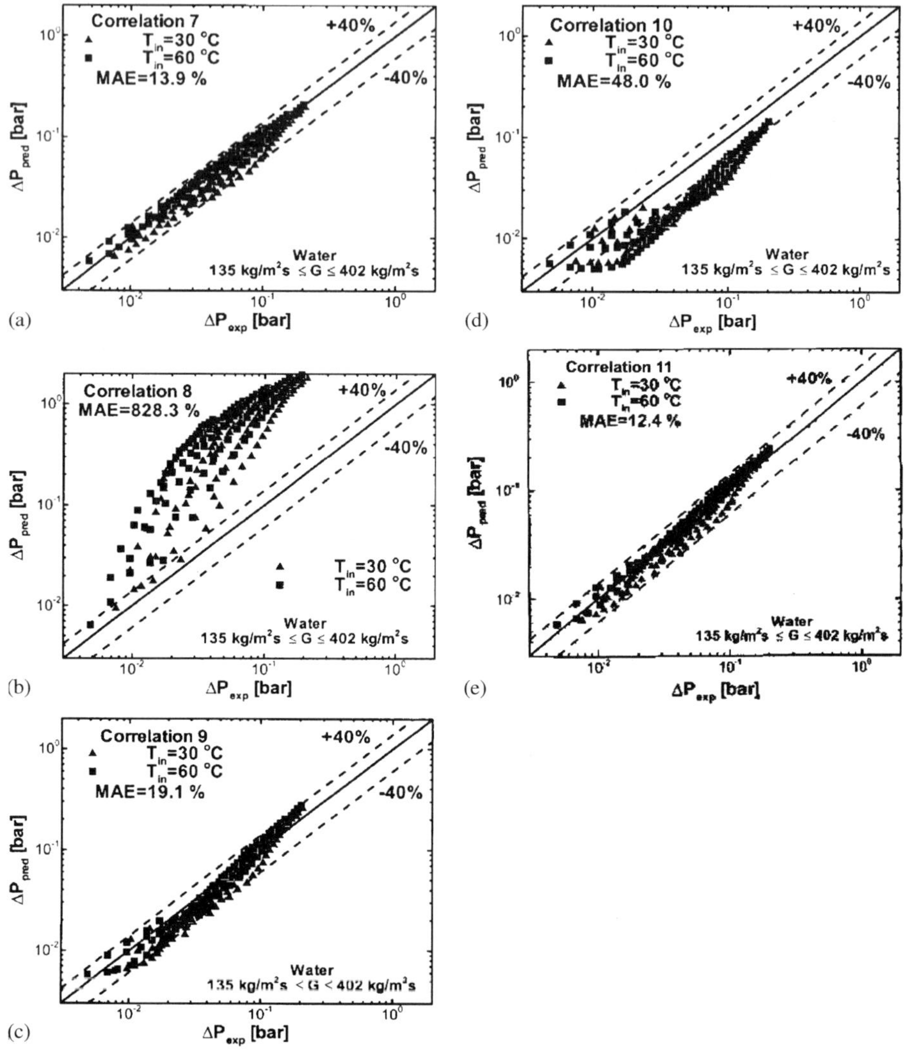

Fig. 6.28 Comparison of pressure drop data with predictions of mini/micro-channel correlations of (**a**) Mishima and Hibiki (1996), (**b**) Tran et al. (2000), (**c**) Lee and Lee (2001a,d) and Yu et al. (2002), (**e**) Qu and Mudawar (2003a). Reprinted from Qu and Mudawar (2003a) with permission

Notice the stronger effect of surface tension where both liquid and vapor are laminar. Figure 6.30a shows good agreement of the pressure drop predictions based on the new correlation with the R-134a data, both in terms of MAE (mean absolute error) and the general trend. The largest deviation is concentrated in the low mass flux and low heat flux region where both the heat loss (which influences the accuracy of the heat flux used in the pressure drop model) and the flow rate measurement uncertainty are greatest. Figure 6.30b shows the present correlation is also very effective at predicting the micro-channel water data of Qu and Mudawar (2003a).

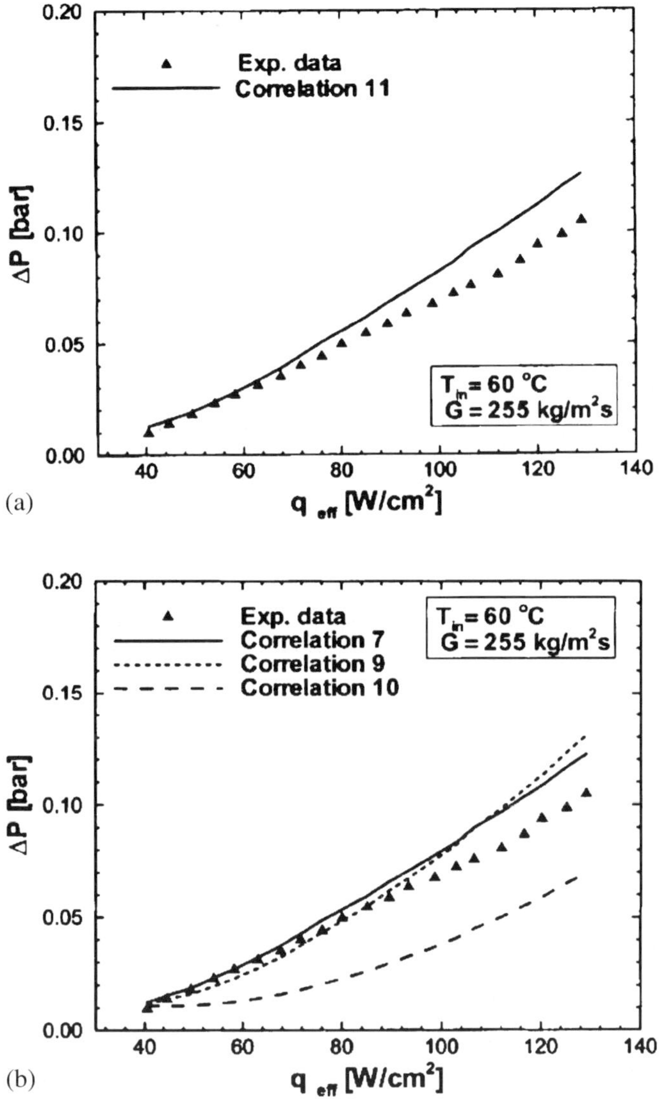

Fig. 6.29 Comparison of pressure drop data with correlation predictions for $T_{\text{in}} = 60\,°\text{C}$ and $G = 255\,\text{kg/m}^2\text{s}$. (**a**) correlation (11), (**b**) correlations (7)–(10). Reprinted from Qu and Mudawar (2003a) with permission

Quiben and Thome (2007a,b) presented an experimental and analytical investigation of two-phase pressure drops during evaporation in horizontal tubes. Experiments were performed under diabatic conditions in tubes of $d = 8$ and 13 mm in the range of vapor quality $x = 0-1$, mass velocity $G = 70-700\,\text{kg/m}^2\text{s}$, heat flux $q = 6.0-57.5\,\text{kW/m}^2$. The test fluids were R-134a, R-22 and R-410A. The results

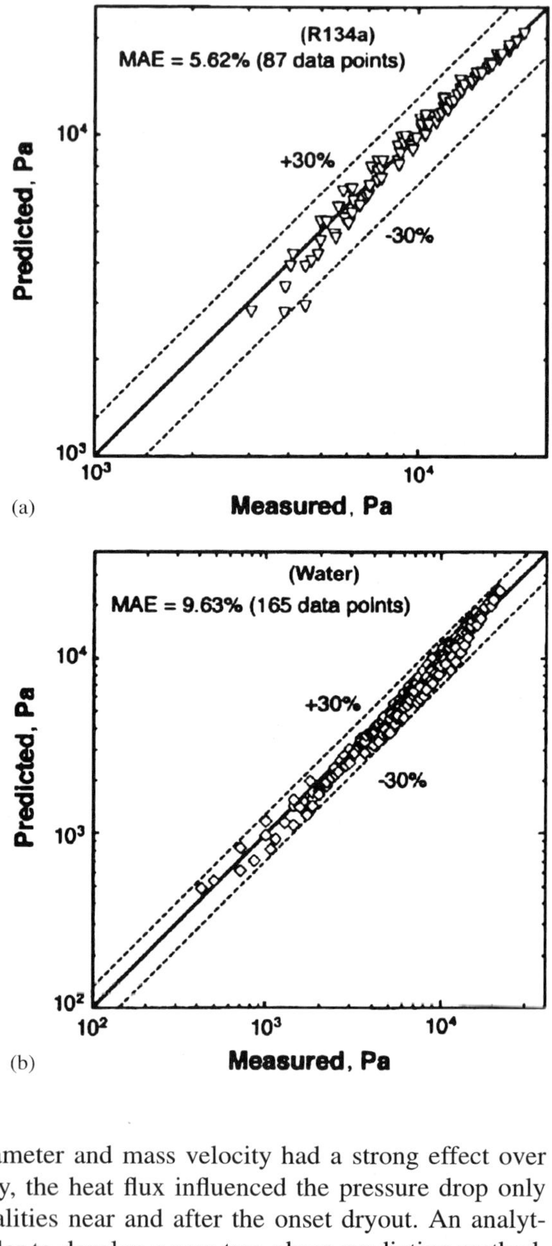

Fig. 6.30 Comparison of correlation predictions with (**a**) R-134a (Lee and Mudawar (2005a)) and (**b**) water Qu and Mudawar's (2003a) micro-channel water data. Reprinted from Lee and Mudawar (2005a) with permission

showed that while the fluid, diameter and mass velocity had a strong effect over the entire range of vapor quality, the heat flux influenced the pressure drop only for a particular range vapor qualities near and after the onset dryout. An analytical study was undertaken in order to develop a new two-phase prediction method. A model has been developed following a phenomenological approach and the interfacial structure between the phases was taken into account. The flow pattern effects are particularly important at low flow rates (stratification effects) and high vapor qualities (dryout effects). This model works well in annular flow and captures the position of the pressure drop peak under condition of dryout at high vapor quality.

6.4.2 Heat Transfer in Two-Phase Flow Boiling

The heat transfer coefficient of boiling flow through a horizontal rectangular channel with low aspect ratio (0.02–0.1) was studied by Lee and Lee (2001b). The mass flux in these experiments ranged from 50 to 200 $\mathrm{kg/m^2\,s}$, maximum heat flux was 15 $\mathrm{kW/m^2}$, and the quality ranged from 0.15 to 0.75, which corresponds to annular flow. The experimental data showed that under conditions of the given experiment, forced convection plays a dominant role.

The detail experimental study of flow boiling heat transfer in two-phase heat sinks was performed by Qu and Mudawar (2003b). It was shown that the saturated flow boiling heat transfer coefficient in a micro-channel heat sink is a strong function of mass velocity and depends only weakly on the heat flux. This result, as well as the results by Lee and Lee (2001b), indicates that the dominant mechanism for water micro-channel heat sinks is forced convective boiling but not nucleate boiling.

Heat transfer characteristics for saturated boiling were considered by Yen et al. (2003). From this study of convective boiling of HCFC123 and FC72 in micro-tubes with inner diameter 190, 300 and 510 μm one can see that in the saturated boiling regime, the heat transfer coefficient monotonically decreased with increasing vapor quality, but independent of mass flux.

The convective and nucleate boiling heat transfer coefficient was the subject of experiments by Grohmann (2005). The measurements were performed in micro-tubes of 250 and 500 μm in diameter. The nucleate boiling metastable flow regimes were observed. Heat transfer characteristics at the nucleate and convective boiling in micro-channels with different cross-sections were studied by Yen et al. (2006). Two types of micro-channels were tested: a circular micro-tube with a 210 μm diameter, and a square micro-channel with a 214 μm hydraulic diameter. The heat transfer coefficient was higher for the square micro-channel because the corners acted as effective nucleation sites.

Several popular macro-channel correlations and recently recommended small-channel correlations were examined by Lee and Mudawar (2005b). Predictions were adjusted for the three-sided wall heating and rectangular geometry using the following relation:

$$h_{\mathrm{tp}} = h_{\mathrm{tp,cor}} \frac{\mathrm{Nu}_3}{\mathrm{Nu}_4} \tag{6.53}$$

where $h_{\mathrm{tp,cor}}$ is the value predicted from a correlation for uniform circumferential heating, and Nu_3 and Nu_4 are the single-phase Nusselt numbers for laminar flow with three-sides and four-sides wall heating, respectively (Shah and London 1978).

$$\mathrm{Nu}_3 = 8.235(1 - 1.883\beta + 3.767\beta^2 - 5.814\beta^3 + 5.361\beta^4 - 2.0\beta^5) \tag{6.54}$$

and

$$\mathrm{Nu}_4 = 8.235(1 - 2.042\beta + 3.085\beta^2 - 2.477\beta^3 + 1.058\beta^4 - 0.186\beta^5) \tag{6.55}$$

where β is the ratio of the channel depth to width.

Figure 6.31 compares the measured heat transfer coefficient by Lee and Mudawar (2005b) in two-phase flow of R-134a to predictions based on previous studies. The predictive accuracy of a correlation was measured by the mean absolute error, defined as

$$\mathrm{MAE} = \frac{1}{N}\sum\left[\frac{|h_{\mathrm{tp,pred}} - h_{\mathrm{tp,exp}}|}{h_{\mathrm{tp,exp}}} \times 100\%\right]. \tag{6.56}$$

Figure 6.31 shows all correlations yield poor predictions evidenced by their large MAE values.

Experiments by Lee and Mudawar (2005b) reveal the range of parameters at which heat transfer is controlled by nucleate boiling or annular film evaporation. The first of these processes occurs only at low qualities ($x < 0.05$) corresponding to very low heat fluxes; the second one at moderate ($0.05 < x < 0.55$) or high ($x > 0.55$) qualities that correspond to high enough heat fluxes. New correlations were suggested by Lee and Mudawar (2005b). They are based on the Martinelli parameter X and account for micro-channel effects not represented in the prior correlations.

Table 6.8 summarizes the new correlations for the three quality regions. The low and high-quality regions are based solely on the Martinelli parameter while the mid-range includes the effects of Bo and $\mathrm{We_{fo}}$ as well. Overall, convection to liquid is important for both the low and mid-quality regions, while convection to vapor becomes important for the high-quality region. For the latter, the low viscosity of R-134a vapor yields vapor Reynolds numbers corresponding to turbulent flow at high-heat flux conditions despite the small hydraulic diameter of the micro-channel. Thus, the single-phase vapor term in the high quality correlation must allow for both laminar or turbulent vapor flow.

Table 6.8 shows that the effect of the Martinelli parameter is important for each of the three quality ranges. The present correlations show the heat transfer coefficient is

Table 6.8 Two-phase flow boiling in micro-channels. Heat transfer coefficient. Reprinted from Lee and Mudawar (2005b) with permission

x_e	Correlation	Data	MAE (%)
0–0.05	$h_{\mathrm{tp}} = 3.856X^{0.267}h_{\mathrm{sp,L}}$ $X^2 = \frac{(\mathrm{d}p/\mathrm{d}z)_\mathrm{L}}{(\mathrm{d}p/\mathrm{d}z)_\mathrm{G}}, \quad h_{\mathrm{sp,L}} = \frac{\mathrm{Nu}_3 k_\mathrm{L}}{d_\mathrm{h}}$ $X_{\mathrm{vv}} = \left(\frac{\mu_\mathrm{L}}{\mu_\mathrm{G}}\right)^{0.5}\left(\frac{1-x_\mathrm{e}}{x_\mathrm{e}}\right)^{0.5}\left(\frac{v_\mathrm{L}}{v_\mathrm{G}}\right)^{0.5}$ $X_{\mathrm{vt}} = \left(\frac{f_\mathrm{L}\mathrm{Re}_\mathrm{G}^{0.25}}{0.079}\right)^{0.5}\left(\frac{1-x_\mathrm{e}}{x_\mathrm{e}}\right)^{0.5}\left(\frac{v_\mathrm{L}}{v_\mathrm{G}}\right)^{0.5}$ $\mathrm{Re}_\mathrm{G} = \frac{Gx_\mathrm{e}d_\mathrm{h}}{\mu_\mathrm{G}}$	50 Water data points	11.6
0.05–0.55	$h_{\mathrm{tp}} = 436.48\mathrm{Bo}^{0.522}\mathrm{We}_\mathrm{L}^{0.351}X^{0.665}h_{\mathrm{sp,L}}$ $\mathrm{Bo} = \frac{q}{Gh_{\mathrm{LG}}}, \quad \mathrm{We_{fo}} = \frac{v_\mathrm{L}G^2 d_\mathrm{h}}{\sigma}$	83 R-134a data points 157 Water data points	11.9
0.55–1.0	$h_{\mathrm{tp}} = \max\left\{(108.6X^{1.665}h_{\mathrm{sp,G}}), h_{\mathrm{sp,G}}\right\}$ $h_{\mathrm{sp,G}} = \frac{\mathrm{Nu}_3 k_\mathrm{G}}{d_\mathrm{h}}$ for laminar gas flow $h_{\mathrm{sp,G}} = 0.023\mathrm{Re}_\mathrm{G}^{0.8}\mathrm{Pr}_\mathrm{G}^{0.4}$ for turbulent gas flow	28 R-134a data points	16.1

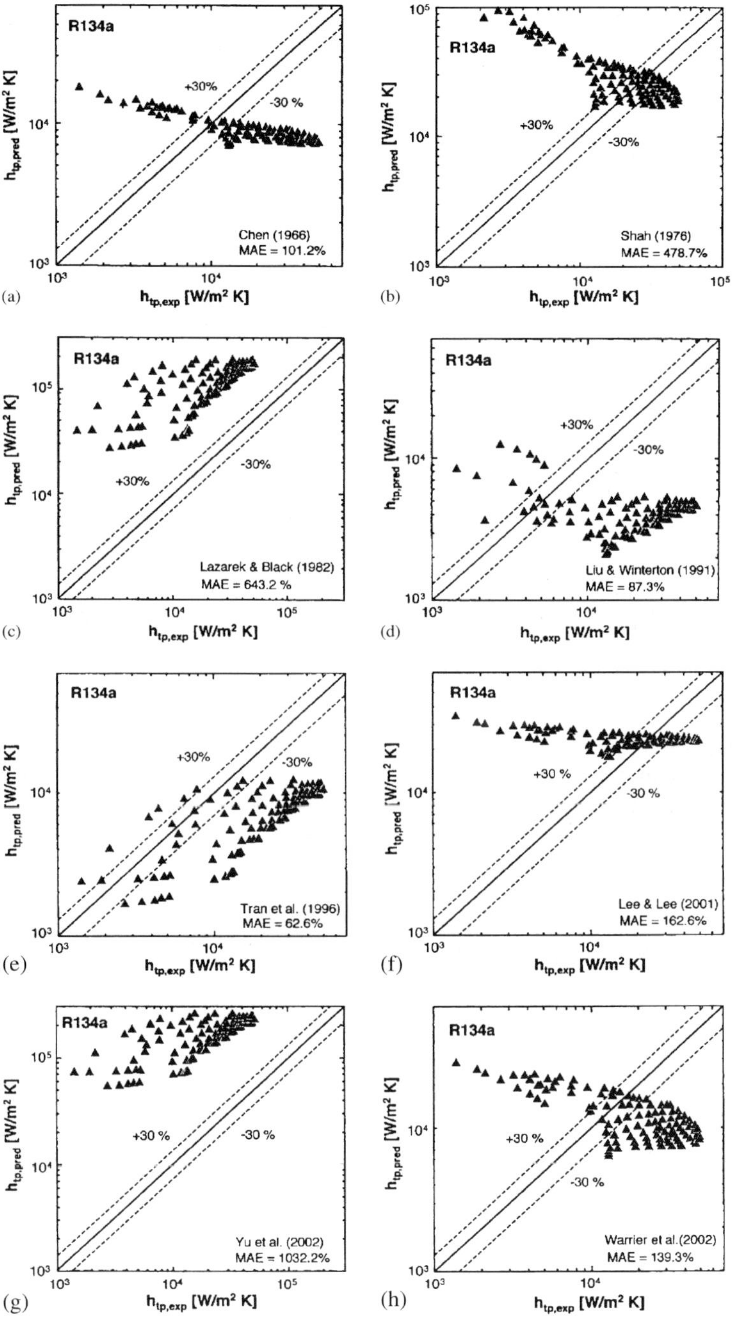

Fig. 6.31 Comparison of R-134a heat transfer coefficient data with predictions based on Chen (1966), Shah (1982), Lazarek and Black (1982), Liu and Winterton (1991), Tran et al. (1996), Lee et al. (2001b), Yu et al. (2002) and Warrier et al. (2002). Reprinted from Lee and Mudawar (2005b) with permission

proportional to the Martinelli parameter raised to a positive exponent, whereas prior macro-channel correlations employ a negative exponent for the same parameter.

In this table the parameters are defined as follows: Bo is the boiling number, d_h is the hydraulic diameter, f is the friction factor, h is the local heat transfer coefficient, k is the thermal conductivity, Nu is the Nusselt number, Pr is the Prandtl number, q is the heat flux, v is the specific volume, X is the Martinelli parameter, X_{vt} is the Martinelli parameter for laminar liquid–turbulent vapor flow, X_{vv} is the Martinelli parameter for laminar liquid–laminar vapor flow, x_e is thermodynamic equilibrium quality, z is the streamwise coordinate, μ is the viscosity, ρ is the density, σ is the surface tension; the subscripts are L for saturated fluid, LG for property difference between saturated vapor and saturated liquid, G for saturated vapor, sp for single-phase, and tp for two-phase.

Figure 6.32 illustrates good agreement between predictions based on the correlation scheme suggested by Lee and Mudawar (2005b) and both the R-134a and water data. An overall MAE of 12.26% indicates good predictive capability, especially with most of the data falling within a ±30% error range and capturing the correct data trend.

The work by Steinke and Kandlikar (2004b) focused on the obtaining heat transfer data during flow boiling in micro-channels. An experimental investigation was performed for flow boiling using water in six parallel, horizontal micro-channels

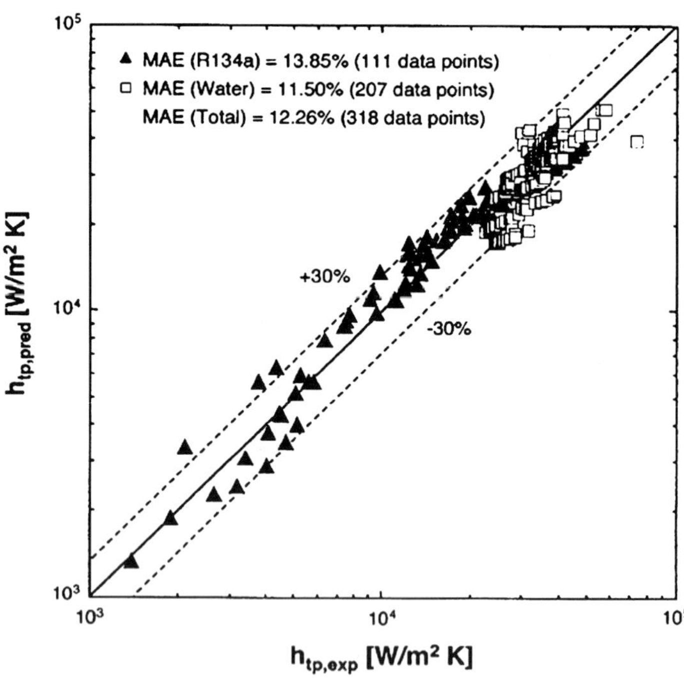

Fig. 6.32 Comparison of heat transfer coefficient data for R-134s and water with predictions based on Lee and Mudawar correlation. Reprinted from Lee and Mudawar (2005b) with permission

with a hydraulic diameter of 207 μm. The channels had a slightly trapezoidal cross-section, with the top and bottom widths differing by about 15 μm. The average channel dimensions are 214 μm wide by 200 μm deep and 57.15 mm long. The range of parameters are mass flux from 157 to 1,782 kg/m^2 s, heat flux from 5 to 930 kW/m^2, inlet temperature of 22 °C, quality from subcooled to 1.0, and atmospheric pressure at the exit.

$$h_{\mathrm{tp}} = 0.6683\mathrm{Co}^{-0.2}(1-x)^{0.8} f_2 h_{\mathrm{L}} + 1{,}058\mathrm{Bo}^{0.7}(1-x)^{0.8} F \cdot h_{\mathrm{L}} \tag{6.57}$$

$$\mathrm{Co} = \left(\frac{\rho_{\mathrm{G}}}{\rho_{\mathrm{L}}}\right)^{0.5} \left(\frac{1-x}{x}\right)^{0.8} \tag{6.58}$$

$$\mathrm{Bo} = \frac{q}{G h_{\mathrm{LG}}} \tag{6.59}$$

where Co is the convection number given in Eq. (6.58), Bo is the boiling number given in Eq. (6.59), f_2 is the multiplier, h_{L} is the heat transfer coefficient with all liquid flow, and x is the quality. The F number for water is 1.0 and the f_2 is multiplier is 1.0 for micro-channel flow.

The empirical correlation (6.57) predicts a heat transfer coefficient about two times higher than that measured by Qu and Mudawar (2003b) during flow boiling of water ($x < 0.15$) and during flow boiling of R-134a ($x = 0.4$–0.8). Correlation (6.57) also overpredicts the experimental data obtained by Yen et al. (2003) during convective boiling of HCFC123 and FC72 in $d = 190$ μm tubes in the range of $x = 0.4$–0.9.

6.4.3 Critical Heat Flux of Flow Boiling

Available data sets for flow boiling critical heat flux (CHF) of water in small-diameter tubes are shown in Table 6.9. There are 13 collected data sets in all. Only taking data for tube diameters less than 6.22 mm, and then eliminating duplicate data and those not meeting the heat balance calculation, the collected database included a total of 3,837 data points (2,539 points for saturated CHF, and 1,298 points for subcooled CHF), covering a wide range of parameters, such as outlet pressures from 0.101 to 19.0 MPa, mass fluxes from 5.33 to 1.34×10^5 kg/m^{2}s, critical heat fluxes from 0.094 to 276 MW/m^2, hydraulic diameters of channels from 0.330 to 6.22 mm, length-to-diameter ratios from 1.00 to 975, inlet qualities from −2.35 to 0, and outlet thermal equilibrium qualities from −1.75 to 1.00.

Zhang et al. (2006) compared some correlation with each database. The comparison results are tabulated in Table 6.10. Mean deviation is defined as $(1/N) \times \sum \left| (q_{\mathrm{crit,exp}} - q_{\mathrm{crit,cal}})/q_{\mathrm{crit,exp}} \right| \times 100\%$, a bold font in Table 6.10 denoting the smallest of mean deviations predicted by four correlations including a new correlation, and an underlined font being the smallest except for the correlation by Zhang et al. (2006).

Figure 6.33 compared graphically the predicted CHF values by the Bowring (1972) correlation, the Katto (1984) correlation and the Shah (1987) correlation

Table 6.9 Collected database for flow boiling CHF of water in small-diameter tubes. Reprinted from Zhang et al. (2006) with permission

Symb.	Reference	d_h [mm]	l/d_h	P_0 [MPa]	G [kg/m^2 s]	x_{in}[%]	x_{out}[%]	$q_{crit,exp}$ [MW/m^2]
○	Thompson and Macbeth (1964)	1.02–5.74	11.7–792	0.103–19.0	$13.0–1.57\times10^4$	−235 to −0.031	−44.8 to 99.9	0.113–21.4
△	Lowdermilk et al. (1958)	1.30–4.78	25.0–250	0.101–0.690	$27.1–3.42\times10^4$	−29.1 to −0.032	−3.02 to 99.1	0.167–41.6
▽	Becker et al. (1965)	3.93–6.07	164.8–382	1.13–6.97	$470–5.45\times10^3$	−67.5 to −23.4	0.086–9,607	1.59–5.66
◫	Griffel (1965)	6.22	147	6.89–10.3	$1.85\times10^3–1.39\times10^4$	−91.4 to −6.38	−7.07 to 30.4	2.60–7.81
⬡	Nariai et al. (1987, 1989)	1.00–3.00	3.33–50.0	0.101–1.05	$4.30\times10^3–2.99\times10^4$	−31.5 to −3.07	−18.3 to 5.55	4.50–66.1
⬡	Inasaka and Nariai (1989)	3.00	33.3	0.290–1.05	$4.30\times10^3–2.99\times10^4$	−31.5 to −12.5	−18.3 to −3.99	7.30–44.5
◁	Inasaka (1993)	1.00–3.00	3.17–50.4	0.101	$6.71\times10^3–2.09\times10^4$	−15.6 to −6.66	−13.2 to 1.10	4.64–67.0
□	Celata et al. (1993)	2.5	40	0.585–2.61	$1.12\times10^4–4.00\times10^4$	−46.1 to −18.7	−35.6 to 0.6	1.21–60.6
⬠	Vandervort et al. (1994)	0.330–2.67	1.66–26.2	0.131–2.28	$5.03\times10^3–4.18\times10^4$	−28.2 to −1.91	−22.6 to 28.4	4.60–124
⦶	Lezzi et al. (1994)	1.00	239–975	1.90–7.20	$7.76\times10^2–2.74\times10^4$	−64.4 to 0	64.3–0.976	0.285–2.36
◇	Kureta (1997)	1.00–6.00	1.00–113	0.101	$5.33–1.91\times10^4$	−17.2 to 0.032	−14.6 to 99.5	0.0935–158
⟠	Roach et al. (1999)	1.13–1.45	110–141	0.336–1.04	$256–1.04\times10^4$	−27.9 to −13.1	36.2–97.4	0.860–3.70
▷	Mudawar and Bowers (1999)	0.406–2.54	2.36–34.2	0.250–17.2	$5.00\times10^3–1.34\times10^5$	−189 to −11.7	−175 to −6.23	9.40–276
	Total (13 databases)	0.330–6.22	1.00–975	0.101–19.0	$5.33–1.34\times10^5$	−235 to 0	−175 to 99.9	0.0935–276

Table 6.10 Assessment of CHF correlations for flow boiling of water. Reprinted from Zhang et al. (2006) with permission

Reference	Mean deviation [%]						
	Correlations for saturated CHF				Correlations for subcooled CHF		
	Bowring (1972)	Katto and Ohno (1984)	Shah (1987)	Zhang et al. (2006)	Inasaka and Nariai (1987)	Celata et al. (1993)	Hall and Mudawar (2000)
Thompson and Macbeth (1964)	**12.3**	15.2	12.6	17.8	18.7	21.8	7.68
Lowdermilk et al. (1958)	32.4	21.9	15.7	**11.2**	30.3	13.2	32.7
Becker et al. (1965)	**5.51**	7.14	10.0	11.5			
Griffel (1965)	5.61	5.47	**4.43**	17.2	4.48	21.3	5.97
Nariai et al. (1987, 1989)	73.8	35.7	**21.8**	29.0	16.4	31.4	17.8
Inasaka and Nariai (1989)	81.2	37.1			9.25	17.1	7.48
Inasaka (1993)			**14.3**	38.3	18.9	28.0	21.7
Celata et al. (1993)					23.2	28.7	14.8
Vandervort et al. (1994)	82.5	16.9	42.9	**16.7**	17.3	24.6	29.6
Lezzi et al. (1994)	32.3	**7.84**	15.3	9.40			
Kureta (1997)	52.5	51.9	37.9	**20.2**	42.7	47.0	35.6
Roach et al. (1999)	18.6	33.5	**16.4**	18.0			
Mudawar and Bowers (1999)					83.5	38.7	18.8
Total (13 databases)	29.3	26.4	20.6	**16.8**	30.5	30.1	19.2

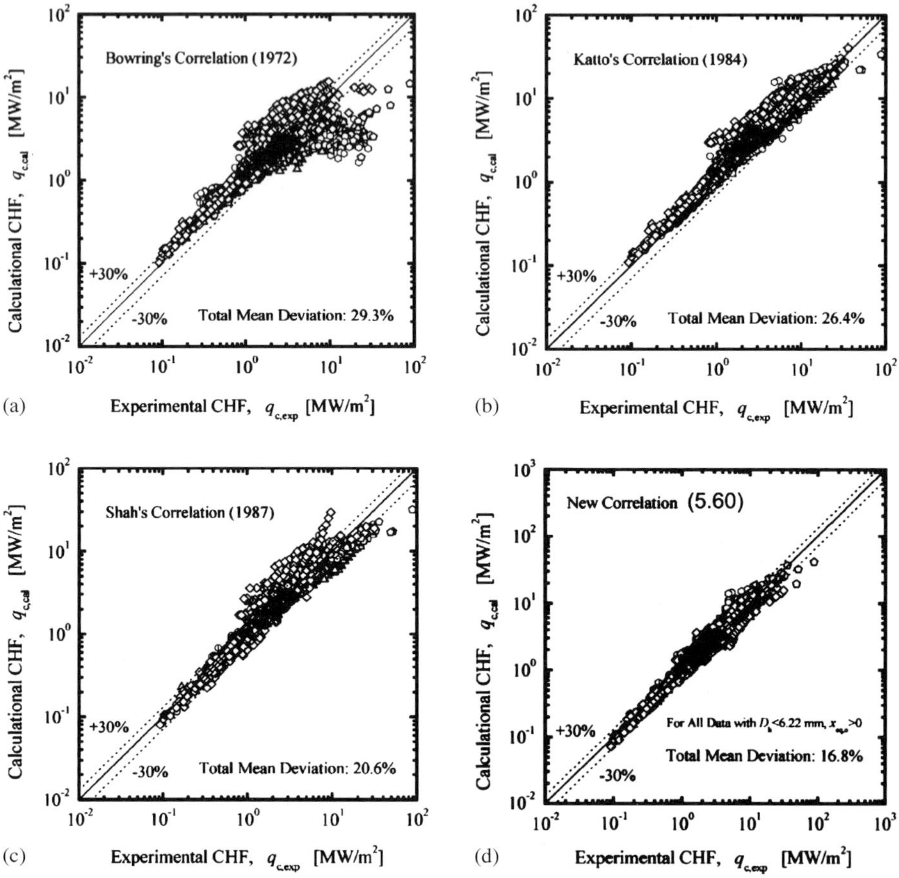

Fig. 6.33 Evaluation of correlations with saturated CHF data. (**a**) Bowring correlation (1972), (**b**) Shah correlation (1987), (**c**) Katto correlation (1984), (**d**) Zhang et al. correlation (2006). Reprinted from Zhang et al. (2006) with permission

with the experimental values, respectively. Symbols for each data set are depicted in Table 6.9. For low CHF values, shown in Fig. 6.33a, the Bowring (1972) relation correlated generally well with the experimental CHF values, although there is a slight overprediction. For high CHF values, however, the Bowring (1972) relation fails to correlate the experimental CHF values. A large scatter of predictions can be observed and many of them tend to deviate systematically from the experimental values. Figure 6.33b shows the behavior of the Katto (1984) correlation. Although the relation correlates all the data successfully within a relatively small scatter, it seems that there exist some slight overpredictions for most of the CHF data. Figure 6.33c shows the evaluation of the Shah (1987) correlation. For low CHF values, the predictions by the Shah (1987) correlation have a very small scatter and most are centered within the error band of $\pm 30\%$. For high CHF values, the Shah (1987) correlation tends to predict CHF with some scatter. The best agreement with experimental data on saturated CHF is the correlation by Zhang et al. (2006) (Fig. 6.33d):

$$\mathrm{Bo} = 0.0352\left[\mathrm{We} + 0.0119\left(\frac{L}{d_\mathrm{h}}\right)^{2.31}\left(\frac{\rho_\mathrm{G}}{\rho_\mathrm{L}}\right)^{0.361}\right]^{-0.295} \times\left[\left(\frac{L}{d_\mathrm{h}}\right)^{-0.311}\left(2.05\left(\frac{\rho_\mathrm{L}}{\rho_\mathrm{L}}\right)^{0.170} - x_\mathrm{in}\right)\right] \tag{6.60}$$

where $\mathrm{Bo} = q_\mathrm{crit}/(h_\mathrm{LG}G)$ is the boiling number, $\mathrm{We} = (G^2 d_\mathrm{h})/(\sigma\rho_\mathrm{L})$ is the Weber number, q_crit is the critical heat flux, h_LG is the latent heat of evaporation, G is the mass flux, x_in is the thermodynamical equilibrium quality at inlet, ρ_G and ρ_L are the density of saturated vapor and liquid, respectively, d_h is the hydraulic diameter, L is the heated length, and σ is the surface tension.

Hall and Mudawar (2000) provided a comprehensive review of the current state of the knowledge of subcooled CHF for water flow boiling in channels, and designed a statistical correlation with five parameters based on almost all available subcooled CHF databases in the literature:

$$\mathrm{Bo} = \frac{C_1 \mathrm{We}^{C_2}(\rho_\mathrm{L}/\rho_\mathrm{G})^{C_3}[1 - C_4(\rho_\mathrm{L}/\rho_\mathrm{G})^{C_5} x_\mathrm{in}]}{1 + 4C_1 C_4 \mathrm{We}^{C_2}(\rho_\mathrm{L}/\rho_\mathrm{G})^{C_3+C_5}(L/d_\mathrm{h})} \tag{6.61}$$

where the Weber number $\mathrm{We} = G^2 d_\mathrm{h}/(\rho_\mathrm{L}\sigma)$, $C_1 = 0.0722$, $C_2 = -0.312$, $C_3 = -0.644$, $C_4 = 0.900$, and $C_5 = 0.724$. The correlation was developed using a total of 4,860 data points and predicted CHF with a rms error of 14.3% in the following parametric ranges: $0.1 < P_0 < 20\,\mathrm{MPa}$, $0.25 < d_\mathrm{h} < 15.0\,\mathrm{mm}$, $2 < L/d_\mathrm{h} < 200$, $300 < G < 30{,}000\,\mathrm{kg/m^2\,s}$, $-2.00 < x_\mathrm{in} < 0.00$, and $-1.00 < x_\mathrm{out} < 1.00$.

A theoretical model for the prediction of the critical heat flux of refrigerants flowing in heated, round micro-channels has been developed by Revellin and Thome (2008). The model is based on the two-phase conservation equations and includes the effect of the height of the interfacial waves of the annular film. Validation has been carried out by comparing the model with experimental results presented by Wojtan et al. (2006), Qu and Mudawar (2004), Bowers and Mudawar (1994), Lazareck and Black (1982). More than 96% of the data for water and R-113, R-134a, R-245fa were predicted within ±20%.

6.5 Explosive Boiling of Water in Parallel Micro-Channels

The thermohydrodynamical processes of boiling in a micro-channel heat sink were subject to a number of experimental investigations performed during the last decade. Periodic wetting and rewetting phenomena were observed by Hetsroni et al. (2003, 2005), Zhang et al. (2002), Steinke and Kandlikar (2004a), and Kandlikar and Balasubramanian (2004). Flow boiling of water in parallel silicon micro-channels with trapezoidal cross-sectional area with hydraulic diameters of 158.8 and 82.8 μm, was studied by Wu and Cheng (2003a,b, 2004). Long period fluctuations in fluid pressure, fluid temperature, and fluid mass flux were measured.

6.5.1 Quasi-Periodic Boiling in a Certain Single Micro-Channel of a Heat Sink

In the study by Hetsroni et al. (2006b) the test module was made from a square-shaped silicon substrate 15 × 15 mm, 530 μm thick, and utilized a Pyrex cover, 500 μm thick, which served as both an insulator and a transparent cover through which flow in the micro-channels could be observed. The Pyrex cover was anodically bonded to the silicon chip, in order to seal the channels. In the silicon substrate parallel micro-channels were etched, the cross-section of each channel was an isosceles triangle. The main parameters that affect the explosive boiling oscillations (EBO) in an individual channel of the heat sink such as hydraulic diameter, mass flux, and heat flux were studied. During EBO the pressure drop oscillations were always accompanied by wall temperature oscillations. The period of these oscillations was very short and the oscillation amplitude increased with an increase in heat input. This type of oscillation was found to occur at low vapor quality.

Period between successive events

Figure 6.34 shows the dependence of the dimensionless period of phase transformations (i.e., the time between bubble venting), t^*, on boiling number Bo ($t^* = t/Ud_h$,

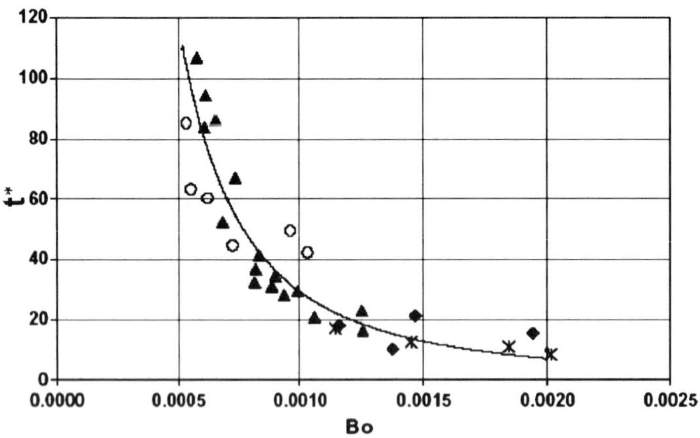

Fig. 6.34 Dependence of dimensionless time interval between cycles on boiling number: circles (○) represent $d_h = 100$ μm, water, triangles (▲) represent $d_h = 130$ μm, water, diamonds (◇) represent $d_h = 220$ μm, water, star (∗) represents $d_h = 220$ μm, ethanol. Reprinted from Hetsroni et al. (2006b) with permission

$Bo = q/Gh_{LG}$, where t is the period between successive events, U is the mean velocity of single-phase flow in the micro-channel, d_h is the hydraulic diameter of the channel, q is heat flux, m is mass flux, h_{LG} is the latent heat of vaporization). The dependence t^* on Bo can be approximated, with a standard deviation of 16%, by

$$t^* = 0.000030 Bo^{-2} . \tag{6.62}$$

6.5.2 The Initial Thickness of the Liquid Film

The term initial liquid film thickness is defined as the average thickness of fluid, evenly distributed during period t, over the surface of the circular micro-channel, after venting of the elongated bubble. This surface is located downstream of the ONB and may be characterized by the heated length L, and hydraulic diameter d_h, assuming that during the period t the liquid film has disappeared due to evaporation. The heat removed from the wall surface is the same as that required for the liquid film evaporation during the period t. The heat balance is:

$$\pi \delta d_h L \rho_L = q \pi d_h L t / h_{LG} \tag{6.63}$$

where ρ_L is the liquid density.

The average liquid thickness δ, can be calculated as:

$$\delta = qt/\rho_L h_{LG} . \tag{6.64}$$

Figure 6.35 shows dependence of the dimensionless initial liquid thickness of water and ethanol δ^*, on the boiling number Bo, where $\delta^* = \delta U/\nu$, U is the mean velocity of single-phase flow in the micro-channel, and ν is the kinematic viscosity of the

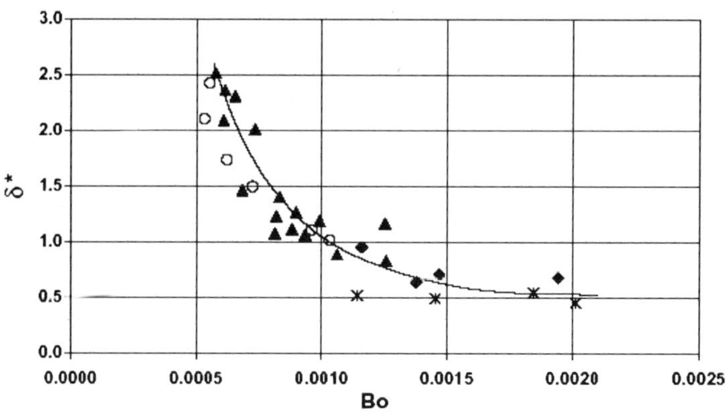

Fig. 6.35 Dependence of dimensionless initial film thickness on boiling number: *circles* (○) represent $d_h = 100\ \mu m$, water, *triangles* (▲) represent $d_h = 130\ \mu m$, water, *diamonds* (♦) represent $d_h = 220\ \mu m$, water, *star* (∗) represents $d_h = 220\ \mu m$, ethanol. Reprinted from Hetsroni et al. (2006b) with permission

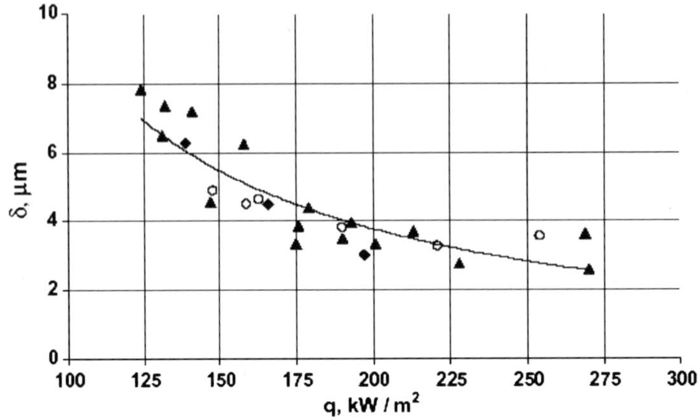

Fig. 6.36 Variation of initial film thickness for water versus heat flux: *circles* (○) represent d_h = 100 μm, water, *triangles* (▲) represent d_h = 130 μm, water, *diamonds* (♦) represent d_h = 220 μm, water. Reprinted from Hetsroni et al. (2006b) with permission

liquid at saturation temperature. The dependence of δ^* on Bo can be approximated, with a standard deviation of 18%, by

$$\delta^* = 0.00015\text{Bo}^{-1.3}\,. \tag{6.65}$$

The initial thickness of the liquid film is a key parameter of the explosive boiling. This point may be discussed in some detail with regard to the beginning of the critical heat flux (CHF) regime. The variation of the initial thickness of the film of water versus the heat flux is provided in Fig. 6.36. For explosive boiling the film thickness decreases with increasing heat flux from 125 to 270 kW/m^2 from about 8 to 3 μm. This range of values is on the same order of magnitude as those given by Moriyama and Inoue (1996) and by Thome et al. (2004) for R-113 in small spaces (100–400 μm). Decreasing liquid film thickness with increasing heat flux is a distinct feature of dryout during explosive boiling. Under these conditions at which the instantaneous temperature of the heater surface exceeds 125 °C, the value of δ was in the range of (3 ± 0.6) μm. This value may be considered as minimum initial film thickness. If the liquid film reached the minimum initial film thickness δ_{min}, CHF regime occurred. According to Thome et al. (2004) δ_{min} is assumed to be on the same order of magnitude as the surface roughness. The values of the minimum initial film thickness calculated by Thome et al. (2004) for R-113 at saturation temperature 47.2 °C was in the range of 1.5–3.5 μm.

6.5.3 System that Contains a Number of Parallel Micro-Channels

Hetsroni et al. (2006b) also studied the effect of EBO in individual channels on the average characteristics of the whole heat sink: total pressure drop and temperature fluctuations on the heater, and the heat transfer coefficient. The high-frequency oscillations in individual micro-channels are superimposed and lead to total low-frequency pressure drop and temperature oscillations of the system.

Fluctuation of pressure drop, fluid and heated wall temperatures

The experimental investigations of boiling instability in parallel micro-channels have been carried out by simultaneous measurements of temporal variations of pressure drop, fluid and heater temperatures. The channel-to-channel interactions may affect pressure drop between the inlet and the outlet manifold as well as associated temperature of the fluid in the outlet manifold and heater temperature. Figure 6.37 illustrates this phenomenon for pressure drop in the heat sink that contains 13 micro-channels of $d_h = 220\,\mu m$ at mass flux $G = 93.3\,kg/m^2\,s$ and heat flux $q = 200\,kW/m^2$. The temporal behavior of the pressure drop in the whole boiling system is shown in Fig. 6.37a. The considerable oscillations were caused by the flow pattern alternation, that is, by the liquid/two-phase alternating flow in the micro-channels. The pressure drop FFT is presented in Fig. 6.37b. Under

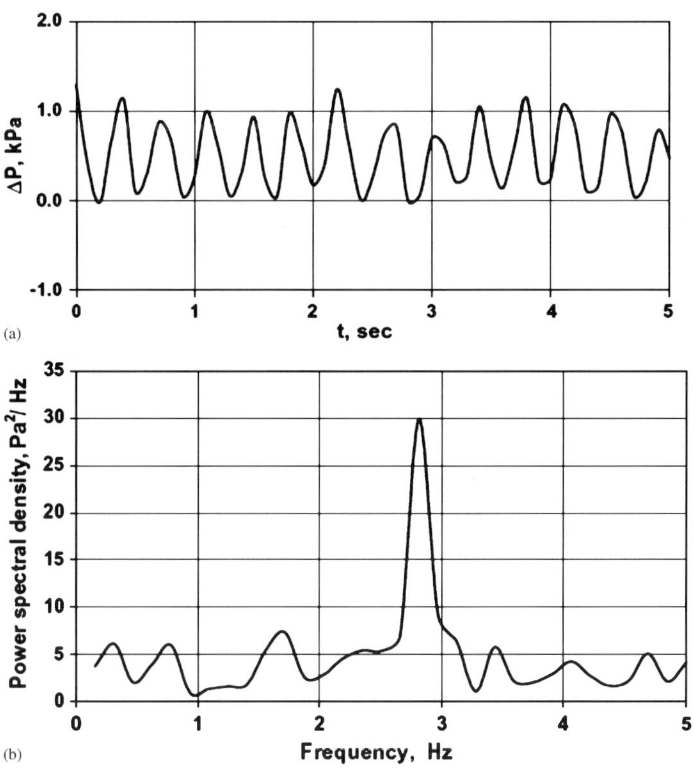

Fig. 6.37 Time variation of pressure drop at $q = 200\,kW/m^2$: (**a**) pressure drop fluctuations, (**b**) pressure drop amplitude spectrum. Reprinted from Hetsroni et al. (2006b) with permission

condition of the given experiment the period of pressure drop fluctuation is about $t = 0.36$ s. The results differ significantly from those reported by Wu and Cheng (2004). In the range of average values of mass flux $G = 112\text{–}146\,\text{kg/m}^2\text{s}$ and heat flux $q = 135\text{–}226\,\text{kW/m}^2$ these authors observed much longer oscillation period (from $t = 15.4$ to 202 s). In experiments conducted by Wu and Cheng (2004), the water in the pressure tank was moved by the compressed nitrogen gas to the test section. According to the authors when the boiling occurred in the test section, the pressure drop across the test was suddenly increased due to generation of vapor bubbles. This increase in pressure drop caused a decrease in mass flux. The long period pressure drop fluctuations may be connected to the period of increasing or decreasing of incoming mass flux. The former depends not only on boiling in the micro-channels of heat sinks, but also on the nitrogen pressure in the water tank and the total length of the pipe connecting the water tank to the test section. In our ex-

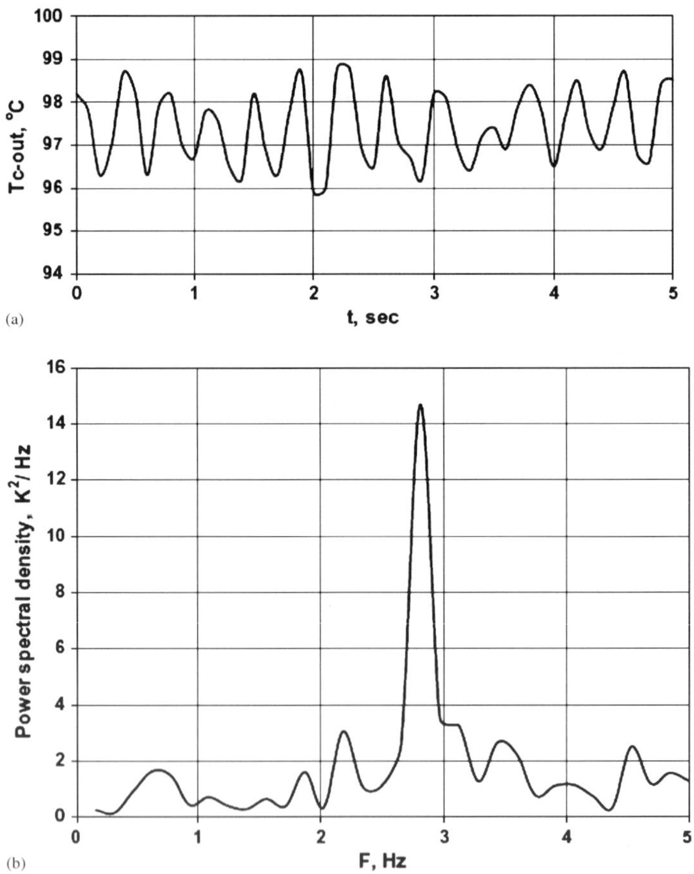

Fig. 6.38 Time variation of fluid temperature at the outlet manifold $q = 200\,\text{kW/m}^2$: (**a**) temperature fluctuations, (**b**) temperature amplitude spectrum. Reprinted from Hetsroni et al. (2006b) with permission

periments, the mass flow rate was independent of pressure drop fluctuations, and the oscillation periods are very much different from those recorded by Wu and Cheng (2004).

The pressure drop fluctuation provides insight into the temperature behavior of the fluid in the outlet manifold. The pressure drop fluctuation frequency is representative of the oscillations in the system. Figure 6.38a,b shows time variation and FFT of the fluctuation component of the fluid temperature. From Fig. 6.38a one can see that the average fluid temperature at the outlet manifold is less than the saturation temperature. This results in the fact that only single liquid comes to the outlet manifold through some of the parallel micro-channels.

The time variation of the mean and maximum heater temperature is presented in Fig. 6.39. The mean heater temperature (i.e., the average temperature of the whole heater) changed in the range of $\Delta T_{av} = 10\,K$. The maximum heater temperature changed in the range of $\Delta T_{max} = 6\,K$. Comparison between Figs. 6.37, 6.38 and 6.39, shows that the time period (frequency) is the same for the pressure drop, the fluid temperature at the outlet manifold, and the mean and maximum heater temperature fluctuations. It also allows one to conclude that these fluctuations are in phase.

When the heat flux is increased, at constant value of mass flux, the oscillation amplitudes of the pressure drop, the fluid and the heater temperatures also increase.

6.5.4 Average Heat Transfer Coefficient

It was observed that at the same boiling number and inlet temperature, an increase in diameter shifts the ONB further from the inlet. The region of the local dryout decreases and the average heated surface temperature decreases as well. Under this condition the heat transfer coefficient increases with increased hydraulic diameter.

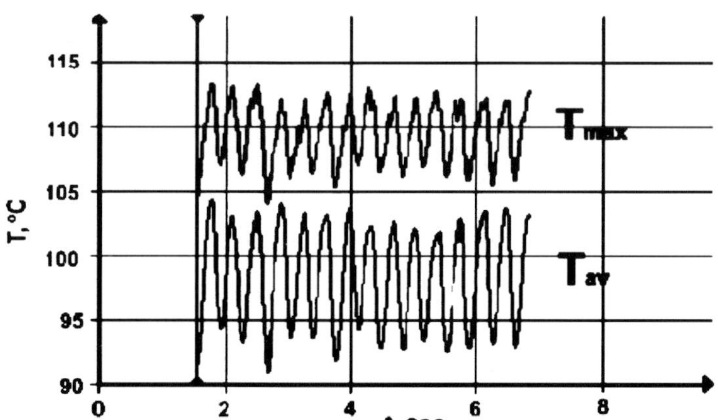

Fig. 6.39 Time variation of average and maximum heater temperature at $q = 200\,kW/m^2$. Reprinted from Hetsroni et al. (2006b) with permission

In order to take into account the effect of surface tension and micro-channel hydraulic diameter, we have applied the Eotvos number $\mathrm{Eo} = g(\rho_L - \rho_G)d_h^2/\sigma$. Figure 6.40 shows the dependence of the Nu/Eo on the boiling number Bo, where $\mathrm{Nu} = hd_h/k_L$ is the Nusselt number, h is the heat transfer coefficient, and k_L is the thermal conductivity of fluid. All fluid properties are taken at the saturation temperature. This dependence can be approximated, with a standard deviation of 18%, by the relation:

$$\mathrm{Nu}/\mathrm{Eo} = 0.030\mathrm{Bo}^{-1.5} \tag{6.66}$$

Simultaneous measurements of temporal variations of pressure drop, fluid and heater temperatures show the boiling instability in parallel micro-channels. The channel-to-channel interactions may affect pressure drop between the inlet and the outlet manifold, as well as associated temperature of the fluid in the outlet manifold and the temperature of the heater. The frequency is the same for the pressure drop, the fluid temperature at the outlet manifold, and for the fluctuations of the mean and maximum temperature of the heater. All these fluctuations are in phase. When the heat flux increases, at a constant value of mass flux, the oscillation amplitudes of the pressure drop, the fluid and the heater temperatures also increase.

The large heated wall temperature fluctuations are associated with the critical heat flux (CHF). The CHF phenomenon is different from that observed in a single channel of conventional size. A key difference between micro-channel heat sink and a single conventional channel is the amplification of the parallel channel instability prior to CHF. As the heat flux approached CHF, the parallel channel instability, which was moderate over a wide range of heat fluxes, became quite intense and should be associated with a maximum temperature fluctuation of the heated surface. The dimensionless experimental values of the heat transfer coefficient may be correlated using the Eotvos number and boiling number.

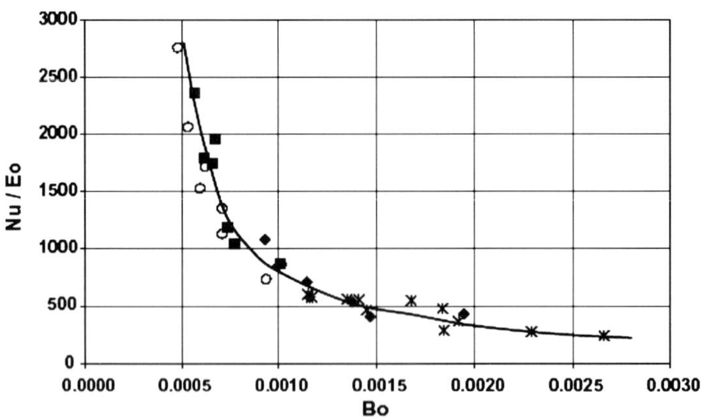

Fig. 6.40 Dependence of Nu/Eo on Bo: *circles* (○) represent $d_h = 100\,\mu m$, water, *triangles* (▲) represent $d_h = 130\,\mu m$, water, *diamonds* (♦) represent $d_h = 220\,\mu m$, water, *star* (∗) represents $d_h = 220\,\mu m$, ethanol. Reprinted from Hetsroni et al. (2006b) with permission

Summary

1. Incipience boiling

Wall superheat

Significant differences in boiling inception phenomena have been reported in micro-channels compared to conventional size channels. Phase-change processes, particularly boiling incipience, are important in the micro-scale as well in the macro-scale, but relatively little work has been done to compare the experimental results in this area. Most of the studies on this phenomenon in conventional size channels ($d_h = 1-20$ mm) were carried out in a single channel. Heat exchangers with micro-channels (d_h less than about 0.5 mm) are generally subject to two problems. The first of these is flow distribution among parallel channels. The second is conjugate effects, circumferential and axial heat conduction in the material forming the channel, so that the actual heat flux and temperature distributions at the channel wall are difficult to estimate. Typical micro-channels studied in the literature had one unheated side wall (usually a transparent cover plate in the experimental apparatus) and heat was applied from the bottom, with the result that the heat flux and wall temperature distribution may be circumferentially non-uniform.

There is significant disagreement between experimental results for conventional size channels due to different experimental conditions. For example, the wall has to be superheated to a relatively great extent to initiate the nucleate boiling in the experiments by Hapke et al. (2000) compared to those reported by Sato and Matsumura (1964) and by Bergles and Rohsenow (1964). Moreover, a mass flux dependence of the wall superheat was reported by Hapke et al. (2000). Measurements of incipient boiling heat flux performed by Qu and Mudawar (2002) in a heat sink containing rectangular parallel micro-channels showed that $q_{\text{ONB}} \sim \Delta T_{\text{S,ONB}}$ whereas measurements by Bergles and Rohsenow (1964) showed that heat flux depends on the wall superheat as $q_{\text{ONB}} \sim \Delta T^2_{\text{S,ONB}}$ and is independent of mass flux.

For qualitative analysis of the conditions at which the boiling incipience was studied experimentally the parameter $D = \Delta T_{\text{sub,ONB}}/T_{\text{S}}$ may be used. Depending on the value of D, the channels can be subdivided into two groups: $D < 1$ and $D \ll 1$. When $D < 1$ ($D = 0.125-0.25$) the onset of nucleate boiling occurred at a bulk temperature significantly less than saturation. When $D \ll 1$ ($D = 0.011-0.043$) the onset of nucleate boiling occurred at values of the bulk temperature close to saturation.

Incipient boiling heat flux

When $D < 1$ ($T_{\text{in}} \ll T_{\text{S}}$) incipient boiling heat flux increases with increasing mass velocity. When $D \ll 1$ ($T_{\text{in}} \sim T_{\text{S}}$) incipient boiling heat flux weakly depends on mass velocity. For micro-channels boiling incipient heat flux may weakly depend on inlet temperature. This case corresponds to flow boiling in parallel micro-channels, in which vapor penetrates the inlet manifold.

Surfactant solutions

For some kind of surfactant solutions boiling incipience was accompanied with hysteresis and the wall superheat up to 24 K was observed.

Effect of dissolved gases

Desorption of the dissolved gases formed bubbles of gas and a limited amount of bubbles containing gas–water vapor mixture. Under these conditions, during flow boiling of water boiling incipience occurred at channel wall temperatures below that of saturation temperature. Addition of surfactants led to an increase in wall temperature. In this case the boiling occurred in the range of heat flux of 5.2–21 $\mathrm{W/cm^2}$ at wall temperatures of 107–121 °C.

2. Bubble dynamics in a confined space

Dimensional analysis shows that the behavior of the bubble radius with time depends on the parameter $\Pi = q/(\rho_L U c_{pL} \Delta T_S)$. In the range of $\Pi = 0.0079$–0.026, linear behavior was observed, and when $\Pi > 0.026$ exponential bubble growth took place.

3. Pressure drop

The complexity of steam–liquid flow in a micro-channel is the result of interactions between liquid inertia, the liquid viscous force, and surface tension. The approach developed by Lee and Mudawar (2005a) may be used to calculate the pressure drop in micro-channels. Two key measures of these interactions are the Reynolds and Weber numbers based on liquid properties.

4. Heat transfer in two-phase flow boiling

The extent to which an incoming liquid will be vaporized is a design variable that depends on the intended application. In micro-scale refrigeration systems, the change in vapor quality may be substantial, on the order of 0.8 for example. In electronics cooling applications, the equilibrium vapor quality may remain at 0, or be very small; in those designs, the aim is to capture the high-heat transfer coefficients of subcooled flow boiling, without the added complexities of net vapor generation (e.g., the need to incorporate a condenser). The distinction between low and high-quality outflow affects heat transfer coefficients. Experiments by Lee and Mudawar (2005b) revealed the range of parameters at which heat transfer was controlled by nucleate boiling or annular film evaporation. The first of this process occurred only at low qualities ($x < 0.05$), the second one at moderate ($0.05 < x < 0.55$) or high

($x > 0.55$) qualities. New correlations were suggested by Lee and Mudawar (2005b). They are based on the Martinelli parameter and may be used for two-phase flow boiling of water and refrigerant R-134a.

5. Critical heat flux

Several thousand CHF points have been reported in the boiling literature of the past 50 years. Most of these data were obtained with stable flow in single conventional size circular tubes and conventional size channels. The tubes were usually of uniform wall thickness, and direct electrical heating was utilized to simulate the constant heat flux. The correlation suggested by Zhang et al. (2006) gives the best agreement with experimental data on saturated CHF in the channels of 0.33–6.22 mm. As discussed in detail in Chap. 2, there is significant difference between experimental results of CHF obtained in a single micro-channel of $d_h = 0.5$ mm and in the block that contained twenty-one 0.215×0.21 mm channels.

The study of CHF in micro-channels has not received much attention in the literature. Single-tube CHF data are not available for micro-channels with hydraulic diameters less than 0.3 mm. Under conditions of explosive boiling, Hetsroni et al. (2006b) suggested that the initial thickness of the liquid film may by considered as a key parameter that affects CHF in the system containing a number of parallel micro-channels of $d_h = 0.1-0.22$ mm. New experiments should be performed to validate the dimensionless groups caused by the evaporating interface near the heated wall and CHF for single micro-channels, as well as for blocks that contain parallel micro-channels.

6. Flow instability

The channel-to-channel interactions may affect pressure drop between the inlet and the outlet manifold, as well as associated temperature of the fluid in the outlet manifold and temperature of the heater. The frequency and the phase are the same for all these fluctuations. They increase at a constant value of mass flux with increasing heat flux. The large heated wall temperature fluctuations are associated with the CHF. As the heat flux approached CHF, the parallel-channel instability, which was moderate over a wide range of heat fluxes, became quite intense and should be associated with maximum temperature fluctuation of the heated surface.

References

Agostini B, Revellin R, Thome J (2008) Elongated bubbles in micro-channels. Part I: Experimental study and modeling of elongated bubble velocity. Int. J. Multiphase Flow 34:590–601

Bankoff SG, Haute T (1957) Ebullition from solid surfaces in the absence of pre-existing gaseous phase. Trans ASME 79:735–740

Becker KM, Hernborg G, Bode M, Eriksson O (1965) Burnout data for flow of boiling water in vertical round ducts, annuli and rod clusters, AE-177. Aktiebolaget Atomenergi, Stockholm, Sweden

Becker R, Doring W (1935) The kinetic treatment of nuclear formation in superheated vapors. Ann Phys 24:719–752

Bergles AE, Rohsenow WW (1964) The Determination of Forced-Convection Surface-Boiling Heat Transfer. Trans ASME J Heat Transfer 86:365–372

Blander M, Katz JL (1975) Bubble nucleation in liquids. AIChE J 21:833–848

Bowers MB, Mudawar I (1994) High flux boiling in low flow rate, low pressure drop mini-channel and micro-channel heat sinks. Int. J. Heat Mass Transfer 37:321–332

Bowring RW (1972) A simple but accurate round tube uniform heat flux, dryout correlation over the pressure range 0.7–17 MN/m2 (100–2500 psia), AEEW-R 789. United Kingdom Atomic Energy Authority, Harwell

Brutin B, Tadrist L (2004) Pressure drop and heat transfer analysis of flow boiling in micro-channel: influence of the inlet condition on two-phase flow stability. Int J Heat Mass Transfer 47:2367–2377

Carey van P (1992) Liquid–vapor phase-change phenomena. An introduction to the thermophysics of vaporization and condensation processes in heat transfer equipment. Hemisphere, New York

Celata GP, Cumo M, Mariani A (1997) Experimental evaluation of the onset of subcooled flow boiling at high liquid velocity and subcooling. Int J Heat Mass Transfer 40:2979–2885

Celata GP, Cumo M, Mariani A (1993) Burnout in highly subcooled water flow boiling in small diameter tubes. Int J Heat Mass Transfer 36:1269–1285

Chen JC (1966) Correlation for boiling heat transfer to saturated fluids in convective flow. Ind Eng Chem Process Des Develop 5:322–329

Cole R (1974) Boiling nucleation. In: Advances in Heat Transfer 10. Academic, New York, pp 86–166

Cole R, Shulman HL (1966) Bubble growth rates at high Jacob numbers Int J Heat Mass Transfer 9:1377–1390

Collier G, Thome J (1994) Convective Boiling and Condensation, 3rd edn. Oxford University Press, Oxford

Davis EJ, Anderson GH (1966) The incipience of nucleate boiling in forced convection flow. AIChE 12:774–780

Dhir VK (1998) Boiling heat transfer. Ann Rev Fluid Mech 30:365–401

Farkas L (1927) The velocity of nucleus formation in superheated vapors. Z Phys Chem 125:236–240

Fritz WW (1936) Ende über den Verdampfungsvorgang nach kinematographischen Aufnahmen an Dampfblasen. Phys Z 37:391–401

Geld C (2004) Prediction of dynamic contact angle histories of a bubble growing at a wall. Int J Heat Fluid Flow 25:74–80

Ghiaasiaan SM, Chedester RC (2002) Boiling incipience in microchannels. Int J Heat Mass Transfer 45:4599–4606

Griffel J (1965) Forced convection boiling burnout for water in uniformly heated tubular test sections. USAEC Report NYO 187-7, TID-4500, Columbia University

Grohmann S (2005) Measurement and modeling of single-phase and flow-boiling heat transfer in micro-tubes. Int J Heat Mass Transfer 48:4072–4089

Hall DD, Mudawar I (2000) Critical heat flux (CHF) for water flow in tubes – II. Subcooled CHF correlations. Int J Heat Mass Transfer 43:2605–2640

Han CH, Griffith P (1965) The mechanism of heat transfer in nucleate pool boiling. Part I: Bubble initiation, growth and departure Int J Heat Mass Transfer 8:887–904

Hapke I, Boye H, Schmidt J (2000) Onset of nucleate boiling in micro-channels. Int J Therm Sci 39:505–513

Helden W, Geld C, Boot P (1995) Forces on bubbles growing and detaching in flow along a vertical wall. Int J Heat Mass Transfer 38:2075–2088

Hetsroni G, Mosyak A, Pogrebnyak E, Sher I, Segal Z (2006a) Bubble growth in saturated pool boiling in water and surfactant solution. Int J Multiphase Flow 22:159–182
Hetsroni G, Mosyak A, Pogrebnyak E, Segal S (2006b) Periodic boiling in parallel micro-channels at low vapor quality. Int J Multiphase Flow 32:1141–1159
Hetsroni G, Mosyak A, Pogrebnyak E, Segal S (2007) Natural convection boiling of water and surfactant in narrow horizontal annular channels. Int J Multiphase Flow 33:469–483
Hetsroni G, Mosyak A, Pogrebnyak E, Segal Z (2005) Explosive boiling of water in parallel micro-channels. Int J Multiphase Flow 31:371–392
Hetsroni G, Mosyak A, Segal Z, Pogrebnyak E (2003) Two-phase flow pattern in parallel micro-channels. Int J Multiphase Flow 29:344–360
Hetsroni G, Mosyak A, Segal Z, Ziskind G (2002a) A uniform temperature heat sink for cooling of electronic devices. Int J Heat Mass Transfer 45:3275–3286
Hetsroni G, Gurevich M, Mosyak A, Rozenblit R, Yarin LP (2002b) Subcooling boiling of surfactant solutions. Int J Multiphase Flow 28:347–361
Hino R, Ueda T (1985) Studies on heat transfer and flow characteristics in subcooled flow boiling. Part 1: Boiling characteristics. Int J Multiphase Flow 11:269–281
Hsu YY (1962) On size range of active nucleation cavities on a heating surface. J Heat Transfer 84:207–216
Hsu YY, Graham RW (1961) An analytical and experimental study of the thermal boundary layer and ebullition cycle in nucleate boiling. NASA TN D-594
Hwan YW, Kim MS (2006) The pressure drop in micro-tubes and correlation development. Int J Heat Mass Transfer 49:1804–1812
Inasaka F (1993) Critical heat flux of subcooled flow boiling in water under uniform heating conditions. Papers Ship Res Inst 30(4):1–69
Inasaka F, Nariai H (1987) Critical heat flux and flow characteristics of subcooled flow boiling in narrow tubes. JSME Int J 30:1595–1600
Inasaka F, Nariai H (1989) Critical heat flux of subcooled flow boiling with water. Proc NURETH-4 1:115–120
Jens WH, Lottes PA (1951) Analysis of heat transfer burnout drop and density data for high pressure water. US ANL Rep 4627
Kakac S, Bon B (2008) A review of two-phase flow dynamic instabilities in tube boiling systems. Int. J. Heat Mass Transfer 51:399–433
Kandlikar SG (2002) Fundamental issues related to flow boiling in mini-channels and micro-channels. Exp Thermal Fluid Sci 26:389–407
Kandlikar SG, Steinke ME, Tian S, Campbell LA (2001) High speed photographic observation of flow boiling of water in parallel mini-channels. In: 35th Proceeding of National Heat Transfer Conference, ASME, New York
Kandlikar SG, Balasubramanian P (2004) An extension of the flow boiling correlation to transition, laminar and deep laminar flows in mini-channels and micro-channels. Heat Transfer Eng 25:86–93
Kandlikar SG, Mizo V, Cartwright M, Ikenze E (1997) Bubble nucleation and growth characteristics in subcooled flow boiling water. National Heat Transfer Conference HTD-342. ASME, New York, pp 11–18
Kandlikar SG (2006) Nucleation characteristics and stability considerations during flow boiling in micro-channels. Exp. Thermal and Fluid Science 30:441–447
Katto Y, Ohno H (1984) An improved version of the generalized correlation of critical heat flux for the forced convective boiling in uniformly heated vertical tubes. Int J Heat Mass Transfer 27:1641–1648
Kays WM, Krawford ME (1993) Convective heat and mass transfer. McGraw-Hill, New York
Kennedy JE, Roach GM, Dowling ME, Abdel-Khalik SI, Chiaasiaan SM, Jeter SM, Quershi ZH (2000) The onset of flow instability in uniformly heated horizontal micro-channels. Trans ASME J Heat Transfer 122:118–125
Klausner JF, Mei R, Bernard D, Zeng L (1993) Vapor bubble departure in forced convection boiling. Int J Heat Mass Transfer 36:651–661

Klein D, Hetsroni G, Mosyak A (2005) Heat transfer characteristics of water and APG surfactant solution in a micro-channel heat sink. Int J Multiphase Flow 31:393–415

Kosar, A, Kuo, CJ, Peles, Y (2005) Boiling heat transfer with reentrant cavities. Int. J. Heat Mass Transfer 48:4867–4886

Kuo, CJ, Peles, Y (2007) Local measurements of flow boiling in structured surface micro-channels. Int. J. Heat Mass Transfer 50:4513–4526.

Kureta M (1997) Critical heat flux for flow boiling of water in small diameter tubes under atmospheric pressure. Dissertation, Kyoto University (in Japanese)

Kutateladze SS (1963) Fundamentals of heat transfer. Edward Arnold, London

Lazarek GM, Black SH (1982) Evaporative heat transfer, pressure drop and critical heat flux in a small vertical tube with R-113. Int J Heat Mass Transfer 25:945–959

Lee HC, Oh BD, Bae SW, Kim MH (2003) Single bubble growth in saturated pool boiling on a constant wall temperature surface. Int J Multiphase Flow 29:1857–1874

Lee HJ, Lee SY (2001a) Pressure drop correlations for two-phase flow within horizontal rectangular channels with small heights. Int J Multiphase Flow 27:782–796

Lee HJ, Lee SY (2001b) Heat transfer correlation for boiling flows in small rectangular horizontal channels with low aspect ratios. Int J Multiphase Flow 27:2043–2062

Lee J, Mudawar I (2005a) Two-phase flow in high-heat-flux micro-channel heat sink for refrigeration cooling applications. Part I: pressure drop characteristics. Int J Heat Mass Transfer 48:928–940

Lee J, Mudawar I (2005b) Two-phase flow in high-heat-flux micro-channel heat sink for refrigeration cooling applications. Part II: heat transfer characteristics. Int J Heat Mass Transfer 48:941–955

Lee PC, Tseng FC, Pan C (2004) Bubble dynamics in microchannels. Part 1: single microchannel. Int J Heat Mass Transfer 47:5575–5589

Lezzi AM, Niro A, Beretta GP (1994) Experimental data on CHF for forced convection water boiling in long horizontal capillary tubes. In: Hewitt GF (ed) Heat Transfer 1994: Proceedings of the 10th International Heat Transfer Conference, vol 7, Institution of Chemical Engineers, Rugby, pp 491–496

Li HY, Tseng FC, Pan C (2004) Bubble dynamics in micro-channels. Part II: two parallel micro-channels. Int J Heat Mass Transfer 47:5591–5601

Li J, Cheng P (2004) Bubble cavitation in a micro-channel. Int J Heat Mass Transfer 47:2689–2698

Liu D, Lee PS, Garimella SV (2005) Prediction of the onset of nucleate boiling in microchannel flow. Int J Heat Mass Transfer 48:5134–5149

Liu Z, Winterton RHS (1991) A general correlation for saturated and subcooled flow boiling in tubes and annuli, based on a nucleate pool boiling equation. Int J Heat Mass Transfer 34:2759–2766

Lockhart RW, Martinelli RC (1949) Proposed correlation of data for isothermal two-phase two-component flow in pipes. Chem Eng Prog 45:39–48

Lowdermilk WH, Lanzo CD, Siegel BL (1958) Investigation of boiling burnout and flow stability for water flowing in tubes, NACA TN 4382. National Advisory Committee for Aeronautics, Washington, DC

Mei R, Chen W, Klausner JF (1995a) Vapor bubble growth in heterogeneous boiling Int J Heat Mass Transfer 38:909–919

Mei R, Chen W, Klausner JF (1995b) Vapor bubble growth in heterogeneous boiling. Part I: growth rate and thermal fields. Int J Heat Mass Transfer 38:921–934

Mikic BB, Rohsenow WM, Griffith P (1970) On bubble growth rates. Int J Heat Mass Transfer 13:657–666

Mishima K, Hibiki T (1996) Some characteristics of air–water two-phase flow in small diameter vertical tubes. Int J Multiphase Flow 22:703–712

Moriyama K, Inoue A (1996) Thickness of the liquid film formed by growing bubble in a narrow gap between two horizontal plates. J Heat Transfer Trans ASME 118:132–139

Mudawar I, Bowers MB (1999) Ultra-high critical heat flux (CHF) for subcooled water flow boiling. I: CHF data and parametric effects for small diameter tubes. Int J Heat Mass Transfer 42:1405–1428

Nakoryakov VE, Pokusaev BG, Shreiberg IR (2000) Wave propagation in gas–liquid media, 2nd edn. CRC, Boca Raton

Nariai H, Inasak F, Uehara K (1989) Critical heat flux in narrow tubes with uniform heating. Heat Transfer Jpn Res 18:21–30

Nariai H, Inasaka F, Shimura T (1987) Critical heat flux of subcooled flow boiling in narrow tube. ASME JSME Thermal Eng Joint Conf Proc 5:455–462

Nigmatulin IR (1991) Dynamics of multiphase media and 2. Hemisphere, London

Ory E, Yuan H, Prosperetti A, Popinet S, Zaleski S (2000) Growth and collapse of vapor bubble in a narrow tube. Phys Fluids 12:1268–1277

Ozawa M, Akagawea K, Sakaguchi T (1989) Flow instabilities in parallel-channel flow systems of gas-liquid two-phase mixtures. Int J Multiphase Flow 15:639–657

Peles YP (1999) VLSI chip cooling by boiling-two-phase flow in micro-channels. Dissertation, Faculty of Mechanical Engineering, Technion–Israel Institute of Technology, Haifa

Peles YP, Yarin LP, Hetsroni G (2001) Steady and unsteady flow in heated capillary. Int J Multiphase Flow 22:577–598

Peng XF, Hu HY, Wang BX (1997) Boiling nucleation during liquid flow in micro-channels. Int J Heat Mass Transfer 41:101–106

Plesset MS, Zwick SA (1954) The growth of vapor bubbles in superheated. J Appl Phys 25:474–478

Prodanovic V, Fraser D, Salcudean M (2002) On transition from partial to fully developed subcooled flow boiling. Int J Heat Mass Transfer 45:4727–4738

Qu W, Mudawar I (2003a) Measurement and prediction of pressure drop in two-phase micro-channel heat sinks. Int J Heat Mass Transfer 46:2737–2753

Qu W, Mudawar I (2003b) Flow boiling heat transfer in two-phase micro-channel heat sink. I: Experimental investigation and assessment of correlation methods. Int J Heat Mass Transfer 46:2755–2771

Qu W, Mudawar I (2002) Prediction and measurement of incipient boiling heat flux in micro-channel heat sinks. Int J Heat Mass Transfer 45:3933–3945

Qu W, Mudawar I (2004) Measurement and correlation of critical heat flux in two-phase micro-channel heat sinks. Int J Heat Mass Transfer 47:2045–2059

Quiben JM, Thome JR (2007a) Flow pattern based two-phase pressure drop model for horizontal tubes. Part I. Diabatic and adiabatic experimental study. Int. J. Heat and Fluid Flow. 28(5):1049–1059

Quiben JM, Thome JR (2007b) Flow pattern based two-phase pressure drop model for horizontal tubes. Part II. New phenomenological model. Int. J. Heat and Fluid Flow. 28(5):1060–1072

Rayleigh JWS (1917) On the pressure developed in a liquid during the collapse of a spherical cavity. Phil Mag 34:94–98

Revellin R, Thome J. (2008) A theoretical model for the prediction of the critical hat flux in heated micro-channel. Int. J. Heat and Mass Transfer 51:1216–1225

Roach GM, Abdel-Khalik SI, Ghiaasiaan SM, Dowling MF, Jeter SM (1999) Low-flow critical heat flux in heated microchannels. Nucl Sci Eng 131:411–425

Robinson AJ, Judd RL (2001) Bubble growth in a uniform and spatially distributed temperature field. Int J Heat Mass Transfer 44:2699–2710

Sato T, Matsumura H (1964) On the conditions of incipient subcooled-boiling with forced convection. Bull Jpn Soc Mech Eng 7:392–398

Sedov LI (1993) Similarity and dimensional methods in mechanics, 10th edn. CRC, Boca Raton

Shah MM (1982) Chart correlation for saturated boiling heat transfer: equation and further study. ASHRAE Trans 88:185–196

Shah MM (1987) Improved general correlation for critical heat flux during upflow in uniformly heated vertical tubes. Int J Heat Fluid Flow 8:326–335

Shah RK, London AL (1978) Laminar flow forced convection in ducts: a source book for compact heat exchanger analytical data. Advances in Heat Transfer, suppl 1. Academic, New York

Sher I, Hetsroni G (2002) An analytical model for nucleate pool boiling with surfactant additives. Int J Multiphase Flow 28:699–706

Shuai J, Kulenovic R, Droll M (2003) Heat transfer and pressure drop for flow boiling of water in narrow vertical rectangular channels. In: Proceedings for 1st International Conference on Micro-channels, Rochester, New York, 24–25 April 2003, ICMM 2003-1084

Staniszewski BE (1959) Nucleate boiling bubble growth and departure MIT DSR Project N7-7673, Technical Report N16

Steinke ME, Kandlikar SG (2004a) Control and effect of dissolved air in water during flow boiling in micro-channels. Int J Heat Mass Transfer 47:1925–1935

Steinke M, Kandlikar SG (2004b) An experimental investigation of flow boiling characteristics of water in parallel micro-channels. Trans ASME J Heat Transfer 126:518–526

Stoddard RM, Blasick AM, Ghiaasiaan SM, Abdel-Khalik SI, Jeter SM, Dowling MF (2002) Onset of flow instability and critical heat flux in thin horizontal annuli. Exp Thermal Fluid Sci 26:1–14

Su S, Huang S, Wang X (2001) Study of boiling incipience and heat transfer enhancement in forced flow through narrow channels. Int J Heat Mass Transfer 31:253–260

Revellin R, Agostini B, Thome J (2008) Elongated bubbles in micro-channels Part II: Experimental study and modeling of bubble collisions. Int. J. Multiphase Flow 34:602–613

Thom JRS, Walker WM, Fallon TA, Reising GFS (1965) Boiling in subcooled water during flow up heated tubes or annuli. In: Symposium on Boiling Heat Transfer in Steam Generating Units and Heat Exchangers, Manchester, 15–16 September 1965. IMechE London

Thome JR, Dupont V, Jacobi AM (2004) Heat transfer model for evaporation in micro-channels. Part I: Comparison with database. Int J Heat Mass Transfer 47:3375–3385

Thompson B, Macbeth RV (1964) Boiling water heat transfer – burnout in uniformly heated round tubes: a compilation of world data with accurate correlations, AEEW-R 356. United Kingdom Atomic Energy Authority, Winfrith, UK

Thorncroft GE, Klausner JF, Mei R (1998) An experimental investigation of bubble growth and detachment in vertical upflow and downflow boiling. Int J Heat Mass Transfer 41:3857–3871

Thorncroft GE, Klausner JF, Mei R (2001) Bubble forces and detachment models Multiphase Sci Technol 13:35–76

Tran TN, Chyu M-C, Wambsganss MW, France DM (2000) Two-phase pressure drop of refrigerants during flow boiling in small channels: an experimental investigation and correlation development. Int J Multiphase Flow 26:1739–1754

Tran TN, Wambsganss MW, France DM (1996) Small circular and rectangular channel boiling with two refrigerants. Int J Multiphase Flow 22:485–498

Unal HC (1975) Determination of the initial point of net vapor generation in flow boiling system. Int J Heat Mass Transfer 18:1095–1099

van Stralen SJD (1966) The mechanism of nucleate boiling in pure liquids and in binary mixtures. Int J Heat Mass Transfer 9:995–1046

Vandervort CL, Bergles AE, Jensen MK (1994) An experimental study of critical heat flux in very high heat flux subcooled boiling. Int J Heat Mass Transfer 37:161–173

Volmer M, Weber A (1926) Nucleus formation in superheated systems. Z Phys Chem 119:277–281

Wang G, Cheng P, Bergles AE 2008 Effects of inlet outlet configurations on flow boiling instability in parallel micro-channels. Int. J. Mass Transfer 51:2267–2281

Warrier GR, Pan T, Dhir VK (2002) Heat transfer and pressure drop in narrow rectangular channels. Exp Therm Fluid Sci 26:53–64

Wojtan L, Revellin R, Thome J (2006) Investigation of critical heat flux in single uniformly heated micro-channels. Exp. Therm, Fluid Sci. 30:765–774

Wu HY, Cheng P (2004) Boiling instability in parallel silicon micro-channels at different heat flux. Int J Heat Mass Transfer 47:3631–3641

Wu HY, Cheng P (2003a) Liquid/two-phase/vapor alternating flow during boiling in micro-channels at high heat flux. Int Comm Heat Transfer 39:295–302

Wu HY, Cheng P (2003b) Visualization and measurements of periodic boiling in silicon micro-channels. Int J Heat Mass Transfer 46:2603–2614
Wu WT, Yang YM (1992) Enhanced boiling heat transfer by surfactant additives. In: Proceedings Engineering Foundation Conference On Pool and External Flow Boiling, Santa Barbara, California, 1992, pp 361–366
Yang YM, Maa JR (2003) Boiling heat transfer enhancement by surfactant additives. In: Proceedings of the 5th International Conference Boiling Heat Transfer, ICBHT, Montego Bay, Jamaica, 4–8 May 2003
Ye H, Naguib N, Gogotsi Y (2004) TEM study of water in carbon nanotubes. JEOL News 39:2–7
Yen T-H, Shji M, Takemura F, Suzuki Y, Kasagi N (2006) Visualization of convective boiling heat transfer in single micro-channels with different shaped cross-sections. Int J Heat Mass Transfer 49:3884–3894
Yen T-H, Kasagi N, Suzuki Y (2003) Forced convective boiling heat transfer in micro-tubes at low mass and heat fluxes. Int J Multiphase Flow 29:1771–1792
Yu W, France DM, Wambsganss MW, Hull JR (2002) Two-phase pressure drop, boiling heat transfer, and critical heat flux to water in a small-diameter horizontal tube. Int J Multiphase Flow 28:927–941
Zel'dovich YaB (1943) On the theory of new phase formation: cavitation. Acta Physicochim URSS 18:1–7
Zeng LZ, Klausner JF, Mei R (1993) A unified model for the prediction on bubble detachment diameters in boiling system. Part I: Pool boiling. Int J Heat Mass Transfer 36:2261–2270
Zhang W, Hibiki T, Mishima K, Mi Y (2006) Correlation of critical heat flux for flow boiling of water in mini-channels. Int J Heat and Mass Transfer 49:1058–1072
Zhang J, Manglik RM (2005) Additive absorption and interfacial characteristics of nucleate pool boiling in aqueous surfactant solutions. J Heat Transfer 127: 684–691
Zhang L, Koo J-M, Jiang L, Asheghi M, Goodson KE, Santiago JK (2002) Measurements and modeling of two-phase flow in micro-channels with nearly constant heat flux boundary conditions. J Microelectromech Syst 11: 12–19
Zuber N (1961) The dynamics of vapor bubbles in non-uniform temperature fields. Int J Heat Mass Transfer 2:83–98

Nomenclature

a	Half of channel height
b	Half of channel width
c_p	Heat capacity at constant pressure
C	Parameter C (Eqs. 6.39, 6.50)
Co	Convection number (Eqs. 6.57, 6.58)
d	Diameter
d_0	Initial diameter of bubble
d_*	Characteristic size
D	Subcooling parameter at ONB point
G	Mass flux, mass velocity of liquid plus vapor
g	Acceleration due to gravity
h	Heat transfer coefficient
h_{LG}	Latent heat
k	Thermal conductivity, Boltzman constant
k_s	Surface roughness

L	Channel length
m	Mass flow rate
M	Mass, constant (Eqs. 6.9, 6.13)
MAE	Mean absolute error
No	Molecular number per unit volume
P	Pressure
P_s^*	Pressure 1 bar
P_s^+	Dimensionless pressure
q	Heat flux
r	Bubble radius
r_b	Radius of bubble embryo
r_c	Radius of cavity
r_{max}^*	Largest cavity radius
t	Time
t^*	Dimensionless period
T	Temperature
T_f	Fluid temperature in annular space
T_∞	Temperature of surrounding fluid
u	Streamwise velocity
U	Average flow velocity in micro-channel, uncertainty
U_b	Velocity of bubble tail
U_{LS}	Superficial liquid velocity
U_{GS}	Superficial gas velocity
v	Spanwise velocity
x	Vapor quality
X	Martinelli parameter
x, y, z	Cartesian coordinates
$\mathrm{Bn} = \delta\sqrt{\frac{g(\rho_L - \rho_G)}{\sigma}}$	Bond number
$\mathrm{Bo} = \frac{q}{Gh_{LG}}$	Boiling number
$\mathrm{Eo} = \frac{\delta^2 g(\rho_L - \rho_G)}{\sigma}$	Eotvos number
$\mathrm{Fo} = \frac{2\alpha t}{\delta^2}$	Fourier number
$\mathrm{Ja} = \frac{(T_\infty - T_s)\rho_L c_{pL}}{\rho_G h_{LG}}$	Jacob numer
$\mathrm{Nu} = \frac{hd_h}{k}$	Nusselt number
$\mathrm{Po} = \lambda \cdot \mathrm{Re}$	Poiseuille number
Pr	Prandtl number

$\mathrm{Re} = \dfrac{U d_{\mathrm{h}}}{\nu}$	Reynolds number
$\mathrm{St}_{\mathrm{ONB}} = \dfrac{q_{\mathrm{ONB}}}{\rho_{\mathrm{L}} u_{\mathrm{in}} c_{\mathrm{p}} (T_{\mathrm{s}} - T_{\mathrm{in}})}$	Stanton number at ONB point
$\mathrm{We} = \dfrac{G^2 d_{\mathrm{h}}}{\sigma \rho}$	Weber number

Greek symbols

α	Thermal Diffusivity, exponent
β	Ratio of channel depth to width, exponent
ΔP	Pressure drop
ΔT	Temperature difference
$\Delta T^{*}_{\mathrm{ONB}}$	Wall superheat at pressure of 1 bar
δ	Gap size, liquid thickness, thermal boundary layer
δ^{*}	Dimensionless liquid thickness
η	Dimensionless distance from the wall
θ	Temperature difference, contact angle
μ	Dynamic viscosity
ν	Kinematic viscosity
Π	Dimensionless parameter (Eqs. 6.43, 6.44)
ρ	Density
σ	Surface tension
τ_{w}	Wall shear stress
Φ_{L}	Two-phase pressure drop multiplier
ψ	Parameter (Eqs. 6.5, 6.7)

Subscripts

av	Average
B	Bubble, bulk
c	Cavity, circular, center
CHF	Critical heat flux
CMC	Critical micelle concentration
crit	Critical
exp	Experimental
G	Gas
h	Hydraulic, heated
heat	Heat
het	Heterogeneous
hom	Homogeneous

in	Inlet
L	Liquid
LG	Liquid/gas
max	Maximum
min	Minimum
n	Spanwise
out	Outlet
ONB	Onset of nucleate boiling
p	Streamwise
pred	Predicted
s	Saturation
s,B	Saturation in the bubble
sub	Subcooled
sub,ONB	Subcooled at ONB point
tp	Two-phase
tp,a	Acceleration component
tp,cor	Uniform circumferential heating
tp,f	Friction component
vt	Laminar liquid–turbulent vapor
vv	Laminar liquid–laminar vapor
w	Wall
w,ONB	Wall at ONB point
3	Three sides heating
4	Four sides heating

Chapter 7
Design Considerations

The details of the specific features of the heat transfer coefficient, and pressure drop estimation have been covered throughout the previous chapters. The objective of this chapter is to summarize important theoretical solutions, results of numerical calculations and experimental correlations that are common in micro-channel devices. These results are assessed from the practical point of view so that they provide a sound basis and guidelines for the evaluation of heat transfer and pressure drop characteristics of single-phase gas–liquid and steam–liquid flows.

7.1 Single-Phase Flow

Fully developed flows in smooth channels

The Poiseuille number in laminar flows is constant but dependent upon the flow passage geometry (Shah and London 1978). The heat transfer rate in laminar channel flow is very sensitive to the thermal boundary conditions. Hence, it is essential to carefully identify the thermal boundary conditions in laminar flow. A systematic analysis of thermal boundary conditions for internal flow is given by Shah and London (1978). Three important thermal boundary conditions T, H1, and H2 are described; they are pertinent to most micro-devices. The T boundary condition refers to the constant wall temperature T_w both axially and peripherally throughout the passage length. The wall heat transfer rate is constant in the axial direction while the wall temperature at any cross-section is constant in the peripheral direction for the H1 boundary condition. The wall heat transfer rate is constant in the axial direction, as well as in the peripheral direction for the H2 boundary condition. The H1 boundary condition is realized at $q_w = \text{const.}$ for highly conductive materials for which the temperature gradients in the peripheral direction are minimum; the H2 boundary condition is realized at $q_w = \text{const.}$ for very low conductive materials for which temperature gradients exist in the peripheral direction. For intermediate values of thermal conductivity, the boundary condition at $q_w = \text{const.}$ will be between H1 and

H2. It may be noted that the H1 and H2 boundary conditions for the symmetrically heated passages with no sharp corners (e.g., circular, flat, and concentric annular channels) are identical; they are simply designated as H.

The values of the Nusselt and the Poiseuille numbers for heat transfer and friction for fully developed laminar flows through specified channels are presented in Table 7.1 (Shah and London 1978).

Correlations concerning symmetrically heated channels should be adjusted for non-uniform circumferential heating. An example is given in Sect. 6.4.2, Eqs. (6.53), (6.54), and (6.55).

Theoretical solutions have been verified by many researchers for flow friction and heat transfer in a single channel. Hence, these results provide a valuable guideline for devices that may employ many such channels in parallel. However, passage-

Table 7.1 The Nusselt and the Poiseuille numbers for fully developed laminar flow

Geometry ($L/d_h > 100$)		Nu_{H1}	Nu_{H2}	Nu_T	Po
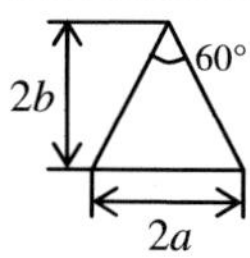	$\frac{2b}{2a} = \frac{\sqrt{3}}{2}$	3.111	1.892	2.47	53.332
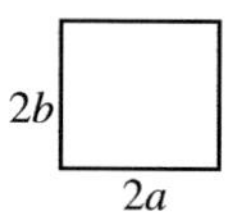	$\frac{2b}{2a} = 1$	3.608	3.091	2.976	56.908
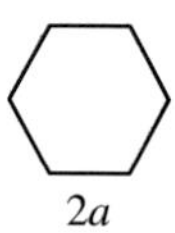		4.002	3.862	3.34	60.216
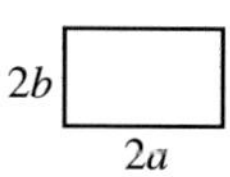	$\frac{2b}{2a} = \frac{1}{2}$	4.123	3.017	3.391	62.192
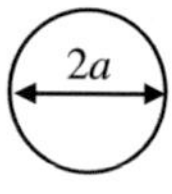		4.364	4.364	3.657	64
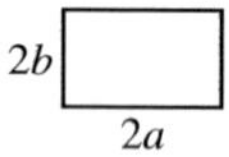	$\frac{2b}{2a} = \frac{1}{4}$	5.331	2.94	4.439	72.932
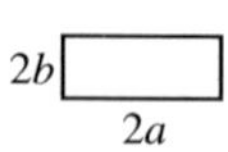	$\frac{2b}{2a} = \frac{1}{4}$	6.049	2.93	5.137	78.808
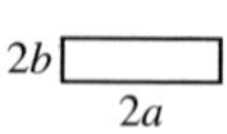	$\frac{2b}{2a} = \frac{1}{8}$	6.490	2.94	5.597	82.340
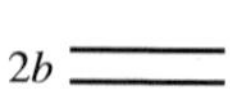	$\frac{2b}{2a} \to 0$	8.235	8.235	7.541	96.00

to-passage flow non-uniformity could result in significant deviations in Nu and Po from the analytical predictions. Also, actual thermal boundary conditions for heat transfer may not correspond to any previously described boundary conditions. In addition, the developing flow effect may be present if the flow passage is not long enough. These and other effects such as fouling, fluid property variations, etc., could affect the actual Nu and Po as presented in Table 7.1. For all these reasons, when accurate Nu and Po are needed for specific applications, they should be obtained experimentally or from numerical calculations even for simple passage geometries.

For smooth micro-channels the transition from laminar to turbulent flow occurs at Re = 1,500–2,200. For turbulent flows the friction factor maybe calculated as

$$\lambda = 0.316 \cdot \mathrm{Re}^{0.25} . \tag{7.1}$$

Direct measurements of turbulent heat transfer in smooth pipes led to the correlation known as the Dittus–Boelter equation

$$\mathrm{Nu} = 0.023 \cdot \mathrm{Re}^{0.8}\, \mathrm{Pr}^{n} \tag{7.2}$$

where n is 0.4 and 0.3 for heating and cooling, respectively, Re is the Reynolds number, and Pr is the Prandtl number. The analytical solutions, e.g., Eq. (6.36), based on the thermal boundary approach appears to provide a reasonable basis for understanding the mechanism of heat transfer.

Entrance effects

For most channels, the mean Nusselt number and friction factor will be within 10% of the fully developed value if $L/d_h > 0.2\mathrm{Re}\,\mathrm{Pr}$. If $L/d_h < 0.2\mathrm{Re}\,\mathrm{Pr}$, the fully developed analytical solutions may not be adequate, since Nu and λ are higher in the developing flow region. However, if the passage-to-passage non-uniformity exists in a micro-device, it reduces Nu substantially, and also reduces λ slightly (and it could be neglected for practical purposes).

Effect of roughness

For channels above $d_h = 1$ mm the surface roughness generally does not affect Nu and λ as long as the height of the surface roughness is negligible compared to the channel hydraulic diameter (i.e., $k_s/d_h < 0.01$). For small diameter tubes ($d_h < 0.6$ mm) the relative roughness of $k_s/d_h > 0.003$ increases heat transfer up to 25–30% in the range of Re = 1,000–2,200. This effect is more pronounced at higher values of k_s/d_h and Re (Kandlikar et al. 2003).

Axial conduction in the walls

For high values of the Reynolds number, the mean value of the Nusselt number does not differ significantly from the theoretical value for fully developed flow. On the contrary, at low Re the effects of conjugate heat transfer on the mean value of the Nusselt number are very important because conduction along the channel walls becomes a competitive mechanism to heat transfer through internal convection.

The mean heat transfer coefficient calculated using average wall and fluid temperatures (assuming a linear wall and fluid temperature distribution) is not consistent with mechanisms of heat transfer. This is also an explanation for the dependence of the Nu on Re found in experiments. This effect will decrease as the convective term of heat transfer increases with respect to the conductive term in the channel wall (Hetsroni et al. 2004; Tiselj et al. 2004; Maranzana et al. 2004). This is evidenced through numerical simulation, and when the effects of conjugate heat transfer dominate, the temperature distribution along the micro-channel is not linear. For a circular tube, the region of significance of axial conduction in the walls in the conjugated heat transfer problem, used also by Celata et al. (2005) may be calculated as:

$$(k_{\mathrm{w}}/k_{\mathrm{L}})\left((d_{\mathrm{o}}^2-d_{\mathrm{in}}^2)/d_{\mathrm{in}}L\right)/\mathrm{Re}\,\mathrm{Pr} > 10^{-2} \tag{7.3}$$

where k_{w} and k_L is the thermal conductivity of wall and liquid, respectively, d_{o} and d_{in} is the outer and inner tube diameter, and L is the tube length.

The dimensionless quantity on the left-hand side of Eq. (7.3) labeled as M by the above-cited authors, allows for the comparison of heat transfer by axial conduction in the wall to the convective heat transfer in the flow.

7.2 Gas–Liquid Flow

Void fraction

The void fraction data was obtained in micro-channels and showed significant differences from conventional size channels, depending on the channel cross-section and inlet geometry. For the micro-channels with a diameter of 100 µm, the effects of inlet geometry and the gas–liquid mixing method on the void fraction were seen to be quite strong, while the conventional size channels have shown a much smaller effect of inlet geometry on the void fraction. The homogeneous flow model and the Armand (1946) correlation, $\alpha=\beta$, where $\beta=U_{\mathrm{GS}}/(U_{\mathrm{GS}}+U_{\mathrm{LS}})$, or the correlation $\alpha=0.83\beta$ recommended by Ali et al. (1993) may be used for narrow channels with $d_{\mathrm{h}}\sim 1$ mm.

Kawahara et al. (2002) presented void fraction data obtained in a 100 µm micro-channel connected to a reducing inlet section and T-junction section. The superficial velocities are $U_{\mathrm{GS}}=0.1-60\,\mathrm{m/s}$ for gas, and $U_{\mathrm{LS}}=0.02-4\,\mathrm{m/s}$ for liquid. The void fraction data obtained with a T-junction inlet showed a linear relationship between the void fraction and volumetric quality, in agreement with the homogeneous model predictions. On the contrary, the void fraction data from the reducing section inlet experiments showed a non-linear void fraction-to-volumetric quality relationship:

$$\alpha=0.03\beta^{0.5}/(1-0.97\beta^{0.5})\,. \tag{7.4}$$

Pressure drop

This matter was discussed in Sect. 5.8. For channels of $d_h = 0.9–3.2$ mm, the two-phase pressure drop can be calculated using the Lockhart–Martinelli model with parameter C, ranging from 5 to 20. The parameter C decreases when the hydraulic diameter decreases (Zhao and Bi 2001). For channels of $d_h = 100\,\mu$m, (Kawahara et al. 2002) two-phase pressure drop can be correlated within an accuracy of $\pm 10\%$ using the Lockhart–Martinelli model with $C = 0.24$.

Heat transfer

There is a lack of information on the effect of superficial liquid and gas velocities on heat transfer in micro-channels. We studied this problem in the test section that contained 21 parallel triangular micro-channels of $d_h = 130\,\mu$m.

The heat transfer coefficient h increases with increasing superficial liquid velocity U_{LS}. Enhancement of heat transfer is more pronounced for low values of superficial liquid velocity. An increase in superficial gas velocity U_{GS} to the same value of the superficial liquid velocity led to a decrease in the heat transfer coefficient. The Nusselt number may be calculated as:

$$\mathrm{Nu_L} = 0.044\mathrm{Re}_{LS}^{0.96}\mathrm{Re}_{GS}^{-0.18} \quad \text{for} \quad \mathrm{Re_{GS}} = 4.7\text{–}270\,, \quad \mathrm{Re_{LS}} = 4.0\text{–}8.0 \qquad (7.5)$$

$$\mathrm{Nu_L} = 0.13\mathrm{Re}_{LS}^{0.96}\mathrm{Re}_{GS}^{-0.40} \quad \text{for} \quad \mathrm{Re_{GS}} = 4.7\text{–}270\,, \quad \mathrm{Re_{LS}} = 8.0\text{–}56 \qquad (7.6)$$

where $\mathrm{Nu_L} = hd_h/k_L$, $\mathrm{Re_{LS}} = U_{LS}d_h/\nu_L$, $\mathrm{Re_{GS}} = U_{GS}d_h/\nu_G$, h is the heat transfer coefficient, k_L is the liquid thermal conductivity, U_{LS} and U_{GS} are the liquid and gas superficial velocities, respectively, and ν_L and ν_G are the liquid and gas kinematic viscosity, respectively.

7.3 Boiling in Micro-Channels

7.3.1 Boiling Incipience

Wall superheat

Experimental and analytical studies showed that wall superheat significantly depends on the heat flux. This dependence is close to $\Delta T_{S,ONB} \sim q_{ONB}^{0.5}$. Wall superheat corresponding to nucleate boiling may be calculated using Eq. (6.9).

Effect of pressure

Empirical correlation (6.14) by Bergles and Rohsenow (1964) is recommended when taking the dependence of wall superheat on pressure into account. It agrees fairly well with the prediction of theoretical analysis based on the Hsu (1962) model.

Liquid subcooling at ONB point

For qualitative analysis of the conditions at which the ONB phenomenon takes place the parameter $D = \Delta T_{\mathrm{sub,ONB}}/T_{\mathrm{S}}$ may be used. Depending on the value of D, the channels can be subdivided into two groups: (1) $D < 1$ (parameter is in the range of $D = 0.1-0.5$), and (2) $D \ll 1$ (parameter is in the range of $D = 0.01-0.1$). For the first group, defined as "short channels," the relative heated length is $L_{\mathrm{ONB}}/d_{\mathrm{h}} < 100$. For the second group, defined as "long channels," the relative heated length is $L_{\mathrm{ONB}}/d_{\mathrm{h}} > 100$. When the value of D lies in the range of 0.125–0.25, as in experiments by Bergles and Rohsenow (1964), the onset of nucleate boiling occurred at values of the bulk temperature $T_{\mathrm{B,ONB}}$, significantly less than the saturation temperature. When the parameter was in the range of $D = 0.011-0.074$, as in the experiments by Celata et al. (1997) and by Kennedy et al. (2000), the ONB occurred at values of the bulk temperature that were close to saturation. Equations (6.32), (6.33) and (6.34) are suggested for calculation of L_{ONB} at given values of q_{ONB}, inlet velocity, and the difference between saturated and inlet temperature.

Effect of mass flux

Heat flux q_{ONB} is essentially independent of mass flux in the case that the inlet fluid temperature does not differ significantly from the saturation temperature, i.e., $T_{\mathrm{S}} - T_{\mathrm{in}}$ is about 10–20 K. Heat flux q_{ONB} is in excess of 10–20% for the case of $T_{\mathrm{S}} - T_{\mathrm{in}} > 20$ K.

Effect of inlet temperature

Incipient boiling heat flux q_{ONB} decreases with an increase in the inlet temperature. It was shown in Sect. 6.1.4 that an increase in the inlet flow temperature may lead to a drastic decrease in q_{ONB}. As a consequence of the above discussion the fact is that at the same value of T_{in} the value of q_{ONB} increases with increasing inlet velocity (see Eqs. 6.32, 6.33, and 6.34).

Boiling incipience in surfactant solutions

Under some conditions boiling incipience in surfactant solutions may be quite different from that in Newtonian fluids. For some kinds of degraded solutions (i.e., solutions that were used after 6–10 runs) boiling occurred at wall superheat higher than that observed in fresh solutions or water. Hysteresis was observed during boiling of degraded solutions. It is speculated that molecules of degraded surfactant are more amenable to the formation of a surfactant monolayer, which renders the interface less flexible and results in the dampening of interfacial motion.

Effect of dissolved gases

Desorption of the dissolved bubbles formed bubbles of gas and a limited amount of bubbles containing gas–water vapor mixture. As a result, boiling incipience occurs at a channel wall temperature below that of the saturation temperature. Onset of nucleate boiling in surfactant solutions that contain dissolved gases takes place at surface temperatures significantly higher than the saturation temperature (Klein et al. 2005).

Dynamics of vapor bubble

In micro-channels the maximum length of the bubble in the streamwise direction is about eight times larger than that in the spanwise direction (Hetsroni et al. 2003b). The parameter $\Pi = q/(\rho_L U c_{pL} \Delta T_s)$ may be used to distinguish linear and exponential regimes of bubble growth. Bubble formation causes pressure and temperature fluctuations.

7.3.2 Flow Boiling: Pressure Drop Characteristics

Predictions of the homogeneous equilibrium flow model and previous separated flow models yielded relatively poor predictions of pressure drop. The mean absolute error (Eq. 6.48) is in the range of MAE $= \pm 40\%$ (Qu and Mudawar 2003a). A new approach was developed to improve the accuracy of pressure drop prediction in two-phase micro-channels (Lee and Mudawar 2005a). Since the bubbly and churn flow patterns are rarely detected in high-flux micro-channel flow, the separated flow model was deemed more appropriate than the homogeneous. In both the slug and annular regimes, mass transfer by liquid break-up and deposition is highly influenced by surface tension. It is assumed that the added complexity of two-phase flow in a micro-channel is the net result of interactions between liquid inertia, the liquid viscous force, and surface tension. Two key measures of these interactions are the Reynolds number (Eq. 6.47) and Weber number (Eq. 6.48) based on liquid properties. Equations (6.50), (6.51), and (6.52) are suggested to evaluate the pressure drop in a copper micro-channel heat sink. These correlations were examined experimentally by measurements of the pressure drop in water and refrigerant R-134a in a two-phase micro-channel heat sink containing parallel $231 \times 713\,\mu\text{m}$ channels. The operating conditions for water were as follows: inlet temperature of $T_{in} = 30.0$ or $60.0\,°\text{C}$, mass flux of $G = 135–400\,\text{kg/m}^2\text{s}$, heat flux of $q = 40.0–130\,\text{W/cm}^2$, and outlet pressure of $P = 2\,\text{bar}$. Experiments performed with refrigerant R-134a spanned the following conditions: inlet pressure of $P_{in} = 1.44–6.60\,\text{bar}$, mass flux of $G = 127–654\,\text{kg/m}^2\text{s}$, inlet quality of $x_{in} = 0.001–0.25$, outlet quality of $x_{in} = 0.001–0.25$, and heat flux of $q = 31.6–93.8\,\text{W/cm}^2$. Equations (6.50), (6.51), and (6.52) predict the pressure drop with the mean absolute error of MAE $= \pm 30\%$. New experiments should be performed in micro-channels of different geometry and a comprehensive method should be developed for determining micro-channel heat sink pressure drop for coolants with different thermophysical properties.

7.3.3 Flow Boiling: Heat Transfer

The correlations considered below may be used only in the range of experimental conditions at which these correlations were obtained. More local data for different channel geometries should be obtained for further validating its applicability to different flow regimes, aspect ratios, sizes and geometries.

Nucleate boiling and annular film evaporation

The study by Steinke and Kandlikar (2004) was performed for water flow boiling in six parallel, horizontal, lightly trapezoidal micro-channels of $d_h = 207\,\mu m$. The ranges of parameters are: mass flux from 157 to $1{,}782\,kg/m^2\,s$, heat flux from 5 to $930\,kW/m^2$, inlet temperature of $22\,°C$, quality from sub-cooled to 1.0, and atmospheric pressure at the exit. Equations (6.57), (6.58) and (6.59) were suggested to calculate the heat transfer coefficient. The optimal range of the correlations is between qualities of 0.2 to 0.8. The correlation underpredicts the heat transfer coefficients at the ONB condition. In a recent publication, Kandlikar (2004) presented two new non-dimensional groups that are thought to be important in micro-channel flows. The groups are based on the surface tension and momentum change due to evaporation, as well as the viscous shear force and inertia force. This and prior studies by other authors pointed to annular flow as the dominant two-phase flow pattern in micro-channels at moderate to high heat fluxes. These observations lend credence to the hypothesis that the dominant heat transfer mechanism for two-phase heat sinks is forced convective boiling and not nucleate boiling. It should be stressed that the empirical correlations (6.57), (6.58) and (6.59) predict a heat transfer coefficient about twice that measured by Qu and Mudawar (2003b) during flow boiling of water ($x < 0.15$) and during flow boiling of R-134a ($x = 0.4–0.8$). These correlations also overpredict the experimental data obtained by Yen et al. (2003) for convective boiling of HCFC 123 and FC72 in $d = 190\,\mu m$ tubes in the range of $x = 0.4–0.9$.

Experiments by Lee and Mudawar (2005b) revealed a range of parameters in which heat transfer is controlled by nucleate boiling or annular film evaporation. The micro-channels were formed by cutting 53 of $231\,\mu m$ wide and $713\,\mu m$ deep micro-slots into the $25.3 \times 25.3\,mm^2$ top surface on an oxygen-free copper block. A transparent cover plate formed a top-insulating surface for the micro-channels. Experimental operating conditions spanned the following ranges: inlet quality of $x_{e,in} = 0.001–0.25$, outlet quality of $x_{e,out} = 0.49$ to superheat, mass flux of $G = 127–654\,kg/m^2 s$, heat flux of $q = 159–938\,kW/m^2$, and inlet pressure of $P_{in} = 1.44–6.60\,bar$.

Two-phase heat transfer in micro-channel heat sink was associated with different mechanisms for low, medium, and high-quality flows. Bubble flow and nucleate boiling occur only at low qualities ($x_e < 0.05$) corresponding to very low heat fluxes. High fluxes produce medium-quality ($x_e = 0.05–0.55$) or high-quality ($x_e = 0.55–1.0$) flows depending on the flow rate, where heat transfer is dominated by annular film evaporation. Due to the large differences in heat transfer mechanisms between the three quality regions, Lee and Mudawar (2005b) reported that better predictions are possible by dividing the quality range into smaller ranges corresponding to these flow transitions.

Table 6.8 summarizes the new correlations for the three quality regions. The low and high-quality regions are based solely on the Martinelli parameter, while the mid-range includes the effects of Bo and We_L as well. Overall, convection to liquid is important for both the low and mid-quality regions, while convection to vapor becomes important for the high-quality region. For the latter, the low viscosity of R-134a vapor yields vapor Reynolds numbers corresponding to turbulent flow at high-heat flux conditions despite the small hydraulic diameter of the micro-channel. Thus, the single-phase vapor term in the high-quality correlation must allow for both laminar and turbulent vapor flow. The present correlations show that the heat transfer coefficient is proportional to the Martinelli parameter raised to a positive exponent, whereas prior macro-channel correlations employ a negative exponent for the same parameter.

Critical heat flux in flow boiling

Available CHF databases in the literature for flow boiling of water in small-diameter tubes were collected by Zhang et al. (2006). Three correlations by Bowring, Katto and Shah were evaluated with the CHF data for saturated flow boiling, and three correlations by Inasaka and Nariai, Celata et al. and Hall and Mudawar were evaluated with the CHF data for subcooled flow boiling. The results obtained by Zhang et al. (2006) are presented in Sect. 6.4.3. The collected database included 2,539 points for saturated CHF and 1,298 points for subcooled CHF, covering a wide range of parameters, such as outlet pressures from 0.101 to 19.0 MPa, mass fluxes from 5.33 to $1.34 \times 10^5\ \mathrm{kg/m^2s}$, critical heat fluxes from 0.094 to $276\ \mathrm{MW/m^2}$, hydraulic diameters of channels from 330 to 6.22 mm, length-to-diameter ratios from 1.00 to 975, inlet qualities from -2.35 to 0, and outlet thermal equilibrium qualities from -1.75 to 1.00. A new simple correlation (6.59) was developed by Zhang et al. (2006) for saturated CHF in small-diameter tubes. In contrast to other correlations, this correlation consists only of a single equation and predicts the experimental data with mean deviation of 16.8%. The authors concluded that the correlation may be used to predict CHF of saturated flow boiling in small-diameter tubes. However, results presented by Zhang et al. (2006) consider the macro-scale flow boiling methods adapted to micro-channel data without recognizing any new heat transfer mechanisms. Such an approach cannot explain the disagreement between experimental results discussed above in Chap. 2. In particular, there is significant scatter between results obtained by Wojtan et al. (2006) and Qu and Mudawar (2004). Hence, additional phenomena, channel geometry, surface roughness, instability effects, additives to fluid, and heat transfer mechanisms must have an effect on micro-channel boiling.

The Hall–Mudawar correlation (6.61) seem to be the most reliable tools for CHF prediction in subcooled flow boiling regions.

Table 7.2 Physical properties of saturated water

T [°C]	$p\,(10^{-2})$ [kPa]	ρ [kg/m^3]	h [kJ/kg]	c_p [kJ/kg K]	$k\,(10^2)$ [W/m K]	$\alpha\,(10^8)$ [m^2/s]	$\nu\,(10^6)$ [m^2/s]	$\sigma\,(10^4)$ [N/m]	Pr
0	1.013	999.9	0	4.212	55.1	13.1	1.789	756.4	13.67
10	1.013	999.7	42.04	4.191	57.4	13.7	1.306	741.6	9.52
20	1.013	998.2	83.91	4.183	59.9	14.3	1.006	726.9	7.02
30	1.013	995.7	125.7	4.174	61.8	14.9	0.805	712.2	5.42
40	1.013	992.2	167.5	4.174	63.5	15.3	0.659	696.5	4.31
50	1.013	988.1	209.3	4.174	64.8	15.7	0.556	676.9	3.54
60	1.013	983.1	251.1	4.179	65.9	16.0	0.478	662.2	2.98
70	1.013	977.8	293.0	4.187	66.8	16.3	0.415	643.5	2.55
80	1.013	971.8	355.0	4.195	67.4	16.6	0.365	625.9	2.21
90	1.013	965.3	377.0	4.208	68.0	16.8	0.326	607.2	1.95
100	1.013	958.4	419.1	4.220	68.3	16.9	0.295	588.6	1.75
110	1.43	951.0	461.4	4.233	68.5	17.0	0.272	569.0	1.60
120	1.98	943.1	503.7	4.250	68.6	17.1	0.252	548.4	1.47
130	2.70	934.8	546.4	4.266	68.6	17.2	0.233	528.8	1.36
140	3.61	926.1	589.1	4.287	68.5	17.2	0.217	507.2	1.26
150	4.76	917.0	632.2	4.313	68.4	17.3	0.203	486.6	1.17
160	6.18	907.0	675.4	4.346	68.3	17.3	0.191	466.0	1.10
170	7.92	897.3	719.3	4.380	67.9	17.3	0.181	443.4	1.05
180	10.03	886.9	763.3	4.417	67.4	17.2	0.173	422.8	1.00
190	12.55	876.0	807.8	4.459	67.0	17.1	0.165	400.2	0.96
200	15.55	863.0	852.5	4.505	66.3	17.0	0.158	376.7	0.93
210	19.08	852.8	897.7	4.555	65.5	16.9	0.153	354.1	0.91
220	23.20	840.3	943.7	4.614	64.5	16.6	0.18	331.6	0.89
230	27.98	827.3	990.2	4.681	63.7	16.4	0.145	310.0	0.88
240	33.48	813.6	1037.5	4.756	62.8	16.2	0.141	285.5	0.87
250	39.78	799.0	1085.7	4.844	61.8	15.9	0.137	261.9	0.86
260	46.94	784.0	1135.7	4.949	60.5	15.6	0.135	237.4	0.87
270	55.05	767.9	1185.7	5.070	59.0	15.1	0.133	214.8	0.88
280	64.19	750.7	1236.8	5.230	57.4	14.6	0.131	191.3	0.90
290	74.45	732.3	1290.0	5.485	55.8	13.9	0.129	168.7	0.93
300	85.92	712.5	1344.9	5.736	54.0	13.2	0.128	144.2	0.97
310	98.70	691.1	1402.2	6.071	52.3	12.5	0.128	120.7	1.03
320	112.90	667.1	1462.1	6.574	50.6	11.5	0.128	98.10	1.11
330	128.65	640.2	1526.2	7.244	48.4	10.4	0.127	76.71	1.22
340	146.08	610.1	1594.8	8.165	45.7	9.17	0.127	56.70	1.39
350	165.37	574.4	1671.4	9.504	43.0	7.88	0.126	38.16	1.60
360	186.74	528.0	1761.5	13.984	39.5	5.36	0.126	20.21	2.35
370	210.53	450.5	1892.5	40.321	33.7	1.86	0.126	4.709	6.79

7.3.4 Natural Convection Boiling

Natural convection boiling of water and surfactants at atmospheric pressure in narrow horizontal annular channels was studied experimentally by Hetsroni et al. (2007) in the range of Bond numbers of 0.185–1.52. The channel length was 24 and 36 mm, the gap size was 0.45, 1.2, 2.2 and 3.7 mm. Thc heat flux was in the range of 20–500 kW/m^2, the concentration of surfactant solutions was varied from 10 to 600 ppm. Effect of restriction and channel length on boiling curves is presented in Chap. 2. For water boiling at Bond numbers less than unity, the CHF in restricted space is lower than that in unconfined space. This effect increases with increasing the channel length. For water at Bond number of 1.52, boiling can almost be considered as unconfined. The addition of surfactant led to enhancement of heat transfer compared to water boiling in the same gap size; however, this effect decreased with decreasing gap size. Hysteresis was observed for boiling in degraded surfactant solutions. For the same gap size, CHF in surfactant solutions was significantly lower than that in water.

7.3.5 Explosive Boiling

Hetsroni et al. (2003b, 2003, 2005a,b) observed the explosive saturated boiling regime in micro-channels, which exists before the annular flow regime. The test

Table 7.3 Physical properties of R-11

T [°C]	$p(10^{-2})$ [kPa]	ρ' [kg/m^3]	ρ'' [kg/m^3]	h [kJ/kg]	c_p [kJ/kg K]	$k(10^2)$ [W/m K]	$\alpha(10^7)$ [m^2/s]	$\nu(10^6)$ [m^2/s]	$\sigma(10^4)$ [N/m]	Pr
−40	0.05093	1620.7	0.3619	204.06	0.84	0.107	0.783	0.605	264.8	7.72
−30	0.09206	1600.8	0.6285	200.69	0.85	0.104	0.772	0.510	253.0	6.60
−20	0.1576	1579.0	1.037	197.29	0.86	0.101	0.761	0444	240.3	5.83
−10	0.2573	1554.7	1.633	193.84	0.86	0.098	0.747	0.394	224.6	5.27
0	0.4030	1534.4	2.477	190.34	0.87	0.095	0.733	0.354	212.8	4.83
10	0.6083	1511.5	3.630	186.79	0.88	0.093	0.716	0.324	201.0	4.52
20	0.8891	1487.9	5.165	183.17	0.88	0.090	0.697	0.300	186.3	4.30
30	1.263	1463.7	7.158	179.45	0.89	0.87	0.678	0.277	173.6	4.09
40	1.748	1439.0	9.709	175.60	0.90	0.084	0.658	0.260	160.8	3.95
50	2.366	1413.8	12.900	171.53	0.91	0.081	0.636	0.245	149.1	3.85
60	3.138	1387.9	16.858	167.35	0.92	0.078	0.614	0.232	137.3	3.78
70	4.088	1361.3	21.697	162.87	0.93	0.075	0.591	0.222	123.6	3.76
80	5.240	1333.9	27.579	158.09	0.95	0.073	0.566	0.212	111.8	3.74
100	8.253	1275.8	43.196	147.46	0.98	0.0670	0.539	0.200	89.2	3.72
120	12.393	1211.8	65.660	135.03	1.02	0.061	0.505	0.189	66.7	3.74
140	17.896	1138.3	98.619	120.22	1.07	0.056	0.466	0.183	45.1	3.92

modules had 13, 21 and 26 triangular micro-channels with hydraulic diameters of 220, 130, and 100 μm, respectively. Tests were performed in the range of mass flux 32–200 kg/m^{2}s, heat flux 120–270 kW/m^2, and vapor quality at the outlet manifold 0.01–0.08. The matter is discussed in Sect. 6.5.1. During quasi-periodic boiling in a certain single micro-channel of a heat sink the pressure drop oscillations were always accompanied by wall temperature oscillations. The period of these oscillations was very short and the oscillation amplitude increased with an increase in heat input. The dependence of the dimensionless time interval between cycles in a given channel on boiling number is shown in Eq. (6.62). The initial liquid film thickness may be calculated using Eq. (6.65). The high-frequency oscillations in individual micro-channels are superimposed and lead to total low frequency (2–5 Hz) pressure drop and temperature oscillations of the whole heat sink. Temporal variations of pressure drop, fluid temperature at the outlet manifold and heater temperature are periodic with the same oscillation frequency. All these fluctuations are in phase. The CHF phenomenon is different from that observed in channels of conventional size. A key difference between micro-channel heat sink and conventional channels is the amplification of instability prior to CHF. The heat transfer coefficient may be calculated with standard deviation of 18% by Eq. (6.66).

7.4 Selected Properties of Liquids Used for Cooling Micro-Devices

The data of physical properties of saturated water are presented in Table 7.2.

Table 7.4 Physical properties of R-12

T [°C]	$p\,(10^{-2})$ [kPa]	ρ' [kg/m^3]	ρ'' [kg/m^3]	h [kJ/kg]	c_p [kJ/kg K]	$k\,(10^2)$ [W/m K]	$\alpha\,(10^7)$ [m^2/s]	$\nu\,(10^6)$ [m^2/s]	$\sigma\,(10^4)$ [N/m]	Pr
−40	0.6423	170.95	1515	4.096	0.8834	0.1000	0.747	0.280	180.44	3.79
−30	1.0052	167.43	1486	6.201	0.8960	0.0954	0.716	0.254	165.73	3.55
−20	1.5102	163.62	1457	9.039	0.9085	0.0907	0.685	0.236	152.98	3.44
−10	2.1927	159.48	1426	12.80	0.9211	0.0861	0.656	0.220	137.29	3.36
0	3.0881	154.95	1394	17.66	0.9337	0.0814	0.625	0.211	124.05	3.38
10	4.2933	150.05	1361	23.80	0.9504	0.0768	0.594	0.204	110.81	3.44
20	5.6731	144.70	1327	31.52	0.9672	0.0721	0.562	0.199	98.06	3.55
30	7.4451	138.75	1292	41.16	0.9839	0.0675	0.531	0.194	85.32	3.66
40	9.5947	132.14	1254	53.12	1.0006	0.0628	0.500	0.191	71.59	3.82
50	12.1463	124.56	1213	68.56	1.0844	0.0582	0.442	0.186	61.19	4.12
60	15.1814	116.90	1167	85.69	1.1179	0.0535	0.410	0.184	42.76	4.49
70	18.7265	108.02	1108	108.81	1.1597	0.0477	0.368	0.183	32.07	4.97
80	22.8393	97.68	1064	138.31	1.2225	0.0419	0.322	0.182	24.12	5.65

Refrigerants R-N ($N = 11, 12, \ldots$)

The same data on physical properties of liquid refrigerants R-N (R-11, R-12, R-13, R-21, R-22, R-113) and their vapor are presented in Tables 7.3–7.8. The detailed data on thermophysical properties of different refrigerants (density, enthalpy, heat capacity, viscosity, thermal conductivity and diffusivity) are found in books by Platzer et al. (1990), Andersen (1959), and Danilova et al. (1976).

Many micro-channel blocks have a high thermal conductivity and conjugate effects may become important. In this chapter we considered analytical and experi-

Table 7.5 Physical properties of R-13

T [°C]	$p\,(10^{-2})$ [kPa]	ρ' [kg/m^3]	ρ'' [kg/m^3]	h [kJ/kg]	c_p [kJ/kg K]	$k\,(10^2)$ [W/m K]	$\alpha\,(10^7)$ [m^2/s]	$\nu\,(10^6)$ [m^2/s]	$\sigma\,(10^4)$ [N/m]	Pr
−110	0.161	162.57	1634	1.257	0.816	0.114	0.855	0.408	204.0	4.77
−100	0.333	158.47	1597	2.457	0.854	0.109	0.799	0.340	186.0	4.26
−90	0.628	154.00	1558	4.427	0.879	0.104	0.760	0.298	168.0	3.92
−80	1.098	149.26	1520	7.452	0.908	0.0984	0.712	0.261	151.0	3.67
−70	1.805	144.19	1481	11.848	0.929	0.0933	0.677	0.240	134.0	3.55
−60	2.817	138.62	1439	18.044	0.971	0.0882	0.631	0.221	117.9	3.50
−50	4.204	132.43	1395	26.497	1.009	0.0831	0.590	0.204	101.3	3.46
−40	6.051	125.56	1349	37.850	1.042	0.0780	0.555	0.199	85.5	3.59
−30	8.42	117.98	1300	52.938	1.072	0.0729	0.523	0.194	69.9	3.72
−20	11.43	109.53	1247	72.833	1.105	0.0678	0.493	0.192	54.9	3.90

Table 7.6 Physical properties of R-21

T [°C]	$p\,(10^{-2})$ [kPa]	ρ' [kg/m^3]	ρ'' [kg/m^3]	h [kJ/kg]	c_p [kJ/kg K]	$k\,(10^2)$ [W/m K]	$\alpha\,(10^7)$ [m^2/s]	$\nu\,(10^6)$ [m^2/s]	$\sigma\,(10^4)$ [N/m]	Pr
−40	0.093	1514.1	0.4989	267.1	1.001	0.123	0.814	0.415	277.5	5.10
−30	0.162	1492.7	0.8563	262.1	1.009	0.119	0.795	0.372	261.8	4.68
−20	0.277	1470.9	1.395	256.7	1.017	0.116	0.775	0.337	247.1	4.35
−10	0.454	1448.5	2.180	252.0	1.026	0.113	0.758	0.312	231.4	4.11
0	0.706	1425.6	3.276	246.7	1.034	0.109	0.742	0.288	216.7	3.88
10	1.06	1402.2	4.756	241.7	1.042	0.105	0.725	0.271	202.0	3.74
20	1.53	1378.2	6.708	236.3	1.051	0.102	0.705	0.256	187.3	3.63
30	2.15	1353.8	9.229	230.9	1.059	0.098	0.690	0.244	173.6	3.54
40	2.95	1328.8	12.43	224.4	1.068	0.095	0.672	0.233	159.8	3.47
50	3.96	1302.4	15.87	218.1	1.072	0.091	0.658	0.224	146.1	3.40
60	5.216	1277.4	21.50	211.2	1.080	0.087	0.642	0.216	133.4	3.37
70	6.76	1250.7	27.03	203.8	1.089	0.083	0.622	0.210	119.6	3.38
80	–	1225.4	47.62	196.3	1.097	0.080	0.597	0.205	109.8	3.44

Table 7.7 Physical properties of R-22

T [°C]	$p\,(10^{-2})$ [kPa]	ρ' [kg/m^3]	ρ'' [kg/m^3]	$h\,(10^{-3})$ [kJ/kg]	c_p [kJ/kg K]	$k\,(10^2)$ [W/m K]	$\alpha\,(10^7)$ [m^2/s]	$\nu\,(10^6)$ [m^2/s]	$\sigma\,(10^4)$ [N/m]	Pr
−70	2048	1489	1.064	250.58	0.9504	0.1344	0.879	0.434	230.48	3.94
−60	0.3746	1465	1.869	245.05	0.9835	0.1198	0.831	0.323	215.25	3.88
−50	0.6472	1439	3.096	239.48	1.0161	0.1163	0.795	0.275	201.03	3.46
−30	1.6465	1382	7.407	227.55	1.0819	0.1082	0.724	0.232	168.67	3.20
−20	2.4614	1350	10.76	220.94	1.1145	0.1035	0.688	0.218	152.00	3.17
−10	3.5598	1318	15.29	214.36	1.1476	0.1000	0.661	0.210	136.31	3.18
0	5.0013	1285	21.23	206.95	1.1803	0.0954	0.629	0.204	120.13	3.25
10	6.8547	1249	28.90	198.29	1.2129	0.0907	0.599	0.199	103.95	3.32
20	9.1691	1213	38.76	188.41	1.2460	0.0872	0.577	0.197	90.12	3.41
30	12.0228	1176	51.55	177.27	1.2786	0.0826	0.549	0.196	75.90	3.55
40	15.4845	1132	67.57	164.75	1.3117	0.0790	0.532	0.196	60.21	3.67
50	19.6424	1084	88.50	155.33	1.3444	0.0744	0.510	0.196	46.68	3.78
60	24.4378	1032	111.48	149.13	1.3732	0.0709	0.500	0.202	33.69	3.92
70	30.0765	969	146.18	142.00	1.4068	0.0669	0.491	0.208	21.28	4.11
80	36.5979	895	196.35	134.50	1.4403	0.0628	0.487	0.219	11.18	4.41

Table 7.8 Physical properties of R-113

T [°C]	$p\,(10^{-2})$ [kPa]	ρ' [kg/m^3]	ρ'' [kg/m^3]	h [kJ/kg]	c_p [kJ/kg K]	$k\,(10^2)$ [W/m K]	$\alpha\,(10^7)$ [m^2/s]	$\nu\,(10^6)$ [m^2/s]	$\sigma\,(10^4)$ [N/m]	Pr
0	0.148	1626	1.23	160.56	0.9127	0.0766	0.517	0.573	215.25	11.15
10	0.236	1600	1.90	157.97	0.9295	0.0743	0.500	0.506	202.01	10.10
20	0.364	1580	2.85	154.97	0.9462	0.0722	0.483	0.447	192.21	9.23
30	0.541	1560	4.12	152.95	0.9630	0.0702	0.467	0.394	182.40	8.45
40	0.783	1530	5.59	148.84	0.9797	0.0681	0.456	0.359	168.18	7.88
50	1.101	1510	8.00	146.12	0.9965	0.0661	0.439	0.323	158.86	7.35
60	1.512	1480	10.8	145.70	1.0090	0.0640	0.428	0.302	144.16	7.06
70	2.033	1460	14.2	138.88	1.0258	0.0620	0.414	0.278	136.31	6.71
80	2.677	1430	18.5	135.11	1.0425	0.0598	0.403	0.254	124.54	6.30
90	3.472	1400	23.7	130.71	1.0593	0.0578	0.389	0.236	111.79	6.07
100	4.421	1380	30.0	126.11	1.0760	0.0556	0.375	0.218	102.97	5.82

mental correlations for pressure drop and heat transfer, suggested for use in single-phase and two-phase flow in micro-channels, as well as in conventional size channels. It is recommended that such correlations be used in the range of boundary conditions and experimental parameters under which they were obtained.

References

Ali MI, Sadatomi M, Kawaji M (1993) Two-phase flow in narrow channels between two plates. Can J Chem Eng 71:657–666

Andersen SA (1959) Automatic refrigeration. McLaren, Glasgow

Armand AA (1946) The resistance during the movement of a two-phase system in horizontal pipes. Izv Vses Teplotekh Inst 1:16–23 (AERE-Lib/Trans 828)

Bergles AE, Rohsenow WW (1964) The determination of forced-convection surface-boiling heat transfer. Trans ASME J Heat Transfer 86:365–372

Celata GP, Cumo M, Marconi V, McPhail SJ, Zummo Z (2005) Micro-tube heat transfer scaling effects: an experimental validation. In: ECI Conference on Heat Transfer and Fluid Flow in Microscale II Ciocco, Castelvecchio Pascoili, 25–30 September 2005

Danilova GN, Filatkin VN, Scherbov MG, Buchko NA (1976) The book of heat and mass transfer problems for food and refrigerating industry. Food Industry, Moscow

Hetsroni G, Gurevich M, Mosyak A, Rozenblit R (2004) Drag reduction and heat transfer of surfactants flowing in a capillary tube. Int J Heat Mass Transfer 47:3797–3809

Hetsroni G, Mosyak A, Pogrebnyak E, Segal Z (2007) Natural convection boiling of water and surfactants in narrow horizontal channels. Int J Multiphase Flow 33:469–483

Hetsroni G, Mosyak A, Pogrebnyak E, Yarin LP (2005a) Heat transfer in micro-channels: Comparison of experiments with theory and numerical results. Int J Heat Mass Transfer 48:5580–5601

Hetsroni G, Mosyak A, Segal Z, Pogrebnyak E (2005b) Explosive boiling of water in parallel micro-channels. Int J Multiphase Flow 31:371–392

Hetsroni G, Mosyak A, Segal Z, Pogrebnyak E (2003b) Two-phase flow pattern in parallel micro-channels. Int J Multiphase Flow 29:344–360

Hetsroni G, Mosyak A, Segal Z, Pogrebnyak E (2003) Two-phase flow pattern in parallel micro-channels. Int J Multiphase Flow 29:344–360

Hsu YY (1962) On size range of active nucleation cavities on a heating surface. J Heat Transfer 84:207–216

Kandlikar SG (2004) Heat transfer mechanisms during flow boiling in micro-channels. ASME J Heat Transfer 126:8–16

Kandlikar SG, Joshi S, Tian S (2003) Effect of surface roughness on heat transfer and fluid flow characteristics at low Reynolds numbers in small diameter tubes. Heat Transfer Eng 24:4–16

Kawahara A, Chung PM, Kawaji M (2002) Investigation of two-phase flow pattern, void fraction and pressure drop in a micro-channel. Int J Multiphase Flow 28:1411–1435

Kennedy JE, Roach GM, Dowling ME, Abdel-Khalik SI, Ghiaasiaan SM, Jeter SM, Quershi ZH (2000) The onset of flow instability in uniformly heated horizontal micro-channels. Trans ASME J Heat Transfer 122:118–125

Klein D, Hetsroni G, Mosyak A (2005) Heat transfer characteristics of water and APG surfactant solution in micro-channel heat sink. Int J Multiphase Flow 31:393–415

Lee J, Mudawar I (2005a) Two-phase flow in high-heat-flux micro-channel heat sink for refrigeration cooling applications. Part I: pressure drop. Int J Heat Mass Transfer 48:928–940

Lee J, Mudawar I (2005b) Two-phase flow in high-heat-flux micro-channel heat sink for refrigeration cooling applications. Part II: heat transfer characteristics. Int J Heat Mass Transfer 48:941–955

Maranzana G, Perry I, Maillet D (2004) Mini and Micro-channels: Influence of axial conduction in the walls. Int J Heat Mass Transfer 47:3993–4004

Platzer B, Plot A, Maurer G (1990) Thermophysical properties of refrigerants. Springer, Berlin Heidelberg New York

Qu W, Mudawar I (2003a) Measurement and prediction of pressure drop in two-phase micro-channel heat sinks. Int J Heat Mass Transfer 46:2737–2753

Qu W, Mudawar I (2003b) Flow boiling heat transfer in two-phase micro-channel heat sinks. I. Experimental investigation and assessment of correlation methods. Int J Heat Mass Transfer 46:2755–2771

Qu W, Mudawar W (2004) Measurement and correlation of critical heat flux in two-phase micro-channel heat sinks. Int J Heat Mass Transfer 47:2045–2059
Shah RK, London AL (1978) Laminar flow forced convection in ducts: a source book for compact heat exchanger analytical data. Advances in Heat Transfer, suppl 1. Academic, New York
Steinke ME, Kandlikar SG (2004) An experimental investigation of flow boiling characteristics of water in parallel micro-channels. Trans ASME J Heat Transfer 126:518–526
Tiselj I, Hetsroni G, Mavko B, Mosyak A, Pogrebnyak E, Segal Z (2004) Effect of axial conduction on heat transfer in micro-channels. Int J Heat Mass Transfer 47:2551–2565
Wojtan L, Revellin R, Thome IR (2006) Investigation of saturated critical heat flux in a single, uniformly heated micro-channel. Exp Thermal Fluid Sci 30:765–774
Yen TH, Kasagi N, Suzuki Y (2003) Forced convective boiling heat transfer in micro-tubes at low mass and heat fluxes. Int J Multiphase Flow 29:1771–1792
Zhang W, Hibiki T, Mishima K, Mi Y (2006) Correlation of critical heat flux for flow boiling of water in mini-channels. Int J Heat Mass Transfer 49:1058–1072
Zhao TS, Bi QC (2001) Pressure drop characteristics of gas–liquid two-phase flow in vertical miniature triangular channels. Int J Heat Mass Transfer 44:2523–2534

Nomenclature

a	Half length of channel bottom
b	Half length of channel side
d	Diameter
c_{p}	Heat capacity at constant pressure
C	Concentration
G	mass flux
H	Heat transfer coefficient
k	Thermal conductivity
k_{s}	Average height of surface roughness
L	Length
n	Exponent (Eq. 7.2)
P	Pressure
q	Heat flux
T	Temperature
x	Mass quality
$D = \dfrac{\Delta T_{\mathrm{sub,ONB}}}{T_{\mathrm{S}}}$	Parameter
$\mathrm{Nu} = \dfrac{hd_{\mathrm{h}}}{k}$	Nusselt number
Po	Poiseuille number
$\mathrm{Pr} = \dfrac{\nu}{\alpha}$	Prandtl number
$\mathrm{Re} = \dfrac{Ud_{\mathrm{h}}}{\nu}$	Reynolds number

Greek symbols

α	Void fraction
β	Homogeneous (volumetric)
$\Delta T_{\mathrm{sub,ONB}} = T_{\mathrm{S}} - T_{\mathrm{B,ONB}}$	
λ	Friction factor
ρ	Density
$\Pi = \frac{q}{\rho_{\mathrm{L}} U C_{\mathrm{pL}} \Delta T_{\mathrm{s}}}$	

Subscripts

B	Bulk
crit	Critical
e	Equilibrium
exp	Experimental
f	Fluid
G	Gas, mass flux
GS	Gas superficial
h	Hydraulic
H1, H2	Thermal boundary condition at constant wall heat flux
in	Inner, inlet
L	Liquid
LS	Liquid superficial
max	Maximum
o	Out
ONB	ONB point
s	Saturation
T	Thermal boundary condition at constant wall pressure
w	Wall, water

Part II
Special Topics

Chapter 8
Capillary Flow with a Distinct Interface

The quasi-one-dimensional model of laminar flow in a heated capillary is presented. In the frame of this model the effect of channel size, initial temperature of the working fluid, wall heat flux and gravity on two-phase capillary flow is studied. It is shown that hydrodynamical and thermal characteristics of laminar flow in a heated capillary are determined by the physical properties of the liquid and its vapor, as well as the heat flux on the wall.

8.1 Preliminary Remarks

For a while now, the problem of flow and heat transfer in heated capillaries has attracted attention from a number of research groups, with several applications to engineering. The knowledge of the thermohydrodynamic characteristics of capillary flow with evaporative meniscus allows one to elucidate the mechanism of heat and mass transfer in porous media, to evaluate the efficiency of cooling system of electronic devices with high power density, as well as to optimize MEMS.

In the earliest research on micro-scale flow heat transfer, Tuckerman and Pease (1981) studied the single-phase fluid flow and heat transfer characteristics in micro-channels and demonstrated that electronic chips could effectively be cooled by means of the forced convection flow of water through micro-channels made from silicon. Following Tuckerman's (1984) research, some other researchers (Wu and Little 1984; Weisberg et al. 1992; Peng and Peterson 1995) have contributed to a better understanding of the fluid flow and heat transfer mechanism occurring in micro-channels with single-phase flow. An extensive review of the available cooling data for single-phase micro-channel flow has written by Bailey et al. (1995). Although single-phase coolants can effectively cool miniature devices, they present some inherent disadvantages like large pressure drops and streamwise increases in the heat sink temperature. However, two-phase heat dissipation can achieve very high heat fluxes for a constant flow rate while maintaining a relatively constant surface temperature.

Bowers and Mudawar (1994a) performed an experimental study of boiling flow within mini-channel (2.54 mm) and micro-channel ($d = 510\,\mu m$) heat sink and demonstrated that high values of heat flux can be achieved. Bowers and Mudawar (1994b) also modeled the pressure drop in the micro-channels and mini-channels, using the Collier (1981) and Wallis (1969) homogenous equilibrium model, which assumes the liquid and vapor phases form a homogenous mixture with equal and uniform velocity, and properties were assumed to be uniform within each phase.

Landerman (1994) developed an analytical model for two-phase boiling heat transfer in a high aspect ratio rectangular channel. The flow regimes in the channel were mapped and then the heat transfer and wall temperature were evaluated, using heat transfer coefficients taken from the literature.

Wayner et al. (1976) developed a simple procedure to obtain the heat transfer coefficient for the interline region of an adsorption controlled wetting film. Xu and Carey (1990) developed an analytical model to predict the heat transfer characteristics of film evaporating on a microgroove surface.

A different analytical approach to the pressure drop of boiling two-phase flow in extremely narrow channels (35–110 μm between plates) was suggested by Morijama and Inoue (1992). The momentum equations for the liquid and the vapor were introduced in order to evaluate the pressure loss along the gap for slug flow and film flow regimes, assuming equal and constant liquid film thickness on the upper and lower wall. The modern state of the problem is found in recent experimental data and corresponding bibliography, discussed in Chaps. 3–6.

The capillary flows where the working liquid undergoes a change of phase on the evaporating meniscus were studied (applied for the problems of grooved heat pipes) by Potast and Wayner (1972), Xu and Carey (1990), Stephan and Busse (1992), Khrustalev and Faghri (1994, 1995), Peterson and Ma (1996), Ha and Peterson (1998), Peterson and Ha (1998), and Weislogel and Lichter (1998). The detail theoretical investigation of the evaporative meniscus of a perfect wetting liquid in a heated channel with constant wall temperature was performed by Morris (2003). A mathematical model of evaporative liquid–vapor meniscus in a capillary slot has been developed by Khrustalev and Faghri (1996). The model includes two-dimensional steady-state momentum conservation and energy equations for both the vapor and liquid phase, while the liquid–vapor interface curvature was constant along the interface.

The flow in a heated capillary depends on a number of parameters including the channel geometry, physical properties of the liquid and the heat flux. An immediate consequence of the liquid heating and evaporation is convective motion of both phases. The latter leads to a velocity and temperature field transformation and a change in the meniscus shape.

Peles et al. (2000) elaborated on a quasi-one-dimensional model of two-phase laminar flow in a heated capillary slot due to liquid evaporation from the meniscus. Subsequently this model was used for analysis of steady and unsteady flow in heated micro-channels (Peles et al. 2001; Yarin et al. 2002), as well as the study of the onset of flow instability in heated capillary flow (Hetsroni et al. 2004).

Below we consider a quasi-one-dimensional model of flow and heat transfer in a heated capillary, with hydrodynamic, thermal and capillarity effects. We estimate the influence of heat transfer on steady-state laminar flow in a heated capillary, on the shape of the interface surface and the velocity and temperature distribution along the capillary axis.

The present model takes into account how capillary, friction and gravity forces affect the flow development. The parameters which influence the flow mechanism are evaluated. In the frame of the quasi-one-dimensional model the theoretical description of the phenomena is based on the assumption of uniform parameter distribution over the cross-section of the liquid and vapor flows. With this approximation, the mass, thermal and momentum equations for the average parameters are used. These equations allow one to determine the velocity, pressure and temperature distributions along the capillary axis, the shape of the interface surface for various geometrical and regime parameters, as well as the influence of physical properties of the liquid and vapor, micro-channel size, initial temperature of the cooling liquid, wall heat flux and gravity on the flow and heat transfer characteristics.

Chapter 8 consists of the following: in Sect. 8.2 the physical model of the process is described. The governing equations and conditions of the interface surface are considered in Sects. 8.3 and 8.4. In Sect. 8.5 we present the equations transformations. In Sect. 8.6 we display equations for the average parameters. The quasi-one-dimensional model is described in Sect. 8.7. Parameter distribution in characteristic zones of the heated capillary is considered in Sect. 8.8. The results of a parametrical study on flow in a heated capillary are presented in Sect. 8.9.

8.2 The Physical Model

A capillary system is said to be in a steady-state equilibrium position when the capillary forces are equal to the hydrostatic pressure force (Levich 1962). The heating of the capillary walls leads to a disturbance of the equilibrium and to a displacement of the meniscus, causing the liquid–vapor interface location to change as compared to an unheated wall. This process causes pressure differences due to capillarity and the hydrostatic pressures exiting the flow, which in turn causes the meniscus to return to the initial position. In order to realize the above-mentioned process in a continuous manner it is necessary to carry out continual heat transfer from the capillary walls to the liquid. In this case the position of the interface surface is invariable and the fluid flow is stationary. From the thermodynamical point of view the process in a heated capillary is similar to a process in a heat engine, which transforms heat into mechanical energy.

The idealized picture of the flow in a heated micro-channel is shown in Fig. 8.1a. Such flow possesses a number of specific properties due to its unique structure, which forms because of liquid evaporation and the interaction of pure vapor and liquid flows separated by the interface surface. The latter has an infinitely thin surface with a jump in pressure and velocity, while the temperature is equal. One can

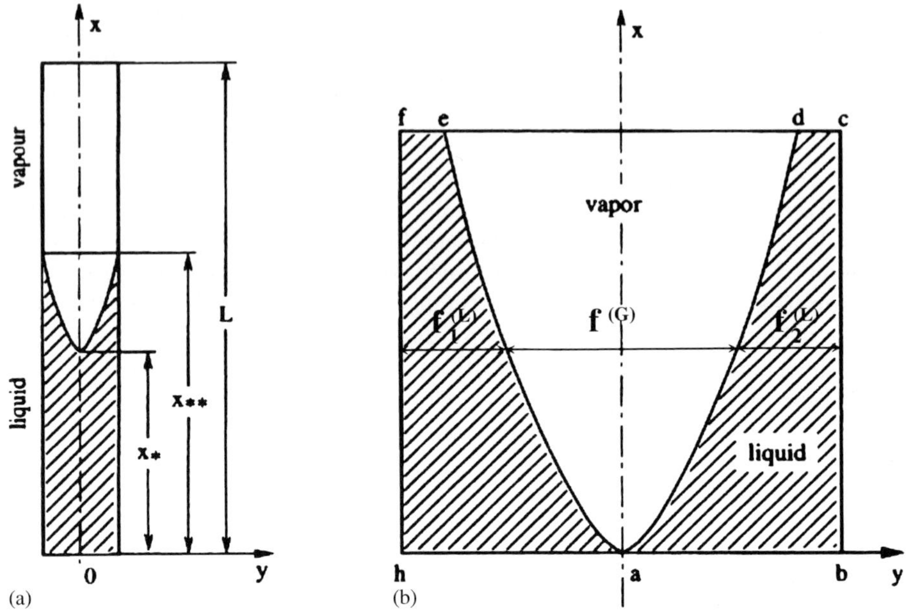

Fig. 8.1a,b The structure of flow in a heated capillary. Reprinted from Peles et al. (2000) with permission. (**a**) A heated capillary scheme: (1) heated ($0 \le x < x_*$), (2) evaporation ($x_* < x \le x_{**}$), and (3) superheated ($x > x_{**}$) regions. (**b**) The evaporation regions: *I* a – b, *II* b – c, *III* c – d, *IV* d – a, *V* a – e, *VI* e – f, *VII* f – h, *VIII* h – a, *I′* a – d, *II′* d – e, *III′* e – a. The terms $f^{(\mathrm{G})}$ and $f^{(\mathrm{L})}$ are the cross-section of the vapor and liquid flows, $f^{(\mathrm{L})} = f_1^{(\mathrm{L})} + f_2^{(\mathrm{L})}$

see that in a heated capillary there are three distinct regions (heated, evaporation, superheated) corresponding to different types of flow. Within the first and the third, single-phase liquid (vapor) flow occurs. Heat transfer from the wall to the fluid is accompanied by an increase of the liquid (vapor) temperature and velocity downstream. In the second region liquid–vapor flow interaction takes place. Heat flux causes the liquid to progress downstream and to evaporate.

8.3 Governing Equations

Consider the mass, thermal and momentum balance equations. The key assumption of the present analysis is that the Knudsen number of the flow in the capillary is sufficiently small. This allows one to use the continuum model for each phase. Due to the moderate flow velocity, the effects of compressibility of the phases, as well as mechanical energy, dissipation in the phases are negligible. Assuming that thermal conductivity and viscosity of vapor and liquid are independent of temperature and pressure, we arrive at the following equations:

$$\operatorname{div} \rho^{(\alpha)} \mathbf{v}^{(\alpha)} = 0 \tag{8.1}$$

$$\rho^{(\alpha)}(\mathbf{v}^{(\alpha)}\cdot\nabla)h^{(\alpha)} = k^{(\alpha)}\nabla^2 T^{(\alpha)} \tag{8.2}$$

$$\rho^{(\alpha)}(\mathbf{v}^{(\alpha)}\cdot\nabla)\mathbf{v}^{(\alpha)} = -\nabla P^{(\alpha)} + \mu^{(\alpha)}\nabla^2\mathbf{v}^{(\alpha)} + \frac{\mu^{(\alpha)}}{3} grad\, div\mathbf{v}^{(\alpha)} + \rho^{(\alpha)}\mathbf{g} \tag{8.3}$$

where ρ, $\mathbf{v}$, T and h are the density, velocity, temperature and enthalpy (bold letters denote vectors). The term $\mathbf{v}$ has components u, v, w, which are directed along the x, y and z-axes, respectively, $\mathbf{g}$ is the acceleration due to gravity, P is the pressure, and k and μ are the thermal conductivity and viscosity, respectively. The parameters ∇ and ∇^2 are the gradient and the Laplacian operator. The superscript $\alpha = \mathrm{G}$ and L corresponds to vapor ($\alpha = \mathrm{G}$) or liquid ($\alpha = \mathrm{L}$). The term that represents the so-called second viscosity is omitted in Eq. (8.3).

In order to close the system of Eqs. (8.1–8.3) it is necessary to supplement the gas equation of state and the equation determining the dependence of liquid density on temperature

$$P^{(\mathrm{G})} = P^{(\mathrm{G})}(\rho^{(\mathrm{G})}T^{(\mathrm{G})}) \tag{8.4}$$

$$\rho^{(\mathrm{L})} = \rho^{(\mathrm{L})}(T^{(\mathrm{L})}) \tag{8.5}$$

The system of Eqs. (8.1–8.5) should also be supplemented by a correlation determining the dependence of enthalpy on temperature: $h^{(\alpha)} = h^{(\alpha)}(T^{(\alpha)})$.

8.4 Conditions at the Interface Surface

We now describe the conditions that correspond to the interface surface. For stationary capillarity flow, these conditions can be expressed by the equations of continuity of mass, thermal fluxes on the interface surface and the equilibrium of all acting forces (Landau and Lifshitz 1959). For a capillary with evaporative meniscus the balance equations have the following form:

$$\sum_{\alpha=1}^{2} \rho^{(\alpha)}\mathbf{v}^{(\alpha)}\mathbf{n}_i^{(\alpha)} = 0 \tag{8.6}$$

$$\sum_{\alpha=1}^{2} \left(\rho^{(\alpha)}\mathbf{v}^{(\alpha)}h^{(\alpha)} - k^{(\alpha)}\frac{\partial T^{(\alpha)}}{\partial x_\ell}\right)\mathbf{n}_i^{(\alpha)} = 0 \tag{8.7}$$

$$\sum_{\alpha=1}^{2} (P^{(\alpha)} + \rho^{(\alpha)}v_i^{(\alpha)}v_k^{(\alpha)})n_i^{(\alpha)} = (\tau_{ik}^{(\mathrm{G})} - \tau_{ik}^{(\mathrm{L})})n_k + \sigma(r_1^{-1} + r_2^{-1})n_i^{(\mathrm{G})} + \frac{\partial\sigma}{\partial x_\ell} \tag{8.8}$$

where σ is the surface tension, τ_{ik} is the tensor of viscous tension, $\mathbf{v}^{(\alpha)}\mathbf{n}_i^{(\alpha)}$ and $(\partial T^{(\alpha)}/\partial x_\ell)\mathbf{n}_i^{(\alpha)}$ are the normal components of the velocity vector and temperature interface surface gradient, respectively, r_1 and r_2 are the general radii of curvature of the interface surface, and $\mathbf{n}_i$ and $\mathbf{n}_k$ correspond to the normal and the tangent directions. The term $\mathbf{n}_i^{(\mathrm{G})} = -\mathbf{n}_i^{(\mathrm{L})}$, $\mathbf{n}_i^{(\mathrm{G})}$ is directed inside the first fluid (vapor) $l = 1, 2, 3$, $\Sigma_{\alpha=1}^{2}$ is the sum of terms related to vapor ($\alpha = 1 = G$) and liquid ($\alpha = 2 = L$).

When the interface surface is expressed by a function $x = \varphi(y,z)$ the general radii of curvature are found from the equation (Smirnov 1964):

$$Ar^2 + Br + c = 0 \tag{8.9}$$

where $A = (a_1a_2 - a_3^2)$; $B = a_4[2a_3a_5a_6 - (1 + a_5^2)a_2 - (1 + a_6^2)a_1]$; $c = (1 + a_5^2 + a_6^2)^2$; $a_1 = (\partial^2x/\partial y^2)_\mathrm{b}$; $a_2 = (\partial^2x/\partial z^2)_\mathrm{b}$; $a_3 = \left(\partial^2x/(\partial y\partial z)\right)_\mathrm{b}$; $a_4 = [1 + (\partial x/\partial y)_\mathrm{b}^2 + (\partial x/\partial z)_\mathrm{b}^2]^{1/2}$; $a_5 = (\partial x/\partial y)_\mathrm{b}$; $a_6 = (\partial x/\partial z)_\mathrm{b}$. The index b corresponds to the interface surface.

For the flow in a slot (the plane problem: $x = \varphi(y)$) the general radii of curvature equals

$$r_1 = \infty, r_2 = \frac{(1 + f^{(\mathrm{G})\prime 2})^{3/2}}{f^{(\mathrm{G})\prime\prime}} = \psi^{-1} \tag{8.10}$$

where $f^{(\mathrm{G})\prime}$ and $f^{(\mathrm{G})\prime\prime}$ equal $(\mathrm{d}f^{(\mathrm{G})}/\mathrm{d}x)_\mathrm{b}$ and $(\mathrm{d}^2f^{(\mathrm{G})}/\mathrm{d}x^2)_\mathrm{b}$, respectively; $f^{(\mathrm{G})} = |k|\,y_\mathrm{b}$ is half the vapor flow cross-section area and k is a unit vector in the z-direction. The vapor pressure on the interface surface may be found from the Clausius–Clapeyron equation.

8.5 Equation Transformation

As already mentioned, the system of Eqs. (8.1–8.5) is supplemented by the Clausius–Clapeyron equation, as well as by the correlation that determines the dependence of enthalpy on temperature and describes the thermohydrodynamical characteristics of flow in a heated capillary. It is advantageous to analyze parameters of such flow to transform the system of governing equations to the form that is convenient for significant simplification of the problem.

8.5.1 Equation for Pressure and Temperature at Interface Surface

The vapor and liquid densities, pressure and temperature on the interface surface are connected by the following equation (Carey 1992):

$$\frac{1}{\rho^{(\mathrm{G})}}\frac{\mathrm{d}P^{(\mathrm{G})}}{\mathrm{d}T} + \frac{1}{\rho^{(\mathrm{L})}}\frac{\mathrm{d}P^{(\mathrm{L})}}{\mathrm{d}T} = \frac{h_\mathrm{LG}}{T}. \tag{8.11}$$

On the meniscus surface the deviation of vapor pressure $P^{(\mathrm{G})}$ from the saturation pressure P_sat depends on the surface tension σ, liquid density $\rho^{(\mathrm{L})}$, gas constant R, temperature T, and radii of curvature r. When $P^{(\mathrm{G})} - P_\mathrm{sat}(T) \ll (2\sigma/r_2)$ (which is usually the case), $P^{(\mathrm{G})}$ can be approximated for most systems (Carey 1992) by the following expression:

$$P^{(\mathrm{G})} = P_\mathrm{sat}\exp(-(2\sigma/RT\rho^{(\mathrm{L})}r_2)). \tag{8.12}$$

When the ratio $2\sigma/RT\rho^{(L)}r_2 \ll 1$ it is possible to assume $P^{(G)} = P_{sat}$. For example, for realistic conditions: $T = 300\,K$, $\rho^{(L)} = 10^3\,kg/m^3$, $r_2 = 5\times10^{-4}\,m$, $R = 462\,J/kg\,K$, $\sigma = 0.0727\,N/m$, the vapor pressure $P^{(G)} = P_{sat}\exp(-2\times10^{-6})$, i.e., $P^{(G)} = P_{sat}$. In this case vapor pressure on the interface may be found from the Clausius–Clapeyron equation

$$\frac{dP^{(G)}}{dT^{(G)}} = \frac{h_{LG}\rho^{(G)}}{T^{(G)}}\,. \tag{8.13}$$

Assuming the vapor is an ideal gas

$$P^{(G)} = R\rho^{(G)}T^{(G)} \tag{8.14}$$

and combining Eqs. (8.12) and (8.13), we arrive (after integration) at the dependence of the vapor pressure on the temperature at the interface surface

$$P^{(G)} = \tilde{P}\exp(-h_{LG}/RT^{(G)}) \tag{8.15}$$

where $\tilde{P} = P'\exp(h_{LG}/RT')$; P' and T' are some values of the pressure and temperature on the saturation line.

8.5.2 Transformation of the Mass, Momentum and Energy Equations

To transform the mass energy and momentum equations, we use Eqs. (8.1–8.3) in the following form:

$$\frac{\partial}{\partial x_\ell}(\rho^{(\alpha)}v_\ell^{(\alpha)}) = 0 \tag{8.16}$$

$$\frac{\partial}{\partial x_\ell}\left[\rho^{(\alpha)}v_\ell^{(\alpha)}h^{(\alpha)} - k^{(\alpha)}\frac{\partial T^{(\alpha)}}{\partial x_\ell}\right] = 0 \tag{8.17}$$

$$\frac{\partial}{\partial x_k}(\Pi_{ik}^{(\alpha)}) = 0 \tag{8.18}$$

where $\Pi_{ik}^{(\alpha)} = P^{(\alpha)}\delta_{ik} + \rho^{(\alpha)}v_i^{(\alpha)}v_k^{(\alpha)} - \tau_{ik}^{(\alpha)} - X_k^{(\alpha)}$, δ_{ik} is the Kronecker delta; $\delta_{ik} = 1$ for $i = k$, $\delta_{ik} = 0$ for $i \neq k$; $i,k = 1,2,3$; $\tau_{ik}^{(\alpha)} = \mu^{(\alpha)}((\partial v_i^{(\alpha)}/\partial x_k) + (\partial v_k^{(\alpha)}/\partial x_i) - (2/3)\delta_{ik}(\partial v_\ell^{(\alpha)}/\partial x_\ell))$; $X_k^{(\alpha)} = \int \rho^{(\alpha)}\mathbf{g}_k\,dx$, $\mathbf{g}_k(g,0,0)$, g is the component of vector $\mathbf{g}$ directed along x-axis, $k = 1,2,3$.

Let us transform Eqs. (8.16–8.18) using Gauss's theorem

$$\int\limits_{V^{(\alpha)}} \frac{\partial}{\partial x_\ell}(\rho^{(\alpha)}v_\ell^{(\alpha)})\,dV = \int\limits_{\Sigma^{(\alpha)}} \rho^{(\alpha)}\mathbf{v}^{(\alpha)}\,d\mathbf{S} \tag{8.19}$$

$$\int_{V^{(\alpha)}} \frac{\partial}{\partial x_\ell}\left[(\rho^{(\alpha)} v_\ell^{(\alpha)} h^{(\alpha)}) - k^{(\alpha)}\frac{\partial T^{(\alpha)}}{\partial x_\ell}\right] \mathrm{d}V = \int_{\Sigma^{(\alpha)}} \left[(\rho^{(\alpha)} \mathbf{v}^{(\alpha)} h^{(\alpha)}) - k^{(\alpha)}\frac{\partial T^{(\alpha)}}{\partial x_\ell}\right] \mathbf{dS} \tag{8.20}$$

$$\int_{V^{(\alpha)}} \frac{\partial}{\partial x_\ell}(\Pi_{ik}^{(\alpha)})\,\mathrm{d}V = \int_{\Sigma^{(\alpha)}} \Pi_{ik}^{(\alpha)}\,\mathbf{dS} \tag{8.21}$$

where $\Sigma^{(\alpha)}$ and $V^{(\alpha)}$ are the surface and the volume of the vapor ($\alpha = \mathrm{G}$) and liquid ($\alpha = \mathrm{L}$) media, respectively; $\mathbf{S} = \mathbf{n}S$, $\mathbf{n}$ is the external normal to that surface. The integrals $\int_{\Sigma^{(\mathrm{G})}} A^{(\mathrm{G})}\,\mathbf{dS}$ and $\int_{\Sigma^{(\mathrm{G})}} A^{(\mathrm{L})}\,\mathbf{dS}$, $A^{(\alpha)} = \rho^{(\alpha)}\mathbf{v}^{(\alpha)}, \rho^{(\alpha)}\mathbf{v}^{(\alpha)}h^{(\alpha)}, k^{(\alpha)}\nabla T_\ell^{(\alpha)}, \Pi_{ik}^{(\alpha)})$ are equal (see Fig. 8.1b):

$$\int_{\Sigma^{(\mathrm{L})}} \mathbf{A}^{(\mathrm{L})}\,\mathbf{dS} = \sum_{k=\mathrm{I}}^{\mathrm{VIII}} \int_{S_k^{(\mathrm{L})}} \mathbf{A}^{(\mathrm{L})}\,\mathbf{dS} \tag{8.22}$$

$$\int_{\Sigma^{(\mathrm{G})}} \mathbf{A}^{(\mathrm{G})}\,\mathbf{dS} = \sum_{k=I'}^{III'} \int_{S_k^{(\mathrm{G})}} \mathbf{A}^{(\mathrm{G})}\,\mathbf{dS} \tag{8.23}$$

where k is the number of restricted surface elements.

Summing integrals (8.22) and (8.23), one also accounts for:

1. Impenetrability of the capillary walls

$$\int_{S_{\mathrm{II}}} \rho^{(\mathrm{L})}\mathbf{v}^{(\mathrm{L})}\,\mathbf{dS} = \int_{S_{\mathrm{VII}}} \rho^{(\mathrm{L})}\mathbf{v}^{(\mathrm{L})}\,\mathbf{dS} = 0\,;$$

$$\int_{S_{\mathrm{II}}} \rho^{(\mathrm{L})}\mathbf{v}^{(\mathrm{L})}h^{(\mathrm{L})}\,\mathbf{dS} = \int_{S_{\mathrm{VII}}} \rho^{(\mathrm{L})}\mathbf{v}^{(\mathrm{L})}h^{(\mathrm{L})}\,\mathbf{dS} = 0$$

2. Adhesion of fluid to the wall

$$\int_{S_{\mathrm{II}}} \rho^{(\mathrm{L})}\mathbf{v}^{(\mathrm{L})}\,\mathbf{dS} = \int_{S_{\mathrm{VII}}} \rho^{(\mathrm{L})}\mathbf{v}^{(\mathrm{L})}\,\mathbf{dS} = 0$$

3. Equality

$$\int_{S_{\mathrm{II}}} P^{(\mathrm{L})}\,\mathbf{dS} + \int_{S_{\mathrm{VII}}} P^{(\mathrm{L})}\,\mathbf{dS} = 0$$

4. Opposed signs of the normal to the interface surface on the liquid side and on the vapor side ($\mathbf{n}_i^{(\mathrm{G})} = -\mathbf{n}_i^{(\mathrm{L})}$), as well as conditions (8.6–8.8) and correlation (8.10) we obtain the following equations (for the plane problem):

$$\sum_{\alpha=1}^{2} \int_{f^{(\alpha)}} \rho^{(\alpha)}\mathbf{v}^{(\alpha)}\,\mathbf{dS} = c_1 \tag{8.24}$$

$$\sum_{\alpha=1}^{2}\int_{f^{(\alpha)}}\left(\rho^{(\alpha)}\mathbf{v}^{(\alpha)}h^{(\alpha)}-k^{(\alpha)}\frac{\partial T^{(\alpha)}}{\partial x_\ell}\right)\mathrm{d}\mathbf{S}=Q_{\mathrm{ext}}+c_2 \tag{8.25}$$

$$\begin{aligned}&\left(\sum_{\alpha=1}^{2}\int_{f^{(\alpha)}}(P^{(\alpha)}\delta_{ik}+\rho^{(\alpha)}v_i^{(\alpha)}v_k^{(\alpha)}-\tau_{ik}^{(\alpha)}-X_k^{(\alpha)})\,\mathrm{d}\mathbf{S}\right)\\&+\int_{f_{\mathrm{int}}}\left[\sigma\psi\mathbf{n}_i^{(\mathrm{G})}+\frac{\partial\sigma}{\partial x_\ell}\right]\mathrm{d}\mathbf{S}=F_{\mathrm{ext}}+c_3\end{aligned} \tag{8.26}$$

where $Q_{\mathrm{ext}}=2\int\limits_{S_{\mathrm{II}}}k^{(\mathrm{L})}(\partial T^{(\mathrm{L})}/\partial x_\ell)\,\mathrm{d}\mathbf{S}$, $F_{\mathrm{ext}}=2\int\limits_{S_{\mathrm{II}}}\tau_{ik}^{(\alpha)}\,\mathrm{d}\mathbf{S}$, $S_{\mathrm{II}}=|k|\int\limits_{x_{\mathrm{in}}}^{x}\mathrm{d}x$, $f^{(\alpha)}$ is the cross-section of vapor ($\alpha=\mathrm{G}$) or liquid ($\alpha=\mathrm{L}$) flows, $f_{\mathrm{int}}=2|k|\int\limits_{x_{\mathrm{in}}}^{x}\left[1+(f^{(\mathrm{G})'})^2\right]^{1/2}\mathrm{d}x$ is the interface surface, constants c_1, c_2 and c_3 correspond to $2\int\limits_{S_{\mathrm{I}}}\mathbf{A}^{(\mathrm{L})}\,\mathrm{d}\mathbf{S}$, and x_{in} is the initial value of x.

In addition we write the balance equation for each phase:

$$\int_{f^{(\alpha)}}\rho^{(\alpha)}\mathbf{v}^{(\alpha)}\,\mathrm{d}\mathbf{S}+\int_{f_{\mathrm{int}}}\rho^{(\alpha)}\mathbf{v}^{(\alpha)}\,\mathrm{d}\mathbf{S}=c_4^{(\alpha)} \tag{8.27}$$

$$\begin{aligned}&\int_{f^{(\alpha)}}\left(\rho^{(\alpha)}\mathbf{v}^{(\alpha)}h^{(\alpha)}-k^{(\alpha)}\frac{\partial T^{(\alpha)}}{\partial x}\right)\mathrm{d}\mathbf{S}\\&+\int_{f_{\mathrm{int}}}\left(\rho^{(\alpha)}\mathbf{v}^{(\alpha)}h^{(\alpha)}-k^{(\alpha)}\frac{\partial T^{(\alpha)}}{\partial x_\ell}\right)\mathrm{d}\mathbf{S}=Q_{\mathrm{ext}}+c_5^{(\alpha)}\end{aligned} \tag{8.28}$$

$$\begin{aligned}&\int_{f^{(\alpha)}}(P^{(\alpha)}\delta_{ik}+\rho^{(\alpha)}v_i^{(\alpha)}v_k^{(\alpha)}-\tau_{ik}^{(\alpha)}-X_k^{(\alpha)})\,\mathrm{d}\mathbf{S}\\&+\int_{f_{\mathrm{int}}}(P^{(\alpha)}\delta_{ik}+\rho^{(\alpha)}v_i^{(\alpha)}v_k^{(\alpha)}-\tau_{ik}^{(\alpha)}-X_k^{(\alpha)})\,\mathrm{d}\mathbf{S}=F_{\mathrm{ext}}^{(\alpha)}+c_6^{(\alpha)}\,.\end{aligned} \tag{8.29}$$

The system of Eqs. (8.24–8.29) allows to determine the component of velocity u, v, temperature T and pressure P within the regions of liquid and vapor flows, as well as the coordinates of interface x_{b}, y_{b}. The geometrical correlation for the cross-section of the liquid and vapor flows

$$\sum_{\alpha=1}^{2}f^{(\alpha)}=f \tag{8.30}$$

is valid at any value of operating parameters (here f is the cross-section of capillary).

It should be noticed that Eq. (8.30) can be used together with any five equations of the system (8.24–8.29) for calculating flow in heated micro-channel.

8.6 Equations for the Average Parameters

Let us introduce average (over the flow cross-section) parameters as:

$$\left\langle J^{(\alpha)}\right\rangle = f^{(\alpha)^{-1}} \int\limits_{f^{(\alpha)}} J^{(\alpha)}\,\mathrm{d}\mathbf{S} \tag{8.31}$$

where J is the parameter under examination and $\langle\,\rangle$ is the operator indicating averaging over the cross-section. Then Eqs. (8.24–8.26), and (8.27–8.29) transform to the following form:

$$\sum_{\alpha=1}^{2} f^{(\alpha)}\left\langle \rho^{(\alpha)}\mathbf{v}^{(\alpha)}\right\rangle = c_1 \tag{8.32}$$

$$\sum_{\alpha=1}^{2} f^{(\alpha)}\left\langle \rho^{(\alpha)}\mathbf{v}^{(\alpha)}h^{(\alpha)} - k^{(\alpha)}\frac{\partial T^{(\alpha)}}{\partial x_\ell}\right\rangle = Q_{\text{ext}} + c_2 \tag{8.33}$$

$$\begin{aligned}&\left(\sum_{\alpha=1}^{2} f^{(\alpha)}\left\langle P^{(\alpha)}\delta_{ik} + \rho^{(\alpha)}v_i^{(\alpha)}v_k^{(\alpha)} - \tau_{xk}^{(\alpha)} - X_k^{(\alpha)}\right\rangle\right)\\ &+\int\limits_{f_{\text{int}}}\left[\sigma\psi n_i^{(\mathrm{G})} + \frac{\partial\sigma}{\partial x_\ell}\right]\mathrm{d}\mathbf{S} = F_{\text{ext}} + c_3\end{aligned} \tag{8.34}$$

$$f^{(\alpha)}\left\langle \rho^{(\alpha)}\mathbf{v}^{(\alpha)}\right\rangle + \int\limits_{f_{\text{int}}} \rho^{(\alpha)}\mathbf{v}^{(\alpha)}\,\mathrm{d}\mathbf{S} = c_4^{(\alpha)} \tag{8.35}$$

$$\begin{aligned}&f^{(\alpha)}\left\langle \rho^{(\alpha)}\mathbf{v}^{(\alpha)}h^{(\alpha)} - k^{(\alpha)}\frac{\partial T^{(\alpha)}}{\partial x_\ell}\right\rangle\\ &+\int\limits_{f_{\text{int}}}\left\langle \rho^{(\alpha)}\mathbf{v}^{(\alpha)}h^{(\alpha)} - k^{(\alpha)}\frac{\partial T^{(\alpha)}}{\partial x_\ell}\right\rangle \mathrm{d}\mathbf{S} = Q_{\text{ext}}^{(\alpha)} + c_5^{(\alpha)}\end{aligned} \tag{8.36}$$

$$\begin{aligned}&f^{(\alpha)}\left\langle P^{(\alpha)}\delta_{ik} + \rho^{(\alpha)}v_i^{(\alpha)}v_k^{(\alpha)} - \tau_{xk}^{(\alpha)} - X_x^{(\alpha)}\right\rangle\\ &+\int\limits_{f_{\text{int}}}\left\langle P^{(\alpha)}\delta_{ik} + \rho^{(\alpha)}v_i^{(\alpha)}v_k^{(\alpha)} - \tau_{xk}^{(\alpha)} - X_x^{(\alpha)}\right\rangle \mathrm{d}\mathbf{S} = F_{\text{ext}}^{(\alpha)} + c_6^{(\alpha)}\,.\end{aligned} \tag{8.37}$$

These equations, supplemented by the expression for the liquid density and vapor pressure, may be integrated into the general case only numerically. However, for some important particular cases, reasonable approximations can be introduced which simplify the system of equations for the average parameters to a form that can be integrated analytically. This approach, developed below, yields expressions for a set of first-order integral equations of the average parameters.

8.7 Quasi-One-Dimensional Approach

Significant simplification of the governing equations may be achieved by using a quasi-one-dimensional model for the flow. Assume that: (1) the ratio of meniscus depth to its radius is sufficiently small, (2) the velocity, temperature and pressure distributions in the cross-section are close to uniform, and (3) all parameters depend on the longitudinal coordinate. Differentiating Eqs. (8.32–8.35) and (8.37) we reduce the problem to the following dimensionless equations:

$$\frac{\mathrm{d}}{\mathrm{d}\overline{x}}\left(\sum_{\alpha=1}^{2}\overline{\rho}^{(\alpha)}\overline{u}^{(\alpha)}\overline{f}^{(\alpha)}\right)=0 \tag{8.38}$$

$$\frac{\mathrm{d}}{\mathrm{d}\overline{x}}\left(\sum_{\alpha=1}^{2}\overline{\rho}^{(\alpha)}\overline{u}^{(\alpha)}\overline{h}^{(\alpha)}\overline{f}^{(\alpha)}\right)=\vartheta+\mathrm{Pe}^{-1}\sum_{\alpha=1}^{2}\overline{k}^{(\alpha)}\frac{\mathrm{d}}{\mathrm{d}\overline{x}}\left(\overline{f}^{(\alpha)}\frac{\mathrm{d}\overline{T}^{(\alpha)}}{\mathrm{d}\overline{x}}\right) \tag{8.39}$$

$$\begin{aligned}&\mathrm{Eu}\frac{\mathrm{d}}{\mathrm{d}\overline{x}}\left(\sum_{\alpha=1}^{2}\overline{f}^{(\alpha)}\overline{P}^{(\alpha)}\delta_{ik}\right)+\frac{\mathrm{d}}{\mathrm{d}\overline{x}}\left(\sum_{\alpha=1}^{2}f^{(\alpha)}\overline{\rho}^{(\alpha)}\overline{u}^{(\alpha)2}\right)\\&=\frac{\mathrm{d}\overline{F}_{\mathrm{ext}}}{\mathrm{d}\overline{x}}+\mathrm{Re}^{-1}\frac{d}{d\overline{x}}\left(\sum_{\alpha=1}^{2}\overline{\mu}^{(\alpha)}\overline{\tau}_{ik}^{(\alpha)}\overline{f}^{(\alpha)}\right)-\mathrm{We}^{-1}\frac{\mathrm{d}}{\mathrm{d}\overline{x}}\left(\int_{\overline{f}_{\mathrm{int}}}\overline{\sigma}\overline{\psi}n_i^{(1)}\,\mathrm{d}\overline{S}\right)\\&-\mathrm{We}^{-1}\frac{\mathrm{d}}{\mathrm{d}\overline{x}}\left(\int_{\overline{f}_{\mathrm{int}}}\frac{\partial\overline{\sigma}}{\partial\overline{x}_\ell}\,\mathrm{d}\overline{S}\right)+\mathrm{Fr}^{-1}\frac{\mathrm{d}}{\mathrm{d}\overline{x}}\left(\sum_{\alpha=1}^{2}\overline{f}^{(\alpha)}\overline{X}_k^{(\alpha)}\right)\end{aligned} \tag{8.40}$$

$$\frac{\mathrm{d}}{\mathrm{d}\overline{x}}\left(\overline{\rho}^{(\alpha)}\overline{u}^{(\alpha)}\overline{f}^{(\alpha)}\right)=\frac{\mathrm{d}}{\mathrm{d}\overline{x}}\overline{G}_{\mathrm{int}}^{(\alpha)} \tag{8.41}$$

$$\begin{aligned}&\mathrm{Eu}\frac{\mathrm{d}}{\mathrm{d}\overline{x}}\left(\overline{f}^{(\alpha)}\overline{P}^{(\alpha)}\delta_{ik}\right)+\frac{\mathrm{d}}{\mathrm{d}\overline{x}}\left(\overline{f}^{(\alpha)}\overline{\rho}^{(\alpha)}\overline{u}^{(\alpha)2}\right)\\&=\frac{\mathrm{d}\overline{F}_{\mathrm{ext}}^{(\alpha)}}{\mathrm{d}\overline{x}}+\mathrm{Re}^{-1}\frac{\mathrm{d}}{\mathrm{d}\overline{x}}\left(\overline{\mu}^{(\alpha)}\overline{\tau}_{ik}^{(\alpha)}\overline{f}^{(\alpha)}\right)-\frac{\mathrm{d}\overline{F}_{\mathrm{int}}^{(\alpha)}}{\mathrm{d}\overline{x}}+\mathrm{Fr}^{-1}\frac{\mathrm{d}}{\mathrm{d}\overline{x}}\left(\overline{f}^{(\alpha)}\overline{X}^{(\alpha)}\right)\end{aligned} \tag{8.42}$$

$$\sum_{\alpha=1}^{2}\overline{f}^{(\alpha)}=1 \tag{8.43}$$

where the characteristic scales pressure P_{L0}, density ρ_{20}, velocity u_{20}, temperature T_{L0}, heat capacity c_{p20}, viscosity μ_{20}, thermal conductivity k_{20}, surface tension σ_{L0} (σ_{L0} corresponds to T_{L0}), area f, and length d (d is half the capillary width) are used to define the following dimensionless parameters: $\overline{P}^{(\alpha)}=P^{(\alpha)}/P_{\mathrm{L0}}$; $\overline{\rho}^{(\alpha)}=\rho^{(\alpha)}/\rho_{\mathrm{L0}}$; $\overline{u}^{(\alpha)}=u^{(\alpha)}/u_{\mathrm{L0}}$; $\overline{T}^{(\alpha)}=T^{(\alpha)}/T_{\mathrm{L0}}$; $\overline{c}_{\mathrm{p}}^{(\alpha)}=c_{\mathrm{p}}^{(\alpha)}/c_{\mathrm{pL0}}$; $\overline{\mu}^{(\alpha)}=\mu^{(\alpha)}/\mu_{\mathrm{L0}}$; $\overline{k}^{(\alpha)}=k^{(\alpha)}/k_{\mathrm{L0}}$; $\overline{\sigma}^{(\alpha)}=\sigma^{(\alpha)}/\sigma_{\mathrm{L0}}$; $\overline{f}^{(\alpha)}=f^{(\alpha)}/f$; $\overline{S}=S/f$; $\overline{x}=x/d$; $\overline{F}=F/\rho_{\mathrm{L0}}u_{\mathrm{L0}}^2f$; $\overline{G}=G/\rho_{\mathrm{L0}}u_{\mathrm{L0}}f$; $\overline{\psi}=\psi d$; $\overline{Q}_{\mathrm{ext}}=Q_{\mathrm{ext}}/\rho_{\mathrm{L0}}u_{\mathrm{L0}}c_{\mathrm{p_{L0}}}h_{\mathrm{L0}}$; $\overline{X}=X/\rho_{\mathrm{L0}}u_{\mathrm{L0}}^2$.

The terms $\mathrm{Re}=u_{\mathrm{L0}}d/\nu_{\mathrm{L0}}$; $\mathrm{Eu}=P_{\mathrm{L0}}/\rho_{\mathrm{L0}}u_{\mathrm{L0}}^2$; $\mathrm{Fr}=u_{\mathrm{L0}}^2/gd$; $\mathrm{Pe}=u_{\mathrm{L0}}d\rho_{\mathrm{L0}}\cdot c_{\mathrm{pL0}}/k_{\mathrm{L0}}$; $\mathrm{We}=d\rho_{20}u_{20}^2/\sigma_{20}$ are the Reynolds, the Euler, the Froude, the Weber and the Peclet numbers; $\vartheta = q/\rho_{\mathrm{L0}}u_{\mathrm{L0}}c_{\mathrm{pL0}}T_{\mathrm{L0}}$, $Q_{\mathrm{ext}} = 2\int_{S_{\mathrm{II}}} q\,\mathrm{d}\mathbf{S}$, q is the heat flux on the wall, $F_{\mathrm{ext}} = \int_{S_{\mathrm{II}}} t\,\mathrm{d}S$, $t = \xi\rho^{(\alpha)}\left(u^{(\alpha)}\right)^2$ is the friction stress, ξ is the drag coefficient, $G_{\mathrm{int}} = \int_{f_{\mathrm{int}}} \rho^{(\alpha)}\mathbf{v}^{(\alpha)}\,\mathrm{d}\mathbf{S}$, $F_{\mathrm{int}} = \int_{f_{\mathrm{int}}} \left(P^{(\alpha)}\delta_{\mathrm{ik}} + \rho^{(\alpha)}v_i^{(\alpha)}v_k^{(\alpha)} - \tau_{ik}^{(\alpha)} - X_k^{(\alpha)}\right)\mathrm{d}\mathbf{S}$, and subscript 0 refers to the initial state.

From the frame of the quasi-one-dimensional model it is possible to determine the hydrodynamic and thermal characteristics of the flow in a heated capillary, accounting for the influence of the capillary force.

The dimensionless forms of Eqs. (8.5), (8.14), (8.15) and (8.43) are

$$\overline{\rho}^{(\mathrm{L})} = \overline{\rho}^{(\mathrm{L})}(\overline{T}^{(\mathrm{L})}) \tag{8.44}$$

$$\mathrm{Eu}\,\overline{P}^{(\mathrm{G})} = \gamma\overline{\rho}^{(\mathrm{G})}\overline{T}^{(\mathrm{G})} \tag{8.45}$$

$$\mathrm{Eu}\,\overline{P}^{(\mathrm{G})} = \Omega\exp(-\omega\overline{T}^{(\mathrm{G})}) \tag{8.46}$$

$$\sum_{\alpha=1}^{2}\overline{f}^{(\alpha)} = 1 \tag{8.47}$$

where $\gamma = RT_{\mathrm{L0}}/u_{\mathrm{L0}}^2$; $\omega = h_{\mathrm{LG}}/RT_{\mathrm{L0}}$; $\Omega = \tilde{P}/\rho_{\mathrm{L0}}u_{\mathrm{L0}}^2$.

For typical conditions of the capillarity flow, the non-dimensional groups have the following orders: $\mathrm{Re} \sim 1$, $\mathrm{Pe} \sim 10$, $\mathrm{Eu} \sim 10^7$, $\mathrm{Fr} \sim 10^{-3}$, $\mathrm{We} \sim 10^{-4}$, $\vartheta \sim 1$.[1] So $\overline{\lambda}^{(\mathrm{G})} \sim 10^{-1}$, $\overline{k}^{(\mathrm{L})} = 1$, $\overline{\mu}^{(\mathrm{G})} \sim 10^{-2}$, $\overline{\mu}^{(\mathrm{L})} = 1$, $\overline{u}^{(\mathrm{G})} \sim 10^3$, $\overline{P}^{(\alpha)} \sim 1$, $\overline{T} \sim 1$, $\overline{f}^{(\alpha)} \sim 1$, $\overline{x} \sim 1$, and the order of magnitude of the derivatives in Eqs. (8.39–8.42) is 1.[2] Accordingly, it is possible to omit the first term on the left-hand side of Eqs. (8.40) and (8.42) and the second term on the right-hand side. At moderate and large heat fluxes on the wall ($q > 10\,\mathrm{W/m^2}$), the first terms on the left-hand and right-hand side of Eq. (8.39) are of the same order. In this case it is possible to omit the second term on the right-hand side of Eq. (8.39).

8.8 Parameters Distribution in Characteristic Zones

Consider now the case where the axial heat transfer due to the temperature gradient is negligible compared to the heat transfer from the capillary wall and the friction caused by the velocity gradient in the x-direction is negligible compared to the momentum losses at the fluid–wall interface.

Liquid heating region: $0 < \overline{x} < \overline{x}_*$, $(\overline{f}^{(G)} = 0, \overline{f}^{(L)} = 1)$

Equations (8.38–8.40) contain the following integrals:

---

[1] For water flow in a heated capillary with $d = 5 \times 10^{-4}$ m at $u = 10^{-3}$–10^{-2} m/s and $q = 10^6\ \mathrm{Wm^2}$.

[2] Within the heating and superheat regions $\overline{x} > 1$ derivatives in Eqs. (8.39–8.42) are $O(1)$.

$$\overline{\rho}^{(L)}\overline{u}^{(L)} = c_1 \quad (8.48)$$

$$\overline{\rho}^{(L)}\overline{u}^{(L)}h^{(L)} = \vartheta\overline{x} + c_2 \quad (8.49)$$

$$\overline{\rho}^{(L)}\overline{u}^{(L)^2} + \mathrm{Eu}\overline{P}^{(L)} = \overline{F}_{\mathrm{ext}} + \mathrm{Fr}^{-1}\overline{\rho}^{(L)}\overline{x} + c_3 \,. \quad (8.50)$$

The constants c_1, c_2 and c_3 are found from conditions[3]

$$\overline{x} = 0\,, \quad \overline{\rho}^{(L)} = 1\,, \quad \overline{u}^{(L)} = 1\,, \quad h^{(L)} = 1\,. \quad (8.51)$$

To determine $\overline{F}_{\mathrm{ext}}$ in Eq. (8.50) we use the relation $F_{\mathrm{ext}} = \int\limits_{S_{\mathrm{II}}} t\,d\mathbf{S}$ and $S_{\mathrm{II}} = |k| \int\limits_{x_{\mathrm{in}}}^{x} dx$. Assuming $\zeta = 32/\mathrm{Re}$ we obtain

$$\overline{F}_{\mathrm{ext}} = 32/\mathrm{Re}\cdot\overline{x}\,. \quad (8.52)$$

The system of Eqs. (8.48–8.50) may be written as

$$\overline{\rho}^{(L)}\overline{u}^{(L)} = 1 \quad (8.53)$$

$$\overline{T}^{(L)} = 1 + \vartheta\overline{x} \quad (8.54)$$

$$(\overline{u}^{(L)} - 1) + \mathrm{Eu}(\overline{P}^{(L)} - 1) = -\overline{x}(32/\mathrm{Re} + \overline{\rho}^{(L)}/\mathrm{Fr})\,. \quad (8.55)$$

The heating region length is determined from the conditions

$$\overline{x} = \overline{x}_* \quad \overline{P} = \overline{P}_s \quad \overline{T} = \overline{T}_s\,. \quad (8.56)$$

Since the dependence $\rho^{(L)} = f(T^{(L)})$ is very weak, it is possible to neglect the liquid density $\rho^{(L)}$ variation within the heating region. Then Eqs. (8.54) and (8.55) may be transformed to:

$$\overline{P}^{(L)} = 1 - (\overline{T}^{(L)} - 1)\Lambda \quad (8.57)$$

where $\Lambda = \frac{32/\mathrm{Re} + \overline{\rho}^{(L)}/\mathrm{Fr}}{\vartheta \mathrm{Eu}}$.

The dependences (8.56) and (8.57) determine the parameters at the end of the heating region. The Λ parameter corresponding to micro-cooling system conditions is on the order of 10^{-2}. In this case P^* and T^* practically do not depend on Λ. Accordingly, the heating region length depends on the non-dimensional group ϑ only.

$$\overline{x}_* = (\overline{T}_* - 1)/\vartheta \quad (8.58)$$

Evaporation region: $\overline{x}_* \leq \overline{x} \leq \overline{x}_{**}$

Equations (8.38–8.42) have the following integrals

$$\sum_{\alpha=1}^{2}\left(\overline{\rho}^{(\alpha)}\overline{u}^{(\alpha)}\overline{f}^{(\alpha)}\right) = c_4 \quad (8.59)$$

[3] The constants $c_1 = 1$, $c_2 = 1$, $c_3 = 1 + \mathrm{Eu}$.

$$\sum_{\alpha=1}^{2}\left(\overline{\rho}^{(\alpha)}\overline{u}^{(\alpha)}\overline{f}^{(\alpha)}\overline{h}^{(\alpha)}\right)=\vartheta\overline{x}+c_5 \tag{8.60}$$

$$\sum_{\alpha=1}^{2}\left(\overline{\rho}^{(\alpha)}\ \overline{u}^{(\alpha)^2}\overline{f}^{(\alpha)}\right)+\mathrm{Eu}\sum_{\alpha=1}^{2}\left(\overline{P}^{(\alpha)}\ \overline{f}^{(\alpha)}\right)=\overline{F}_{\mathrm{ext}}-\mathrm{We}^{-1}\int_{\overline{f}_{\mathrm{int}}}\overline{\sigma\psi n_i}\mathrm{dS}$$
$$-\mathrm{We}^{-1}\int_{\overline{f}_{\mathrm{int}}}\frac{\partial\overline{\sigma}}{\partial\overline{x}_\ell}\mathrm{dS}+\mathrm{Fr}^{-1}\int_{\overline{x}_*}^{\overline{x}}\left(\sum_{\alpha=1}^{2}\overline{\rho}^{(\alpha)}\overline{f}^{(\alpha)}\right)\mathrm{d}\overline{x}+c_6 \tag{8.61}$$

$$\overline{\rho}^{(\mathrm{G})}\overline{u}^{(\mathrm{G})}\overline{f}^{(\mathrm{G})}=-\int_{\overline{x}*}^{\overline{x}}\overline{\rho}^{(\mathrm{G})}\overline{u}^{(\mathrm{G})}\sqrt{1+\left(f^{(\mathrm{G})'}\right)^2}\,\mathrm{d}\overline{x}+c_7 \tag{8.62}$$

$$\overline{\rho}^{(\mathrm{G})}\overline{u}^{(\mathrm{G})}\overline{f}^{(\mathrm{G})}+\mathrm{Eu}\overline{P}^{(\mathrm{G})}\overline{f}^{(\mathrm{G})}$$
$$=-\int_{\overline{x}*}^{\overline{x}}\left(\mathrm{Eu}\overline{P}^{(\mathrm{G})}+\overline{\rho}^{(\mathrm{G})}\overline{u}^{(\mathrm{G})^2}\right)\sqrt{1+\left(f^{(\mathrm{G})'}\right)^2}\,\mathrm{d}\overline{x}+\mathrm{Fr}\int_{\overline{x}*}^{\overline{x}}\overline{\rho}^{(\mathrm{G})}\overline{f}^{(\mathrm{G})}\,\mathrm{d}\overline{x}+c_8\,. \tag{8.63}$$

The constants c_4, c_5, c_6, c_7 and c_8 are found from the conditions[4]

$$\overline{x}=\overline{x}_*,\ \overline{f}^{(\mathrm{G})}=0,\ \overline{f}^{(\mathrm{L})}=1,\ \overline{\rho}^{(\alpha)}=\overline{\rho}_*^{(\alpha)},\ \overline{u}^{(\alpha)}=\overline{u}_*^{(\alpha)},\ \overline{P}^{(\alpha)}=\overline{P}_*^{(\alpha)},\ \overline{h}^{(\alpha)}=\overline{h}_*^{(\alpha)}\,.$$

Equations (8.59) and (8.63) may be written as

$$\overline{\rho}^{(\mathrm{G})}\overline{u}^{(\mathrm{G})}\overline{f}^{(\mathrm{G})}+\overline{\rho}^{(\mathrm{L})}\overline{u}^{(\mathrm{L})}\overline{f}^{(\mathrm{L})}=1 \tag{8.64}$$

$$\overline{\rho}^{(\mathrm{G})}\overline{u}^{(\mathrm{G})}\overline{f}^{(\mathrm{G})}\overline{h}^{(\mathrm{G})}+\overline{\rho}^{(\mathrm{L})}\overline{u}^{(\mathrm{L})}\overline{f}^{(\mathrm{L})}\overline{h}^{(\mathrm{L})}=\overline{h}_*^{(\mathrm{G})}+\vartheta(\overline{x}-\overline{x}_*) \tag{8.65}$$

$$\left(\overline{\rho}^{(\mathrm{G})}\overline{u}^{(G)^2}\overline{f}^{(\mathrm{G})}+\overline{\rho}^{(\mathrm{L})}\overline{u}^{(L)^2}\overline{f}^{(\mathrm{L})}-\overline{\rho}_*^{(\mathrm{L})}\overline{u}_*^{(L)^2}\right)$$
$$+\mathrm{Eu}\left(\overline{P}^{(\mathrm{G})}\overline{f}^{(\mathrm{G})}+\overline{P}^{(\mathrm{L})}\overline{f}^{(\mathrm{L})}-\overline{P}_*^{(\mathrm{L})}\right)$$
$$=\mathrm{We}^{(-1)}\int_{f_{\mathrm{int}}}\overline{\sigma\psi n_i}\,\mathrm{d}\overline{S}-\mathrm{We}^{(-1)}\int_{f_{\mathrm{int}}}\frac{\partial\overline{\sigma}}{\partial\overline{x}_\ell}\,\mathrm{d}\overline{S} \tag{8.66}$$
$$-(32/\mathrm{Re})(\overline{x}-\overline{x}_*)-\mathrm{Fr}^{-1}\int_{\overline{x}_*}^{\overline{x}}\left(\textstyle\sum_{\alpha=1}^{2}\overline{\rho}^{(\alpha)}\overline{f}^{(\alpha)}\right)\mathrm{d}\overline{x}$$

$$\overline{\rho}^{(\mathrm{G})}\overline{u}^{(\mathrm{G})}\overline{f}^{(\mathrm{G})}=\int_{\overline{x}_*}^{\overline{x}}\overline{\rho}^{(\mathrm{G})}\overline{u}^{(\mathrm{G})}\sqrt{1+(f^{(\mathrm{G})'})^2}\,\mathrm{d}\overline{x} \tag{8.67}$$

$$\overline{\rho}^{(\mathrm{G})}\overline{u}^{(G)^2}\overline{f}^{(\mathrm{G})}+\mathrm{Eu}\overline{P}^{(\mathrm{G})}\overline{f}^{(\mathrm{G})}$$
$$=\int_{\overline{x}_*}^{\overline{x}}\left(\mathrm{Eu}\overline{P}^{(\mathrm{G})}+\overline{\rho}^{(\mathrm{G})}\overline{u}^{(G)^2}\right)\sqrt{1+\left(f^{(\mathrm{G})'}\right)^2}\,\mathrm{d}\overline{x}-\mathrm{Fr}^{-1}\int_{\overline{x}_*}^{\overline{x}}\overline{\rho}^{(\mathrm{G})}\overline{f}^{(\mathrm{G})}\,\mathrm{d}\overline{x}\,. \tag{8.68}$$

---

[4] The constants $c_4=1$, $c_5=h_*^{(\mathrm{G})}-\vartheta\overline{x}_*$, $c_6=\mathrm{Eu}\overline{P}_*^{(\mathrm{L})}+\overline{\rho}_*^{(\mathrm{L})}u_*^{(L)^2}-(32/\mathrm{Re})\overline{x}_*$, $c_7=0$, $c_8=0$

The left-hand side of Eq. (8.65) can be presented as:

$$\begin{aligned}&\overline{\rho}^{(G)}\overline{u}^{(G)}\overline{f}^{(G)}\overline{h}^{(G)}-\overline{\rho}^{(G)}\overline{u}^{(G)}\overline{f}^{(G)}\overline{h}^{(L)}+\overline{\rho}^{(G)}\overline{u}^{(G)}\overline{f}^{(G)}\overline{h}^{(L)}+\overline{\rho}^{(L)}\overline{u}^{(L)}\overline{f}^{(L)}\overline{h}^{(L)}\\&=\overline{\rho}^{(G)}\overline{u}^{(G)}\overline{f}^{(G)}\left(\overline{h}^{(G)}-\overline{h}^{(L)}\right)+\overline{h}^{(L)}\left(\overline{\rho}^{(G)}\overline{u}^{(G)}\overline{f}^{(G)}+\overline{\rho}^{(L)}\overline{u}^{(L)}\overline{f}^{(2)}\right)\\&=\overline{\rho}^{(G)}\overline{u}^{(G)}\overline{f}^{(G)}\left(\overline{h}^{(G)}-\overline{h}^{(L)}\right)+\overline{h}^{(L)}\,.\end{aligned}$$

Then Eq. (8.65) has the form

$$\overline{\rho}^{(G)}\overline{u}^{(G)}\overline{f}^{(G)}\left(\overline{h}^{(G)}-\overline{h}^{(L)}\right)+\overline{h}^{(L)}=\overline{h}_*^{(L)}+\vartheta(\overline{x}-\overline{x}_*)\,. \tag{8.69}$$

If $\overline{h}^{(\alpha)}=\overline{c}_{\mathrm{p}}^{(\alpha)}\overline{T}$ ($\overline{c}_{\mathrm{p}}^{(L)}=1$) then[5]

$$\overline{T}=\left(1+h_{\mathrm{LG}}\overline{\rho}^{(G)}\overline{u}^{(G)}\overline{f}^{(G)}\right)^{-1}\left[\overline{T}_*+\vartheta(\overline{x}-\overline{x}_*)\right] \tag{8.70}$$

where $\overline{h}_{\mathrm{LG}}=h_{\mathrm{LG}}/c_{\mathrm{p}_{20}}T_{20}$.

The location of $\overline{x}_{**}$ can be determined from the conditions[6]

$$\overline{x}=\overline{x}_{**},\quad \overline{f}^{(G)'}=\tan\theta,\quad \overline{f}^{(G)}=1 \tag{8.71}$$

where θ is the contact angle.

At a given Weber number, condition (8.71) may be satisfied at some values of the Euler number, which play the role of eigenvalues.

The second integral on the right-hand side of Eq. (8.66) may be presented as

$$\int\limits_{\overline{f}_{\mathrm{int}}}\frac{\partial\overline{\sigma}}{\partial\overline{x}_\ell}\,\mathrm{d}\overline{S}=\int\limits_{\overline{f}_{\mathrm{int}}}\frac{\partial\overline{\sigma}}{\partial\overline{T}}\cdot\frac{\partial\overline{T}}{\partial\overline{x}_\ell}\,\mathrm{d}\overline{S} \tag{8.72}$$

where $\left(\frac{\partial\overline{\sigma}}{\partial\overline{T}}\right)<0$ for most known liquids.

In the partial case when $\overline{\sigma}$ is constant the first integral on the right-hand side of Eq. (8.66) may be expressed as

$$\int\limits_{\overline{f}_{\mathrm{int}}}\overline{\sigma}\psi n_i\,\mathrm{d}\overline{S}=\left(\mathrm{arctg}\overline{f}^{(G)'}-\pi/2\right)\,. \tag{8.73}$$

Superheat region: ($\overline{f}^{(G)}=1$, $\overline{f}^{(L)}=0$)

Equations (8.38–8.40) have the following integrals

$$\overline{\rho}^{(G)}\overline{u}^{(G)}=c_9 \tag{8.74}$$

$$\overline{\rho}^{(G)}\overline{u}^{(G)}\overline{h}^{(G)}=\overline{\vartheta}\overline{x}+c_{10} \tag{8.75}$$

---

[5] Equation (8.70) may be presented in the following form $\overline{\rho}^{(G)}\overline{u}^{(G)}\overline{f}^{(G)}=(1+\vartheta\overline{x}-\overline{T}^{(L)})/\overline{h}_{\mathrm{LG}}$.

[6] The solution of Eq. (8.66) must satisfy the following condition: $\overline{x}=\overline{x}_*$ $\overline{f}^{(1)'}=\infty$ ($\overline{f}^{(G)}=0$), $\overline{x}=\overline{x}_{**}$ $\overline{f}^{(G)'}=\tan\theta$ ($\overline{f}^{(G)}=1$).

$$\overline{\rho}^{(\mathrm{G})}\overline{u}^{(G)^2}+\mathrm{Eu}\overline{P}^{(\mathrm{G})}=\overline{F}_{\mathrm{ext}}+\mathrm{Fr}^{-1}\overline{\rho}^{(\mathrm{G})}\overline{x}+c_{11} \tag{8.76}$$

where the constants c_9, c_{10} and c_{11} are found from the conditions $\overline{x}=\overline{x}_{**}$, $\overline{\rho}^{(\mathrm{G})}=\overline{\rho}^{(\mathrm{G})}_{**}$, $\overline{u}^{(\mathrm{G})}=\overline{u}^{(\mathrm{G})}_{**}$, $\overline{P}^{(\mathrm{G})}=\overline{P}^{(\mathrm{G})}_{**}$, $\overline{T}^{(\mathrm{G})}=\overline{T}^{(\mathrm{G})}_{**}$, $(c_9=1,\ c_{10}=\overline{c}_{\mathrm{p}}^{(\mathrm{G})}\overline{T}^{(\mathrm{G})}_{**}-\vartheta\overline{x}_{**}$, $c_{11}=\overline{u}^{(\mathrm{G})}_{**}+\mathrm{Eu}\overline{P}^{(\mathrm{G})}_{**}-32\overline{x}_{**}/\mathrm{Re}-\mathrm{Fr}^{-1}\overline{\rho}^{(\mathrm{G})}_{**}\overline{x}_{**})$.

Then,

$$\overline{\rho}^{(\mathrm{G})}\overline{u}^{(\mathrm{G})}=1 \tag{8.77}$$

$$\overline{T}^{(\mathrm{G})}=\overline{T}^{(\mathrm{G})}_{**}+\vartheta/\overline{c}_{\mathrm{p}}^{(\mathrm{G})}(\overline{x}-\overline{x}_{**}) \tag{8.78}$$

$$\left(\overline{u}^{(\mathrm{G})}-\overline{u}^{(\mathrm{G})}_{**}\right)+\mathrm{Eu}\left(\overline{P}^{(\mathrm{G})}-\overline{P}^{(\mathrm{G})}_{**}\right)=-(32/\mathrm{Re})\left(\overline{x}-\overline{x}_{**}\right)-\mathrm{Fr}^{-1}\left(\overline{\rho}^{(\mathrm{G})}\overline{x}-\overline{\rho}^{(\mathrm{G})}_{**}\overline{x}_{**}\right). \tag{8.79}$$

The systems of Eqs. (8.56–8.58), (8.64–8.66), and (8.77–8.79) allow us to find the density, velocity, temperature and pressure distributions along the capillary axis, as well as the interface surface shape.

8.9 Parametrical Study

8.9.1 Thermohydrodynamic Characteristics of Flow

The numerical solution of Eqs. (8.38–8.43) was carried out for laminar flow, for a set of non-dimensional groups varying within the limits: $5\times10^8\leq\mathrm{Eu}\leq10^{12}$; $1<\vartheta<10^3$; $5\times10^{-10}\leq\mathrm{We}\leq10^{-6}$; $5\times10^{-3}\leq\mathrm{Re}\leq10$; $10^{-8}\leq\mathrm{We}\leq10^{-2}$. The non-dimensional groups were chosen based on water flow in a 500 μm slot, heat flux variation of 1 $(\mathrm{W/cm^2})$ to 100 $(\mathrm{W/cm^2})$ and $10^{-5}(\mathrm{m/s})\leq u_{\mathrm{L0}}\leq2\times10^{-1}(\mathrm{m/s})$ velocity variation.

The calculations show that the liquid pressure monotonically decreases along the heating region. Within the evaporation region a noticeable difference between the vapor and liquid pressures takes place. The latter is connected with the effect of the Laplace force due to the curvature of the interface surface. In the superheated region the vapor pressure decreases downstream.

Figure 8.2a,b shows the character of the liquid and the vapor pressure distribution along the evaporation region. It is found that for the above-mentioned parameters, the vapor pressure is practically independent from x. Accordingly, the vapor temperature, as well as the density, are also approximately constant. The latter makes it possible to reduce the number of equations by three. The remaining five equations consist of four equations that contain only four unknowns $(\overline{u}^{(\mathrm{G})},\overline{u}^{(\mathrm{L})},\overline{f}^{(\mathrm{G})},\overline{f}^{(\mathrm{L})})$ and one equation (the combined vapor–liquid momentum balance), which contains the additional unknown $\overline{P}^{(\mathrm{L})}$. That means that the system of governing equations may be solved successively: at first to obtain $\overline{u}^{(\mathrm{G})}$, $\overline{u}^{(\mathrm{L})}$, $\overline{f}^{(\mathrm{G})}$, $\overline{f}^{(\mathrm{L})}$ and then to obtain

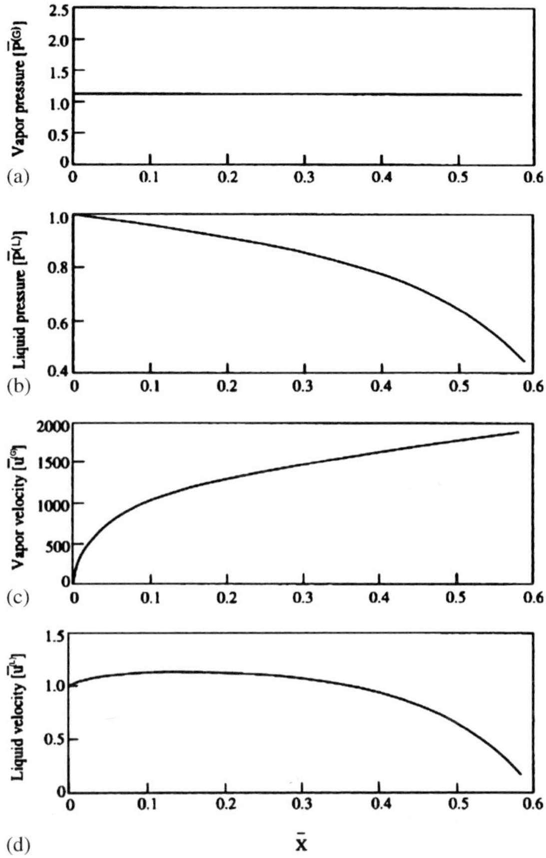

Fig. 8.2a–d The vapor and liquid pressure and velocity distribution within the evaporation region for We $= 10^{-9}$, $\vartheta = 2.5$, Eu $= 1.6 \times 10^{11}$, Re $= 0.04$, Fr $= 1.3 \times 10^{-7}$. Reprinted from Peles et al. (2000) with permission

$\overline{P}^{(L)}$ by solving the equation for the total momentum. Note that $\overline{u}^{(G)}$, $\overline{u}^{(L)}$, $\overline{f}^{(G)}$, $\overline{f}^{(L)}$ depend on ϑ and $\overline{P}^{(L)}$, which is a function of the other non-dimensional groups: $\overline{P}^{(G)} = f(\mathrm{Eu}, \mathrm{Re}, \mathrm{Fr}, \mathrm{We})$. The liquid pressure decreases along the evaporation region and a sharp drop takes place near the top of the evaporation region.

The liquid and vapor velocity versus axial position for the evaporation region are shown in Fig. 8.2c,d. It can be seen that the vapor velocity increases as $\overline{x}^{0.4}$, whereas the liquid velocity is almost constant along a wide range of $\overline{x}$ and decreases very sharply towards the top of the evaporation region.

The effect of various parameters on the difference between vapor and liquid pressure is illustrated in Figs. 8.3 and 8.4. The effect of the Euler and Weber numbers as well as the thermal parameter ϑ is highly noticeable. An increase in Eu, We and ϑ leads to a decrease in $\Delta\overline{P}$, whereas the difference of both phase pressures is practically independent of Reynolds number. An increase in the Froude number is accompanied by an increase in $\Delta\overline{P}$ for a small Fr. At Fr $> 10^{-6}$ the effect of Fr on $\Delta\overline{P}$ is negligible.

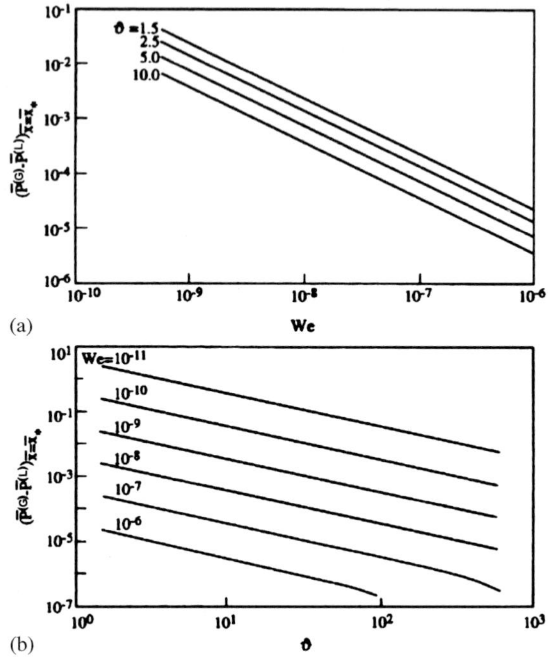

Fig. 8.3a,b The dependence $\Delta\overline{P}(\mathrm{We})$ and $\Delta\overline{P}(\vartheta)$ for different ϑ at $\overline{x}_*$ for $\mathrm{Eu} = 1.6 \times 10^{11}$, $\mathrm{Re} = 0.04$, $\mathrm{Fr} = 10^{-6}$. Reprinted from Peles et al. (2000) with permission

The $\vartheta(\theta)$ correlation corresponding to various Weber numbers is shown in Fig. 8.5. The shape of the interface surface in a capillary flow with phase change is presented in Fig. 8.6. As the calculations show, the curvature of the meniscus is not constant and grows toward the periphery.

8.9.2 The Effect of Regulated Parameters

The regulated parameters of the problem are: the width and length of micro-channel $\overline{d} = dh_{\mathrm{LG}}\rho^{(\mathrm{L})}/\sigma$ and $\overline{L} = Lh_{\mathrm{LG}}\rho^{(\mathrm{L})}/\sigma$, initial temperature of the liquid $\overline{T}_{\mathrm{inl}}^{(\mathrm{L})} = T_{\mathrm{inl}}^{(\mathrm{L})}c_{\mathrm{p}}^{(\mathrm{L})}/h_{\mathrm{LG}}$, gravity acceleration $\overline{g} = g\sigma^3/\nu^{(\mathrm{L})^2}\rho^{(\mathrm{L})^3}h_{\mathrm{LG}}^3$, and heat flux on the wall $\overline{q} = q\sigma/\rho^{(L)^2}h_{\mathrm{LG}}^2\nu^{(\mathrm{L})}$. The effect of these parameters on the flow characteristics, the liquid and vapor velocities $\overline{u}^{(i)} = u^{(i)}\sigma/\nu^{(\mathrm{L})}h_{\mathrm{LG}}\rho^{(\mathrm{L})}$, temperatures $\overline{T}^{(i)} = T^{(i)}c_{\mathrm{p}}^{(\mathrm{L})}/h_{\mathrm{LG}}$, pressures $\overline{P}^{(i)} = P^{(i)}/\rho^{(L)^3}(\nu^{(\mathrm{L})}h_{\mathrm{LG}}/\sigma)^2$ and the length of the heating and evaporation region $\overline{x}_j = x_j h_{\mathrm{LG}}\rho^{(\mathrm{L})}/\sigma$ was studied numerically.

The subsequent calculations were performed for laminar flow of water ($\rho^{(\mathrm{L})} = 10^3\,\mathrm{kg/m^3}$, $c_{\mathrm{p}} = 4.19\,\mathrm{kJ/kg\,K}$, $\sigma = 0.059\,\mathrm{N/m}$, $h_{\mathrm{LG}} = 2{,}256\,\mathrm{kJ/kg}$, $\theta = 0.67\,\mathrm{rad}$) in a vertical slot of height $d = 1.5$ mm. The inlet water temperature $T_{\mathrm{in}}^{(\mathrm{L})}$, gap size d, heat flux q and acceleration due to gravity g were varied within the limits; $273 < T_{\mathrm{in}}^{(\mathrm{L})} < 373$ (K), $1 < d < 500$ (μm), $10 < q < 600$ (W/cm^2), $1 < g < 600$ (m/s^2).

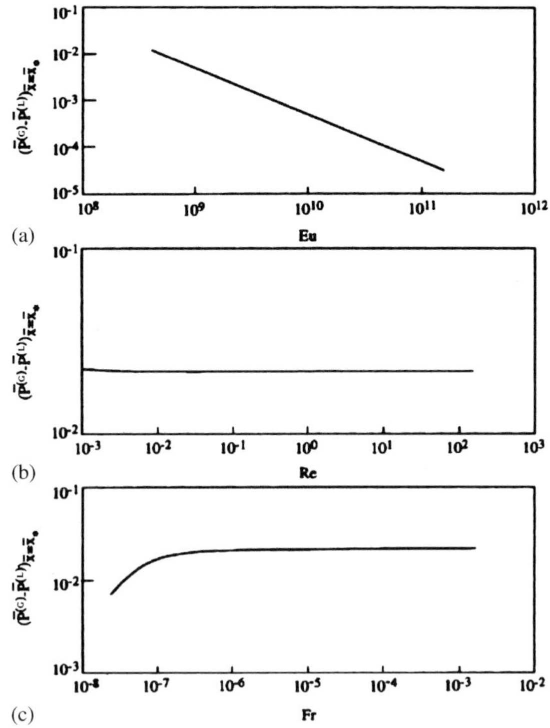

Fig. 8.4 (**a**) The dependence $\Delta\overline{P}(\mathrm{Eu})$ at $\overline{x}_*$ for We $= 10^{-7}$, $\vartheta = 10$, Re $= 0.1$, Fr $= 10^{-7}$. (**b**) The dependence $\Delta\overline{P}(\mathrm{Re})$ at $\overline{x}_*$ for We $= 10^{-7}$, $\vartheta = 10$, Eu $= 2.5 \times 10^8$, Fr $= 10^{-5}$. (**c**) The dependence $\Delta\overline{P}(\mathrm{Fr})$ at $\overline{x}_*$ for We $= 10^{-7}$, $\vartheta = 10$, Eu $= 2.5 \times 10^8$, Re $= 0.1$. Reprinted from Peles et al. (2000) with permission

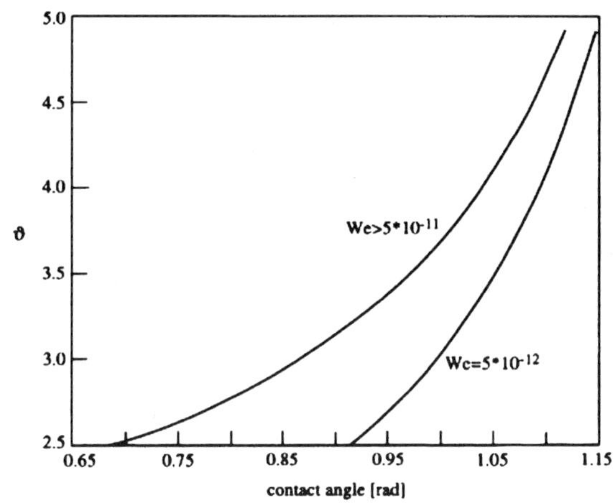

Fig. 8.5 The correlations $\vartheta(\theta)$ at We – var. Reprinted from Peles et al. (2000) with permission

The numerical solution for the entire flow domain (i.e., the liquid single-phase flow, two-phase flow and superheated single-phase vapor flow) begins with the evaluation in the two-phase domain for an initial guess of the vapor pressure at the beginning of this domain ($\overline{x}_*$). The liquid single-phase length $\overline{x}_*$ is obtained based on the inlet temperature, and the inlet pressure is calculated. If it is the inlet pressure, the properties of the single-phase flow in the liquid and vapor domain are calculated and

Fig. 8.6 The shape of interface surface for $\mathrm{We} = 10^{-7}$, $\vartheta = 2.5$, $\mathrm{Eu} = 1.6 \times 10^{11}$, $\mathrm{Fr} = 10^{-7}$, $\mathrm{Re} = 0.04$. Reprinted from Peles et al. (2000) with permission

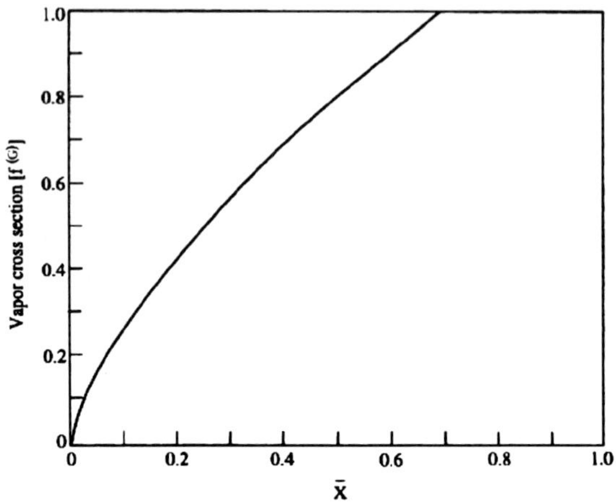

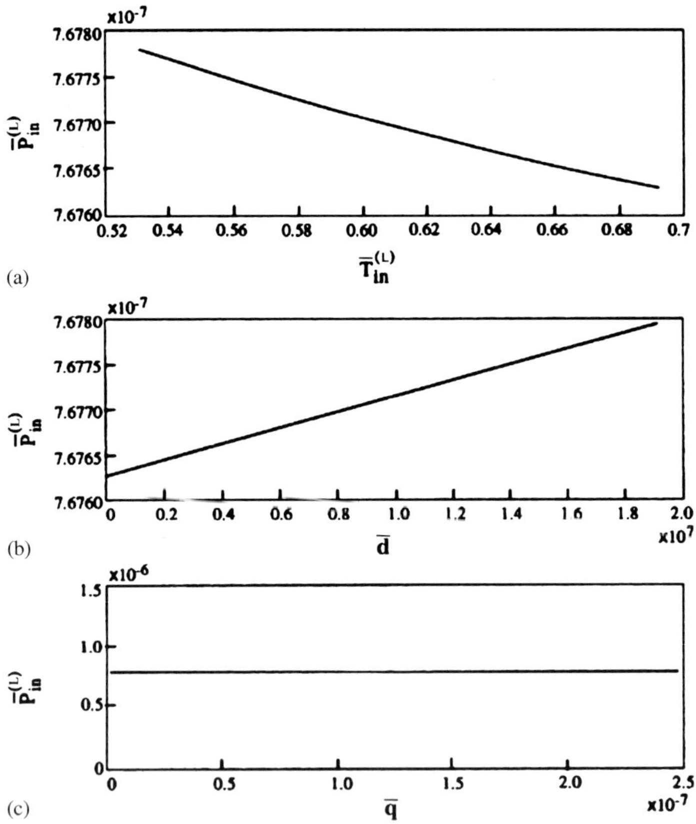

Fig. 8.7a–c The effect of inlet liquid temperature, gap size and wall heat flux from the wall on the pressure in inlet cross-section pressure of the micro-channel. Reprinted from Peles et al. (2000) with permission

the numerical evaluation ends, if not the vapor pressure at $\overline{x}_*$ in the two-phase domain is changed and another iteration takes place. The calculations show that $\overline{P}_{\text{in}}^{(L)}$, for a fixed vapor evaporation pressure, depends very weakly on $\overline{T}_{\text{in}}^{(L)}$, $\overline{d}$ and $\overline{q}$ at large Euler numbers. For example, the variation of $\overline{T}_{\text{in}}^{(L)}$, $\overline{d}$ and $\overline{q}$ within the limits $0.52 < \overline{T}_{\text{in}}^{(L)} < 0.68$, $4\times 10^4 < \overline{d} < 2\times 10^7$; $10^{-8} < \overline{q} < 2.5\times 10^{-7}$ corresponds (at $\text{Eu} > 10^8$) to less than 0.02% change in $\overline{P}_{\text{in}}^{(L)}$ (Fig. 8.7). The temperature in the meniscus symmetry point T_{20} equals the saturation temperature T_{sat}. Since the pressure drop in the liquid region of the capillary flow is small, it is possible to assume that T_{sat} corresponds to $P_{\text{out}}^{(G)}$. The estimations show that such an assumption does not affect practically, the results of the calculations.

The effect of the inlet liquid temperature, the size of the capillary gap, the wall heat flux and gravity on the hydrodynamic and the thermal characteristics of the flow in micro-channel are illustrated in Figs. 8.7–8.15. These data show that the preliminary heating of the liquid (increase of $\overline{T}_{\text{in}}^{(L)}$) is accompanied by displacement of the meniscus toward the inlet of the capillary. In accordance with that, the length of the liquid region of the flow ($\overline{x}_*$) decreases, whereas the length of vapor region increases (Fig. 8.8). It is worth noting that the length of the evaporation region (as well as the shape of interface surface) does not depend on $\overline{T}_{\text{in}}^{(L)}$.

Expansion of the vapor region leads to a vapor temperature and velocity growth at the outlet (Fig. 8.9a–c). The latter is accompanied by a significant change of the micro-channel drag. The calculation has shown that the decrease in the liquid region drag is smaller than the growth of the hydraulic drag at the vapor region. As

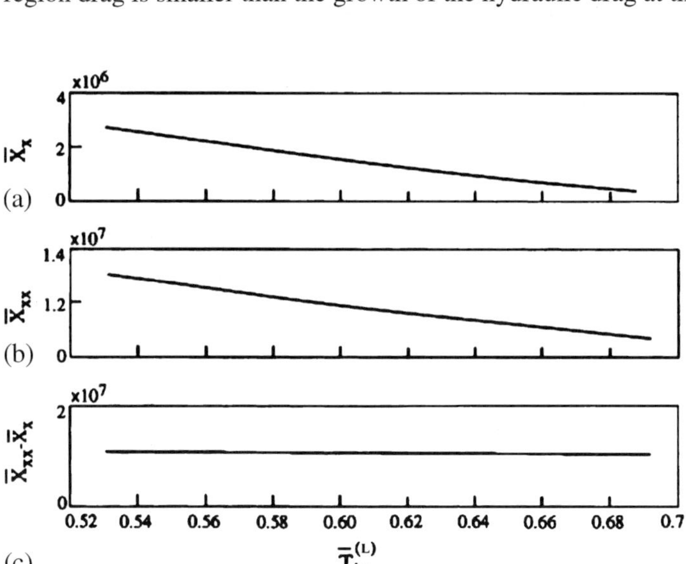

Fig. 8.8a–c The effect of inlet liquid temperature on the meniscus position: (**a**) the dependence of $\overline{x}_*(\overline{T}_{\text{in}}^{(L)})$, (**b**) the dependence of $\overline{x}_{**}(\overline{T}_{\text{in}}^{(L)})$, (**c**) the dependence of $\Delta\overline{x}(\overline{T}_{\text{in}}^{(L)})$, ($\Delta\overline{x} = \overline{x}_{**} - \overline{x}_*$). Reprinted from Peles et al. (2000) with permission

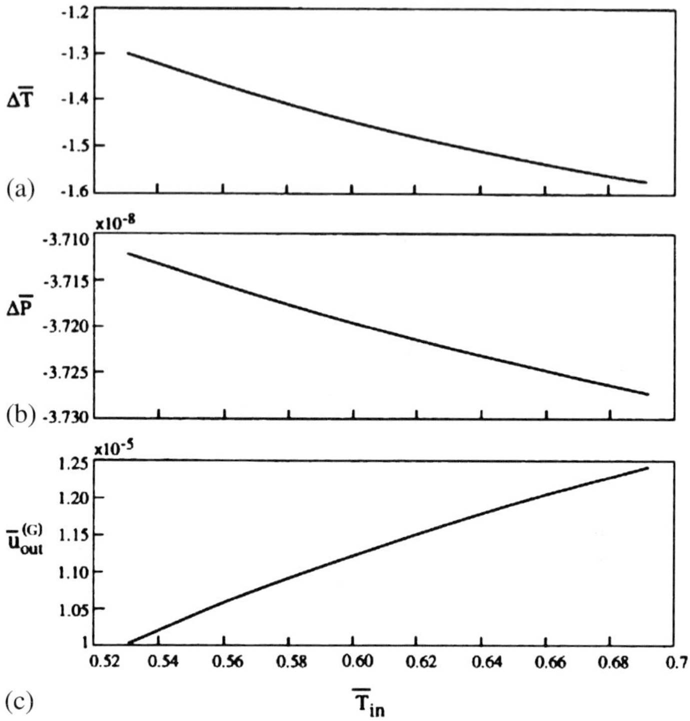

Fig. 8.9a–c The dependencies of vapor temperature, pressure and velocity in the outlet cross-section of the capillary on the inlet liquid temperature: (**a**) the dependence of $\Delta\overline{T}(\overline{T}_{\text{in}}^{(\text{L})})$, (**b**) the dependence of $\Delta\overline{P}(\overline{T}_{\text{in}}^{(\text{L})})$, (**c**) the dependence of $\overline{u}_{\text{out}}^{(\text{G})}(\overline{T}_{\text{in}}^{(\text{L})})$, $(\Delta\overline{T} = \overline{T}_{\text{in}}^{(\text{L})} - \overline{T}_{\text{out}}^{(\text{G})}, \Delta\overline{P} = \overline{P}_{\text{out}}^{(\text{G})} - \overline{P}_{\text{in}}^{(\text{L})})$. Reprinted from Peles et al. (2000) with permission

a consequence the total pressure drop between the inlet and outlet cross-sections of the micro-channel increase as the liquid preliminary heating increases (Fig. 8.9b).

The effect of capillary gap size on the interface surface, vapor velocity and the difference of pressure and temperature between inlet and outlet cross-sections of the micro-channel are illustrated in Figs. 8.10 and 8.11. It is seen that an increase of d leads to the expansion of the liquid region. This effect is explained as follows: at the fixed values of the other parameters an increase of d leads to a growth in the total mass flux of the liquid through the capillary. Since the area of heat transfer surface and wall heat flux are invariable, the energy per unit mass of liquid decreases. Accordingly the heating rate of the liquid decreases as well. The latter is accompanied by a displacement of the meniscus toward the outlet cross-section and therefore $\overline{x}_*$ increases. The length of the evaporation region increases proportionally to the size of the gap, whereas the $(x_{**} - x_*)/d$ ratio does not depend on d. The latter shows that the shape of the interface surfaces are similar for various d. The decrease of the superheat region length with gap size growth leads to a decrease in the outlet vapor velocity and temperature (Fig. 8.11).

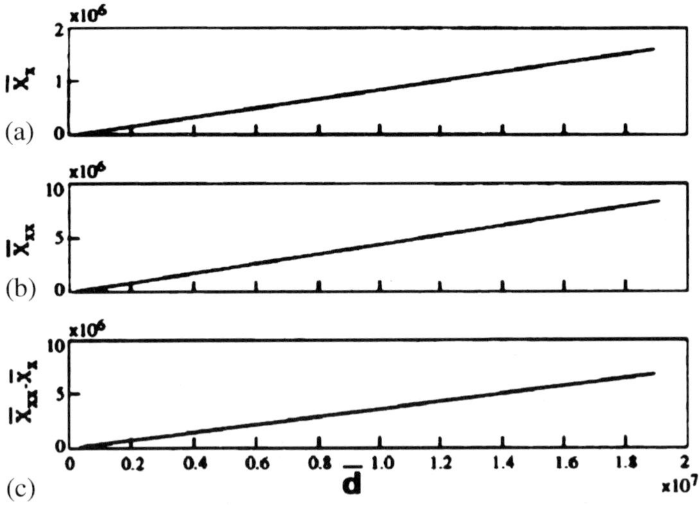

Fig. 8.10a–c The effect of gap size on the meniscus position: (**a**) the dependence of $\overline{x}_*(\overline{d})$, (**b**) the dependence of $\overline{x}_{**}(\overline{d})$, (**c**) the dependence of $\Delta\overline{x}(\overline{d})$, $\Delta\overline{x} = \overline{x}_{**} - \overline{x}_*$. Reprinted from Peles et al. (2000) with permission

The effect of gravity on the liquid and vapor parameters in the inlet and outlet cross-section is presented in Figs. 8.12 and 8.13. It is seen that an increase in the gravity is accompanied by a significant growth of the liquid pressure $\overline{P}_{\text{in}}^{(\text{L})}$ (Fig. 8.12a). At the same time an increase of the vapor pressure in the outlet cross-section is observed. However, the rate of liquid and vapor pressure growth are very different. This causes an increase of the difference $\Delta\overline{P} = \overline{P}_{\text{in}}^{(\text{L})} - \overline{P}_{\text{out}}^{(\text{G})}$ as gravity increases, as well as the sign change at some value of $\overline{g}$. The temperature difference $\Delta\overline{T} = \overline{T}_{\text{in}}^{(\text{L})} - \overline{T}_{\text{out}}^{(\text{G})}$ and vapor velocity in the outlet cross-section of the micro-channels practically do not depend on gravitational acceleration (Fig. 8.12).

The effect of wall heat flux on the length of the heating and evaporation regions, vapor velocity, temperature and pressure in the outlet cross-section is shown in Figs. 8.13, 8.14, and 8.15. These data illustrate some important features of capillary flow at large Euler numbers.

At large Euler number ($\text{Eu} > 10^8$) the mass flux through the micro-channels is directly proportional to the wall heat flux. In this case the energy per unit mass of liquid absorbed from the wall does not depend on the value of the heat flux. As a consequence the length of the heating and evaporation regions, as well as the liquid and vapor temperatures are invariable on heat flux q. This phenomenon, which may be called "the effect of self-regulation," has an important impact when estimating the limiting permissible thermal states of the system with phase change of a cooling liquid. Ha and Peterson (1998) showed that for V-shape micro-grooves the evaporation length decreases slightly as heat flux increases. However, the Euler number in this research is on the order of 10^7 ($\text{Eu} < 10^8$) and the capillary pumping mechanism

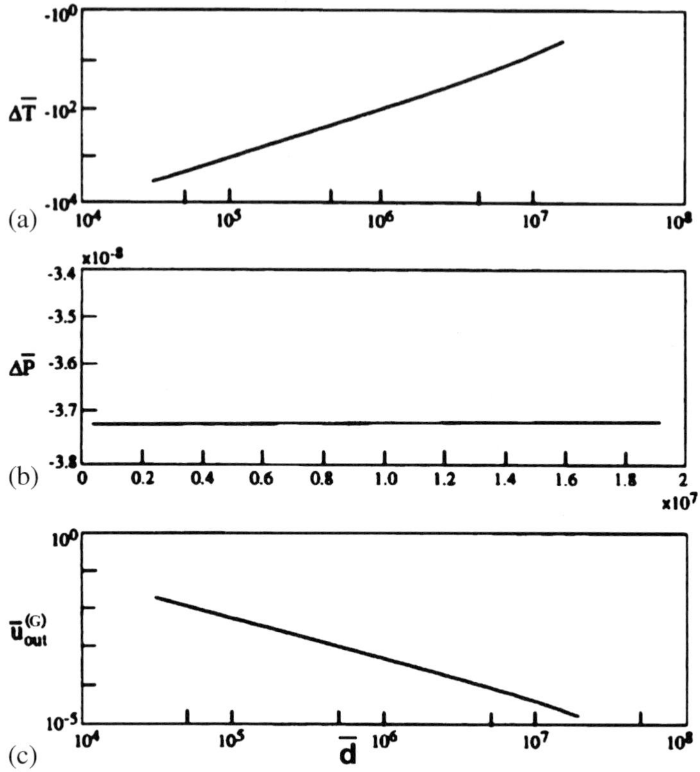

Fig. 8.11a–c The dependence of vapor temperature, pressure and velocity gap size: (**a**) the dependence of $\Delta\overline{T}(\overline{d})$, (**b**) the dependence of $\Delta\overline{P}(\overline{d})$, (**c**) the dependence of $\overline{u}_{\mathrm{out}}^{(\mathrm{G})}(\overline{d})$, ($\Delta\overline{T} = \overline{T}_{\mathrm{in}}^{(\mathrm{L})} - \overline{T}_{\mathrm{out}}^{(\mathrm{G})}$, $\Delta\overline{P} = \overline{P}_{\mathrm{in}}^{(\mathrm{L})} - \overline{P}_{\mathrm{out}}^{(\mathrm{G})}$). Reprinted from Peles et al. (2000) with permission

in this kind of geometry is primarily due to the receding radius of curvature parallel to the flow direction.

At $\mathrm{Eu} > 10^8$ the wall temperature T_{w} depends on the liquid (vapor) temperature, the heat transfer intensity and the wall heat flux.

To estimate the limiting permissible value of the wall heat flux we use the thermal balance equation

$$h\Delta T = q \tag{8.80}$$

where h is the convection heat transfer coefficient, $\Delta T = T_{\mathrm{w}} - T^{(\mathrm{G})}$.

Then we have

$$T_{\mathrm{w}}^{\max} = T_{\mathrm{out}}^{(\mathrm{G})} + q/h \tag{8.81}$$

where $T_{\mathrm{w}}^{\max}$ is the maximum temperature of the wall and $T_{\mathrm{out}}^{(\mathrm{G})}$ is the outlet vapor temperature.

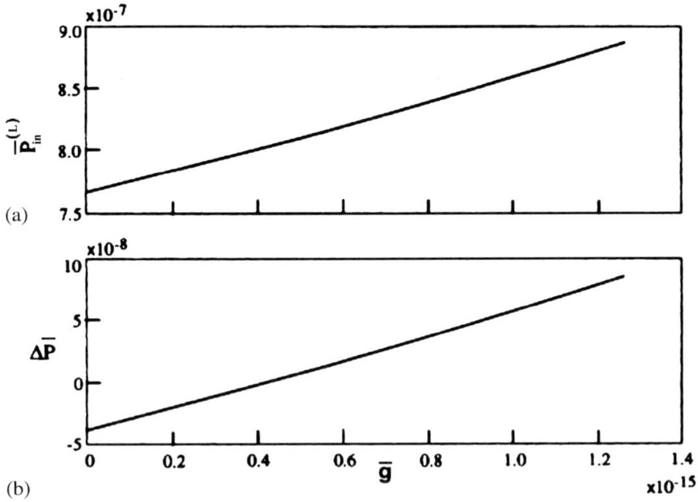

Fig. 8.12a,b The effect of gravity on inlet liquid pressure and pressure difference between the inlet and outlet cross-section of the capillary: (**a**) the dependence of $\overline{P}_{\text{in}}^{(\text{L})}(g)$, (**b**) the dependence of $\Delta\overline{P}(g)$, $(\Delta\overline{P} = \overline{P}_{\text{in}}^{(\text{L})} - \overline{P}_{\text{out}}^{(\text{G})})$. Reprinted from Peles et al. (2000) with permission

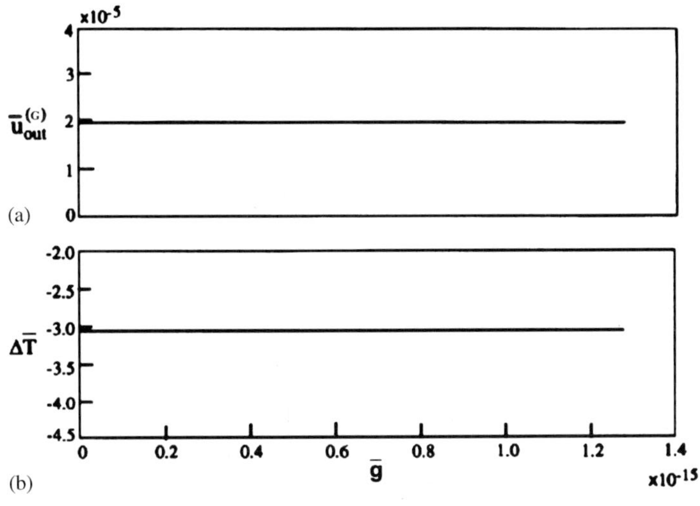

Fig. 8.13a,b The effect of gravity on vapor velocity and temperature in the outlet cross-section of the capillary: (**a**) the dependence of $\overline{u}_{\text{out}}^{(\text{G})}(\overline{g})$, (**b**) the dependence of $\Delta\overline{T}(\overline{g})$, $(\Delta\overline{T} = \overline{T}_{\text{in}}^{(\text{L})} - \overline{T}_{\text{out}}^{(\text{G})})$. Reprinted from Peles et al. (2000) with permission

Recalling that $\text{Nu} = A\text{Re}^n\,\text{Pr}^m$ (Nu and Pr are Nusselt and Prandtl numbers, A, n, m are known constants), we rewrite Eq. (8.80) in the following form

$$T_{\text{w}}^{\text{max}} = T_{\text{out}}^{(\text{G})} + cq \tag{8.82}$$

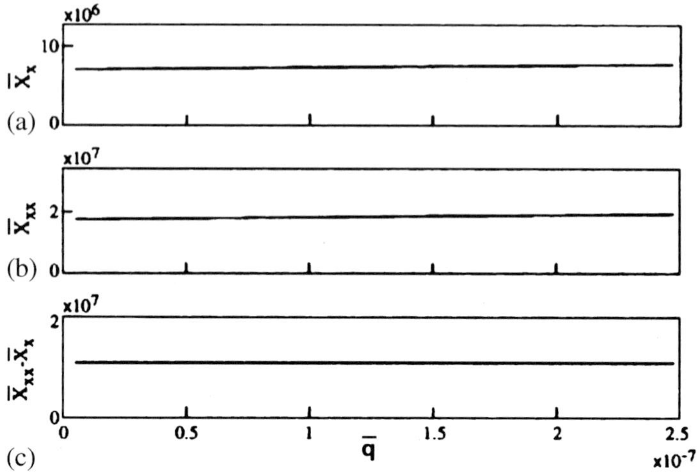

Fig. 8.14a,b The effect of wall heat flux on the meniscus position: (**a**) the dependence of $\overline{x}_*(q)$, (**b**) the dependence of $\overline{x}_{**}(\overline{q})$, (**c**) the dependence of $\Delta\overline{x}_{**}(q)$. Reprinted from Peles et al. (2000) with permission

where $c = (d^{1-n}\nu^n/kA\nu^n\,\mathrm{Pr}^m)$, where k is the thermal conductivity and ν is the kinematic viscosity.

Let the permissible wall temperature equal $T_{\mathrm{w,p}}$. Then we obtain the following estimation for the permissible heat flux on the wall q_{p}

$$cq_{\mathrm{p}} \le T_{\mathrm{w,p}} - T_{\mathrm{out}}^{(\mathrm{G})} \,. \tag{8.83}$$

The inequality (8.83) gives an estimation of the limiting permissible value of heat flux on the wall.

Summary

The quasi-one-dimensional model of flow in a heated micro-channel makes it possible to describe the fundamental features of two-phase capillary flow due to the heating and evaporation of the liquid. The approach developed allows one to estimate the effects of capillary, inertia, frictional and gravity forces on the shape of the interface surface, as well as the on velocity and temperature distributions. The results of the numerical solution of the system of one-dimensional mass, momentum, and energy conservation equations, and a detailed analysis of the hydrodynamic and thermal characteristic of the flow in heated capillary with evaporative interface surface have been carried out.

The following results have been obtained:

1. The velocity, pressure and temperature distribution in a heated capillary with evaporative interface surface are determined by the following parameters ac-

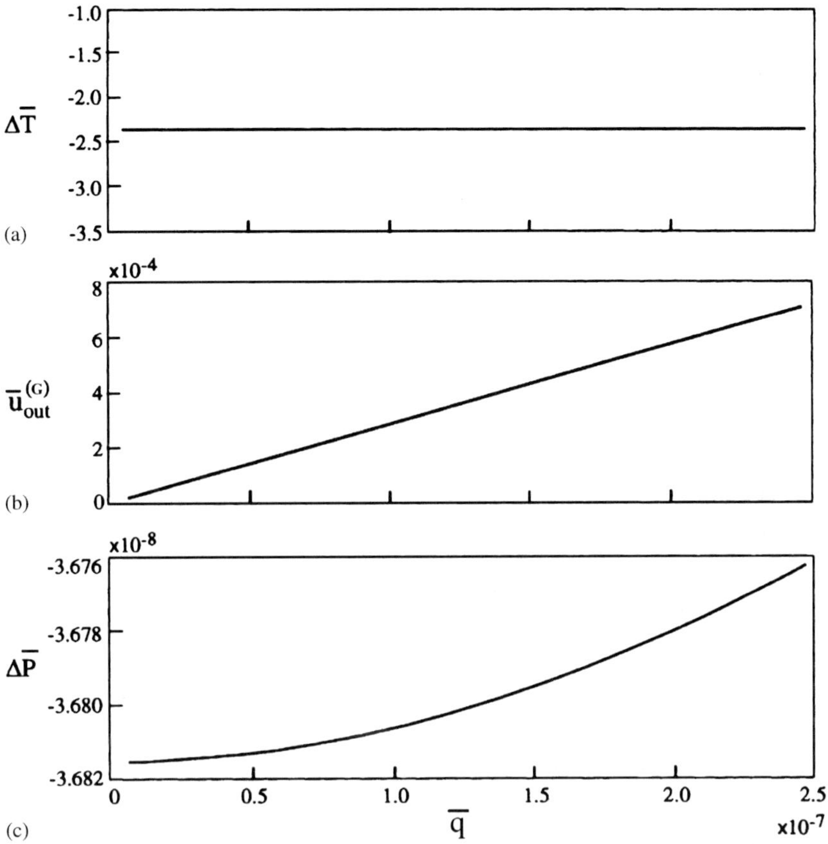

Fig. 8.15a–c The dependencies of vapor temperature, pressure on a velocity in outlet cross-section of capillary on heat flux on the wall: (**a**) the dependence of $\Delta\overline{T}(\overline{q})$, (**b**) the dependence of $\overline{u}^{(G)}_{out}(\overline{q})$, (**c**) the dependence of $\Delta\overline{P}(\overline{q})$, ($\Delta\overline{T} = \overline{T}^{(L)}_{in} - \overline{T}^{(G)}_{out}$, $\Delta\overline{P} = \overline{P}^{(L)}_{in} - \overline{P}^{(G)}_{out}$). Reprinted from Peles et al. (2000) with permission

counting for the physical properties of the liquid and vapor as well as hydrodynamic and thermal conditions of the flow: the Reynolds, Euler, Peclet, Froude and Weber numbers and four non-dimensional groups γ, Ω, ω, and ϑ.

2. The vapor pressure, density and temperature practically do not change along the evaporation region in physically realistic systems. The latter allows one to simplify the system of governing equations and reduce the problem to a successive solution of the shortened system of equations to determine the velocity, liquid pressure and gaseous phases as well as the interface shape in a heated capillary.
3. The difference in pressure between vapor and liquid within the evaporation region depends mainly on the Euler and Weber numbers, as well as on the thermal parameter ϑ. The effect of the Reynolds and Froude numbers on the pressure difference of both phases is negligible.

4. The curvature of the meniscus in a heated capillary with evaporative interface surface is not constant.
5. Increasing the initial liquid temperature is accompanied by a change of the length of the liquid and vapor regions, the increase of the vapor velocity and temperature as well as the total drag in the micro-channel.
6. Heat flux on the wall, gap size of the capillary, as well as gravity affects noticeably the total drag in the micro-channel and temperature difference between the inlet and outlet cross-sections.
7. At Euler number larger than 10^8 a self-regulated regime of flow is realized at which the length of heating, evaporation and superheating regions, as well as liquid and vapor temperatures, do not depend on the wall heat flux. At such flow the wall temperature depends on the liquid (vapor) temperature, heat transfer intensity and wall heat flux.
8. The maximum possible heat flux, which corresponds to the maximum allowed wall temperatures is estimated. This maximum wall heat flux is determined by the difference between the permissible wall temperature and the vapor temperature in the outlet cross-section, which is a function of the Reynolds and Nusselt numbers.

References

Bailey DK, Ameel TA, Warrington RO, Savoie TI (1995) Single-phase forced convection heat transfer in microgeometries: a review. ASME IECEC Paper ES 396:301–310

Bowers MB, Mudawar I (1994a) High flux boiling in low flowrate, low pressure drop mini-channel and micro-channel heat sink. Int J Heat Mass Transfer 37:321–332

Bowers MB, Mudawar I (1994b) Two-phase electronic cooling using mini-channel and micro-channel heat sink. Part 2: Flow rate and pressure drop constraints. J Electron Packag ASME 116:298–305

Carey VP (1992) Liquid–vapour phase-change phenomena. Hemisphere, Washington, DC

Collier SP (1981) Convective boiling and condensation. McGraw-Hill, New York

Ha JM, Peterson GP (1998) Capillary performance of evaporation flow in micro grooves: an analytical approach for very small tilt angles. ASME J Heat Transfer 120:452–457

Hetsroni G, Yarin LP, Pogrebnyak E (2004) Onset of flow instability in a heated capillary tube. Int J Multiphase Flow 30:1424–1449

Khrustalev D, Faghri A (1994) Thermal analysis of microheat pipe. J Heat Transfer ASME 116:189–198

Khrustalev D, Faghri A (1995) Heat transfer during evaporation on capillary grooved structure of heat pipes. J Heat Transfer ASME 117:740–747

Khrustalev D, Faghri A (1996) Fluid flow effect in evaporation from liquid–vapor meniscus. Trans ASME J Heat Transfer 118:725–730

Landau LD, Lifshitz EM (1959) Fluid mechanics, 2nd edn. Pergamon, London

Landerman CS (1994) Micro-channel flow boiling mechanisms leading to Burnout. J Heat Transfer Electron Syst ASME HTD-292:124–136

Levich VG (1962) Physicochemical hydrodynamics. Prentice Hall, London

Morijama K, Inoue A (1992) The thermohydraulic characteristics of two-phase flow in extremely narrow channels (the frictional pressure drop and heat transfer of boiling two-phase flow, analytical model). Heat Transfer Jpn Res 21:838–856

Morris SJS (2003) The evaporating meniscus in the channel. J Fluid Mech 494:297–317
Peles YP, Yarin LP, Hetsroni G (2000) Thermohydrodynamic characteristics of two-phase flow in a heated capillary. Int J Multiphase Flow 26:1063–1093
Peles YP, Yarin LP, Hetsroni G (2001) Steady and unsteady flow in a heated capillary. Int J Multiphase Flow 27:577–598
Peng XF, Peterson GP (1995) The effect of thermofluid and geometrical parameters on convection of liquids through rectangular micro-channels. Int J Heat Mass Transfer 38:755–758
Peterson GP, Ma HB (1996) Theoretical analysis of the maximum heat transport in triangular grooves: a study of idealized et al micro-heat pipes. Trans ASME Heat Transfer 118:731–737
Peterson GP, Ha JM (1998) Capillary performance of evaporation flow in micro grooves: approximate analytical approach and experimental investigation. ASME J Heat Transfer 120:743–751
Potast M, Wayner PC (1972) Evaporation from a two-dimensional extended meniscus. Int J Heat Mass Transfer 15:1851–1863
Smirnov VI (1964) A course of higher mathematics, vol II. Pergamon, London
Stephan PS, Busse CA (1992) Analysis of the heat transfer coefficient of grooved heat pipe evaporator walls. Int J Heat Mass Transfer 35:383–391
Tuckerman D (1984) Heat transfer micro structure for integrated circuits. Dissertation, Stanford University, Stanford
Tuckerman DB, Pease RFW (1981) High-performance heat sinking for VLSI. IEEE Electron Devise Lett EDL-2:126–129
Wallis GB (1969) One-dimensional two-phase flow. McGraw-Hill, New York
Wayner PC, Kao YK, LaCroix LV (1976) The interline heat transfer coefficient of an evaporating wetting film. Int J Heat Mass Transfer 19:487–492
Weisberg A, Bau HH, Zemel JN (1992) Analysis of micro-channels for integrated cooling. Int J Heat Mass Transfer 35:2465–2472
Weislogel MM, Lichter S (1998) Capillary flow in an interior corner. J Fluid Mech 373:349–378
Wu PY, Little WA (1984) Measurement of the heat transfer characteristics of gas flow a fine channels heat exchangers used for microminiature refrigerators. Cryogenics 24:415–420
Xu X, Carey VP (1990) Film evaporation from a micro-grooved surface: an approximate heat transfer model and its comparison with experimental data. J Thermophys 4(4):512–520
Yarin LP, Ekelchik LA, Hetsroni G (2002) Two-phase laminar flow in a heated micro-channels. Int J Multiphase Flow 28:1589–1616

Nomenclature

c_{p}	Thermal conductivity
d	Width of micro-channel
F_{ext}	Drag force
g	Acceleration due to gravity
h	Enthalpy, heat transfer coefficient
h_{LG}	Latent heat of evaporation
q	Heat flux on the wall
Q_{ext}	External heat flux
k	Thermal conductivity
n_i, n_k	Correspond to the normal and tangent direction
P	Pressure
R	Universal gas constant
r_1, r_2	General radii of curvature of the interface

T	Temperature
u, v, w	Components of velocity directed along x, y and z-axis, respectively
$\mathbf{v}$	Velocity (bold letter denote vector)
x, y, z	Cartesian coordinates
$\mathrm{Eu} = \dfrac{P}{\rho u^2}$	Euler number
$\mathrm{Fr} = \dfrac{u^2}{gd}$	Froud number
$\mathrm{Nu} = \dfrac{\alpha d}{k}$	Nusselt number
$\mathrm{Pe} = \dfrac{ud}{\alpha}$	Peclet number
$\mathrm{We} = \dfrac{\rho u^2 d}{\sigma}$	Weber number

Greek symbols

μ	Viscosity
∇	Gradient
∇^2	Laplacian operator
σ	Surface tension
τ_{ik}	Tensor of viscous tension

Subscripts

in	Initial state
w	Wall
ext	External
s	Saturation

Superscript

$\alpha = \mathrm{G}$ or L	Corresponds to vapor ($\alpha = \mathrm{G}$) or liquid ($\alpha = \mathrm{L}$)
max	Maximum

Chapter 9
Steady and Unsteady Flow in a Heated Capillary

The forced fluid flow in heated micro-channels with a distinct evaporation front is considered. The effect of a number of dimensionless parameters such as the Peclet, Jacob numbers, and dimensionless heat flux, on the velocity, temperature and pressure within the liquid and vapor domains has been studied, and the parameters corresponding to the steady flow regime, as well as the domains of flow instability are delineated. An experiment was conducted and demonstrated that the flow in micro-channels appear to have to distinct phase domains one for the liquid and the other for the vapor, with a short section of two-phase mixture between them.

9.1 Introduction

The investigations of fluid flow in micro-channels may be divided in two groups: (1) single-phase flow, and (2) evaporative two-phase flow. The first was intensively investigated beginning from the pioneer work by Tukermann and Pease (1981). Two-phase flow is much less understood.

Hsu (1962) developed a semi-theoretical model that provides considerable insight into the effects of a non-uniform liquid superheat resulting from a transient condition in the liquid during the bubble growth and release process. The model shows that there is a finite range of active cavity sizes on the heating surface, which depends on subcooling, pressure, heat flux and physical properties of the fluid. For water at atmospheric pressure, the temperature difference between the wall and saturation temperature is around 15.6 °C, thermal boundary layer thickness of 76 μm near the nucleation site, and heat flux of 30 W/cm^2; the active nucleation sites are those having a cavity opening of radius 2.5–15 μm. The thermal boundary layer far from the nucleation site was measured by optical means by Yamagata et al. (1955). It was found that the thickness of the thermal boundary layer is on the order of 250 μm. Thus, the boundary layers of an adjacent heated micro-channel wall have coverage in a channel with a hydraulic diameter on the order of 100 μm, causing a higher bulk temperature and therefore a wider range of active nucleation sides. The latter results in higher bubble departure frequency. Furthermore, the bubble departure diameter in channels of regular size is on the order of 1 mm, indicating that the bubble initiation process in micro-channels is strongly controlled by the surrounding walls.

Peng and Wang (1993) investigated the flow boiling through micro-channels with a cross-section of $0.6 \times 0.7\,\text{mm}^2$. They observed that no partial nucleate boiling existed and that the velocity and liquid subcooling have no obvious effect on the flow nucleate boiling fluid. Peng et al. (1994a,b, 1996) conducted additional experimental investigations on flow boiling in micro-channels with rectangular cross-sections ranging from 0.1×0.3 to $0.6 \times 0.7\,\text{mm}^2$. Peng et al. (1998) suggested a dimensionless parameter for nucleate boiling in micro-channels, which agree well with the experimental data.

Bowers and Mudawar (1994) performed an experimental study of pressure drop and critical heat flux (CHF) in mini-channels with circular cross-section ($d = 2.54\,\text{mm}$ and $d = 510\,\mu\text{m}$) using R-113. A CHF correlation proposed by Katto (1978) was presented as $q_{\text{CHF}} = Gh_{\text{LG}} = 0.16\text{We}^{-0.19}(L/d)^{-0.54}$, where q_{CHF} is the critical heat flux, G is the mass velocity, L and d are the channel length and hydraulic diameter, respectively, h_{LG} is the latent heat of evaporation and $\text{We} = (\rho_{\text{L}} u^2 L)/\sigma$ is the Weber number, σ is the surface tension, and u is the inlet velocity. When the channel size decreases, the capillary forces acting in the longitudinal direction become significant. The experiment yields CHF values of about $200\,\text{W}/\text{cm}^2$. However, the pressure drop for the small channels was less than 1 KPa compared to 23 KPa for the micro-channels.

Peles et al. (1998) and Khrustalev and Faghri (1996) considered two-phase laminar flow in a heated micro-channel with distinct evaporating meniscus in the frame of quasi-one-dimensional and two-dimensional models.

Below the system of quasi-one-dimensional equations considered in the previous chapter used to determine the position of meniscus in a heated micro-channel and estimate the effect of capillary, inertia and gravity forces on the velocity, temperature and pressure distributions within domains are filled with pure liquid or vapor. The possible regimes of flow corresponding to steady or unsteady motion of the liquid determine the physical properties of fluid and intensity of heat transfer.

We also consider the conditions corresponding to the existence of steady flow with distinct evaporation front. The knowledge of such conditions is necessary to design cooling systems of electronic devices with high power densities, and for choosing operating parameters ensuring their stability, etc. We use the model by Peles et al. (1998), which is based on a simplified one-dimensional model flow with flat evaporation front dividing the liquid and vapor into two different domains.

Chapter 9 consists of the following: in Sect. 9.2 the physical model of two-phase flow with evaporating meniscus is described. The calculation of the parameters distribution along the micro-channel is presented in Sect. 9.3. The stationary flow regimes are considered in Sect. 9.4. The data from the experimental facility and results related to two-phase flow in a heated capillary are described in Sect. 9.5.

9.2 The Physical Model

Evaporative two-phase flow in a heated micro-channel resembles a two-phase slug flow with distinct domains of liquid and vapor. These domains are divided by the infinitely thin evaporating front, which propagates relatively to the fluid with a velocity u'_f equal (numerically) to the linear rate of liquid evaporation. In the frame of reference associated with micro-channel walls, the velocity of the evaporation front is

$$u_f = |u_L| - |u'_f| \tag{9.1}$$

where u_f and u'_f are the velocity of the evaporating front in the systems of coordinates associated with the micro-channel wall and the fluid flow, respectively, and u_L is the liquid velocity in an inertial system of coordinates.

Depending on the correlation between u'_f and u_L, the evaporation front moves downstream ($|u'_f| < |u_L|$) or upstream ($|u'_f| > |u_L|$) (Fig. 9.1). Essentially, in both cases the flow in the micro-channel is unsteady. At the same time, at a certain combination of governing parameters (initial velocity of liquid, heat flux on the wall, etc.) the regime of flow may be steady. It corresponds to the kinematic condition $u'_f = u_L$, which determines the fixed position of the evaporation front in the micro-channel, as well as invariable length of the "liquid" and "vapor" domains. The velocity, temperature and pressure distributions corresponding to such a flow are plotted in Fig. 9.2. The velocity of the vapor increases monotonically along the channel axis, whereas the liquid velocity is almost constant. The latter is a result of low thermal expansion of the liquid: $\gamma = 0.18 \times 10^{-3}\,\mathrm{K}^{-1}$ for water at $T = 20\,°\mathrm{C}$ (Kuchling 1980) and $\rho_L(T_L) \approx$ const. at $20\,°\mathrm{C} < T < 100\,°\mathrm{C}$. Accordingly, within such a range of temperature the change of liquid velocity does not exceed 4%.

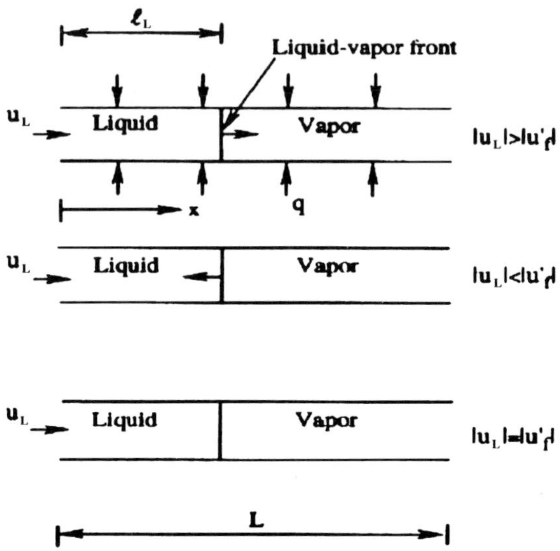

Fig. 9.1 The scheme of the flow in a heated capillary. u_L is the liquid velocity and u'_f is the front velocity relative to the liquid. ℓ_L is the liquid length domain and L is the total channel length. Reprinted from Peles et al. (2001) with permission

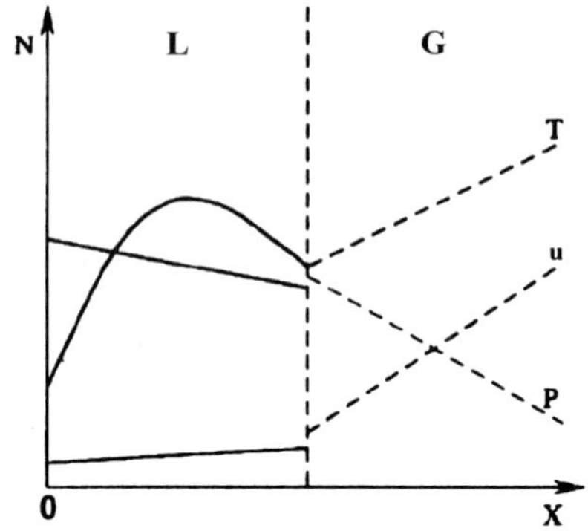

Fig. 9.2 The velocity, temperature and pressure distributions along the axis of a heated capillary ($N = u, T, P$), G and L correspond to vapor and liquid domains, respectively. *Solid line* indicates the liquid domain, and *dotted line* indicates the vapor domain (concave meniscus). Reprinted from Peles et al. (2001) with permission

At the evaporation front there is a jump in the flow velocity, which equals $\Delta u = u_L(\rho_{L.G} - 1)$, where $\rho_{L.G} = \rho_L/\rho_G$, ρ_L and ρ_G are the liquid and vapor densities, respectively. Since $\rho_{L.G} \gg 1$ a jump in the flow velocity is expressed approximately as $\Delta u \simeq u_L \rho_{L.G}$.

The temperature distribution has a characteristic maximum within the liquid domain, which is located in the vicinity of the evaporation front. Such a maximum results from two opposite factors: (1) heat transfer from the hot wall to the liquid, and (2) heat removal due to the liquid evaporation at the evaporation front. The pressure drops monotonically in both domains and there is a pressure jump at the evaporation front due to the surface tension and phase change effect on the liquid–vapor interface.

Taking into account the above-mentioned factors it is possible to present the stationary flow in a heated capillary as a flow of liquid and its vapor divided by an infinitely thin evaporation front. The parameters of these flows are related to each other by the condition of mass, momentum and energy conservation at the evaporation front.

To describe the flow in a horizontal heated capillary we use the mass, momentum and energy balance equations. At moderate velocity, the effects due to compressibility of liquid and vapor, as well as energy dissipation in gaseous and liquid phases are negligible. Assuming that thermal conductivity and viscosity of the vapor and the liquid are independent of temperature and pressure, we arrive at the following system of equations:

$$\mathrm{div}(\rho_i \mathbf{v}_i) = 0 \tag{9.2}$$

$$\rho_i(\mathbf{v}_i \cdot \nabla h_i) = k_i \nabla^2 T_i \tag{9.3}$$

$$\rho_i(\mathbf{v}_i \cdot \nabla)\mathbf{v}_i = -\nabla P_i + \mu_i \nabla^2 \mathbf{v}_i \tag{9.4}$$

where ρ, $\mathbf{v}$, T and h are the density, velocity, temperature and enthalpy (vector $\mathbf{v}$ has components u, v, w which are directed along the x, y, z-axes, respectively), P the pressure, k and μ the thermal conductivity and viscosity, respectively, ∇ and ∇^2 are the gradient and the Laplace operator; from here on the subscript $i = \mathrm{G,L}$ corresponding to vapor and liquid.

The system of Eqs. (9.2–9.4) should be supplemented by relations determining the dependencies $P_G(\rho_G, T_G)$, $\rho_L(T_L)$ and $h_i(T_i)$.

At steady flow in a heated micro-channel the conditions at the evaporation front may be expressed by the continuity of mass, thermal fluxes on the interface surface and the equilibrium of all acting forces (Landau and Lifshitz 1959). With reference to the evaporative meniscus the balance equations have the following form (Peles et al. 1998):

$$\sum_{i=1}^{2} \rho_i \mathbf{v}_i n_i = 0 \tag{9.5}$$

$$\sum_{i=1}^{2} \left(\rho_i \mathbf{v}_i h_i - k_i \frac{\partial T_i}{\partial x_\ell} \right) n_i = 0 \tag{9.6}$$

$$\sum_{i=1}^{2} (P_i + \rho_i v_i v_k) n_i = \left(\tau_{ik}^{(G)} - \tau_{ik}^{(L)} \right) n_k + \sigma (r_1^{-1} + r_2^{-1}) n_i^{(1)} + \frac{\partial \sigma}{\partial x_\ell} \tag{9.7}$$

where σ is the surface tension, τ_{ik} the tensor of viscous tension; $v_i n_i$ and $(\partial T_i/\partial x_i) n_i$; the normal component of the velocity vector and temperature interface surface gradient, respectively, r_1 and r_2 the general curvature radii of the interface surface, n_i and n_k correspond to the normal and tangent directions: $n_i^{(G)} = -n_i^{(L)}$, $x_i = x, y, z$.

The radius of curvature r of the meniscus and its depth H are expressed for a cylindrical micro-channel as

$$r = r_0/\cos\theta \tag{9.8}$$

$$H = r_0(1 - \sin\theta)/\cos\theta \tag{9.9}$$

where r_0 and θ are the radius of the micro-channel and the contact angle, respectively.

When the contact angle is close to $\pi/2$ (for example, for the system water/steel $\theta \sim 0.45\pi$), the ratio H/r_0 is small, we estimate the order of magnitude of the terms in Eqs. (9.2–9.4) and conditions (9.5–9.7). Choosing the meniscus as characteristic scales of length in the longitudinal and transversal directions H and r_0, respectively, we obtain

$$\frac{\partial}{\partial x} \sim \frac{1}{H}, \quad \frac{\partial}{\partial y} \sim \frac{1}{r_0}; \quad \frac{\partial^2}{\partial x^2} \sim \frac{1}{H^2}, \quad \frac{\partial^2}{\partial y^2} \sim \frac{1}{r_0^2}. \tag{9.10}$$

From Eq. (9.10) we obtain

$$\frac{\partial^2}{\partial x^2} \gg \frac{\partial^2}{\partial y^2} \tag{9.11}$$

$$\nabla^2 = \frac{\partial^2}{\partial x^2} + \frac{1}{y} \frac{\partial}{\partial y} \left(y \frac{\partial}{\partial y} \right) \approx \frac{\partial^2}{\partial x^2}. \tag{9.12}$$

The components of velocity at the meniscus surface are connected with each other (at $\rho_G/\rho_L \gg 1$) by relation

$$v/u \approx \tan\varphi \tag{9.13}$$

where $0 < \varphi < \pi/2 - \theta$, $\tan\varphi = \mathrm{d}y^*/\mathrm{d}x^*$, $y^*(x^*)$ is the equation which determines the meniscus shape.

At large contact angle (practically $\theta \geq 0.4\pi$) the following evaluations are valid

$$u > v, \quad u\frac{\partial}{\partial x} \gg v\frac{\partial}{\partial y}. \tag{9.14}$$

Estimates (9.12) and (9.14) effectively reduce the problem of flow in a heated micro-channel to solving a system of one-dimensional mass, momentum and energy balance equations. They have the following form:

$$\frac{\mathrm{d}\rho_i u_i}{\mathrm{d}x} = 0 \tag{9.15}$$

$$\rho_i u_i \frac{\mathrm{d}u_i}{\mathrm{d}x} = -\frac{\mathrm{d}P_i}{\mathrm{d}x} - \frac{\mathrm{d}F_i}{\mathrm{d}x} \tag{9.16}$$

$$\rho_i u_i c_{\mathrm{pi}} \frac{\mathrm{d}T_i}{\mathrm{d}x} = k_i \frac{\mathrm{d}^2 T_i}{\mathrm{d}x^2} + q \tag{9.17}$$

where ρ, u, T and P are the density, velocity, temperature and pressure, F the drag force accounting for the effect of viscosity and c_{p} and k are the heat capacity and thermal conductivity, respectively, q is the heat flux per unit axial length from the wall. The boundary conditions for the present problem are:

$$\text{at } x = 0\,, \quad \rho_{\mathrm{L}} = \rho_{\mathrm{L},0}\,, \quad u_{\mathrm{L}} = u_{\mathrm{L},0}\,, \quad T_{\mathrm{L}} = T_{\mathrm{L},0} \tag{9.18}$$

$$\text{at } x = \ell_{\mathrm{L}}\,, \quad \rho_i = \rho_{\mathrm{i,s}}\,, \quad u_i = u_{\mathrm{i,s}}\,, \quad P_i = P_{\mathrm{i,s}}\,, \quad T_i = T_{\mathrm{s}} \tag{9.19}$$

$$\text{at } x = L\,, \quad \rho_{\mathrm{G}} = \rho_{\mathrm{G},00}\,, \quad u_{\mathrm{G}} = u_{\mathrm{G},00}\,, \quad \frac{\mathrm{d}T_{\mathrm{G}}}{\mathrm{d}x} = 0 \tag{9.20}$$

where $L = \ell_{\mathrm{G}} + \ell_{\mathrm{L}}$ is the total length of the micro-channel, ℓ_{G} and ℓ_{L} the length of the vapor and liquid regions, respectively, subscripts 0 and 00 correspond to the inlet and outlet cross-section, respectively, and s to the evaporation front.

We supplement the system of Eqs. (9.15–9.17) by the equation of state of the vapor and liquid

$$P_{\mathrm{G}} = R\rho_{\mathrm{G}} T_{\mathrm{G}} \tag{9.21}$$

$$\rho_{\mathrm{L}} = \rho_{\mathrm{L}}(T_{\mathrm{L}}) \tag{9.22}$$

by the equation for the vapor pressure at the interface

$$P_{\mathrm{s}} = P_{\mathrm{s}}(T_{\mathrm{s}}) \tag{9.23}$$

and by the mass and thermal balances at the evaporation front

$$u_{\mathrm{G,s}} = u_{\mathrm{L,s}}\rho_{\mathrm{G.L}} \tag{9.24}$$

$$-k_{\mathrm{L}}\left(\frac{\mathrm{d}T_{\mathrm{L}}}{\mathrm{d}x}\right)_{x=\ell_{\mathrm{L}}} + k_{\mathrm{G}}\left(\frac{\mathrm{d}T_{\mathrm{G}}}{\mathrm{d}x}\right)_{x=\ell_{\mathrm{L}}} = h_{\mathrm{LG}}\rho_{\mathrm{L,s}}u_{\mathrm{L,s}} \tag{9.25}$$

where R is the gas constant.

The momentum balance equation at the evaporation front has (neglecting the effect of viscous tension and changing surface tension along of meniscus) the following form:

$$\Delta P = \frac{\sigma}{2r_0}\cos\theta + \rho_{\mathrm{L,s}} v_{\mathrm{L,s}}^2 (1 - \rho_{\mathrm{L,s}}/\rho_{\mathrm{G,s}}) \tag{9.26}$$

where $\Delta P = P_{\mathrm{G,s}} - P_{\mathrm{L,s}}$, and $v_{\mathrm{L,s}}$ is the linear rate of evaporation.

The linear rate of evaporation (cylindrical micro-channel), may be estimated as follows:

$$v_{\mathrm{L,s}} = \frac{4q^*}{h_{\mathrm{LG}}\rho_{\mathrm{L,s}}}\frac{L}{d} \tag{9.27}$$

where q^* is the heat flux at the wall, and h_{LG} is the latent heat of evaporation.

Using Eqs. (9.25) and (9.26) we estimate the contribution of the surface tension in the pressure jump at the evaporation front, for the following parameters: $L/d = 10^2$, $d = 2\times10^{-4}$ m, $\rho_{\mathrm{L}} = 10^3$ kg/m^3, $\rho_{\mathrm{L}}/\rho_{\mathrm{G}} = 10^3$, $h_{\mathrm{LG}} = 2.26\times10^6$ J/kg, $\theta = 80°$, $\sigma = 500\times10^{-4}$ N/m by calculating the ratio of the first term on the right-hand side of Eq. (9.25) to the second one at $q^* = (3;30;300)\times10^4$ W/m^2. As a result we find that the ratio $\sigma\cos\theta/2r_0\rho_{\mathrm{L}}v_{\mathrm{L}}^2(\rho_{\mathrm{L}}/\rho_{\mathrm{G}})$ equals 1.39, 0.0139, 0.000139, respectively, for $q^* = (3;30;300)\times10^4$ W/m^2. It is seen that the pressure jump in a flow in a micro-channel with evaporation front is determined mainly by the phase change. The effect of surface tension shows only at small heat fluxes at the wall. The latter is typical to any fluid/wall system, in particular, for a water/glass system with $\theta = 0$. In this case the ratio $\sigma\cos\theta/2r_0\rho_{\mathrm{L}}v_{\mathrm{L}}^2(\rho_{\mathrm{L}}/\rho_{\mathrm{G}})$ equals 8.93; 0.0893; 0.000893 for $q^* = (3;30;300)\times10^4$ W/m^2.

The above-mentioned estimates show that at high power densities, characteristic for cooling systems of electronic devices, it is possible to neglect the effects due to curvature of the interface and present it as a flat front.

We add to Eqs. (9.15–9.17) and (9.21–9.25) the total mass and energy balances to determine the vapor velocity and the temperature at the outlet cross-section

$$\rho_{\mathrm{G,00}}u_{\mathrm{G,00}} = \rho_{\mathrm{L,0}}u_{\mathrm{L,0}} \tag{9.28}$$

$$Q_0 + \rho_{\mathrm{L,0}}u_{\mathrm{L,0}}c_{\mathrm{pL}}(T_{\mathrm{s}} - T_{\mathrm{L,0}}) + \rho_{\mathrm{L,0}}u_{\mathrm{L,0}}h_{\mathrm{L,G}} + \rho_{\mathrm{L,0}}u_{\mathrm{L,0}}c_{\mathrm{pL}}(T_{\mathrm{G,00}} - T_{\mathrm{s}}) = qL \tag{9.29}$$

where $Q_0 = k_{\mathrm{L}}(\mathrm{d}T/\mathrm{d}x)|_{x=0}$ is the upstream heat losses from the liquid in the inlet cross-section.

9.3 Parameters Distribution Along the Micro-Channel

Using the system (9.15–9.17) we determine the distribution of velocity, temperature and pressure within the liquid and vapor domains. We render the equations dimensionless by the following characteristic scales: $u_{\mathrm{L,0}}$ for velocity, $T_{\mathrm{L,0}}$ for temperature, $\rho_{\mathrm{L,0}}$ for density, $P_{\mathrm{L,0}}$ for pressure, $\rho_{\mathrm{L,0}}u_{\mathrm{L,0}}^2$ for the drag force and L for length

$$\begin{aligned} &\overline{u}_i = u_i/u_{\mathrm{L,0}}\,, \quad \overline{T}_i = T_i/T_{\mathrm{L,0}}\,, \quad \overline{P}_i = P_i/P_{\mathrm{L,0}} \\ &\overline{\rho}_i = \rho_i/\rho_{\mathrm{L,0}}\,, \quad \overline{F}_i = F_i/\rho_{\mathrm{L,0}}u_{\mathrm{L,0}}^2\,, \quad \overline{x} = x/L\,. \end{aligned} \tag{9.30}$$

The dimensionless equations take the following form:

$$\frac{d\overline{\rho}_i\overline{u}_i}{d\overline{x}} = 0 \tag{9.31}$$

$$\frac{d\overline{\rho}_i\overline{u}_i^2}{d\overline{x}} = -\mathrm{Eu}\frac{d\overline{P}_i}{d\overline{x}} - \frac{d\overline{F}_i}{d\overline{x}} \tag{9.32}$$

$$\frac{d\overline{T}_i}{d\overline{x}} = \frac{1}{\mathrm{Pe}_i}\frac{d^2\overline{T}_i}{d\overline{x}^2} + \vartheta_i \tag{9.33}$$

where $\mathrm{Eu} = P_{L,0}/\rho_{L,0}u_{L,0}^2$ is the Euler number, $\mathrm{Pe}_i = u_{L,0}L/\alpha_i$ the Peclet number, $\alpha_i = k_i/\rho_{L,0}c_{pi}$ the thermal diffusivity, and $\vartheta_i = qL/\rho_{L,0}u_{L,0}c_{pi}T_{L,0}$ is the dimensionless heat flux at the wall.

The boundary conditions for system (9.31–9.33) are

$$\text{at } \overline{x}, \overline{\rho}_L = 1\,, \quad \overline{u}_L = 1\,, \quad \overline{T}_L = 1 \tag{9.34}$$

$$\text{at } \overline{x} = \overline{\ell}_L\,, \quad \overline{\rho}_i = \overline{\rho}_{i,s}\,, \quad \overline{u}_i = \overline{u}_{i,s}\,, \quad \overline{T}_i = \overline{T}_s\,, \quad \overline{P}_i = \overline{P}_{i,s} \tag{9.35}$$

$$\text{at } \overline{x} = 1\,, \quad \overline{\rho}_G = \overline{\rho}_{G,00}\,, \quad \overline{u}_G = \overline{u}_{G,00}\,, \quad \frac{d\overline{T}_G}{d\overline{x}} = 0\,. \tag{9.36}$$

The additional relations of Eqs. (9.21) and (9.23) and conditions (9.24) and (9.29) take the following dimensionless form:

$$\overline{P}_i = R_*\overline{\rho}_G\overline{T}_G \tag{9.37}$$

$$\overline{P_S} = \overline{P}_s(\overline{T}_s) \tag{9.38}$$

$$\rho_L = \rho_L(\overline{T}_L) \tag{9.39}$$

$$\overline{u}_{G,s} = \overline{u}_{L,s}\rho_{L.1} \tag{9.40}$$

$$-\left(\frac{d\overline{T}_2}{d\overline{x}}\right)_{\overline{\ell}_L} + k_{G.L}\left(\frac{d\overline{T}_L}{d\overline{x}}\right)_{\overline{\ell}_L} = \mathrm{Ja}\cdot\mathrm{Pe}_L \tag{9.41}$$

$$\mathrm{Eu}\Delta\overline{P} = \mathrm{We}^{-1}\cos\theta + (1-\overline{\rho}) \tag{9.42}$$

$$\overline{\rho}_{G,00}\overline{u}_{G,00} = 1 \tag{9.43}$$

$$Q_0 + (\overline{T}_s - 1) + \mathrm{Ja} + c_{pG.L}(\overline{T}_{G,00} - \overline{T}_s) = \vartheta_L \tag{9.44}$$

where $\overline{Q}_0 = Q_0/(\rho_{L,0}u_{L,0}c_{p,L}T_{L,0})$, $\mathrm{Ja} = h_{LG}/c_{pL}T_{L,0}$ is the Jacob number, $\vartheta_L = qL/(\rho_{L,0}u_{L,0}c_{p,L}T_{L,0})$, $R_* = R\frac{\rho_{L,0}T_{L,0}}{P_{L,0}}$, $c_{pL.G} = c_{pG}/c_{pL}$, $k_{G.L} = k_G/k_L$, $\overline{\rho} = \rho_L/\rho_G$, and $\mathrm{We} = \frac{2r_0\rho_L v_L^2}{\sigma}$.

At large Euler numbers when $\Delta\overline{P} \ll 1$, the vapor pressure may be calculated by the Clausius–Clapeyron equation. In this case $\overline{P}_s$ and $\overline{T}_s$ in Eq. (9.38) correspond to the saturation parameters.

Integrating (9.31) and (9.33) we obtain:

$$\overline{\rho}_i\overline{u}_i = 1 \tag{9.45}$$

$$\overline{T}_i = C_1^{(i)} + \vartheta_i(\overline{x} + \mathrm{Pe}_i^{-1}) + C_2^{(i)}\exp(\mathrm{Pe}_i\overline{x}) \tag{9.46}$$

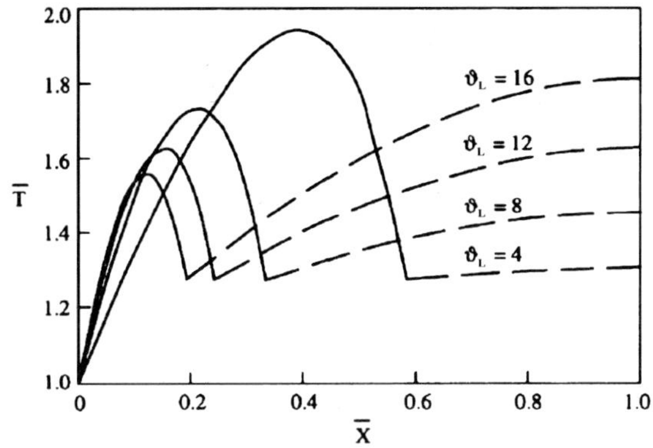

Fig. 9.3 Temperature distribution in a heated capillary. *Solid line*: liquid, *dotted* – vapor (water and atmospheric pressure, Ja $= 1.82$ and $\text{Pe}_L = 6$). Reprinted from Peles et al. (2001) with permission

from the walls and its temperature increases. The liquid temperature reaches a maximum in front of the interface. Then (within $\bar{x}_m < \bar{x} < \bar{\ell}_L$) it decreases to the saturation temperature. Heat flux from the region of high temperature to meniscus spends to liquid evaporation. An increase of ϑ_L leads to the shortening of the liquid domain, as well as to a displacement of the point corresponding to the maximum temperature towards the inlet cross-section (Fig. 9.4). This process is accompanied by a decrease of the maximum liquid temperature and an increase of the heat losses in the inlet cross-section (Fig. 9.5). The effect of Pe_L on the length of the liquid domain is illustrated in Fig. 9.6. An increase of Pe_L leads to a decrease of $\bar{\ell}_L$. The liquid–vapor heat flux ratio at the evaporation front is shown in Fig. 9.7. An increase of the Peclet number is accompanied by an increase of $(q_L/q_G)_{\bar{x}=\bar{\ell}_L}$. This is due to the fact that when the vapor–liquid Peclet number increases two events occur: (1) more heat flux at the front is needed to evaporate a larger quantity of liquid, and (2) its influence on the temperature field, and thus on the temperature slope, diminishes. Since the vapor Peclet number is very large, compared to the liquid Peclet number, the contribution of the vapor domain to the front heat flux increases much slower than that of the liquid as the Peclet number increases, thus $(q_L/q_G)_{\bar{x}=\bar{\ell}_L}$ must increase.

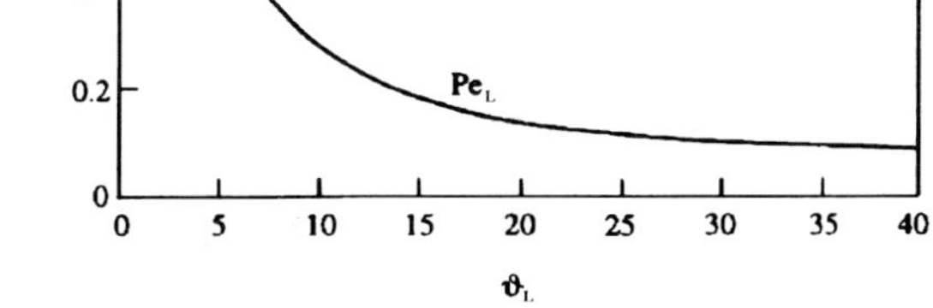

Fig. 9.4 The effect of ϑ_L on the length of liquid domain $\bar{\ell}_L$, for water and atmospheric pressure, Ja $= 1.82$ and $\text{Pe}_L = 6$. Reprinted from Peles et al. (2001) with permission

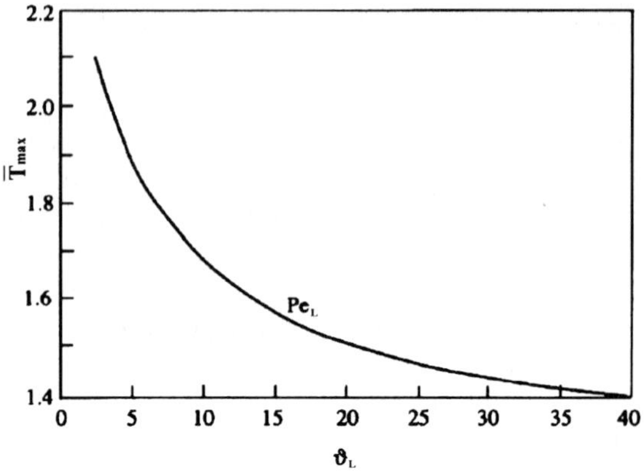

Fig. 9.5 Dependence of the maximal temperature $\overline{T}_{max}$ on ϑ_L, for water at atmospheric pressure, Ja $= 1.82$ and Pe$_L =$ 6. Reprinted from Peles et al. (2001) with permission

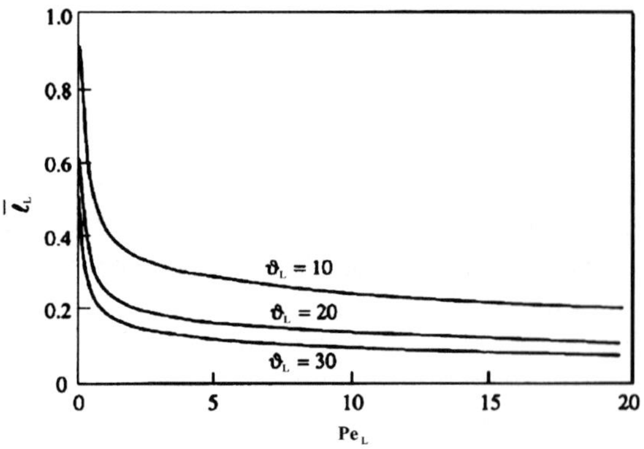

Fig. 9.6 The length of liquid domain as a function of Pe$_L$. Reprinted from Peles et al. (2001) with permission

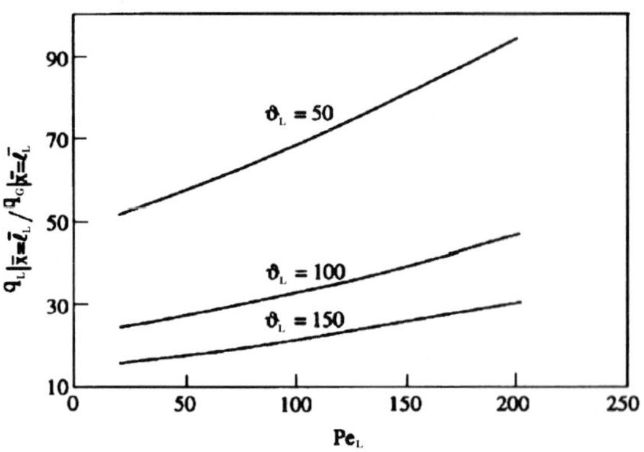

Fig. 9.7 The ratio $(q_L/q_G)_{\overline{x}=\overline{\ell}_L}$ vs. the Peclet number (water at atmospheric pressure). Reprinted from Peles et al. (2001) with permission

Equation (9.62) contains seven dimensionless parameters accounting for the hydrodynamic and thermal effects. The regimes in which stable flows in a heated capillary are possible correspond to the following interval of the length $\overline{\ell}_L$: $0 < \overline{\ell}_L < 1$. The latter allows us to use Eq. (9.62) to define the domains of the existence of stable and unstable flow regimes. In the multi-dimensional parametric space ($\overline{\ell}$, ϑ_L Pe$_L$ Ja $\overline{T}_s$ $k_{L.G}$ $c_{pL.G}$), the limiting values of these parameters correspond to a surface subdividing the parametric space into domains, which correspond to various flow regimes.

The solution of Eq. (9.62) may be represented as 35 spatial (or 21 planar) diagrams which establish correlations between any three (or two, for the planar diagram) parameters when all the other parameters are fixed. These diagrams allow us to outline the domains of stable flow regimes and to determine the values of the parameters corresponding to the change between the stable and unstable flows.

Consider some particular cases, which are of interest to the understanding of the general flow properties in a heated capillary. First, we examine a simple case when the Peclet number is much less than unity. Since $\overline{\ell}_{\mathrm{L}} < 1$, the assumption $\mathrm{Pe}_{\mathrm{L}} \ll 1$ corresponds to the condition $\mathrm{Pe}_{\mathrm{L}}\overline{\ell}_{\mathrm{L}} \ll 1$. In this case Eq. (9.62) reduces to

$$\overline{\ell}_{\mathrm{L}}^2 - K_1\overline{\ell}_{\mathrm{L}} + K_2 = 0 \tag{9.63}$$

where

$$K_1 = (\vartheta_{\mathrm{L}} - \mathrm{Ja})/\vartheta_{\mathrm{L}} \quad K_2 = (\overline{T}_{\mathrm{s}} - 1)/\vartheta_{\mathrm{L}}\mathrm{Pe}_{\mathrm{L}}\,. \tag{9.64}$$

For any finite value of the parameters ϑ_{L} and Pe_{L}, it is seen that the dimensionless length of the liquid domain $\overline{\ell}_{\mathrm{L}}$, cannot be equal to 0 and 1, since the term K_2 and the parameter Ja are positive. To determine a relationship between the parameters corresponding to the stable flow in the case $\mathrm{Pe}_{\mathrm{L}} \ll 1$, Eq. (9.63) should be solved to find $\overline{\ell}_{\mathrm{L}}$

$$\overline{\ell}_{\mathrm{L}} = 0.5\left(K_1 \pm \sqrt{K_1^2 - 4K_2}\right) \tag{9.65}$$

Since $\overline{\ell}_{\mathrm{L}}$ should be positive, the physically realistic solution of Eq. (9.65) is possible only under the following conditions:

$$K_1 > 0 \tag{9.66}$$

$$K_1^2 \geq 4K_2\,. \tag{9.67}$$

From inequality (9.66) it follows that for

$$\vartheta_{\mathrm{L}} > \mathrm{Ja} \tag{9.68}$$

a steady flow is possible.

To estimate the range of possible variations of Pe_{L} and ϑ_{L} for a given $\overline{T}_{\mathrm{s}}$ and Ja inequality (9.67) is presented in the following form:

$$\mathrm{Pe}_{\mathrm{L}} \geq 4\frac{(\overline{T}_{\mathrm{s}} - 1)\vartheta_{\mathrm{L}}}{(\vartheta_{\mathrm{L}} - \mathrm{Ja})^2} \tag{9.69}$$

where the equality corresponds to the low boundary of the stable flow domain.

The dependence $\mathrm{Pe}_{\mathrm{L}} = \mathrm{Pe}_{\mathrm{L}}(\vartheta_{\mathrm{L}})$ is shown in Fig. 9.8a. Each point of this curve corresponds to a certain position of the evaporation front. Extreme points (the upper one $\overline{\ell}_{\mathrm{L}} = 0$ and the lower one $\overline{\ell}_{\mathrm{L}} = 1$) correspond to $\vartheta_{\mathrm{L}} = \mathrm{Ja}$, $\mathrm{Pe}_{\mathrm{L}} \to \infty$ and as $\vartheta_{\mathrm{L}} \to \infty$ the Peclet number Pe_{L} tends to 0, respectively.

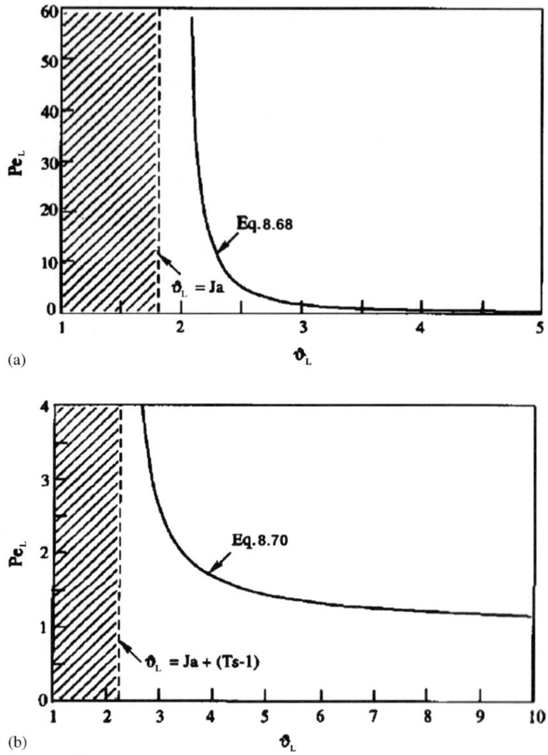

Fig. 9.8a,b The dependence of $\mathrm{Pe_L}(\vartheta_\mathrm{L})$. The *shaded region* indicates unsteady flow: **(a)** $\mathrm{Pe_L} \ll 1$, **(b)** $\mathrm{Pe_L} \gg 1$. Reprinted from Peles et al. (2001) with permission

The domain of the stable flow is located to the right of the boundary $\mathrm{Pe_L}(\vartheta_\mathrm{L})$ (the shaded region in the graph). To the left of this curve is the domain in which stable flows in a heated capillary cannot occur. From the relation between the parameters ϑ_L and Ja, the parametric plane $\mathrm{Pe_L} - \vartheta_\mathrm{L}$ may be subdivided into two domains: (1) $\vartheta_\mathrm{L} < \mathrm{Ja}$, and (2) $\vartheta_\mathrm{L} > \mathrm{Ja}$. Within the first of these the stable flows cannot occur at any values of parameters $\mathrm{Pe_L}$ and $\overline{T}_\mathrm{s}$. At $\vartheta_\mathrm{L} > \mathrm{Ja}$ various flow regimes may be realized: (1) a stable flow at $\mathrm{Pe_L} > \mathrm{Pe_{Lb}}$, and (2) an unstable flow at $\mathrm{Pe_L} < \mathrm{Pe_{Lb}}$ ($\mathrm{Pe_{Lb}}$ is the value of the Peclet number corresponding to the boundary curve).

In the second case when ($\mathrm{Pe_L} \gg 1$, $\mathrm{Pe_L}\overline{\ell}_\mathrm{L} \gg 1$), expression (9.62) reduces to

$$\overline{\ell}_\mathrm{L} = (\mathrm{Pe_L}\vartheta_\mathrm{L})^{-1}\left\{\mathrm{Pe_L}[\mathrm{Ja} + (\overline{T}_\mathrm{s} - 1)] + \vartheta_\mathrm{L}(1 - k_\mathrm{G.L}c_\mathrm{pL.G})\right\}. \tag{9.70}$$

Assuming $\overline{\ell}_\mathrm{L} \leq 1$ in Eq. (9.70) we obtain

$$\mathrm{Pe_L} \leq \frac{\vartheta_\mathrm{L}(1 - k_\mathrm{G.L}c_\mathrm{pG.L})}{\vartheta_\mathrm{L} - [\mathrm{Ja} + (\overline{T}_\mathrm{s} - 1)]}. \tag{9.71}$$

The dependence $\mathrm{Pe_L} = \mathrm{Pe_L}(\vartheta_\mathrm{L})$ is shown in Fig. 9.8b. The line $\vartheta_\mathrm{L} = \mathrm{Ja} + (\overline{T}_\mathrm{s} - 1)$ divides the parametric plane $\mathrm{Pe_L} - \vartheta_\mathrm{L}$ into two domains: (1) an unstable flow regime domain for any value of $\mathrm{Pe_L}[\vartheta_\mathrm{L} < \mathrm{Ja} + (\overline{T}_\mathrm{s} - 1)]$, (2) a stable flow regime domain

$[\vartheta_L > \mathrm{Ja} + (\overline{T}_s - 1)$ and $\mathrm{Pe}_L < 1 - k_{G.L} c_{pL.G}]$. Thus, at $\mathrm{Pe}_L \gg 1$ the domain of stable flows in a heated capillary with a distinct evaporation front is restricted by the lines $\vartheta_L = \mathrm{Ja} + (\overline{T}_s - 1)$ and $\mathrm{Pe}_L = 1 - k_{G.L} c_{pG.L}$. Note that the last restriction had already been met by the restriction $\mathrm{Pe}_L \gg 1$.

9.5 Experimental Facility and Experimental Results

The experimental set-up is shown in Fig. 9.9. Prior to the experiment, the water was passed through a $1/2\,\mu$m cartridge filter, to fill up the reservoir. Water was circulated through the flow loop by a peristaltic pump (6). As the fluid leaves the reservoir (10), it passes through a silicon rubber tube and then flows into the test module (1) and leaves it through the outlet silicon rubber tube. A special Teflon cartridge was manufactured to hold the module in place and to move it in the *x*–*y* plane. The temperature and pressure were measured at both the inlet (7) and exit (8) of the test section using a 0.3 mm copper-constantan thermocouple and silicon pressure gauges, respectively. The thermocouples and pressure gauge output voltages were recorded by multi-scan data acquisition module, which in turn was connected to a PC via a data acquisition card. The input voltage was then translated to °C and kPa by a calibration process prior to the test runs.

A DC power supply with an operating range of 0–60 V was used to supply power to the chip aluminum resistor (2). The Teflon module cartridge was placed under-

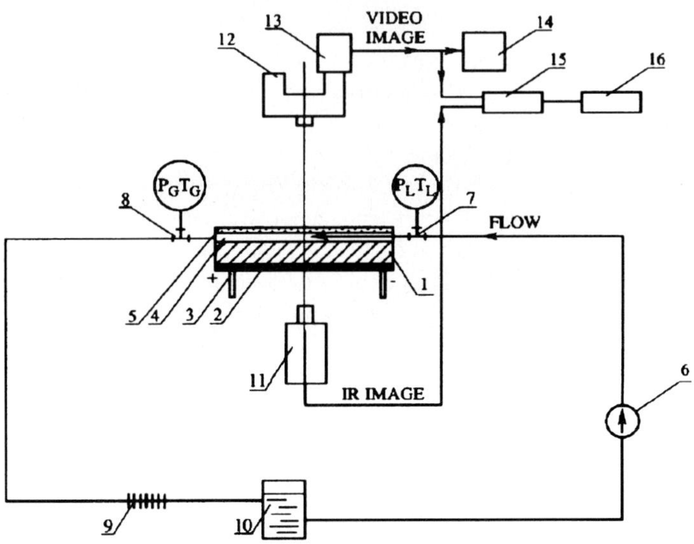

Fig. 9.9 Experimental set-up: *1* test module, *2* heater, *3* electrical contact, *4* micro-channel, *5* Pyrex, *6* peristaltic pump, *7* and *8* pressure and temperature measurements, *9* cooler, *10* reservoir, *11* IR camera, *12* microscope, *13* high-speed video camera, *14* PC, *15* synchronizer, *16* video recorder. Reprinted from Peles et al. (2001) with permission

neath a high-speed CCD camera (13) mounted on a microscope (12), which photographed the boiling process occurring in the micro-channel (4) to the PC (14) and a video recorder (16). At the same time an IR (11) camera measured the resistor's temperature on the other side of the chip. Special software was used to analyze the IR results and to compute the average and maximum temperature on the chip.

The test module is shown in Fig. 9.10. The device of the micro-channel silicon substrate is shown in Fig. 9.11. It consists of some parts: the first and most important is a 525 μm micro-channel silicon substrate (1) bonded to the 500 μm Pyrex cover (5) and the last is the flow manifold. The micro-channel silicon substrate was fabricated by a photolithography process on one side and consists of 16 mm long triangular micro-channels (4) having hydraulic diameter ranging from 50 to 200 μm. On the other side of a $1 \times 1\,\text{cm}^2$ aluminum resistor, silicon substrate was evaporated, forming three distinct channel regions: inlet adiabatic, heat flux dissipation and outlet adiabatic. The channel side was glued to a Pyrex substrate by anodic bonding to form the chip test module.

While the manifolds were fabricated by a plain molding process, the micro-channels substrate fabrication was quite complicated and was achieved by a multi-stage process. The following main stages were used in the process: (1) double side oxidation of a 525 μm (1 0 0) silicon substrate to 1,000 A, (2) single side 1,200 A silicon nitride deposition, (3) silicon nitride channels template opening by reactive

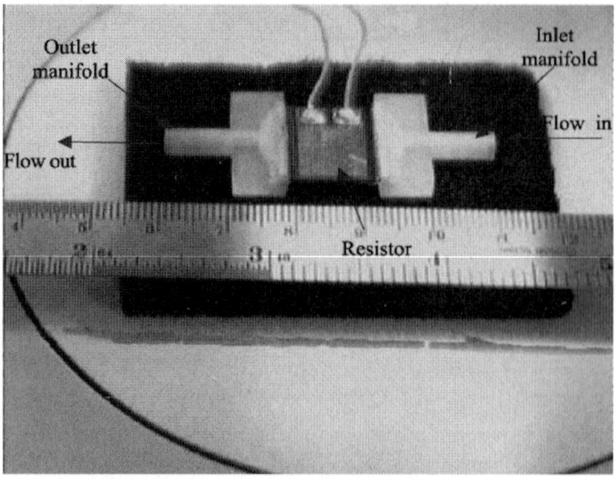

Fig. 9.10 The test module. Reprinted from Peles et al. (2001) with permission

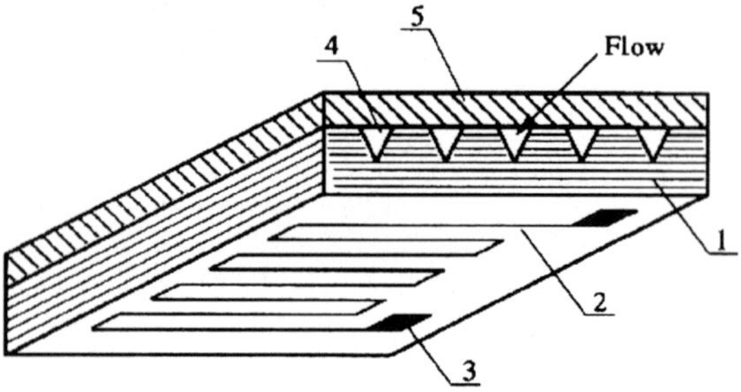

Fig. 9.11 The micro-channel silicon substrate bonded to the 500 μm micro-channel cover Pyrex: *1* test module, *2* heater, *3* electrical contact, *4* micro-channel, *5* Pyrex. Reprinted from Peles et al. (2001) with permission

ion etching (RIE), (4) channels template oxidation buffer oxide etching (BOE), (5) silicon etching by tetramethyl ammonium hydroxide (TMAH), (6) 4,000 A double side oxidation, (7) silicon nitride etching by RIE, (8) 1,000 A oxidation layer etching, (9) silicon/Pyrex anodic bonding, (10) aluminum thin film evaporation, (11) aluminum template etching, and (12) wafer sawing.

Once the micro-channels were fabricated, the manifold were bonded to the substrate by epoxy glue, and the resistor was wired by a silver-based epoxy electric conductor and painted with a 1 pm thin paint to increase the surface emissivity. The triangular micro-channels and the unpainted resistor are shown in Fig. 9.12.

The experiments were conducted on a heat exchanger described above, having 17 triangular micro-channels with hydraulic diameter of 157 μm and mass flow rates of 5.7 ml/min. The applied heat flux ranged from up to 35 W/cm^2. The deviation of q^* from its nominal value does not exceed 0.5%. The characteristic value of the

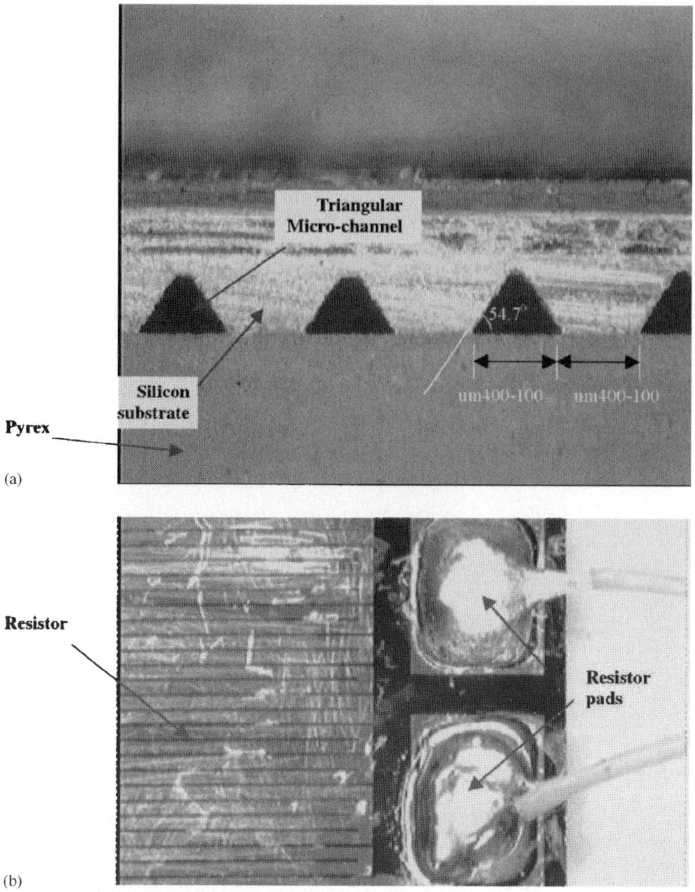

Fig. 9.12 The (**a**) triangular micro-channels, and (**b**) unpainted resistor. Reprinted from Peles et al. (2001) with permission

dimensionless groups at which the observations were carried out are: Eu $\sim 10^6$, $Pe_L \sim 7 \times 10^2$, $\vartheta_L \sim 0.3$, Ja ~ 1.8, and allows us to study the flow in a heated capillary outside the domains of the steady flows.

The experimental results show that the evaporating mechanism in two-phase flows in micro-channels differ considerably from larger sized channels. Two distinct phase domains, one for the liquid and another for the vapor, were observed, with a very short (on the order of the hydraulic diameter) section of two-phase mixture between them. This implies that the outlet vapor mass quality for a steady flow can take only the values of 0 (single-phase liquid flow) or unity (saturated or superheated vapor). The energy required for the flow with zero outlet quality is much lower than the energy required for the quality larger than one. Thus an energy gap is found between those energy levels, for which steady evaporating two-phase flow does not exist. If the applied heat flux is sufficient to initiate evaporation at quality lower than one, an unsteady flow is expected with an outlet phase flow fluctuation corresponding to some time-average mass quality lower than one, as was shown to be the case in the present experiment. This agrees well with the theoretical model results since it lies in the unsteady flow region ($\vartheta_L < \mathrm{Ja} + (\overline{T}_s - 1)$), as can be seen in Fig. 9.8b.

The development of the two-phase flow in a heated capillary at different Peclet number is illustrated in Fig. 9.13. It shows that different mechanisms of two-phase flow formation may occur depending on the value of Pe_L. At small Pe_L the fine bubble formation (on the micro-channel wall) plays a dominant role. Growth of these bubbles leads to a blockage of the micro-channel, to a sharp change of the hydraulic

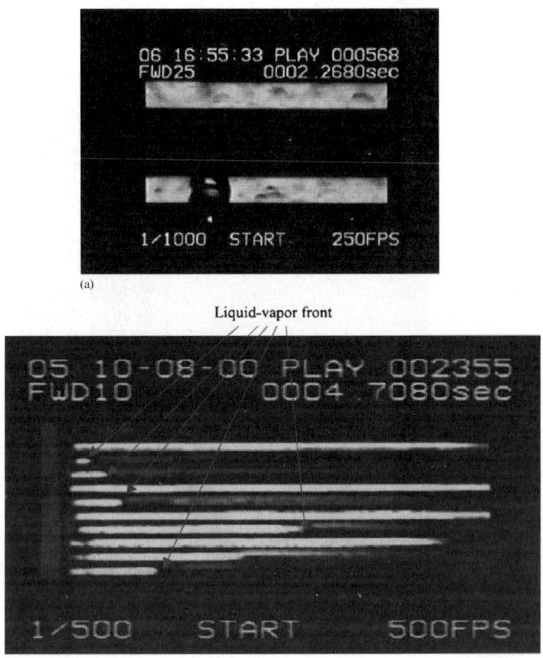

Fig. 9.13a,b Flow types in a heated capillary: (**a**) bubble formation, (**b**) liquid–vapor front. Reprinted from Peles et al. (2001) with permission

resistance and ultimately to an unsteady gas–liquid flow. This effect is negligible at large Peclet number. In this case liquid evaporation is accompanied by formation of the two-phase flow with distinct vapor–liquid evaporation front dividing the domains of the liquid and the vapor. Depending on the values of the governing parameters such flow may be steady or unsteady. A particular flow at $\mathrm{Pe_L} \approx 500$ and $\vartheta_\mathrm{L} \approx 0.3$ (the conditions correspond to boundary of the unsteady flow domains) are shown in Fig. 9.14. A number of successive photos of the flow are presented. They show that the position of the evaporation front changes during the periodical observation. Thus at the given values of parameters the flow in a heated capillary is non-steady. The latter agrees with the estimation of the boundary of steady states given in Sect. 9.4.

The experimental investigation of the flow in a heated capillary shows that, depending on the value of the Peclet number, various types of the process occurred. At small $\mathrm{Pe_L}$, the dominant role is the bubble formation at the channel wall, whereas for $\mathrm{Pe_L} \gg 1$, liquid evaporation leads to formation of an evaporation front.

The experimental results also demonstrated that evaporating two-phase flow in micro-channels with time-average mass quality lower than one is unsteady. This result agrees well with the one-dimensional model prediction. Physically this flow regime can be explained by the special mechanism associated with the micro-channel geometry, which appear to have two distinct phase domains: one for the liquid and the other for the vapor, with a very short section of two-phase mixture between them. It follows that the outlet flow can exhibit only single-phase, either liquid or vapor. Thus, a vapor mass quality lower than 1 points to an alternating outlet phase with a time-average quality lower than one.

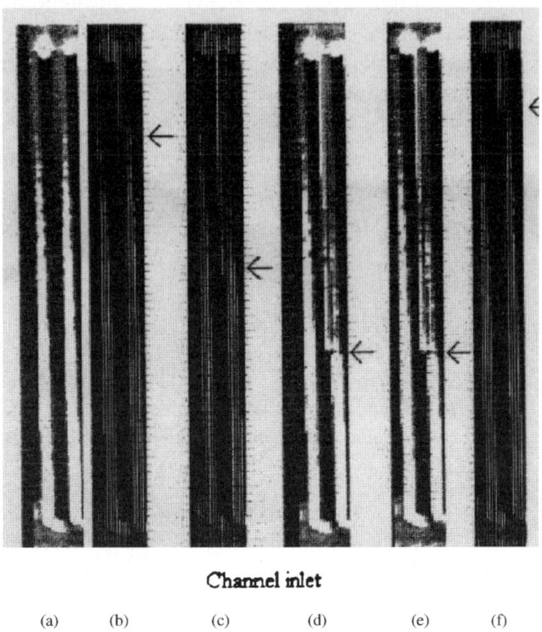

Fig. 9.14 Unsteady flow in a heated capillary: $\vartheta_\mathrm{L} = 0.3$, $\mathrm{Ja} = 1.83$, $\mathrm{Pe_L} = 598$. Reprinted from Peles et al. (2001) with permission

Summary

The quasi-one-dimensional model described in the previous chapter is applied to the study of steady and unsteady flow regimes in heated micro-channels, as well as the boundary of steady flow domains. The effect of a number of dimensionless parameters on the velocity, temperature and pressure distributions within the domains of liquid vapor has been studied. The experimental investigation of the flow in a heated micro-channel is carried out.

The following results have been obtained:

1. The temperature distribution along the micro-channel axis is not monotonic. It has a maximum that is located within the liquid domain. An extraordinary form of the temperature profile is a result of the influence of two opposite factors, namely, absorbs heat from the wall and heat transfer from liquid to the front in order to establish the evaporation process. An increase of heat flux on the wall leads to displacement of the point corresponding to maximum temperature towards the inlet cross-section.
2. An increase of the Peclet number leads to a decrease of the length of liquid domain, as well as an increase of the liquid–vapor heat flux ratio at the evaporating front.
3. The different types of the process can be realized depending on the value of the Peclet number, the dominant role is the bubble formation at the channel wall, whereas for large Pe, liquid evaporation leads to formation of an evaporation front.
4. The experimental results also demonstrated that evaporating two-phase flow in micro-channels with time-average mass quality lower than one is unsteady. This result agrees well with the one-dimensional model prediction. Physically this flow regime can be explained by a special mechanism associated with micro-channel geometry, which appears to have two distinct phase domains: one for the liquid and the other for the vapor, with a very short section of two-phase mixture between them. It follows that the outlet flow can exhibit only single-phase, either liquid or vapor. Thus, a vapor mass quality lower than 1 means alternating outlet phase with a time-average quality lower than one.

References

Bowers MB, Mudawar I (1994) High flux boiling in low flow rate, low pressure drop mini-channel and micro-channel heat sink. Int J Heat Mass Transfer 37:321–332

Hsu YY (1962) On the size range of active nucleating cavities on a heating surface. J Heat Transfer 184:207–213

Katto Y (1978) A generalized correlation for critical heat flux for the forced convection boiling in vertical uniformly heated round tubes. Int J Heat Mass Transfer 21:1527–1542

Khrustalev D, Faghri A (1996) Fluid flow effect in evaporation from liquid-vapor meniscus. ASME J Heat Mass Transfer 118:725–747

Kuchling H (1980) Nachschlagebücher für Grundlagenfächer Physik. Verb Fachbuchverlag, Leipzig

Landau LD, Lifshitz EM (1959) Fluid Mechanics, 2nd edn. Pergamon, London
Peles YP, Yarin LP, Hetsroni G (1998) Heat transfer of two phase flow in a heated capillary, Heat Transfer 1998. In: Proceedings of the 11th International Heat Transfer Conference, Kyongju, Korea, 23–28 August 1998, vol 2
Peles YP, Yarin LP, Hetsroni G (2001) Steady and unsteady flow in a heated capillary. Int J Multiphase Flow 27:577–598
Peng XF, Hu HY, Wang BX (1998) Boiling nucleation during liquid flow in micro-channels. Int J Heat Mass Transfer 41:101–106
Peng XF, Peterson GP, Wang BX (1996) Flow boiling in binary mixtures in micro-channels plates. Int J Heat Mass Transfer 39:1257–1263
Peng XF, Wang BX (1993) Forced convection and flow boiling through micro-channels Int J Heat Mass Transfer 36:3421–3427
Peng XF, Wang BX (1994a) Cooling characteristics with micro-channeled structures. J Enhanced Heat Transfer 1:315–326
Peng XF, Wang BX, Peterson GP, Ma HB (1994b) Experimental investigation of heat transfer in flat plates with rectangular micro-channels. Int J Heat Mass Transfer 37:127–137
Tukermann DB, Pease RFW (1981) High performance heat sink for VLSI. IEEE Electron Device Lett EDI 2:126–129
Yamagata K, Hironai F, Mishikawa K, Matsouka H (1955) Nucleate boiling of water on the horizontal heating surface. Memoirs of the Faculty of Engineering, vol 15. Kyushu University, Japan, p 98

Nomenclature

c_{p}	Specific heat
d	Hydraulic diameter
F	Drag force
H	Depth of meniscus
k	Thermal conductivity
L	Total channel length
ℓ_{l}	Liquid length domain
n	Normal to interface surface
P	Pressure
q	Heat flux permit axial length
Q_0	Heat losses from the liquid in the inlet cross-section
R	Gas constant
r	Radius of curvature of the meniscus
r_0	Radius of micro-channel
T	Temperature
u	Inlet velocity
u_{f}	Velocity of the evaporating front in the system of coordinates associated with the micro-channel walls
u'_{f}	Velocity of the evaporating front in the system of coordinates associated with the fluid flow
$\mathrm{Eu} = \dfrac{P}{\rho u^2}$	Euler number

$\mathrm{Ja} = \dfrac{h_{\mathrm{LG}}}{C_{\mathrm{p}}T}$	Jacob number
$\mathrm{Pe} = \dfrac{uL}{\alpha}$	Peclet number
$\mathrm{Re} = \dfrac{\rho u L}{\mu}$	Reynolds number
$\mathrm{We} = \dfrac{L\rho u^2}{\sigma}$	Weber number

Greek symbols

α	Thermal diffusivity
ϑ	Dimensionless heat flux at the wall
θ	Contact angle
μ	Viscosity
ρ	Density
σ	Surface tension

Subscripts

f	Fluid
G	Vapor
L	Liquid
S	Saturated
t	Total
0	Inlet
00	Outlet

Chapter 10
Laminar Flow in a Heated Capillary with a Distinct Interface

The subject of the present chapter is the analysis of possible states of capillary flow with distinct evaporative meniscus. The system of quasi-one-dimensional mass, momentum and energy equations are applied to classify the operating parameters corresponding to various types of flow. The domains of steady and unsteady states are also outlined.

10.1 Introduction

One of the possible ways to enhance the heat transfer in cooling systems of electronic devices, with high power densities, is phase change of the coolant in micro-channels. The possibility of enhancing heat transfer motivated a number of groups to study two-phase boiling heat transfer in mini- and micro-channels (Bowers and Mudawar 1994; Morijama and Inoue 1992; Landerman 1994; Peng et al. 1996; Peng and Wang 1993).

Two-phase flows in micro-channels with an evaporating meniscus, which separates the liquid and vapor regions, have been considered by Khrustalev and Faghri (1996) and Peles et al. (1998, 2000). In the latter a quasi-one-dimensional model was used to analyze the thermohydrodynamic characteristics of the flow in a heated capillary, with a distinct interface. This model takes into account the multi-stage character of the process, as well as the effect of capillary, friction and gravity forces on the flow development. The theoretical and experimental studies of the steady forced flow in a micro-channel with evaporating meniscus were carried out by Peles et al. (2001). These studies revealed the effect of a number of dimensionless parameters such as the Peclet and Jacob numbers, dimensionless heat transfer flux, etc., on the velocity, temperature and pressure distributions in the liquid and vapor regions. The structure of flow in heated micro-channels is determined by a number of factors: the physical properties of fluid, its velocity, heat flux on the wall, etc. At a fixed geometry, the flow pattern in a micro-channel depends, mainly, on the liquid velocity v and heat flux on the wall q_{w} (i.e., the values of the Peclet number and dimensionless heat flux), where different regimes of flow take place. At large q_{w} and small v, bubble nucleation is the dominant factor that determines the flow pattern. In the case of relatively small q_{w} and large v, bubble nucleation is negligible. Under these conditions a two-phase flow with distinct meniscus (dividing the regions of pure liquid and pure vapor flows) forms in the micro-channel. Such flow characterizes small channels with high critical heat flux.

In spite of the fact that for the last decade the flow in micro-channels has attracted significant interest (Incropera 1995; Peng and Wang 1994; Peng et al. 1994; Ha and Peterson 1998; Peterson and Ha 1998; Triplett et al. (part I) 1999a; Triplett et al. (part II) 1999b; Ghiaasiaan and Abdel-Khalik 2001) a number of important problems related to the hydrodynamic and heat transfer in such channels have been less investigated. This particularly concerns the parametrical dependence of the process, its stability, conditions of the existence of stationary flow, etc.

In Chap. 10 the study of complex processes in a heated capillary with evaporating meniscus is presented. The study consists of the formulation of the problem, detailed analysis of the influence of the physical properties of the coolant and wall heat flux on the thermal regime of the flow, hydraulic resistance of the micro-channel, as well as the efficiency of the cooling system on the whole. In order to study the thermohydrodynamic characteristics of a two-phase capillary flow with phase change at the meniscus a quasi-one-dimensional model is used. It takes into account the principal characteristics of the phenomenon, namely, the effects of the inertia, pressure, gravity, friction forces and capillary pressure due to the curvature of the interface surface, as well as the thermal and dynamical interactions of the liquid and vapor phases. To describe the flow outside of the meniscus, in the domains of the pure liquid or vapor, the one-dimensional mass, momentum and energy equations are used. The possible states of the flow are considered, and the domains of steady and unsteady states are outlined. An equation for stationary two-phase flow regimes in heated micro-channels is derived. This equation is applied to classify the operating parameters, corresponding to various types of flow.

Chapter 10 consists in following: the model of the cooling system of the electronic device with high power density is described in Sect. 10.2. Section 10.3 deals with the mathematical formulation of the problem of laminar flow in a heated capillary with phase change. The statement of the dimensionless variables, reformulation of the problem in these variables, its parametrical study as well as the analysis of the existence of steady and unsteady states are presented in Sects. 10.4, 10.5, and 10.6. In Sect. 10.7 we discuss the results of the numerical calculations. The problem of efficiency of the cooling system and the condition of its optimal functioning are presented in Sect. 10.8. The relation between the saturation parameters of the vapor is presented in Sect. 10.9.1. The integral relations are derived in Sect. 10.9.2. The analysis of the solutions of the governing system of equations is presented in Sect. 10.9.3.

10.2 Model of the Cooling System

The elements of a cooling system of electronic devices are micro-heat exchangers with extremely large surface area per unit volume, low thermal resistance and small mass. As a rule, they are systems of parallel channels with hydraulic diameters ranging from 10 μm to 10^3 μm with inlet and outlet manifolds, which connect the channels (Fig. 10.1). A sketch of the cooling system is presented in Fig. 10.2. It consists of a micro-channel (1), condenser (2), heat exchanger (3), pump (4) and two tanks (5) and (6). The coolant is supplied into the micro-channel (by the pump), where it is heated and vaporizes on the meniscus. The vapor enters into the vapor tank (6) and then into the condenser (2) where it condenses. The water from the condenser (2) enters into the heat exchanger (3) and is cooled down to ambient temperature. The pressure $P_{L,0}$ and temperature $T_{L,0}$ at the inlet of the micro-channel can be changed by regulating the heat exchanger and the pump.

The thermohydrodynamic characteristics of the flow in the heated micro-channels depend on the following factors: the heat flux on the wall, which determines the intensity of the vaporization, the location of the meniscus, the difference between the inlet and outlet pressures, the capillary, and mass and friction forces which act on the liquid and vapor.

The processes in a cooling system of electronic devices with high power density can be modeled as follows. The coolant with temperature $T_{2.0}$ and pressure $P_{2.0}$ enters into the micro-channel from the tank (5) (Fig. 10.2). The mass capacity of the liquid in the tank (5) is large enough, therefore the heat flux from the micro-channel

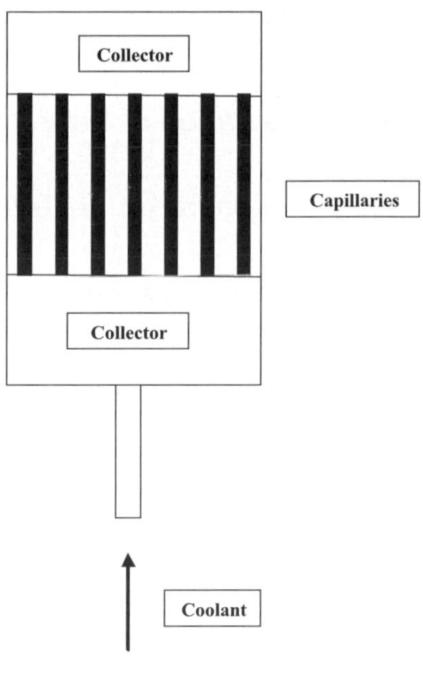

Fig. 10.1 Cooling element of electronic device. Reprinted from Yarin et al. (2002) with permission

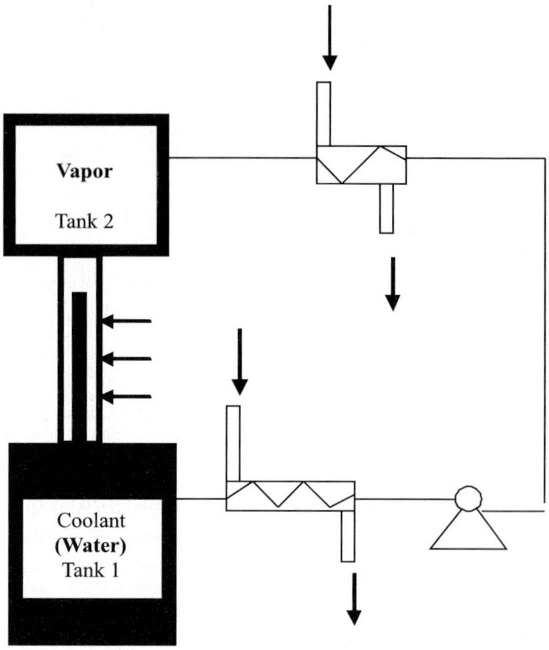

Fig. 10.2 Scheme of a cooling system. Reprinted from Yarin et al. (2002) with permission

does not influence the coolant initial temperature. Phase change occurs directly on the meniscus. Its position depends on the operating parameters and the physical properties of the cooling fluid and its vapor. The upper tank (6) and the condenser (2) ensure a given pressure of the vapor at the outlet of the micro-channel.

10.3 Formulation of the Problem

10.3.1 Conditions on the Interfacial Surface

The existence of the interfacial surface with a finite curvature causes the capillary pressure, which determines the rising of the liquid, and its height in a stationary state (Levich 1962). The latter is determined by the equilibrium of the gravity and capillary forces. The situation changes drastically, when the phase change occurs on the meniscus surface. The liquid evaporation disturbs the equilibrium of the gravity and capillary forces and provides for the motion of both phases and displacement of the interface. In this case the flow field is demarcated by the interface into two domains where pure liquid or vapor flow occurs. Obviously, such structure of flow in a micro-channel is possible when bubble nucleation within the liquid domain is absent. The theoretical analysis by Peng et al. (1998) showed, that the critical value of the heat flux on the wall at which the bubble nucleation is negligible depends on the physical properties of the liquid, and the micro-channel diameter and is expressed by the following inequality

$$\frac{h_{\mathrm{LG}}\alpha}{c\pi(v''-v')d} > q_{\mathrm{cr}} \tag{10.1}$$

where h_{LG} is the latent heat of evaporation, v'' and v' are the vapor and liquid specific volume at saturation, α is the thermal diffusivity of the vapor, d is the micro-channel diameter, c is an empirical constant has order of unity. For water flow in a micro-channel with $d = 10^{-4}$ m ($v'' = 1.673\,\mathrm{m^3/kg}$, $v' = 0.00104\,\mathrm{m^3/kg}$, $h_{\mathrm{LG}} = 2.26\times10^6\,\mathrm{J/kg}$, $\alpha = 18.58\times10^{-8}\,\mathrm{m^2/s}$) the critical heat flux q_{cr} is about $8\times10^4\,\mathrm{W/m^2}$ and increases with decreasing d. In experiments by Peng and Wang (1998), Peng et al. (1994, 1996) vapor bubbles were not observed in water and methanol flows in rectangular micro-channels with a cross-section ranging from 0.1×0.3 mm to 0.6×0.7 mm even when the applied heat flux was much higher than $10^5\,\mathrm{W/m^2}$. Thus, the theoretical estimations, as well as the experiment, show that the considered model of flow in a heated micro-channel is valid for a wide range of heat fluxes, which are of practical interest.

Consider a flow of liquid coolant in a capillary where heating and evaporation occur. The conditions on the interface are expressed by means of the continuity of the mass and thermal fluxes and the equilibrium of all acting forces (Landau and Lifshitz 1959):

$$\sum_{i=1}^{2}(\rho_i \mathrm{v}_i)\mathrm{n}_i = 0 \tag{10.2}$$

$$\sum_{i=1}^{2}(\rho_i \mathrm{v}_i h_i - k_i\nabla T_i)\mathrm{n}_i = 0 \tag{10.3}$$

$$\sum_{i=1}^{2}(P_i + \rho_i v_{\mathrm{in}} v_{\mathrm{in}})\mathbf{n_i} = (\tau_{*G} - \tau_{*L})\mathbf{n}_k + \sigma\left(r_1^{-1} + r_2^{-1}\right)\mathbf{n}_{\mathrm{G}} + \nabla\sigma \tag{10.4}$$

where $\rho, \mathbf{v}, T, h$ and P are the density, velocity, temperature, enthalpy and pressure; σ is the surface tension; k is the thermal conductivity; τ_* is the viscous tension tensor; $v_i = v_i n_i$ and $\nabla T_i n_i$ are the normal components of the velocity vector and the temperature interface surface gradient, respectively; r_1 and r_2 are the general radii of the curvature for the interface; $\mathbf{n}_i$ and $\mathbf{n}_k$ are the normal and tangent directions; $\mathbf{n}_{\mathrm{G}} = -\mathbf{n}_{\mathrm{L}}$ refer to the vapor and the liquid; the bold letters indicate a vector, $\sum_{i=1}^{2}$ is the sum of terms related to vapor ($i = 1 = \mathrm{G}$) and liquid ($i = 2 = \mathrm{L}$).

Consider a liquid–solid system for which the contact angle is close to 90° (for example, the contact angle for the water–steel system is $70° < \theta < 90°$ (Grigoriev and Zorin 1982)). The projections of the velocity vector $\mathbf{v}$ on x-, y-, z-axes are $u = |\mathbf{v}|\sin\theta$, $v = w = |\mathbf{v}|\cos\theta$. For the surface, which is weakly bent, $\cos\theta \ll 1$ and $\sin\theta \sim 1$ we have the following estimate $u \gg v = w$. Analogously, $\partial \mathrm{T}/\partial \mathrm{x} \gg \partial \mathrm{T}/\partial \mathrm{y} = \partial \mathrm{T}/\partial \mathrm{z}$.

Bearing in mind these estimates and the assumptions that $\sigma = \text{const.}$, also that the fluid is incompressible, Eqs. (10.2–10.4) transform as follows:

$$\rho_G \tilde{V}_G = \rho_L \tilde{V}_L \tag{10.5}$$

$$P_G + \rho_G \tilde{V}_G^2 = P_L + \rho_L \tilde{V}_L^2 + f_L \tag{10.6}$$

$$\rho_G \tilde{V}_G h_G - k_G \frac{\partial T_G}{\partial x} = \rho_L \tilde{V}_L h_L - k_L \frac{\partial T_L}{\partial x} \tag{10.7}$$

where $\tilde{V}_i = u_i - V_f$ is the relative velocity, $V_f = \partial x_f / \partial t$ is the velocity of the interface surface, $f_L = 2\sigma/R$ is the capillary pressure, $R = r_1 = r_2$ is the radius of the surface curvature.

Since the coolant and its vapor are conductive fluids, $T_{G,f} = T_{L,f} = T_s$, where the subscripts s and f correspond to the saturation parameters and the interface surface, respectively. The saturation pressure and temperature are weakly connected (Sect. 10.9.1), so that T_s is determined practically by the external pressure $P_{G,00}$.

It should be stressed that Eqs. (10.5) and (10.6) are satisfied for any values of the contact angle, whereas Eq. (10.7) is correct only for θ that is close to 90°.

10.3.2 The Flow Outside of the Interfacial Surface

To calculate the flow fields outside the evaporating meniscus we use the one-dimensional model, developed by Peles et al. (1998, 2000, 2001). Assuming that the compressibility and the energy dissipation are negligible (a flow with moderate velocities), the thermal conductivity and viscosity are independent of the pressure and temperature, we arrive at the following system of equations:

$$\frac{\partial \rho_i}{\partial t} + \frac{\partial \rho_i u_i}{\partial x} = 0 \tag{10.8}$$

$$\rho_i \frac{\partial u_i}{\partial t} + \rho_i u_i \frac{\partial u_i}{\partial x} = -\frac{\partial P_i}{\partial x} - \rho_i g - \frac{\partial F_i}{\partial x} \tag{10.9}$$

$$\rho_i \frac{\partial h_i}{\partial t} + \rho_i u_i \frac{\partial h_i}{\partial x} = \frac{\partial}{\partial x}\left(k_i \frac{\partial T_i}{\partial x}\right) + q \tag{10.10}$$

where F is the specific friction force, q is the specific volumetric rate of heat absorption, and g is the acceleration due to gravity.

The initial and boundary conditions for the problem are

$$t = 0: \ \rho_i = \rho_i(x)\,,\ u_i = u_i(x)\,,\ T_i = T_i(x)\,,\ h_i = h_i(x)\,,\ x_f = x_f^*\,, \tag{10.11}$$

where the x is the longitudinal coordinate; x_f and x_f^* are the actual and initial position of the meniscus:

$$t > 0: x = 0, \rho_L = \rho_{L.0}, u_L = u_{L.0}, T_L = T_{L.0}, h_L = h_{L.0}, P_L = P_{L.0} \tag{10.12}$$

$$t > 0: x = x_f^*, \rho_i = \rho_{i.f}, u_i = u_{i.f}, T_i = T_{i.f}, h_i = h_{i.f}, P_i = P_{i.f} \tag{10.13}$$

$$t > 0: x = L, \rho_G = \rho_{G.00}, u_G = u_{G.00}, \frac{\partial T_G}{\partial x} = 0, \frac{\partial h_G}{\partial x} = 0, P_G = P_{G.00} \tag{10.14}$$

where L is the total length of the micro-channel, and subscripts 0 and 00 are related to the inlet and outlet, respectively (Fig. 10.3).

The boundary conditions (10.12–10.14) correspond to the flow in a micro-channel with a cooled inlet and adiabatic receiver (an adiabatic pipe or tank, which is established at the exit of the micro-channel). Note, that the boundary conditions of the problem can be formulated by another way, if the cooling system has another construction, for example, as follows: $x = 0$, $T_L = T_{L.0}$, $x = L$, $T_G = T_{G.00}$, when the inlet and outlet are cooled; $x = 0$, $\partial T_L/\partial x = 0$, $x = L$, $T_G = T_{G.00}$ in case of the adiabatic inlet and the cooled outlet, etc.

We supplement the system of Eqs. (10.8–10.10) by the equations of state of the vapor and liquid

$$P_G = \rho_G R T_G \tag{10.15}$$

$$\rho_L = \rho_L(T_L) \tag{10.16}$$

and by the equation for the vapor pressure at the interface surface

$$P_{G.f} = P_{G.f}(T_{G.f})\,. \tag{10.17}$$

The quasi-one-dimensional model is based on the system of Eqs. (10.8–10.10) with condition (10.5–10.7) and describes the major features of the flow in the heated

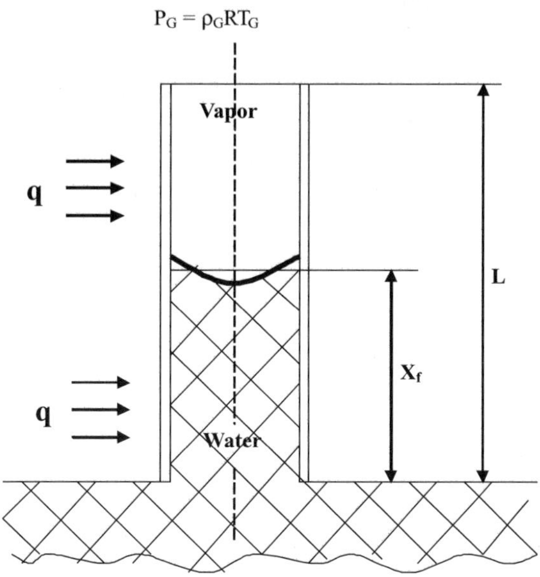

Fig. 10.3 Characteristic domains in heated capillary: *1* flow of vapor, *2* flow of liquid (*bold line* is interface). Reprinted from Yarin et al. (2002) with permission

capillary. This model takes into account the general characteristics of the process due to the curvature of the interfacial surface, as well as the flow of the liquid and vapor, that is caused by evaporation. A similar model was successfully used by Yuan and Prosperetti (1999) for a study on the pumping effect of growing and collapsing bubbles in a tube. It should be noted that the actual position of the meniscus x_{f}, as well as the velocity of the coolant at the inlet $u_{2,0}$, and the vapor temperature at the outlet of the micro-channel $T_{\mathrm{G},00}$ are a priori unknown. In order to determine these parameters it is necessary to supply the system of governing equations, the integral correlations connecting inlet and outlet parameters. They take the following form (Sect. 10.9.2):

$$\frac{\partial}{\partial t}\left(\int_{0}^{x_{\mathrm{f}}}\rho_{\mathrm{L}}\,\mathrm{d}x+\int_{x_{\mathrm{f}}}^{L}\rho_{\mathrm{G}}\,\mathrm{d}x\right)+\left(\rho_{\mathrm{G.00}}u_{\mathrm{G.00}}-\rho_{\mathrm{L.0}}u_{\mathrm{L.0}}\right)=0 \tag{10.18}$$

$$\frac{\partial}{\partial t}\left(\int_{0}^{x_{\mathrm{f}}}\rho_{\mathrm{L}}u_{\mathrm{L}}\,\mathrm{d}x+\int_{x_{\mathrm{f}}}^{L}\rho_{\mathrm{G}}u_{\mathrm{G}}\,\mathrm{d}x\right)=f_{\mathrm{L}}-\left\{\left(\rho_{\mathrm{G}}u_{\mathrm{G}}^{2}\right)_{00}-\left(\rho_{\mathrm{L}}u_{\mathrm{L}}^{2}\right)_{0}\right\}+\left(P_{\mathrm{L.0}}-P_{\mathrm{G.00}}\right)$$

$$-g\left(\int_{0}^{x_{\mathrm{f}}}\rho_{\mathrm{L}}\,\mathrm{d}x+\int_{x_{\mathrm{f}}}^{L}\rho_{\mathrm{G}}\,\mathrm{d}x\right)-\left(F_{\mathrm{L.f}}+F_{\mathrm{G.00}}\right) \tag{10.19}$$

$$\frac{\partial}{\partial t}\left(\int_{0}^{x_{\mathrm{f}}}\rho_{\mathrm{L}}h_{\mathrm{L}}\,\mathrm{d}x+\int_{x_{\mathrm{f}}}^{L}\rho_{\mathrm{G}}h_{\mathrm{G}}\,\mathrm{d}x\right)+\left\{\left(\rho_{\mathrm{G}}u_{\mathrm{G}}h_{\mathrm{G}}\right)_{00}-\left(\rho_{\mathrm{L}}u_{\mathrm{L}}h_{\mathrm{L}}\right)_{0}\right\}$$

$$=-k_{\mathrm{L}}\left(\frac{\partial T_{\mathrm{L}}}{\partial x}\right)_{0}+qL\,. \tag{10.20}$$

Note, that the term $k_{\mathrm{G}}(\partial T_{\mathrm{G}}/\partial x)_{00}$, which belongs to the boundary condition (10.14) is omitted on the right-hand side of Eq. (10.20).

10.4 Non-Dimensional Variables

Assuming steady state in Eqs. (10.8–10.10) and (10.18–10.20), we obtain the system of equations, which determines steady regimes of the flow in the heated micro-channel. We introduce values of density $\rho_{*}=\rho_{\mathrm{L.0}}$, velocity u_{*}, length $\ell_{*}=L$, temperature $T_{*}=T_{L,0,}$ pressure $\Delta P_{*}=P_{L,0}-P_{\mathrm{G},00}$ and enthalpy h_{LG} as characteristic scales. The dimensionless variables are defined as follows:

$$\overline{\rho}=\rho/\rho_{*}\,,\quad \overline{u}=u/u_{*}\,,\quad \overline{x}=x/l_{*}\,,\quad \overline{T}=T/T_{*}$$

$$\overline{h}=h/h_{\mathrm{LG}}\,,\quad \gamma=\frac{kT_{*}}{\rho_{*}u_{*}h_{\mathrm{LG}}l_{*}}\,,\quad \overline{P}=P/\Delta P_{*} \tag{10.21}$$

and we obtain the dimensionless equations

$$\frac{\mathrm{d}\overline{\rho}_i\overline{u}_i}{\mathrm{d}\overline{x}}=0 \tag{10.22}$$

$$\overline{\rho}_i\overline{u}_i\frac{\mathrm{d}\overline{u}_i}{\mathrm{d}\overline{x}}=-\mathrm{Eu}\frac{\mathrm{d}\overline{P}_i}{\mathrm{d}\overline{x}}-\mathrm{Fr}^{-1}\overline{\rho}_i-\frac{\mathrm{d}\overline{F}_i}{\mathrm{d}\overline{x}} \tag{10.23}$$

$$\overline{\rho}_i\overline{u}_i\frac{\mathrm{d}\overline{h}_i}{\mathrm{d}\overline{x}}=\gamma_i\frac{\mathrm{d}^2\overline{T}_i}{\mathrm{d}\overline{x}^2}+\vartheta \tag{10.24}$$

where $\mathrm{Eu}=\Delta P_*/(\rho_* u_*^2)$ and $\mathrm{Fr}=u_*^2/(gL)$ are the Euler and Froude numbers, $\vartheta=qL/(\rho_* u_* h_{\mathrm{LG}})$, $\overline{F}_i=F_i/(\rho_* u_*^2)$, $\gamma_i=\tilde{q}_i/\mathrm{Pe}_i$, $\tilde{q}_i=c_{\mathrm{pi}}T_*/h_{\mathrm{LG}}$, $\mathrm{Pe}_i=u_*L/\alpha_i$ is the Peclet number, $\alpha_i=k_i/\rho_* c_{\mathrm{pi}}$.

Choose the characteristic velocity u_* so, that total heat flux on the wall is fully expended for liquid evaporation (the heating without any losses of heat: $\vartheta\equiv 1$). We conclude that

$$u_*=\frac{qL}{\rho_* h_{\mathrm{LG}}}=\frac{4q_{\mathrm{w}}L}{d\rho_* h_{\mathrm{LG}}}\,. \tag{10.25}$$

The conditions (10.5–10.7) and integral relations (10.18–10.20) become

$$\overline{\rho}_{\mathrm{G}}\overline{\tilde{V}}_{\mathrm{G}}=\overline{\rho}_{\mathrm{L}}\overline{\tilde{V}}_{\mathrm{L}} \tag{10.26}$$

$$\mathrm{Eu}\overline{P}_{\mathrm{G}}+\overline{\rho}_{\mathrm{G}}(\overline{\tilde{V}}_{\mathrm{G}})^2=\mathrm{Eu}\overline{P}_{\mathrm{L}}+\overline{\rho}_{\mathrm{L}}(\overline{\tilde{V}}_{\mathrm{L}})^2+\mathrm{We}^{-1} \tag{10.27}$$

$$\overline{\rho}_{\mathrm{G}}\overline{\tilde{V}}_{\mathrm{G}}\overline{h}_{\mathrm{G}}-\gamma_{\mathrm{G}}\frac{\partial\overline{T}_{\mathrm{G}}}{\partial\overline{x}}=\overline{\rho}_{\mathrm{L}}\overline{\tilde{V}}_{\mathrm{L}}\overline{h}_{\mathrm{L}}-\gamma_{\mathrm{L}}\frac{\partial\overline{T}_{\mathrm{L}}}{\partial\overline{x}} \tag{10.28}$$

and

$$\overline{\rho}_{\mathrm{G.00}}\overline{u}_{\mathrm{G.00}}-\overline{\rho}_{\mathrm{L.0}}\overline{u}_{\mathrm{L.0}}=0 \tag{10.29}$$

$$\mathrm{We}^{-1}-\{(\overline{\rho}_{\mathrm{G}}\overline{u}_{\mathrm{G}}^2)_{00}-(\overline{\rho}_{\mathrm{L}}\overline{u}_{\mathrm{L}}^2)_0\}+\mathrm{Eu}$$
$$-\mathrm{Fr}^{-1}\left(\int_0^{\overline{x}_{\mathrm{f}}}\overline{\rho}_{\mathrm{L}}\,\mathrm{d}\overline{x}+\int_{\overline{x}_{\mathrm{f}}}^{\overline{L}}\overline{\rho}_{\mathrm{G}}\,\mathrm{d}\overline{x}\right)-(\overline{F}_{\mathrm{L.f}}+\overline{F}_{\mathrm{G.00}})=0 \tag{10.30}$$

$$(\overline{\rho}_{\mathrm{G}}\overline{u}_{\mathrm{G}}\overline{h}_{\mathrm{G}})_{00}-(\overline{\rho}_{\mathrm{L}}\overline{u}_{\mathrm{L}}\overline{h}_{\mathrm{L}})_0=-\gamma_{\mathrm{L}}\left(\frac{\partial\overline{T}_{\mathrm{L}}}{\partial\overline{x}}\right)_0+1 \tag{10.31}$$

where $\mathrm{We}=(\rho_* u_*^2 2R)/\sigma$ is the Weber number, $R=d/2\cos\theta$, d is the diameter of the micro-channel.

The system of Eqs. (10.22–10.24) and relations (10.26–10.31) contains five dimensionless parametrical groups Eu, Fr, We, γ_{G} and γ_{L}, which completely determine the problem.

10.5 Parametrical Equation

Now let us transform the integral relations (10.30) and (10.31). For this, we write the solution of Eqs. (10.22) and (10.24). They are (Peles et al. 2001)

$$\overline{\rho}_i \overline{u}_i = \text{const} \tag{10.32}$$

$$\overline{T}_\mathrm{L} = C_1^{(\mathrm{L})} + \vartheta^* [\overline{x} + (\mathrm{Pe}\overline{u}_{\mathrm{L.0}})^{-1}] + C_2^{(\mathrm{L})} \exp(\mathrm{Pe}\overline{x}\overline{u}_{\mathrm{L.0}}) \tag{10.33}$$

where $C_1^{(\mathrm{L})} = (1 - C_2^{(\mathrm{L})}) - \vartheta^*/(\mathrm{Pe}\overline{u}_{\mathrm{L.0}})$, $C_2^{(\mathrm{L})} = ((\overline{T}_\mathrm{s} - 1) - \vartheta^* \overline{x}_\mathrm{f}) / (\exp(\mathrm{Pe}\overline{x}_\mathrm{f}\overline{u}_{\mathrm{L.0}}) - 1)$, $\vartheta^* = h_{\mathrm{LG}}/(c_\mathrm{p} T_* \overline{u}_{\mathrm{L.0}})$, $\overline{T}_\mathrm{s} = T_\mathrm{s}/T_*$, T_s is the temperature on the interface.

We also determine the friction forces $F_{\mathrm{L.f}}$ and $F_{\mathrm{G.00}}$. Assuming that

$$F_i = \int_{x_i'}^{x_i''} \eta_i \frac{\rho_i u_i^2}{2} \frac{\mathrm{d}x}{d} \tag{10.34}$$

and taking into account that $x_\mathrm{L}' = 0$, $x_\mathrm{L}'' = x_\mathrm{f}$, $x_\mathrm{G}' = x_\mathrm{f}$, $x_\mathrm{G}'' = L$ and $\eta_i = 64/\mathrm{Re}_i$ ($\mathrm{Re}_i = u_i d/\nu_i$ where ν_i are the kinematic viscosity) we obtain

$$F_{\mathrm{L.f}} = \frac{8}{r^2} u_\mathrm{L} \mu_\mathrm{L} x_\mathrm{f} \tag{10.35}$$

$$F_{\mathrm{G.00}} = \frac{8}{r^2} u_\mathrm{G} \mu_\mathrm{G} (L - x_\mathrm{f}) \tag{10.36}$$

where r is the micro-channel radius and μ is the dynamic viscosity of the fluid.

Then the last term on the left-hand side of Eq. (10.30) turns out to be

$$(\overline{F}_{\mathrm{L.f}} + \overline{F}_{\mathrm{G.00}}) = \frac{32}{\mathrm{Re_L}} \overline{u}_\mathrm{G} [\overline{x}_\mathrm{f} + \nu_{\mathrm{GL}} (\overline{L} - \overline{x}_\mathrm{f})] \tilde{L} \tag{10.37}$$

where $\mathrm{Re_L} = u_* d/\nu_\mathrm{L}$, $\nu_{\mathrm{G.L}} = \nu_\mathrm{G}/\nu_\mathrm{L}$, $\tilde{L} = L/d$.

Since $\overline{\rho}_\mathrm{G} \overline{u}_\mathrm{G} = \overline{\rho}_\mathrm{L} \overline{u}_\mathrm{L}$, the left-hand side of Eq. (10.31) is

$$(\overline{\rho}_\mathrm{G} \overline{u}_\mathrm{G} \overline{h}_\mathrm{G})_{00} - (\overline{\rho}_\mathrm{L} \overline{u}_\mathrm{L} \overline{h}_\mathrm{L})_0 = \overline{\rho}_\mathrm{L} \overline{u}_\mathrm{L} (\overline{h}_{\mathrm{G.00}} - \overline{h}_{\mathrm{L.0}})\,. \tag{10.38}$$

The right-hand side of Eq. (10.38) transforms to

$$\overline{\rho}_\mathrm{L} \overline{u}_\mathrm{L} (\overline{h}_{\mathrm{G.00}} - \overline{h}_{\mathrm{L.0}}) = \overline{\rho}_\mathrm{L} \overline{u}_\mathrm{L} (1 + J_1 + J_2) \tag{10.39}$$

where $J_1 = (h_{\mathrm{G.00}} - h_{\mathrm{H.f}})/h_{\mathrm{LG}}$, $J_2 = (h_{\mathrm{L.f}} - h_{\mathrm{L.0}})/h_{\mathrm{LG}}$.

Taking into account that at $P = 10^5\,\mathrm{N/m^2}$, $h_{\mathrm{L,f}} = 417.46\,\mathrm{kJ/kg}$, $h_{\mathrm{G,f}} = 2{,}675$ kJ/kg and $h_{\mathrm{LG}} = 2{,}258\,\mathrm{kJ/kg}$ (Johnson 1998), we obtain the following estimates for J_1 and J_2: $0 < J_1 < 0.02$, $0 < J_2 < 0.04$ when $100\,^\circ\mathrm{C} < T_\mathrm{G} < 120\,^\circ\mathrm{C}$ and $80\,^\circ\mathrm{C} < T_\mathrm{L} < 100\,^\circ\mathrm{C}$, respectively. This allows one to assume that $\overline{\rho}_\mathrm{L} \overline{u}_\mathrm{L} (1 + J_1 + J_2) \approx \overline{\rho}_\mathrm{L} \overline{u}_\mathrm{L}$.

The integrals $\int_{\overline{x}_\mathrm{f}}^{\overline{L}} \overline{\rho}_\mathrm{G}\,\mathrm{d}\overline{x}$ and $\int_0^{\overline{x}_\mathrm{f}} \overline{\rho}_\mathrm{L}\,\mathrm{d}\overline{x}$ in Eq. (10.30) have the order of magnitudes

$$\int\limits_{\overline{x}_\mathrm{f}}^{\overline{L}} \overline{\rho}_\mathrm{G}\,\mathrm{d}\overline{x} \sim (\overline{L}-\overline{x}_\mathrm{f})\overline{\rho}_\mathrm{G} \tag{10.40}$$

$$\int\limits_{0}^{\overline{x}_\mathrm{f}} \overline{\rho}_\mathrm{L}\,\mathrm{d}\overline{x} \sim \overline{x}_\mathrm{f}\overline{\rho}_\mathrm{L}\,. \tag{10.41}$$

Since the liquid and vapor densities are estimated as $\overline{\rho}_\mathrm{L} \sim 1$, $\overline{\rho}_\mathrm{G} \sim 10^{-3}$, we have the following estimate for the ratio of the integrals (10.41) and (10.40)

$$\frac{\int\limits_{0}^{\overline{x}_\mathrm{f}} \overline{\rho}_\mathrm{L}\,\mathrm{d}\overline{x}}{\int\limits_{\overline{x}_\mathrm{f}}^{\overline{L}} \overline{\rho}_\mathrm{G}\,\mathrm{d}\overline{x}} \sim \frac{\overline{x}_\mathrm{f}\overline{\rho}_\mathrm{L}}{(\overline{L}-\overline{x}_\mathrm{f})\overline{\rho}_\mathrm{G}}\,. \tag{10.42}$$

When $(\overline{x}_\mathrm{f}\overline{\rho}_\mathrm{L})/\left((\overline{L}-\overline{x}_\mathrm{f})\overline{\rho}_\mathrm{L}\right) \gg 1$ the integral (10.40) is much smaller than the integral (10.41) and it can be neglected. This inequality is fulfilled for the majority of physically realistic conditions when $\overline{x}_\mathrm{f} \gg \overline{L}\cdot 10^{-3}$.

Taking into account the above estimates we write Eq. (10.30) in the following form

$$\overline{u}'_\mathrm{L} = \frac{\mathrm{We}^{-1} + \mathrm{Eu} - \mathrm{Fr}^{-1}\overline{x}_\mathrm{f}}{\frac{32}{\mathrm{Re}_\mathrm{L}}\tilde{L}[\overline{x}_\mathrm{f} + \nu_\mathrm{G.L}(\overline{L}-\overline{x}_\mathrm{f})]} \tag{10.43}$$

where $\overline{u}'_\mathrm{L}$ is the velocity, defined by the momentum equation.

When the heat flux on the wall $q_\mathrm{w} = 0$, the velocity of the fluid also equals zero. Assuming in Eq. (10.43) that $\overline{u}'_\mathrm{L} = 0$, we find the height of the liquid in the microchannel with no heat transfer:

$$\overline{X}_\mathrm{fcap} = \frac{\mathrm{We}^{-1} + \mathrm{Eu}}{\mathrm{Fr}^{-1}}\,. \tag{10.44}$$

In particular, when $\mathrm{Eu} = 0$ (pure capillary height)

$$\overline{X}_\mathrm{fcap} = \frac{\mathrm{Fr}}{\mathrm{We}}\,. \tag{10.45}$$

Using Eq. (10.44), we write Eq. (10.43) as

$$\overline{u}'_\mathrm{L} = A\frac{1-\tilde{x}_\mathrm{f}}{B-\tilde{x}_\mathrm{f}} \tag{10.46}$$

where $A = \mathrm{Re}/(32\mathrm{Fr}\overline{L}(\nu_\mathrm{G.L}-1))$, $B = \tilde{\ell}\nu_\mathrm{G.L}/(\nu_\mathrm{G.L}-1)$, $\tilde{x}_\mathrm{f} = \overline{x}_\mathrm{f}/\overline{X}_\mathrm{fcap}$, $\tilde{\ell} = \overline{L}/\overline{X}_\mathrm{fcap}$.

As seen from Eq. (10.46) the physical realistic values of the liquid velocity $\overline{u}'_\mathrm{L} \geq 0$ correspond to the confined range of the meniscus position $\tilde{x}_\mathrm{f}$: $0 < \tilde{x}_\mathrm{f} \leq 1$. The existence of the upper limit in this inequality may be explained as follows: the heat flux on the wall causes liquid evaporation and determines the displacement of the meniscus. If the heat flux increases, the rate of evaporation also increases. The velocities of the flow of the two phases grow and the interface surface moves towards the inlet of the micro-channel. Thus, the height of the meniscus when $q_w = 0$ is maximum and is equal to $\overline{X}_{fcap}$. Accordingly, the dimensionless meniscus height $\tilde{x}_f = x_f/\overline{X}_{fcap} < 1$ for any value of the heat flux.

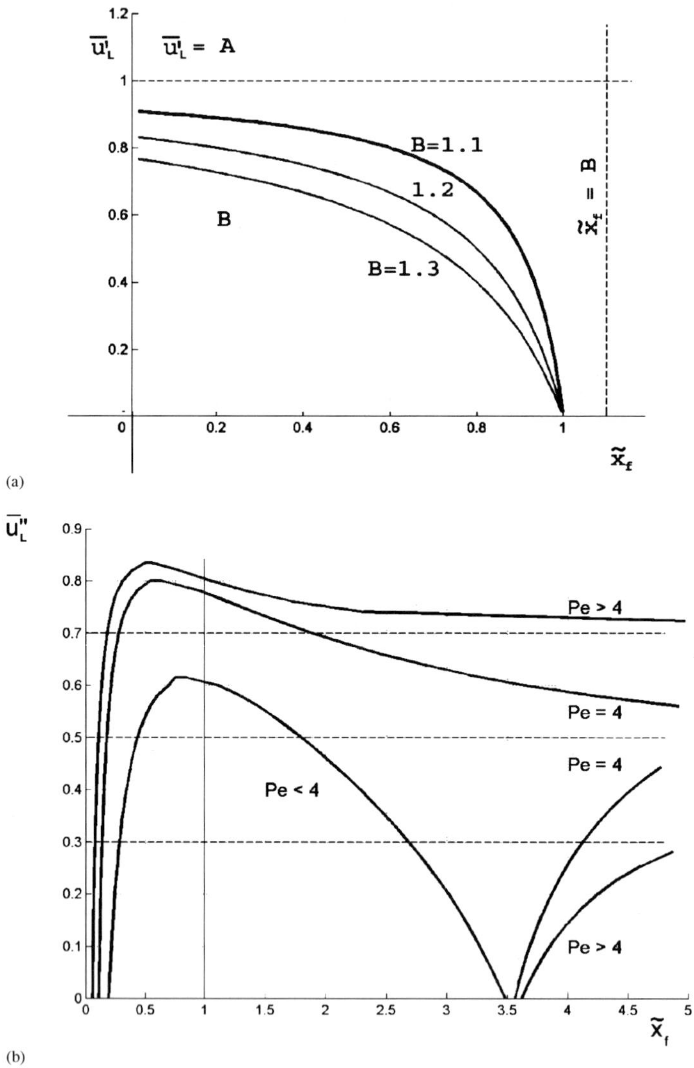

Fig. 10.4a,b Functions $\overline{u}'_L(\tilde{x}_f)$ and $\overline{u}''_L(\tilde{x}_f)$: (**a**) dependence of $\overline{u}'_L(\tilde{x}_f)$ upon various values of parameter B; *arrow* shows an increase of parameter B; (**b**) dependence of $\overline{u}''_L(\tilde{x}_f)$ upon various values of Peclet number. Reprinted from Yarin et al. (2002) with permission

The dependence of $\overline{u}'_{\mathrm{L}}(\tilde{x}_{\mathrm{f}})$ on different values of the parameter B is plotted in Fig. 10.4a (Sect. 10.9.3, Fig. 10.5a–c).

Taking into account Eq. (10.33), we transform Eq. (10.31) into

$$\overline{u}''_{\mathrm{L}} = 1 - \left(\frac{1}{\mathrm{Pe}\overline{u}''_{\mathrm{L}}} + \frac{\overline{q}_{\mathrm{L}}(\overline{T}_S - 1)\overline{u}''_{\mathrm{L}} - \overline{X}_{\mathrm{fcap}}\tilde{x}_{\mathrm{f}}}{\exp(\mathrm{Pe}\overline{X}_{\mathrm{fcap}}\tilde{x}_{\mathrm{f}}\overline{u}''_{\mathrm{L}}) - 1}\right) \tag{10.47}$$

where $\overline{u}''_{\mathrm{L}}$ is the liquid velocity defined from the energy equation, $\overline{q}_{\mathrm{L}} = (c_{\mathrm{pL}}T_*)/h_{\mathrm{LG}}$. The graphs of $\overline{u}''_{\mathrm{L}}(\tilde{x}_{\mathrm{f}})$ for different values of the Peclet number are plotted in Fig. 10.4b.

Now the momentum and energy equations for steady flow in a heated capillary are

$$\overline{u}'_{\mathrm{L}} = \varphi_1(\tilde{x}_{\mathrm{f}}) \tag{10.48}$$

$$\overline{u}''_{\mathrm{L}} = \varphi_2(\tilde{x}_{\mathrm{f}})\,. \tag{10.49}$$

The stationary states correspond to the parametrical equation

$$\overline{u}'_{\mathrm{L}} = \overline{u}''_{\mathrm{L}} \tag{10.50}$$

which includes the following parameters: Re, Pe, Fr, ν_{12}, $\overline{T}_{\mathrm{S}}$, $\overline{L}$, $\tilde{\ell}$ and $\overline{q}_{\mathrm{L}}$.

Equation (10.50) postulates equality of the velocity due to liquid evaporation $\overline{u}''_{\mathrm{L}}$ and the velocity due to the capillary and pressure forces $\overline{u}'_{\mathrm{L}}$.

10.6 Parametrical Analysis

The solution of Eq. (10.50) determines the steady states of the liquid velocity, as well as the position of the meniscus in a heated micro-channel. Equation (10.50) can have one, two or three steady solutions. This depends on the value of the parameter $\tilde{\ell}$ (in the generic case: parameter B), which takes into account the effect of the capillary forces.

Consider the possible regimes of flow corresponding to $\tilde{\ell} \gg 1$ and $\tilde{\ell} \sim 1$. We refer to the first regime as "semi-filled," whereas the second one is "filled."

The change of velocity due to liquid evaporation $\overline{u}''_{\mathrm{L}}$ and influence of the capillary forces $\overline{u}'_{\mathrm{L}}$ versus $\tilde{x}_{\mathrm{f}}$ for $\tilde{\ell} \gg 1$ is illustrated in Fig. 10.6. In the case $\tilde{\ell} \gg 1$ the curves $\overline{u}'_{\mathrm{L}}(\tilde{x}_{\mathrm{f}})$ and $\overline{u}''_{\mathrm{L}}(\tilde{x}_{\mathrm{f}})$ have only one point of intersection, which determines the stationary values of $\overline{u}_{\mathrm{L}} = \overline{u}_{\mathrm{L.st}}$ and $\tilde{x}_{\mathrm{f}} = \tilde{x}_{\mathrm{f.st}}$. It is not difficult to show that this point is stable. Indeed a displacement of the meniscus from its initial position $\tilde{x}_{\mathrm{f.st}}$ to the position $\tilde{x}^{(*)}_{\mathrm{f.st}}$ leads to the situation, when the velocity due to the liquid evaporation $\overline{u}''_{\mathrm{L}}$ exceeds the velocity due to the capillary force $\overline{u}'_{\mathrm{L}}$. This leads to the return of the meniscus to its initial position. If the meniscus displaces to the left, $\overline{u}'_{\mathrm{L}} > \overline{u}''_{\mathrm{L}}$, this also leads to the return of the system to its initial state.

The dependence of the stable values of the liquid velocity $\overline{u}_{\mathrm{L.st}}$ and the meniscus location $\tilde{x}_{\mathrm{f.st}}$ on the Peclet number and upon the parameter A are plotted in Fig. 10.7.

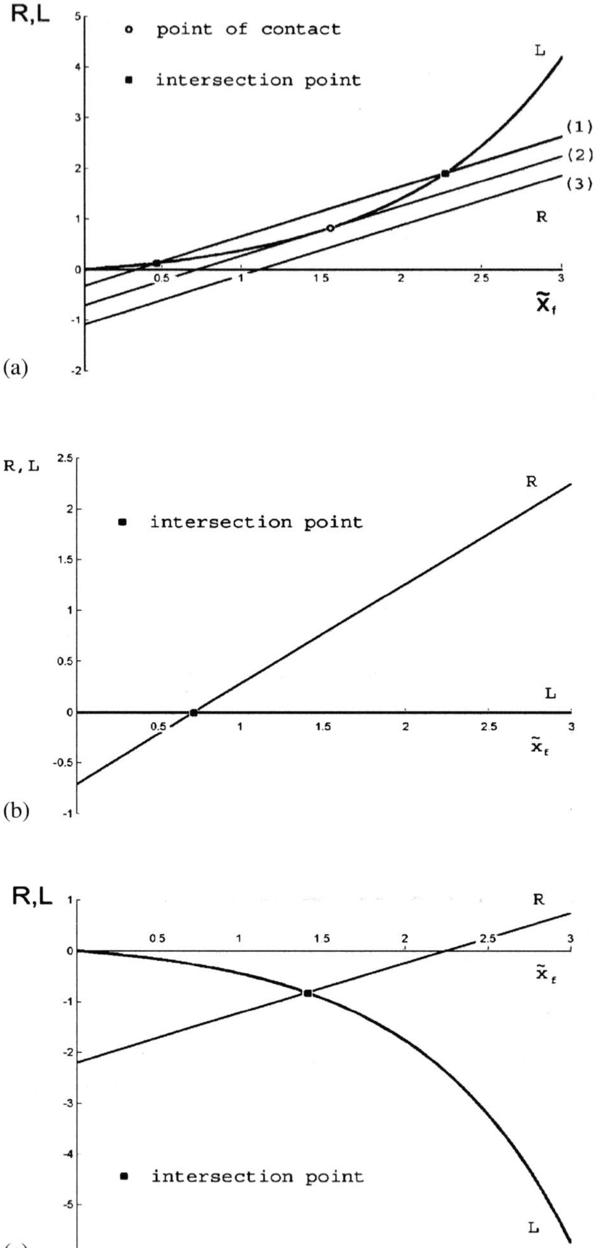

Fig. 10.5a–c Graphical analysis of Eq. (10.44): (**a**) dependence of $L(\tilde{x}_f)$, left-hand side of Eq. (10.44) and $R(\tilde{x}_f)$, right-hand side of Eq. (10.44) at $D < 0$ and any $\overline{u}''_L$ or at $D < 0$ and one of the conditions $0 < \overline{u}''_L < 1/2 - \sqrt{D}$ or $1/2 + \sqrt{D} < \overline{u}''_L < 1$ (D is the discriminate of the equation $\overline{u}''_L - 1 + (1/\mathrm{Pe}\overline{u}''_L) = 0$); 1, 2, 3 correspond to various values of $\overline{u}''_L : \overline{u}''_{L(1)} > \overline{u}''_{L(2)} > \overline{u}''_{L(3)}$. (**b**) The dependence of L and R on $\tilde{x}_f$ at $D = 0$. (**c**) The dependence L and R upon $\tilde{x}_f$ at $D < 0$ and $1/2 + \sqrt{D} < \overline{u}''_L < 1/2 + \sqrt{D}$. Reprinted from Yarin et al. (2002) with permission

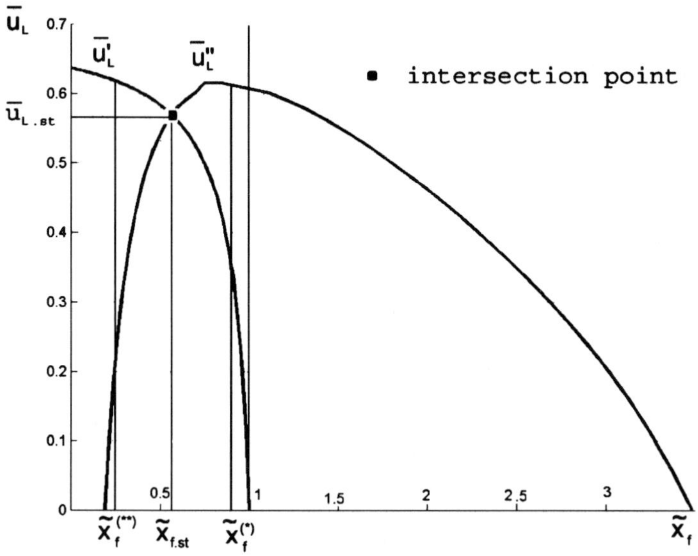

Fig. 10.6 Solution of Eq. (10.47) at $\tilde{\ell} \gg 1$. Reprinted from Yarin et al. (2002) with permission

The effect of the Peclet number on $\overline{u}_{\mathrm{L.st}}$ and $\tilde{x}_{\mathrm{f.st}}$ is shown in Fig. 10.7a. An increase in the Peclet number is accompanied by an increase in $\overline{u}_{\mathrm{L.st}}$ and a decrease in $\tilde{x}_{\mathrm{f.st}}$. The changing of $\overline{u}_{\mathrm{L.st}}$ and $\tilde{x}_{\mathrm{f.st}}$ with A is more complicated (Fig. 10.7b). An increase of A leads to a monotonic increase of $\tilde{x}_{\mathrm{f.st}}$, whereas the dependence of $\overline{u}_{\mathrm{r.st}}(A)$ has a maximum at $A > A_{\mathrm{m}}$, where A_{m} corresponds to an extremum of the curve $\overline{u}''_{\mathrm{L}}(\tilde{x}_{\mathrm{f}})$.

Figure 10.7b illustrates the influence of some physical parameters on the characteristics of the capillary flow. The dimensionless parameter A may be expressed as $A = (1/128)\,(h_{\mathrm{LG}}/q_{\mathrm{w}}L)\,\left(\rho_* g d^3/(\nu_{\mathrm{G}} - \nu_{\mathrm{L}})\right)$. Thus, an increase of the heat flux on the wall or the length of the capillary causes a decrease in the parameter A and is accompanied by the displacement of the meniscus near the inlet of the micro-channel. When the distance $\tilde{x}_{\mathrm{f}}$ between the interface and the inlet is small enough, the heat losses increase and the rate of evaporation u'_{L} decreases.

The possible intersections of curves $\overline{u}'_{\mathrm{L}}(\tilde{x}_{\mathrm{f}})$ and $\overline{u}''_{\mathrm{L}}(\tilde{x}_{\mathrm{f}})$at $\tilde{\ell} \sim 1$ are shown in Fig. 10.8. If the values of A are small enough ($0 < A < A_1$), there is only one intersection point P_1. It is located on the left branch of the curve $\overline{u}''_{\mathrm{L}}(\tilde{x}_{\mathrm{f}})$ and corresponds to the steady regime of the flow. Parameters $A = A_1$ and $A = A_3$ confine the domain within which there are three intersection points of the curves $\overline{u}'_{\mathrm{L}}(\tilde{x}_{\mathrm{f}})$ and $\overline{u}''_{\mathrm{L}}(\tilde{x}_{\mathrm{f}})$. One of these points (intermediate) corresponds to the unstable state. The boundary values $A = A_1$ and $A = A_3$ correspond to two intersection points, one of which is the point of touch. If $A > A_3$ there is only one intersection point (stable). The domains of existence of one, two or three solutions of Eq. (10.50) are shown in Figs. 10.9 and 10.10 in the plane of the parameters A–Pe and the space A–Pe–$\overline{X}_{\mathrm{f.cap}}$. The existence of two stable regimes of flow (with fixed values of the parameters and various meniscus positions) may be explained as follows: when the meniscus is located near

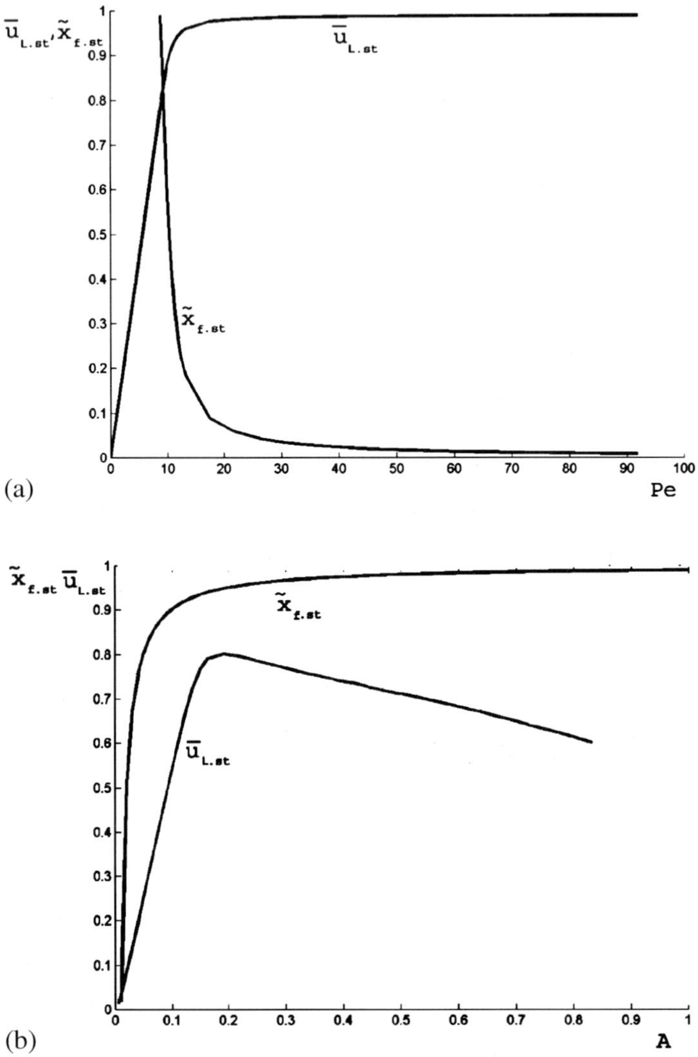

Fig. 10.7a,b Dependence of liquid velocity $\overline{u}_{\mathrm{L.st}}$ and meniscus location $\tilde{x}_{\mathrm{f.st}}$ versus Peclet number and parameter *A* (semi-filled regime): **(a)** $\overline{u}''_{\mathrm{L.st}}(\mathrm{Pe})$; $\tilde{x}_{\mathrm{f.st}}(\mathrm{Pe})$, **(b)** $\overline{u}''_{\mathrm{L.st}}(A)$; $\tilde{x}_{\mathrm{f.st}}(A)$. Reprinted from Yarin et al. (2002) with permission

the outlet, the volume of the capillary is filled by liquid. Since the length of the vapor region is very small, the friction force due to vapor is negligible. The friction force due to liquid is smaller than the gravitational force. Thus, the capillary force is balanced mainly by the gravitational force. At the same time, dynamic equilibrium is possible, when friction forces due to vapor are dominant. This situation appears, when the meniscus approaches the inlet of the micro-channel and the gravity force together with the friction force due to the liquid are negligible.

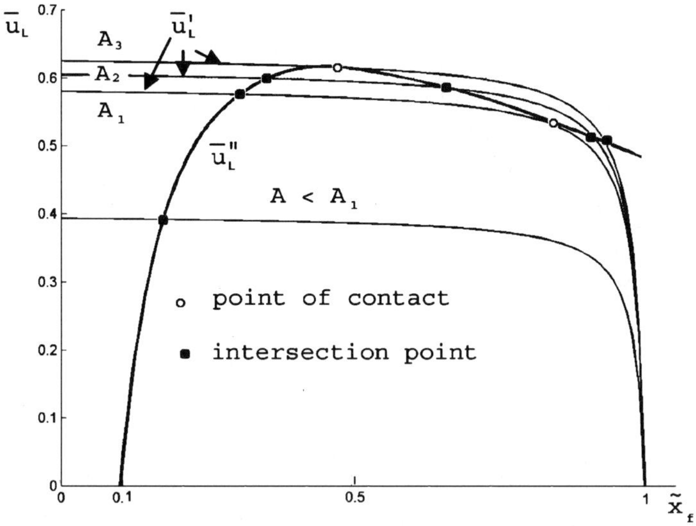

Fig. 10.8 Solution of Eq. (10.47) at $\tilde{\ell} \sim 1$. Reprinted from Yarin et al. (2002) with permission

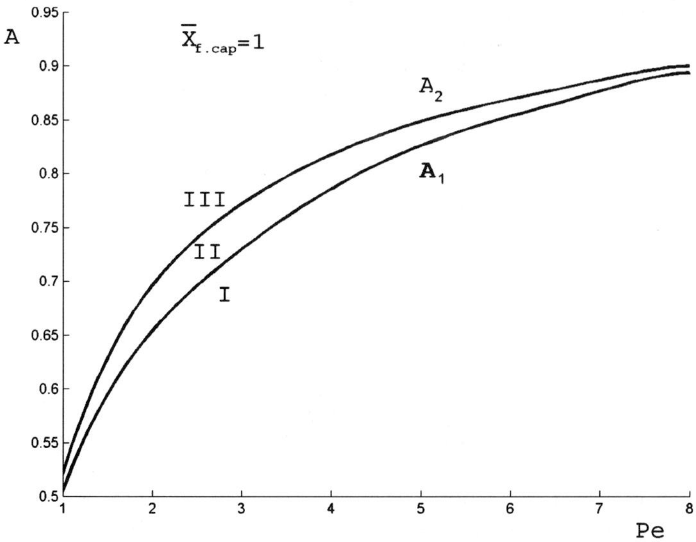

Fig. 10.9 Diagram of steady states: *I* and *III* are domains of existence of single solution, *II* is a domain of existence of three solutions (two stable and one unstable). Lines A_1 and A_2 correspond to two stable solutions. Reprinted from Yarin et al. (2002) with permission

Thus, depending on the values of the parameter $\tilde{\ell}$ we can classify the equilibrium states, using parameter A (Table 10.1). When $A = A_1$ or $A = A_3$ the curves $\overline{u}'_\mathrm{L}$ and $\overline{u}''_\mathrm{L}$ are tangent.

The stable stationary states that correspond to two lower rows in Table 10.1 may be subdivided into two groups: gravity and friction, depending on the dominant

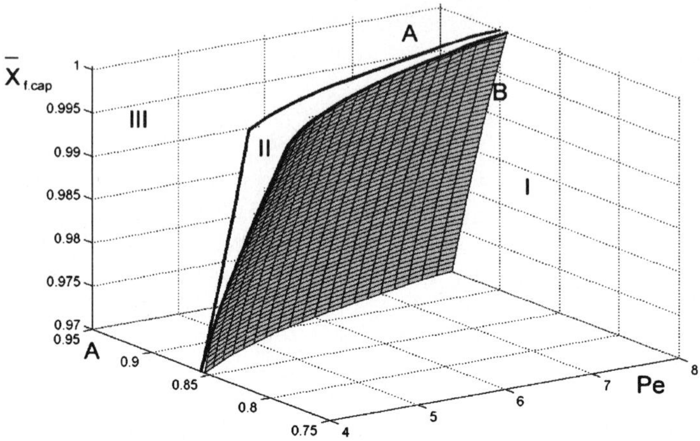

Fig. 10.10 Three-dimensional diagram of steady states: *I* and *III* are domains of existence of single solution, *II* is a domain of existence of three solutions (two stable and one unstable). Boundary surfaces correspond to two stable solutions. Reprinted from Yarin et al. (2002) with permission

Table 10.1 Classification of the equilibrium states using parameter *A*

Regime	$\tilde{\ell}$	Parameter A	Number of stationary states	Stability
Semi-filled	$\tilde{\ell} \gg 1$	$0 < A < \infty$	One	Stable
		$0 < A < A_1$	One	Stable
Filled	$\tilde{\ell} \sim 1$	$A_1 < A < A_3$	Three	Two stable One unstable
		$A = A_1$ $A = A_3$	Two	Stable

factor. The first corresponds to the conditions when the capillary force due to surface tension is compensated mainly by the weight of a liquid column, whereas the second corresponds to the dominant role of the friction force due to the vapor flow.

Table 10.1 shows that only one stationary state is possible, when the meniscus is far enough from the outlet of the capillary. Contrary to this, when the capillary is filled, namely, the interface surface is near the outlet, and two and three stationary states are possible. From the physical point of view, these phenomena may be explained by the different contribution of the friction force due to the vapor. In the first case this force is dominant, whereas in the second its effect is negligible.

10.7 Results and Discussion

The numerical calculations of the liquid velocity and the meniscus position were carried out for the various wall heat flux q_w, the acceleration due to gravity g as well as the length and the diameter of the micro-channel in the ranges $(200 < q_w < 2000)$ W/m^2, $(5 < g < 45)$ m/s^2, $(0.015 < L < 0.05)$ m, and $(0.1 < d < 2)$ 10^{-4} m. The change of the flow velocity and meniscus position versus heat flux is illustrated in Fig. 10.11. An increase in q_w is accompanied by a decrease in the $x_{f.st}$ and growth of the outlet $u_{L.st}$. The sharp change of $x_{f.st}$ and $u_{L.st}$ is observed at relatively small heat fluxes when the meniscus is located away from the inlet and heat losses are negligible. In this case all the energy supplied to the liquid is used for its evaporation so that its velocity is directly proportional to q_w. The character of the process changes qualitatively when q_w is much greater. When the meniscus approaches the inlet cross-section, the heat losses became significant, because they are proportional to $\left((\overline{T}_S - 1)/\tilde{x}_f\right)$. In this case an increase in q_w is accompanied by an increase in the heat losses to the inlet and practically has no effect on the liquid velocity.

The effect of the acceleration due to gravity on the steady-state liquid velocity and the meniscus position is shown in Fig. 10.12. An increase in g is accompanied by the displacement of the meniscus toward the inlet and a decrease in the liquid velocity.

Figure 10.13 illustrates the effect of the micro-channel length on $u_{L.st}$ and $x_{f.st}$. It may be seen that an increase in the micro-channel length is accompanied by a monotonic decrease in $x_{f.st}$, whereas the function $u_{L.st}(L)$ has a maximum. Such a dependence of $u_{L.st}(L)$ is due to opposite features: (a) the growth of the friction forces with L; (b) the growth of the total heat flux to the liquid. When L is relatively small the increase in the total heat flux plays the dominant role, which leads to an increase in the liquid velocity. When L is large enough, the hydraulic resistance of the vapor region is dominant. Its growth leads to the displacement of the meniscus towards the inlet, the increase in the heat losses and the decrease in the liquid velocity.

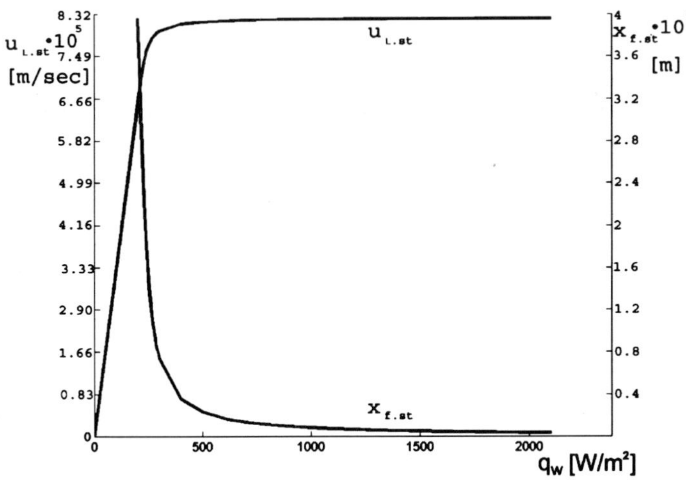

Fig. 10.11 Stable liquid velocity and meniscus location vs. heat flux on the micro-channel wall ($L = 2 \times 10^{-2}$ m, $d = 10^{-4}$ m, $g = 9.8$ m/s^2). Reprinted from Yarin et al. (2002) with permission

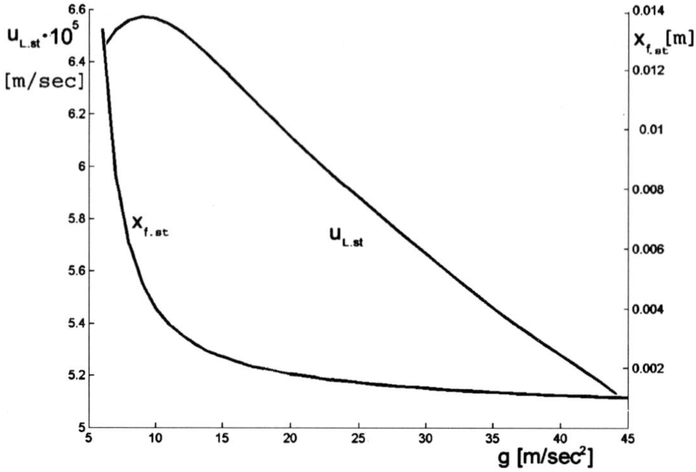

Fig. 10.12 Stable liquid velocity and meniscus location versus gravity acceleration ($L = 2 \times 10^{-2}$ m, $d = 10^{-4}$ m, $q_w = 200\ \mathrm{W/m^2}$). Reprinted from Yarin et al. (2002) with permission

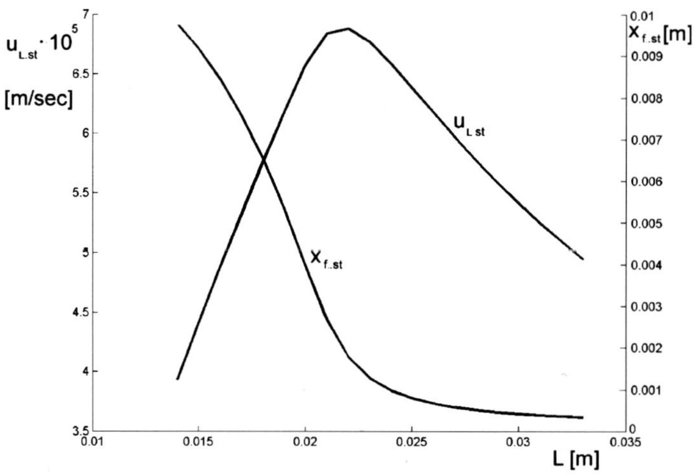

Fig. 10.13 Liquid velocity and the location of the meniscus versus capillary length ($d = 10^{-4}$ m, $g = 9.8\ \mathrm{m/s^2}$, $q_w = 200\ \mathrm{W/m^2}$). Reprinted from Yarin et al. (2002) with permission

The graphs of the functions $u_{L.st}(d)$ and $x_{f.st}(d)$ are plotted in Fig. 10.14. Both curves have rising and falling branches and accordingly a characteristic maximum at a certain (depending on the curve) value of the micro-channel diameter. When $d > 1 \times 10^{-4}$ m the liquid velocity is inversely proportional to d. Within this regime an increase in the micro-channel diameter leads to a significant decrease in the friction force as well as the forces due to the surface tension. Under these conditions the length of the liquid region is shortened. The situation changes qualitatively when

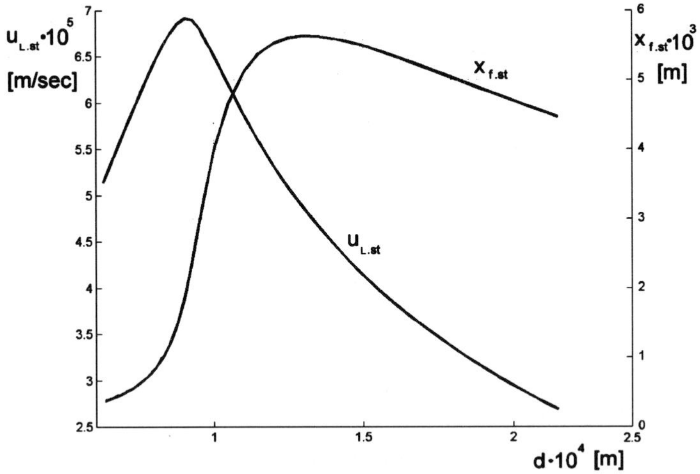

Fig. 10.14 Liquid velocity and meniscus location versus capillary diameter ($L = 2 \times 10^{-2}$ m, $g = 9.8\ \mathrm{m/s^2}$, $q_w = 200\ \mathrm{W/m^2}$). Reprinted from Yarin et al. (2002) with permission

$x_{f.st}$ is small enough and heat losses play a significant role. In this case the growth of $u_{L.st}$ with d takes place.

10.8 Efficiency of the Cooling System

The total energy, which is supplied to the liquid from the wall, is equal to the heat flux q_w, multiplied by the lateral area of the micro-channel: πdLq_w. The energy, which is expended to the liquid vaporization, is equal to the latent heat of evaporation h_{LG}, multiplied by the mass liquid flow rate through a cross-section of the micro-channel, and is equal $\rho_L u_L \left(\pi d^2/4\right) h_{LG}$.

The effect of the heat losses to the inlet on the thermal states of the micro-channel depends mainly on the meniscus position, which is determined by the flow parameters. To characterize this effect, the coefficient of efficiency is introduced; it may be defined as the ratio of the energy expended to the liquid vaporization and the total energy supplied to the micro-channel.

$$\eta = \frac{\rho_L u_L d h_{LG}}{4 q_w L}\,. \tag{10.51}$$

Since the coefficient of efficiency depends on the location of the meniscus, it is different for semi-filled and filled regimes.

In a steady state two-phase flow in a heated capillary, there is the balance of forces, which act on the liquid and its vapor. The analysis of this balance shows that there are two stable states of the flow. They correspond to the different locations of the meniscus, which separates the liquid and the vapor. The existence of such states may be explained by the changes in the values of the different components in the balance equation. When the meniscus is near the outlet, the gravity and the surface tension play the main role, whereas the friction forces of the liquid and the vapor are negligible. In contrast, when the meniscus is near the inlet, the friction and surface tension forces are the main parameters.

The position of the meniscus within the micro-channel defines the type of temperature distribution. In the first case, when the meniscus is near the outlet, the temperature gradient of the vapor region is small. The rate of evaporation is determined mainly by the heat flux in the liquid region. Therefore, the necessary condition of the evaporation consists of the existence of the region (near the meniscus), where the water is overheated (its temperature is higher than the temperature of boiling). The heat losses to the inlet tank cause the existence of the temperature maximum.

When the meniscus is near the inlet of the micro-channel, the heat losses increase, because the gradient of the temperature is greater.

Thus, there are two meniscus positions near the edges of the micro-channel, where the heat losses are significant. Between them there is some position, in which these losses are minimal and the evaporation is the most intense. This corresponds to the optimal efficiency of the cooling system.

The effect of various parameters on the efficiency coefficient is illustrated in Fig. 10.15a–d. It may be seen, that an increase of heat flux on the wall leads to a significant decrease in η. It is due to the meniscus displacement towards the inlet and growth of heat losses into the cooling liquid. The calculations show (Fig. 10.15a–c) that for given values of parameters there is some "optimal" value of a micro-channel's diameter and length, at which the coefficient of efficiency reaches maximum.

The developed theory of two-phase laminar flow with a distinct interface which is based on a one-dimensional approximation, takes into account the major features of the process: the inertia, gravity, surface tension and friction forces and leads to the physically realistic pattern of a laminar flow in a heated micro-channel. This allows one to use the present theory to study the regimes of flow as well as optimizing a cooling system of electronic devices with high power densities.

The obtained solution of the problem shows that the rate of vaporization (the liquid's velocity), the liquid and vapor temperatures, the position of the meniscus in the micro-channel, its hydraulic resistance and the thermal losses are determined by eight non-dimensional groups, accounting for the effects of heat transfer, phase change as well as inertia, friction, surface tension and gravitational forces. The number of such non-dimensional groups may be reduced to four by introducing a general parameter: the capillary height, which depends on the Weber, Froude and Euler numbers. It is shown that in the generic case the governing system of equations that describes capillary flow has three stationary solutions: two stable and one (intermediate) unstable.

An increase in the heat flux is accompanied by an increase in the liquid and vapor velocities, the meniscus displacement towards the outlet cross-section, as well as growth of vapor to liquid forces ratio and heat losses. When q_{w} is large enough, the

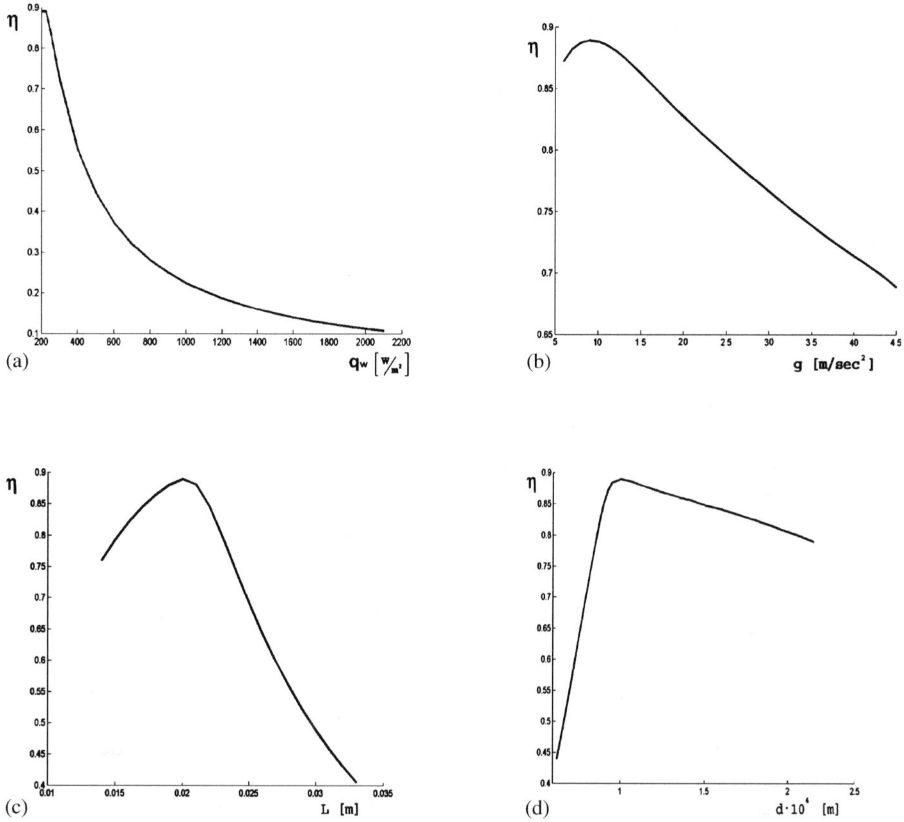

Fig. 10.15 Efficiency of cooling system (friction regime): (**a**) coefficient of efficiency vs. heat flux, (**b**) coefficient of efficiency vs. gravity, (**c**) coefficient of efficiency vs. capillary length, (**d**) coefficient of efficiency vs. capillary diameter. Reprinted from Yarin et al. (2002) with permission

difference between the intensity of heat transfer and heat losses are limited by some final value, which determines the maximum rate of vaporization. Accordingly, when q_{w} is large all characteristic parameters are practically invariable.

The existence of two stable states (at given values of the operating parameters) is due to the dominant role of the gravity or friction forces at the various meniscus positions. A decrease in the gravity leads to the displacement of the meniscus toward the outlet and to a decrease in the heat losses and an increase in the liquid and vapor velocities. A decrease in the micro-channel diameter leads to a monotonic increase in the liquid and vapor velocities, whereas the dependence of the meniscus position versus d has an extremum.

At given values of the parameters, there are optimal values of the micro-channel diameter and length, which correspond to a maximum efficiency coefficient.

10.9 Equation Transformation

10.9.1 The Dependence of the Saturation Pressure and Temperature

The dependence of the saturation pressure P_S and temperature T_S for several liquids is determined by the Antoine equation (Reid et al. 1987). We use the following expression for saturation pressure of the water (Seaver et al. 1989):

$$P_S = a_0 + t_S[a_1 + t_S(a_2 + t_S\{a_3 + t_S[a_4 + t_S(a_5 + a_6 t_S)]\})] \tag{10.52}$$

where

$$a_0 = 6.107799961 \times 10^{-2},\ a_1 = 4.436518521 \times 10^{-3},\ a_2 = 1.428945805 \times 10^{-4},$$
$$a_3 = 2.650648731 \times 10^{-6},\ a_4 = 3.031240396 \times 10^{-8},\ a_5 = 2.034080948 \times 10^{-10},$$
$$a_6 = 6.136820929 \times 10^{-13},$$

t is measured in degrees Celsius.

From Eq. (10.52) it follows that

$$\frac{dT_S}{dP_S} = \frac{1}{a_1 + t_S(2a_2 + t_S\{3a_3 + t_S[4a_4 + t_S(5a_5 + 6a_6 t_S)]\})}\,.$$

For $t_S = 100\,°\mathrm{C}$

$$\frac{dT_S}{dP_S} = 2.6859 \times 10^{-4}\,,$$

for $t_S = 200\,°\mathrm{C}$

$$\frac{dT_S}{dP_S} = 2.4066 \times 10^{-5}\,.$$

This shows that T_S changes weakly with the pressure. This allows us to neglect the dependence of the hydraulic resistance of the vapor on the saturation temperature T_S and assume that it is determined by the external pressure $P_{G,00}$.

10.9.2 Integral Relations

1. Integrating Eq. (10.2) from 0 to $x_{f(-)}$ we obtain

$$\int_0^{x_f}\left(\frac{\partial \rho_L}{\partial t} + \frac{\partial \rho_L u_L}{\partial x}\right)dx = \frac{\partial}{\partial t}\int_0^{x_f}\rho_L\,dx - \rho_{L.f}\frac{dx_f}{dt} + \{(\rho_L u_L)_f - (\rho_L u_L)_0\} = 0\,. \tag{10.53}$$

Integrating Eq. (10.2) from $x_{\mathrm{f}(+)}$ to L yields

$$\int_{x_\mathrm{f}}^{L}\left(\frac{\partial\rho_\mathrm{G}}{\partial t}+\frac{\partial\rho_\mathrm{G}u_\mathrm{G}}{\partial x}\right)\mathrm{d}x=\frac{\partial}{\partial t}\int_{x_\mathrm{f}}^{L}\rho_\mathrm{G}\,\mathrm{d}x+\rho_\mathrm{G.f}\frac{\mathrm{d}x_\mathrm{f}}{\mathrm{d}t}+\{(\rho_\mathrm{G}u_\mathrm{G})_{00}-(\rho_\mathrm{G}u_\mathrm{G})_\mathrm{f}\}=0. \tag{10.54}$$

Summing Eqs. (10.53) and (10.54) we get

$$\frac{\partial}{\partial t}\left(\int_0^{x_\mathrm{f}}\rho_\mathrm{L}\,\mathrm{d}x+\int_{x_\mathrm{f}}^{L}\rho_\mathrm{G}\,\mathrm{d}x\right)+(\rho_\mathrm{G}-\rho_\mathrm{L})_\mathrm{f}\frac{\mathrm{d}x_\mathrm{f}}{\mathrm{d}t}+\{[(\rho_\mathrm{L}u_\mathrm{L})_\mathrm{f}-(\rho_\mathrm{L}u_\mathrm{L})_0] \\ +[(\rho_\mathrm{G}u_\mathrm{G})_{00}-(\rho_\mathrm{G}u_\mathrm{G})_\mathrm{f}]\}=0\,. \tag{10.55}$$

From Eq. (10.11) it follows that

$$(\rho_\mathrm{G}u_\mathrm{G}-\rho_\mathrm{L}u_\mathrm{L})_\mathrm{f}=(\rho_\mathrm{G}-\rho_\mathrm{L})_\mathrm{f}V_\mathrm{f}=(\rho_\mathrm{G}-\rho_\mathrm{L})_\mathrm{f}\frac{\mathrm{d}x_\mathrm{f}}{\mathrm{d}t}\,. \tag{10.56}$$

Then Eq. (10.55) takes the following form

$$\frac{\partial}{\partial t}\left(\int_0^{x_\mathrm{f}}\rho_\mathrm{L}\,\mathrm{d}x+\int_{x_\mathrm{f}}^{L}\rho_\mathrm{G}\,\mathrm{d}x\right)+(\rho_\mathrm{G.00}u_\mathrm{G.00}-\rho_\mathrm{L.0}u_\mathrm{L.0})=0\,. \tag{10.57}$$

2. Multiplying Eq. (10.2) by u_i and summing that equation with Eq. (10.3) we obtain the equation

$$\frac{\partial\rho_iu_i}{\partial t}+\frac{\partial\rho_iu_i^2}{\partial x}=-\frac{\partial P_i}{\partial x}-\rho_ig-\frac{\partial F_i}{\partial x}\,. \tag{10.58}$$

Integrating Eq. (10.58) from 0 to x_f and from x_f to L yields

$$\frac{\partial}{\partial t}\int_0^{x_\mathrm{f}}\rho_\mathrm{L}u_\mathrm{L}\,\mathrm{d}x-(\rho_\mathrm{L}u_\mathrm{L})_\mathrm{f}\frac{\mathrm{d}x_\mathrm{f}}{\mathrm{d}t}+\{(\rho_\mathrm{L}u_\mathrm{L}^2)_\mathrm{f}-(\rho_\mathrm{L}u_\mathrm{L}^2)_0\} \\ =(P_\mathrm{L.0}-P_\mathrm{L.f})-g\int_0^{x_\mathrm{f}}\rho_\mathrm{L}\,\mathrm{d}x-F_\mathrm{L.f} \tag{10.59}$$

$$\frac{\partial}{\partial t}\int_{x_\mathrm{f}}^{L}\rho_\mathrm{G}u_\mathrm{G}\,\mathrm{d}x+(\rho_\mathrm{G}u_\mathrm{G})_\mathrm{f}\frac{\mathrm{d}x_\mathrm{f}}{\mathrm{d}t}+\left\{\left(\rho_\mathrm{G}u_\mathrm{G}^2\right)_{00}-\left(\rho_\mathrm{G}u_\mathrm{G}^2\right)_\mathrm{f}\right\} \\ =(P_\mathrm{G.f}-P_\mathrm{G.00})-g\int_{x_\mathrm{f}}^{L}\rho_\mathrm{G}\,\mathrm{d}x-F_\mathrm{G.00}\,. \tag{10.60}$$

Note that we account for that $F_{L,0} \equiv 0$, $F_{G,f} \equiv 0$ in Eqs. (10.59) and (10.60). Summing Eqs. (10.59) and (10.60), we obtain

$$\frac{\partial}{\partial t}\left(\int_0^{x_f} \rho_L u_L \,dx + \int_{x_f}^{L} \rho_G u_G \,dx\right) + (\rho_G u_G - \rho_L u_L)\frac{dx_f}{dt} + \left\{\left(\rho_L u_L^2\right)_f - \left(\rho_L u_L^2\right)_0 + \left(\rho_G u_G^2\right)_{00} - \left(\rho_G u_G^2\right)_f\right\} = (P_{L.0} - P_{L.f} + P_{G.f} - P_{G.00}) - g\left(\int_0^{x_f} \rho_L \,dx + \int_{x_f}^{L} \rho_G \,dx\right) - (F_{L.f} + F_{G.00})\,. \tag{10.61}$$

Using the conditions (10.12) and (10.13), we transform Eq. (10.61). Taking into account Eq. (10.56), we rewrite Eq. (10.13) as follows:

$$(\rho_G u_G^2 - \rho_L u_L^2)_f = f_L - (P_G - P_L)_f + (\rho_G - \rho_L)_f \left(\frac{dx_f}{dt}\right)^2 . \tag{10.62}$$

Then Eq. (10.61) takes the form

$$\frac{\partial}{\partial t}\left(\int_0^{x_f} \rho_L u_L \,dx + \int_{x_f}^{L} \rho_G u_G \,dx\right) = f_L - [(\rho_G u_G^2)_{00} - (\rho_L u_L^2)_0] + (P_{L.0} - P_{G.00}) - g\left(\int_0^{x_f} \rho_L \,dx + \int_{x_f}^{L} \rho_G \,dx\right) - (F_{L.f} + F_{G.00})\,. \tag{10.63}$$

3. We rewrite Eq. (10.4) in the following form and integrate this equation from 0 to x_f and from x_f to L:

$$\frac{\partial}{\partial t}\int_0^{x_f} \rho_L h_L \,dx - (\rho_L h_L)_f \frac{dx_f}{dt} + \{(\rho_L h_L u_L)_f - (\rho_L h_L u_L)_0\} = k_L\left(\frac{\partial T_L}{\partial x}\right)_f - k_L\left(\frac{\partial T_2}{\partial x}\right)_0 + q x_f \tag{10.64}$$

$$\frac{\partial}{\partial t}\int_{x_f}^{L} \rho_G h_G \,dx + (\rho_G h_G)_f \frac{dx_f}{dt} + \{(\rho_G h_G u_G)_{00} - (\rho_G h_G u_G)_f\} = k_G\left(\frac{\partial T_G}{\partial x}\right)_{00} - k_G\left(\frac{\partial T_G}{\partial x}\right)_f + q(L - x_f)\,. \tag{10.65}$$

Summing Eqs. (10.64) and (10.65) we find

$$\frac{\partial}{\partial t}\left(\int_0^{x_f} \rho_L h_L \,dx + \int_{x_f}^{L} \rho_G h_G \,dx\right) + (\rho_G h_G - \rho_L h_L)_f \frac{dx_f}{dt}$$

$$
\begin{aligned}
&+\{(\rho_L h_L u_L)_f - (\rho_L h_L u_L)_0 + (\rho_G h_G u_G)_{00} - (\rho_G h_G u_G)_f\} \\
&= k_K \left(\frac{\partial T_L}{\partial x}\right)_f - k_L \left(\frac{\partial T_L}{\partial x}\right)_0 + k_G \left(\frac{\partial T_G}{\partial x}\right)_{00} - k_G \left(\frac{\partial T_G}{\partial x}\right)_f + qL\,.
\end{aligned} \tag{10.66}
$$

Using Eq. (10.14) we obtain

$$
\begin{aligned}
&\frac{\partial}{\partial t}\left(\int_0^{x_f} \rho_L h_L \,\mathrm{d}x + \int_{x_f}^{L} \rho_G h_G \,\mathrm{d}x\right) + \{(\rho_G h_G u_G)_{00} - (\rho_L h_L u_L)_0\} \\
&= -k_L \left(\frac{\partial T_L}{\partial x}\right)_0 + k_G \left(\frac{\partial T_G}{\partial x}\right)_{00} + qL\,.
\end{aligned} \tag{10.67}
$$

10.9.3 Analysis of the Equations

1. Let us transform the momentum Eq. (10.43) to the form:

$$
\overline{u}'_2 = A\frac{1-\tilde{x}_f}{B-\tilde{x}_f} \tag{10.68}
$$

where

$$
A = \mathrm{Re}/\left(32\mathrm{Fr}\overline{L}(\nu_{G.L}-1)\right)\,,\quad B = \tilde{\ell}\nu_{G.L}/\left(\nu_{G.L}-1\right),
$$
$$
\tilde{x}_f = \overline{x}_f/\overline{X}_{fcap}\,,\quad \tilde{\ell} = \overline{L}/\overline{X}_{fcap}\,.
$$

If the pressure is moderate ($P < 10^6$ Pa), the ratio of the kinetic viscosities of the vapor and the liquid ν_{12} has the order of about 70 (Johnson 1998). The curve $\overline{u}'_L(\tilde{x}_f)$ is a hyperbola with the horizontal asymptote $\overline{u}'_L = A$ and the vertical $\tilde{x}_f = B$ (Fig. 10.4a). The physical meaning is only the sector of lower branches of the hyperbola, which is between the lines $\tilde{x}_f = 0$ and $\tilde{x}_f = 1$. It corresponds to the position of the meniscus inside the micro-channel.

2. Now let us consider the energy equation of Eq. (10.44). To reveal the shape of the curve $\overline{u}''_L(\tilde{x}_f)$ we consider its intersection with the lines $\overline{u}''_L = \mathrm{const}$. Writing Eq. (10.44) as follows:

$$
\left(\overline{u}''_L - 1 + \frac{1}{\mathrm{Pe}\overline{u}''_L}\right)\left[\exp(\mathrm{Pe}\overline{X}_{fcap}\tilde{x}_f\overline{u}''_L) - 1\right] = \overline{X}_{fcap}\tilde{x}_f - \overline{q}_2\overline{u}''_L(\overline{T}_S - 1)\,. \tag{10.69}
$$

The right-hand side of Eq. (10.69): R is the linear function of $\tilde{x}_f$. An increase of $\overline{u}''_L$ leads to the displacement of these lines downwards. The left-hand side of Eq. (10.69) gives: L as a non-linear function of $\tilde{x}_f$. Its sign depends on the sign of the multiplier $(\overline{u}''_L - 1 + 1/(\mathrm{Pe}\overline{u}''_L))$, i.e., on the determinant $D = 1/4 - 1/\mathrm{Pe}$. There are three cases: $D < 0$, $D = 0$, $D > 0$.

(a) $D < 0$ ($\mathrm{Pe} < 4$). In this case $\overline{u}''_L - 1 + 1/(\mathrm{Pe}\overline{u}''_L) > 0$. Since the expression in square brackets is positive, the left-hand side of Eq. (10.69) is also positive (Fig. 10.5a). The derivatives of functions on left-hand side and right-hand side of Eq. (10.69) at the point $\tilde{x}_{\mathrm{f}} = 0$ equal, respectively,

$$L' = \overline{X}_{\mathrm{fcap}} + \mathrm{Pe}\overline{X}_{\mathrm{fcap}}(\overline{u}''_{\mathrm{L}} - 1)\overline{u}''_{\mathrm{L}} \tag{10.70}$$

$$R' = \overline{X}_{\mathrm{fcap}} \,. \tag{10.71}$$

Since $\overline{u}''_{\mathrm{L}} < 1$, then $R' > L'$. That means that depending on value of $\overline{u}''_2$ there are one or two intersection points of the curves $L(\tilde{x}_{\mathrm{f}})$ and $R(\tilde{x}_{\mathrm{f}})$. Under certain conditions the intersection points are absent. Thus the curve $\overline{u}''_{\mathrm{L}}(\tilde{x}_{\mathrm{f}})$ has the shape as shown in Fig. 10.5a.

(b) $D = 0$ ($\mathrm{Pe} = 4$, $\overline{u}''_{\mathrm{L}} = 1/2$). In this case there is a single intersection point of the curve $\overline{u}''_{\mathrm{L}}$ with the axis of abscissa $\tilde{x}_{\mathrm{f}} = \left(\overline{q}_{\mathrm{L}}(\overline{T}_{\mathrm{S}} - 1)\right) / \left(2\overline{X}_{\mathrm{fcap}}\right)$ (Fig. 10.5b).

(c) $D > 0$ ($\mathrm{Pe} > 4$). When $\overline{u}''_{\mathrm{L}}$ changes within the range $(1/2) - \sqrt{D} \leq \overline{u}''_{\mathrm{L}} \leq (1/2) + \sqrt{D}$, $L(\tilde{x}_{\mathrm{f}})$ decreases with the increasing of $\tilde{x}_{\mathrm{f}}$. In this case there is one intersection point (Fig. 10.5c). When $0 < \overline{u}''_{\mathrm{L}} < (1/2) - \sqrt{D}$ or $(1/2) + \sqrt{D} < \overline{u}''_2 < 1$, i.e., $\overline{u}''_{\mathrm{L}} - 1 + 1/(\mathrm{Pe}\overline{u}''_{\mathrm{L}}) > 0$ and there are two intersection points as it is shown in Fig. 10.5a.

Now we find intersection points of the curve $\overline{u}''_{\mathrm{L}}(\tilde{x}_{\mathrm{f}})$with the abscissa axis. Assuming in Eq. (10.54) $\overline{u}''_{\mathrm{L}} \to 0$ (at finite $\overline{x}_{\mathrm{f}}$), we obtain

$$(\tilde{x}_{\mathrm{f}})_{\mathrm{G.L}} = \frac{1}{\overline{X}_{\mathrm{fcap}}} \pm \frac{1}{\overline{X}_{\mathrm{fcap}}}\sqrt{1 - 2\frac{\overline{q}_{\mathrm{L}}(\overline{T} - 1)}{\mathrm{Pe}}} \,. \tag{10.72}$$

The dependences $\overline{u}''_{\mathrm{L}}(\tilde{x}_{\mathrm{f}})$ for various Pe are plotted in Fig. 10.4b. The shape of these curves significantly depends on the value of the Peclet number. When $\mathrm{Pe} < 4$ the raising and the falling branches of $\overline{u}''_{\mathrm{L}}(\tilde{x}_{\mathrm{f}})$ contain points $(\tilde{x}_{\mathrm{f}})_{\mathrm{G}}$ and $(\tilde{x}_{\mathrm{f}})_{\mathrm{L}}$ on the abscissa axis and form canopy-shaped curves with characteristic maximum depending on Pe. Contrary to that, if $\mathrm{Pe} \geq 4$, the curves $\overline{u}''_{\mathrm{L}}(\tilde{x}_{\mathrm{f}})$ are not continuous. When $\tilde{x}_{\mathrm{f}}$ is large, the upper and lower branches have one ($\mathrm{Pe} = 4$) or two ($\mathrm{Pe} > 4$) asymptotes.

10.10 Two-Dimensional Approach

The quasi-one-dimensional model used in the previous sections for analysis of various characteristics of flow in a heated capillary assumes a uniform distribution of the hydrodynamical and thermal parameters in the cross-section of micro-channel. In the frame of this model, the general characteristics of the flow with a distinct interface, such as position of the meniscus, rate evaporation and mean velocities of the liquid and its vapor, etc., can be determined for given drag and intensity of heat transfer between working fluid and wall, as well as vapor and wall. In accordance with that, the governing system of equations has to include not only the mass, momentum and energy equations but also some additional correlations that determine the drag coefficient and Nusselt number and thereby close the system of governing equations.

In contrast with the one-dimensional model, the two-dimensional model allows to determine the actual parameter distribution in flow fields of the working fluid and its vapor. It also allows one to calculate the drag and heat transfer coefficients by the solution of a fundamental system of equations, which describes the flow of viscous fluid in a heated capillary.

The general features of two-dimensional flow with evaporating liquid–vapor meniscus in a capillary slot were studied by Khrustalev and Faghri (1996). Following this work we present the main results mentioned in their research. The model of flow in a narrow slot is presented in Fig. 10.16. Within a capillary slot two characteristic regions can be selected, where two-dimensional or quasi-one-dimensional flow occurs. Two-dimensional flow is realized in the major part of the liquid domain, whereas the quasi-one-dimensional flow is observed in the micro-film region, located near the wall.

At the assumption that in a narrow slot the gravitational forces are negligible as compared to surface tension, the governing equations describing flow of an incompressible liquid in the central part of flow read as

$$\nabla \cdot \mathbf{v} = 0 \tag{10.73}$$

$$\rho(\mathbf{v} \cdot \nabla)\mathbf{v} = -\nabla P + \mu \nabla^2 \mathbf{v} \tag{10.74}$$

$$\rho c_{\mathrm{p}}(T \cdot \nabla)\mathbf{v} = k\nabla^2 T \tag{10.75}$$

where $\mathbf{v}$ is the vector with components u and v, directed along the x- and y-axis, respectively, P and T are the pressure and temperature, and μ and k are the viscosity and thermal conductivity.

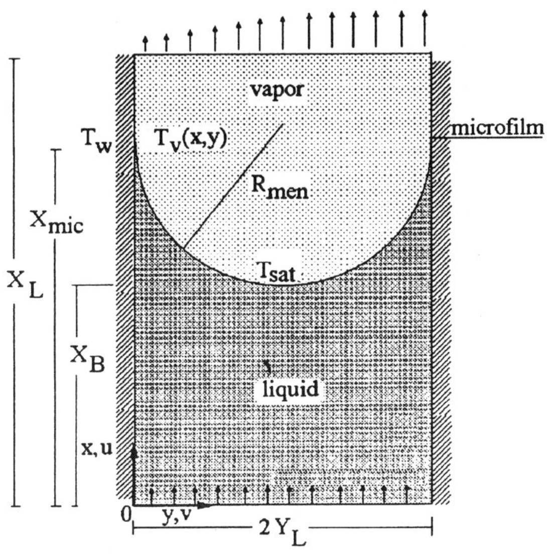

Fig. 10.16 Model of flow in narrow slot. Reprinted from Khrustalev and Faghri (1996) with permission

The boundary conditions for Eqs. (10.73–10.75) are:

1. Non-slip condition on the wall ($u = v = 0$), and parabolic velocity profile u at the inlet
2. Uniform liquid temperature distribution at the inlet, constant wall temperature
3. Symmetry of flow: $v \equiv 0$, $\mathrm{d}u/\mathrm{d}y = 0$, $\mathrm{d}T/\mathrm{d}y = 0$.

At the interface the mass and thermal balance equations are valid. If one assumes that the liquid–vapor interface curvature is constant, accordingly $(P_G - P_L)_{int} = \sigma/R_{men}$, where P_G and P_L are the vapor and liquid pressure at the interface, σ is the surface tension, and R_{men} is the meniscus radius.

To calculate the micro-field, the quasi-one-dimensional model by Khrustalev and Faghri (1994, 1995) is used.

The results of numerical calculations of the velocity distribution within the vapor and liquid domains for two values of the difference between wall and saturation temperatures are shown in Fig. 10.17. It is seen that the vapor velocity reaches 100–150 m/s in the region of micro-film. The liquid velocity is much smaller than those in vapor.

The temperature distribution in the capillary slot is presented in Fig. 10.18. These data show the wall superheat influence on temperature fields in liquid and vapor domains. In these cases, significant heterogeneity of temperature fields is observed.

The results of calculations of the Nusselt number are presented in Fig. 10.19. Here also the data of the calculated heat transfer by the quasi-one-dimensional model by Khrustalev and Faghri (1996) is shown. The comparison of the results related to one and two-dimensional model shows that for relatively small values of wall superheat the agreement between the one and two-dimensional model is good enough (difference about 3%), whereas at large Δt the difference achieves 30%.

Summary

A theory of two-phase laminar flow with a distinct interface has been developed. Although this theory is based on a one-dimensional approximation, it takes into account the major features of the process: the inertia, gravity, surface tension and friction forces. Thus this study may be expected to give the physically realistic pattern of a laminar flow in a heated micro-channel. This allows one to use the present theory to study the regimes of flow, as well as optimizing a cooling system of electronic devices with high power densities.

Partial results are as follows:

1. To calculate the flow parameters under the conditions when the meniscus position and the liquid velocity at the inlet are unknown a priori. The mass, momentum and energy equations are used for both phases, as well as the balance conditions at the interface. The integral condition, which connects flow parameters at the inlet and the outlet cross-sections is derived.

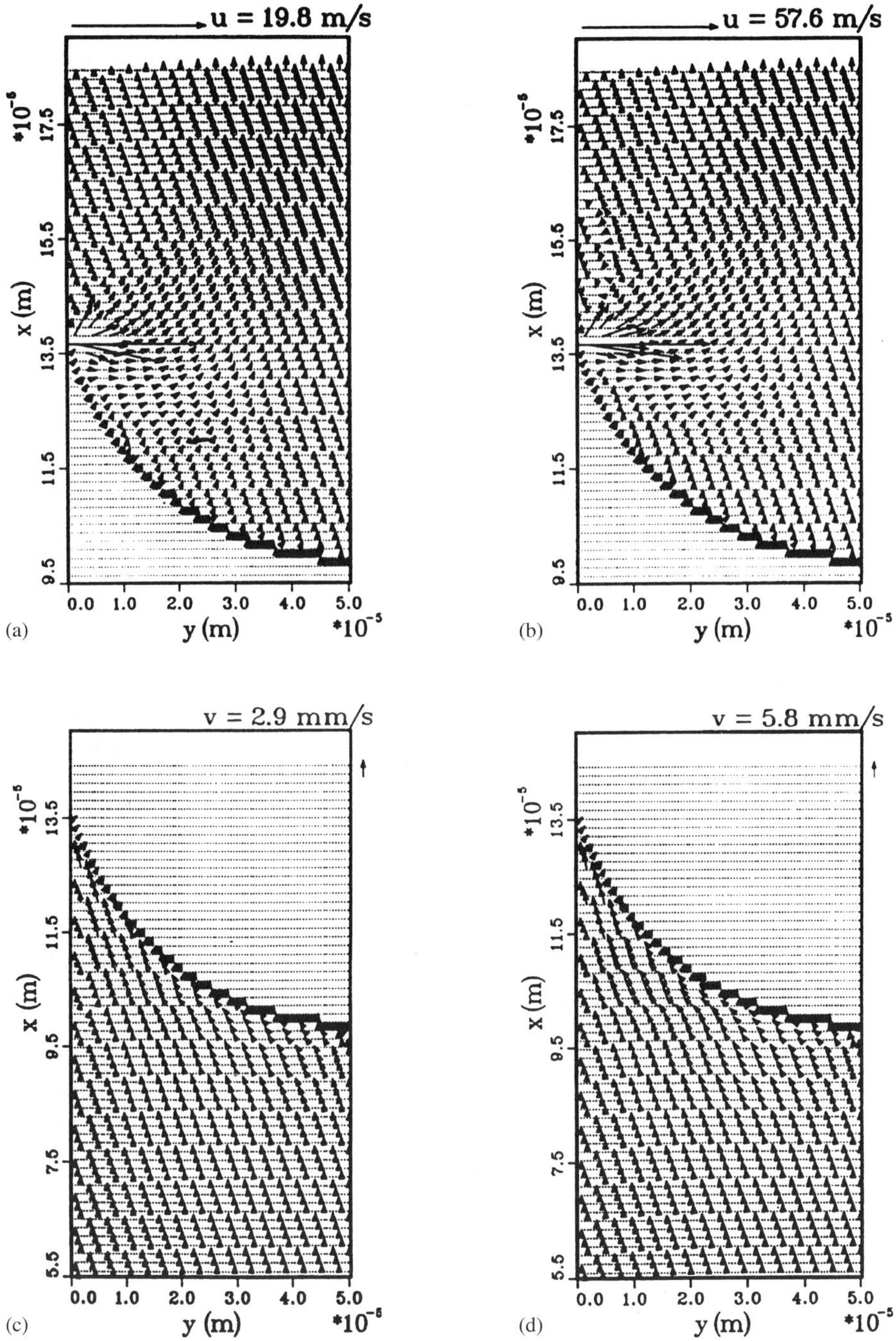

Fig. 10.17a–d Vapor flow over the evaporating meniscus (**a**) $T_w - T_{sat} = 5$ K, (**b**) $T_w - T_{sat} = 10$ K. Liquid flow in the evaporating meniscus (**c**) $T_w - T_{sat} = 5$ K, (**d**) $T_w - T_{sat} = 10$ K. Reprinted from Khrustalev and Faghri (1996) with permission

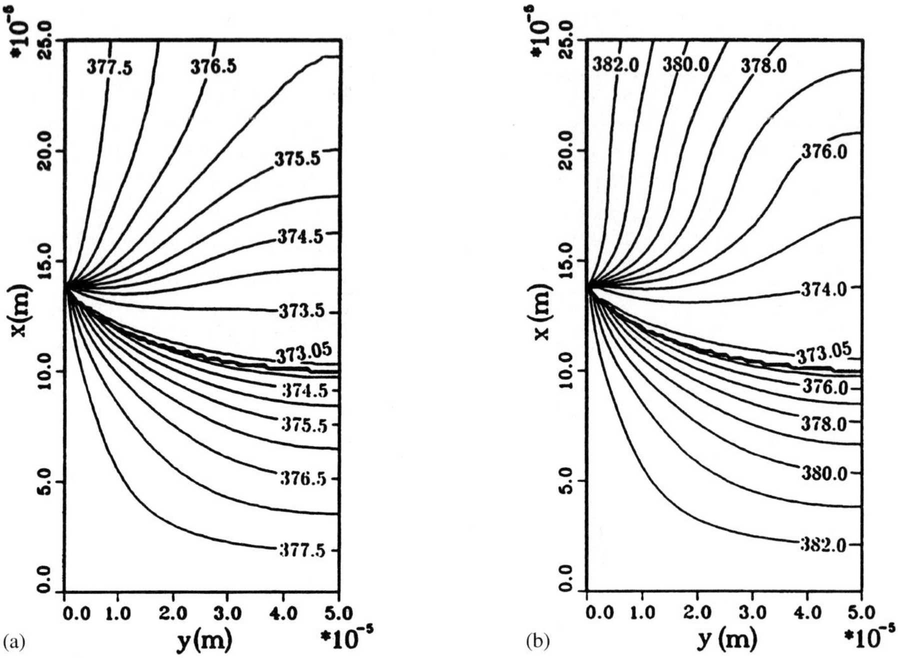

Fig. 10.18a,b Temperature contours in the liquid and vapor: (**a**) $T_w - T_{sat} = 5$ K, (**b**) $T_w - T_{sat} = 10$ K. Reprinted from Khrustalev and Faghri (1996) with permission

Fig. 10.19 Local Nusselt number for the vapor flow along the slot water at atmospheric pressure. Reprinted from Khrustalev and Faghri (1996) with permission

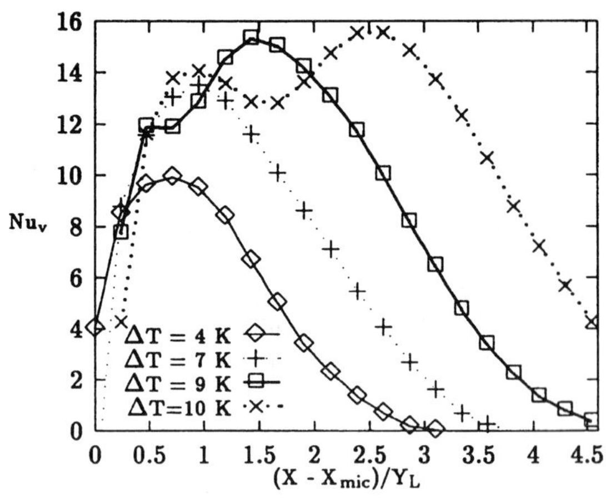

2. It is shown that the rate of vaporization (the liquid's velocity), the liquid and vapor temperatures, the position of the meniscus in the micro-channel, its hydraulic resistance and the thermal losses are determined by eight non-dimensional groups, accounting for the effects of heat transfer, phase change, as well as inertia, friction, surface tension and gravity forces. The number of such non-dimensional groups may be reduced to four by introducing a general

parameter: the capillary height, which depends on the Weber, Froude and Euler numbers.

3. The classification of possible regimes of flow are proposed. It is based on a non-dimensional parameter accounting for the ratio of the micro-channel length to the capillary height. It is shown that in the generic case the governing system of equations, which describes capillary flow, has three stationary solutions: two stable and one (intermediate) unstable.
4. It is shown that an increase in the heat flux is accompanied by an increase in the liquid and vapor velocities, the meniscus displacement towards the outlet cross-section, as well as growth of vapor to liquid forces ratio and heat losses. When q_{w} is large enough, the difference between the intensity of heat transfer and heat losses are limited by some final value, which determines the maximum rate of vaporization. Accordingly, when q_{w} is large all characteristic parameters are practically invariable.
5. It is shown that the existence of two stable states (at given values of the operating parameters) is due to the dominant role of the gravity or friction forces at the various meniscus positions. A decrease in the gravity leads to the displacement of the meniscus toward the outlet and to a decrease in the heat losses and an increase in the liquid and vapor velocities. A decrease in the micro-channel diameter leads to a monotonic increase in the liquid and vapor velocities, whereas the dependence of the meniscus position versus d has an extremum.
6. At given values of the parameters, there are optimal values of the micro-channel diameter and length, which correspond to a maximum efficiency coefficient.

References

Bowers XB, Mudawar I (1994) High flux boiling in low flow rate, low pressure drop mini-channel and micro-channel heat sink. Int J Heat Mass Transfer 37:321–333

Ghiaasiaan SM, Abdel-Khalik SI (2001) Two-phase flow in micro-channels. Advances in Heat Transfer 34:145–253

Grigoriev VA, Zorin VM (eds) (1982) Heat and mass transfer. Thermal experiment reference book. Energoizdat, Moscow (in Russian)

Ha JM, Peterson GP (1998) Capillary performance of evaporating flow in micro-grooves. An approximate analytical approach for very small tilt angles. Trans ASME J Heat Transfer 120:452–457

Incropera FP (1999) Liquid cooling of electronic devices by single-phase convection. Wiley, New York

Johnson RW (ed) (1998) The handbook of fluid dynamics. Appendix C: properties of gases and vapors. CRC, New York

Khrustalev D, Faghri A (1994) Thermal analysis of a microheat pipes. Trans ASME J Heat Transfer 116:189–198

Khrustalev ID, Faghri A (1995) Heat transfer during evaporation on capillary-grooved structures of heat pipes. Trans ASME J Heat Transfer 117:740–747

Khrustalev ID, Faghri A (1996) Fluid flow effect in evaporation from liquid vapor meniscus. Trans ASME J Heat Transfer 118:725–730

Landau LD, Lifshitz EM (1959) Fluid mechanics, 2nd edn. Pergamon, London

Landerman CS (1994) Micro-channel flow boiling mechanisms leading to burnout. J Heat Transfer Electron Syst ASME HTD 292:124–136
Levich VG (1962) Physicochemical hydrodynamics. Prentice Hall, London
Morijama K, Inoue A (1992) The thermohydraulic characteristics of two-phase flow in extremely narrow channels (the frictional pressure drop and heat transfer of boiling two-phase flow, analytical model). Heat Transfer Jpn Res 21:838–856
Peles YP, Yarin LP, Hetsroni G (2001) Steady and unsteady flow in a heated capillary. Int J Multiphase Flow 27:577–598
Peles YP, Yarin LP, Hetsroni G (1998) Heat transfer of two-phase flow in a heated capillary. In: Heat Transfer 1998, Proceedings of the 11th International Heat Transfer Conference, Kyongju, Korea, 23–28 August 1998, vol 2, pp 193–198
Peles YP, Yarin LP, Hetsroni G (2000) Thermohydrodynamic characteristics of two-phase flow in a heated capillary. Int J Multiphase Flow 26:1063–1093
Peng XF, Wang BX (1994) Cooling characteristics with micro-channeled structures. J Enhanced Heat Transfer 1:315–326
Peng XF, Wang BX (1993) Forced convection and flow boiling heat transfer for liquid flowing through micro-channels. Int J Heat Mass Transfer 36:3421–3427
Peng XF, Hu HY, Wang BX (1998) Boiling nucleation during liquid flow in micro-channels. Int J Heat Mass Transfer 41:101–106
Peng XF, Peterson GP, Wang BX (1996) Flow boiling in binary mixtures in micro-channels plates. Int J Heat Mass Transfer 39:1257–1263
Peng XF, Wang BX (1998) Forced-convection and boiling characteristics in micro-channels. In: Heat Transfer 1998, Proceedings of the 11th International Heat Transfer Conference, Kyongju, Korea, 23–28 August 1998
Peng XF, Wang BX, Peterson GP, Ma HB (1994) Experimental investigation of heat transfer on flat plates with rectangular micro-channels. Int J Heat Mass Transfer 37:127–137
Peterson GP, Ha JM (1998) Capillary performance of evaporating flow in micro grooves: an approximate analytical approach and experimental investigation. Trans ASME J Heat Transfer 120:743–751
Reid RC, Prausnitz JM, Poling BE (1987) The properties of gases and liquids. McGraw-Hill, Boston
Seaver M, Galloway A, Manuchia TJ (1989) Acoustic levitation in a free-jet wind tunnel. Rev Sci Instrum 60:3452–3458
Triplett KA, Chiaasiaah SM, Abdel-Khalik SI, LeMouel A, McCord BN (1999b) Gas–liquid two-phase flow in micro-channels. Part 11: void fraction and pressure drop. Int J Multiphase Flow 25:395–410
Triplett KA, Chiaasiaah SM, Abdel-Khalik SI, Sadowski JL (1999a) Gas–liquid two-phase flow in micro-channels. Part 1: two-phase flow patterns. Int J Multiphase Flow 25:377–394
Yuan H, Prosperetti A (1999) The pumping effect of growing and collapsing bubbles in a tube. Micromech Microeng 9:402–413
Yarin LP, Ekelchik LA, Hetsroni G (2002) Two-phase laminar flow in a heated micro-channels. Int J Multiphase Flow 28:1589–1616

Nomenclature

d	Micro-channel diameter
f_L	Capillary pressure
g	Acceleration of gravity
h	Enthalpy
k	Thermal conductivity

L	Total length of micro-channel
$\mathbf{n}, \mathbf{k}$	Normal and tangent direction
P	Pressure
h_{LG}	Latent heat of evaporation
q	Specific volumetric rate of heat absorption
q_{cr}	Critical value of heat flux on the wall at which the bubble nucleation is negligible
q_{w}	Wall heat flux
T	Temperature
$\mathbf{v}$	Velocity (vector $\mathbf{v}$ has components u, v, w, are directed along axis x, y, z, respectively)
$\tilde{V}_{\mathrm{f}}$	Relative velocity
V_{f}	Velocity of interface surface
$\mathrm{Eu} = \dfrac{\Delta P}{\rho u^2}$	Euler number
$\mathrm{Fr} = \dfrac{u^2}{gL}$	Froud number
$\mathrm{Pe} = \dfrac{uL}{\alpha}$	Peclet number
$\mathrm{Re} = \dfrac{uL}{\nu}$	Reynolds number
$\mathrm{We} = \dfrac{\rho u^2}{\sigma}$	Weber number
x, y, z	Cartesian coordinates

Greek symbols

α	Thermal diffusivity
∇	Gradient
θ	Contact angle
ν	Kinematic viscosity
ρ	Density
σ	Surface tension

Subscripts

G, L	Correspond to vapor and liquid, respectively
s	Saturation parameters
0, 00	Inlet and outlet cross-section, respectively
f	Interface surface
cr	Critical

Chapter 11
Onset of Flow Instability in a Heated Capillary

The capillary flow with distinct evaporative meniscus is described in the frame of the quasi-dimensional model. The effect of heat flux and capillary pressure oscillations on the stability of laminar flow at small and moderate Peclet number is estimated. It is shown that the stable stationary flow with fixed meniscus position occurs at low wall heat fluxes (Pe $\ll 1$), whereas at high wall heat fluxes Pe ≥ 1, the exponential increase of small disturbances takes place. The latter leads to the transition from stable stationary to an unstable regime of flow with oscillating meniscus.

11.1 Introduction

Consider the stability of capillary flow when a liquid is heated and evaporated at a meniscus. This problem is important in the context of cooling systems of electronic devices. A growing number of designs in MEMS with high power density require a thorough insight into the mechanism of complex processes in heated microchannels. The latter includes a number of problems related to hydrodynamics of laminar flow developed under conditions of the inertia, friction, gravity and capillary force interactions, heat transfer, as well as phase change. The studies in the last decade concern a wide range of problems connected with stable single-phase flows (Tuckerman and Pease 1984; Tuckerman 1981; Wiesberg et al. 1992; Wang and Peng 1994; Wu and Little 1984; Bailey et al. 1995; Peng et al. 1994; Peng and Peterson 1995, 1996; Adams et al. 1998; Incropera 1999), boiling nucleation and bubble growth in narrow pipes (Peng et al. 1998; Yuan et al. 1999; Ory et al. 2000; Peng et al. 2001), pressure drop and heat transfer in two-phase flows (Morijama and Inoue 1992; Peng and Wang 1993; Bowers and Mudawar 1994; Sobhan and Garimella 2001). At the same time there is a paucity of theoretical studies dealing with capillary flow with a phase change at an evaporative meniscus, in spite of the fact that such flows are interesting in connection with their possible implementation in cooling systems of electronic devices.

The stationary regimes of capillary flows with a distinct meniscus separating the regions of liquid and vapor flows have been considered by Khrustalev and Faghri (1996) and Peles et al. (1998, 2000, 2001). Recently Yarin et al. (2002) investigated in detail the features of two-phase laminar flow in a heated micro-channel and revealed the effect of the inertia, pressure, gravity and friction forces on major flow characteristics. It was shown that, in the general case, the system of equations that describes the capillary flow has three solutions corresponding to stationary regimes of flow. The analysis of stationary states performed in the quasi-stationary approximation (an approach similar to the Semenov's diagram method) showed that two of these states ("upper" and "lower" corresponding to high and low velocities, respectively) are stable whereas the intermediate one is unstable. The stationary or quasi-stationary approximations should be considered as limiting for the solutions of the unsteady problems for infinite time intervals. However, approaches ignoring the dynamics of the transient processes leading to steady states should be supplemented with stability consideration. Indeed, only stable steady states can become attractors of the transient processes. This makes stability studies of the limiting steady states extremely important.

The present chapter deals with the study of the stability of a flow in a heated capillary, with liquid evaporating at a meniscus. The behavior of the vapor–liquid system, which undergoes small perturbations, is analyzed by linear approximation, in the frame of the quasi-one-dimensional model of capillary flow with a distinct interface. The effect of the physical properties of both phases, the wall heat flux and the capillary sizes on the flow stability is studied. The velocity, pressure and temperature oscillations in a capillary tube with constant wall heat flux or constant wall temperature are determined. A scenario of a possible process at small and moderate Peclet numbers corresponding to the flow in capillaries is considered. The boundaries of stability, subdividing the domains of stable and unstable flows, are outlined, and the values of geometrical and operating parameters corresponding to the transition from stable to unstable flow are estimated. The study consists of the problem formulation, analysis of the influence of the physical properties of the liquid and its vapor, and wall heat flux, on velocity, pressure and temperature oscillations in capillary flows, as well as the stability of the flow at small and large Peclet numbers.

Chapter 11 consists of following: Sect. 11.2 deals with the pattern of capillary flow in a heated micro-channel with phase change at the meniscus. The perturbed equations and conditions on the interface are presented in Sect. 11.3. Section 11.4 contains the results of the investigation on the stability of capillary flow at a very small Peclet number. The effect of capillary pressure and heat flux oscillations on the stability of the flow is considered in Sect. 11.5. Section 11.6 deals with the study of capillary flow at a moderate Peclet number.

11.2 Capillary Flow Pattern

We deal here with the stability of flow in a heated capillary tube when liquid is evaporating on the meniscus. The capillary, as shown in Fig. 11.1, is a straight vertical pipe with diameter d and length ℓ. The wall heat flux is uniform: $q_{\mathrm{w}} = \mathrm{const}$. The thermal conditions on the capillary inlet and outlet are:

1. $\overline{T}_{\mathrm{L.in}} = \mathrm{const.}$, the average liquid temperature $\overline{T}_{\mathrm{L}}$ at $x = 0$
2. $\left(\frac{\mathrm{d}\overline{T}_{\mathrm{G}}}{\mathrm{d}x}\right)_{x=\ell} = 0$, the average vapor temperature gradient at $x = \ell$

Hereafter, the subscripts G and L denote vapor and liquid, respectively, and in and 0 denote inlet and outlet of the capillary tube, respectively.

These conditions correspond to a certain design of cooling system, namely, a micro-channel with cooling inlet and adiabatic outlet (Yarin et al. 2002).

The wall heat flux is the cause for the liquid evaporation, and perturbation of equilibrium between the gravity and capillary forces. It leads to the offset of both phases (heated liquid and its vapor) and interface displacement towards the inlet. In this case the stationary state of the system corresponds to an equilibrium between gravity, viscous (liquid and vapor) and capillary forces. Under these conditions the stationary height of the liquid level is less than that in an adiabatic case

$$x_f < x_{\mathrm{f.ad}} = \frac{2\sigma}{r\rho_{\mathrm{L}}g}\cos\theta \tag{11.1}$$

where x_{f} and $x_{\mathrm{f,ad}}$ are the height of the liquid level in a heated and adiabatic capillary tube, respectively, σ is the surface tension coefficient, r is the radius, ρ_{L} is the liquid density, g is the acceleration due to gravity, and θ is the static contact angle as measured from the liquid side of the contact line.

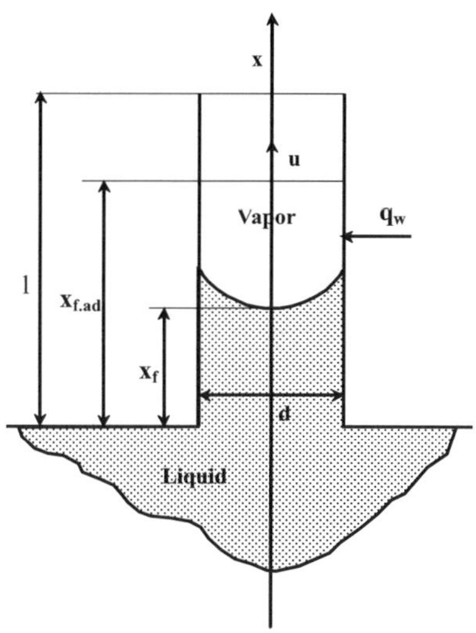

Fig. 11.1 Schematics of a heated micro-channel (*arrows* show flow and heat flux directions). Reprinted from Hetsroni et al. (2004) with permission

Unlike at adiabatic conditions, the height of the liquid level in a heated capillary tube depends not only on σ, r, ρ_{L} and θ, but also on the viscosities and thermal conductivities of the two phases, the wall heat flux and the heat loss at the inlet. The latter affects the rate of liquid evaporation and hydraulic resistance of the capillary tube. The process becomes much more complicated when the flow undergoes small perturbations triggering unsteady flow of both phases. The rising velocity, pressure and temperature fluctuations are the cause for oscillations of the position of the meniscus, its shape and, accordingly, the fluctuations of the capillary pressure. Under constant wall temperature, the velocity and temperature fluctuations promote oscillations of the wall heat flux.

11.3 Equation Transformation

11.3.1 Perturbed Equations

In this section we present the system of quasi-one-dimensional equations, describing the unsteady flow in the heated capillary tube. They are valid for flows with weakly curved meniscus when the ratio of its depth to curvature radius is sufficiently small. The detailed description of a quasi-one-dimensional model of capillary flow with distinct meniscus, as well as the estimation conditions of its application for calculation of thermohydrodynamic characteristics of two-phase flow in a heated capillary are presented in the works by Peles et al. (2000, 2001) and Yarin et al. (2002). In this model the set of equations including the mass, momentum and energy balances is:

$$\frac{\partial \rho_i}{\partial t}+\frac{\partial \rho_i u_i}{\partial x}=0 \tag{11.2}$$

$$\rho_i\frac{\partial u_i}{\partial t}+\rho_i u_i\frac{\partial u_i}{\partial x}=-\frac{\partial P_i}{\partial x}-\rho_i g-\frac{\partial F_i}{\partial x} \tag{11.3}$$

$$\rho_i\frac{\partial h_i}{\partial t}+\rho_i u_i\frac{\partial h_i}{\partial x}=\frac{\partial}{\partial x}\left(k_i\frac{\partial T_i}{\partial x}\right)+q \tag{11.4}$$

where ρ, u, P, h and T are the density, velocity, pressure, enthalpy and temperature, respectively, k is the thermal conductivity, q is the specific rate of volumetric heat absorption, F is the specific friction force, and the subscripts $i=\mathrm{G,L}$ correspond to vapor and liquid, respectively.

The conditions on the interface express the continuity of the mass and heat fluxes and the equilibrium of all acting forces (Landau and Lifshitz 1959). In the frame of reference associated with the interface they are:

$$\rho_{\mathrm{G}}\tilde{V}_{\mathrm{G}}=\rho_{\mathrm{L}}\tilde{V}_{\mathrm{L}} \tag{11.5}$$

$$P_{\mathrm{G}}+\rho_{\mathrm{G}}\tilde{V}_{\mathrm{G}}^2=P_{\mathrm{L}}+\rho_{\mathrm{L}}\tilde{V}_{\mathrm{L}}^2+f_\sigma \tag{11.6}$$

$$\rho_{\mathrm{G}}\tilde{V}_{\mathrm{G}}h_{\mathrm{G}}-k_{\mathrm{G}}\frac{\partial T_1}{\partial x}=\rho_{\mathrm{L}}\tilde{V}_{\mathrm{L}}h_{\mathrm{L}}-k_{\mathrm{L}}\frac{\partial T_{\mathrm{L}}}{\partial x} \tag{11.7}$$

where $V_f = dx_f/dt$ is the velocity of the interface, $\tilde{V}_i = u_i - V_f$ is the velocity relative to the interface, $f_\sigma = 2\sigma/R$ is the capillary pressure, and $R = r/\cos\theta$ is the radius of the interfacial curvature.

In the case when capillary flow undergoes small perturbations, the governing parameters J_j can be presented as a sum of their basic values, corresponding to the stationary flow $\overline{J}_j$, plus small perturbations J'_j

$$J_j = \overline{J}_j + J'_j \tag{11.8}$$

where $J_j = \rho,\ u,\ P,\ T,\ h,\ x_f,\ f_\sigma$ and q. The line over any parameters refers to their average (in time) values. $\partial\overline{J}_j/\partial t = 0,\ \overline{J}'_j = 0$.

In capillary flow with a distinct meniscus separating the regions of pure liquid and pure vapor flows, it is possible to neglect the change in densities of the phases and assume ρ_G and ρ_L are constant. For flow of incompressible fluid ($\rho_i =$ const., $\rho'_i = 0$, $\partial\overline{u}_i/\partial x = 0$) the substitution of (11.8) in Eqs. (11.1–11.3) leads, in a linear approximation, to the following system of equations

$$\overline{\rho}_i \frac{\partial(\overline{u}_i + u'_i)}{\partial x} = 0 \tag{11.9}$$

$$\overline{\rho}_i \frac{\partial u'_i}{\partial t} = -\frac{\partial(\overline{P}_i + P'_i)}{\partial x} - \overline{\rho}_i g - \frac{\partial(\overline{F}_i + F'_i)}{\partial x} \tag{11.10}$$

$$\overline{\rho}_i \frac{\partial h'_i}{\partial t} + \overline{\rho}_i \overline{u}_i \frac{\partial \overline{h}_i}{\partial x} + \overline{\rho}_i u'_i \frac{\partial \overline{h}_i}{\partial x} + \overline{\rho}_i \overline{u}_i \frac{\partial h'_i}{\partial x} = \frac{\partial}{\partial x}\left(k_i \frac{\partial(\overline{T}_i + T'_i)}{\partial x}\right) + \overline{q} + q'\,. \tag{11.11}$$

The equations for stationary flow

$$\frac{d\overline{u}_i}{dx} = 0 \tag{11.12}$$

$$\frac{d\overline{P}_i}{dx} + \overline{\rho}_i g + \frac{d\overline{F}_i}{dx} = 0 \tag{11.13}$$

$$\overline{\rho}_i \overline{u}_i c_{pi} \frac{d\overline{T}_i}{dx} = \frac{d}{dx}\left(k_i \frac{d\overline{T}_i}{dx}\right) + \overline{q}\,. \tag{11.14}$$

We obtain from (11.9–11.11) the equations for small perturbations of velocity, pressure, temperature and enthalpy, as well as the specific volumetric rate of heat absorption. Assuming that $h_i = c_{p_i} T_i$ we arrive at

$$\frac{\partial u'_i}{\partial x} = 0 \tag{11.15}$$

$$\frac{\partial u'_i}{\partial t} = -\frac{1}{\overline{\rho}_i}\left(\frac{\partial P'_i}{\partial x} + \frac{\partial F'_i}{\partial x}\right) \tag{11.16}$$

$$\frac{\partial T'_i}{\partial t} + u'_i \frac{\partial \overline{T}_i}{\partial x} + \overline{u}_i \frac{\partial T'_i}{\partial x} = \frac{\partial}{\partial x}\left(\alpha_i \frac{\partial T'_i}{\partial x}\right) + \tilde{q}'_i \tag{11.17}$$

where $\alpha_i = k_i/(\overline{\rho}_i c_{p_i})$is the thermal diffusivity, $\tilde{q}_i' = q'/(\overline{\rho}_i c_{p_i})$. Substitution of (11.8) in conditions (11.5–11.7) leads (in linear approximations) to the following system of equations:

$$\overline{\rho}_G\left(\overline{u}_G + u_G' - \frac{dx_f'}{dt}\right) = \overline{\rho}_L\left(\overline{u}_L + u_L' - \frac{dx_f'}{dt}\right) \tag{11.18}$$

$$\overline{P}_G + P_G' + \overline{\rho}_G(\overline{u}_G^2 + 2\overline{u}_G u_G') = \overline{P}_L + P_L' + \overline{\rho}_L(\overline{u}_L^2 + 2\overline{u}_L u_L') + \overline{f}_\sigma + f_\sigma' \tag{11.19}$$

$$\begin{aligned}&\overline{\rho}_G\left(\overline{u}_G\overline{h}_G + u_G'\overline{h}_G + \overline{u}_G h_G' - \overline{h}_G\frac{dx_f'}{dt}\right) - k_G\frac{\partial \overline{T}_G}{\partial x} - k_G\frac{\partial T_G'}{\partial x}\\ &= \overline{\rho}_L\left(\overline{u}_L\overline{h}_L + u_L'\overline{h}_L + \overline{u}_L h_L' - \overline{h}_L\frac{dx_f'}{dt}\right) - k_L\frac{\partial \overline{T}_L}{\partial x} - k_L\frac{\partial T_L'}{\partial x}.\end{aligned} \tag{11.20}$$

Here $x_f = \overline{x}_f + x_f'$, x_f is the liquid height in the capillary.

For stationary flow

$$\overline{\rho}_G\overline{u}_G = \overline{\rho}_L\overline{u}_L \tag{11.21}$$

$$\overline{P}_G + \overline{\rho}_G\overline{u}_G^2 = \overline{P}_L + \overline{\rho}_L\overline{u}_L^2 + \overline{f_\sigma} \tag{11.22}$$

$$\overline{\rho}_G\overline{u}_G\overline{h}_G - k_G\frac{\partial \overline{T}_G}{\partial x} = \overline{\rho}_L\overline{u}_L\overline{h}_L - k_L\frac{\partial \overline{T}_L}{\partial x} \tag{11.23}$$

and we obtain from (11.18–11.20) equations for the oscillations at the meniscus surface

$$(\overline{\rho}_G\overline{u}_G' - \overline{\rho}_L\overline{u}_L') = (\overline{\rho}_G - \overline{\rho}_L)\frac{dx_f'}{dt} \tag{11.24}$$

$$(P_G' - P_L') = 2\overline{\rho}_G\overline{u}_L(u_L' - u_G') + f_\sigma' \tag{11.25}$$

$$\begin{aligned}&\overline{\rho}_L\overline{u}_L(h_G' - h_L') + (\overline{\rho}_G\overline{h}_G u_G' - \overline{\rho}_L\overline{h}_L u_L') -\\ &-(\overline{\rho}_G\overline{h}_G - \overline{\rho}_L\overline{h}_L)\frac{dx_f'}{dt} = k_G\frac{\partial T_G'}{\partial t} - k_L\frac{\partial T_L'}{\partial t}\end{aligned} \tag{11.26}$$

where $h = c_p T$, $h_{x=x_f} = c_p T_s$, and T_s is the temperature of the interface that is assumed to be constant and equal to the saturation temperature. The small perturbations of pressure practically does not influence T_s because of the weak dependence of $T_s(P_s)$ (Reid et al. 1987).

The solution of Eqs. (11.15–11.17), subject to the conditions (11.24–11.26), determines the displacement of the interface in time, as well as the evolution of the velocity, pressure and temperature oscillations.

11.3.2 Perturbed Energy Equation for Small Peclet Number

The dimensionless form of Eq. (11.17) is

$$\mathrm{St}\frac{\partial \tilde{T}_i'}{\partial \tilde{t}} + \tilde{u}_i'\frac{\partial \tilde{T}_i}{\partial \tilde{x}} + \tilde{u}\frac{\partial \tilde{T}_i'}{\partial \tilde{x}} = \mathrm{Pe}_i^{*\,-1}\frac{\partial^2 \tilde{T}_i'}{\partial \tilde{x}^2} + \vartheta_i' \tag{11.27}$$

where $\mathrm{St} = \ell_* \psi_* / \overline{u}_*$ and $\mathrm{Pe}_i^* = u_* \ell_* / \alpha_i$ are the Strouhal and Peclet numbers, respectively, and $\vartheta_i' = q' \ell_* / (\rho_i u_* c_{\mathrm{p}_i} T_*)$, $\tilde{u}_i = \overline{u}_i / \overline{u}_*$, $\tilde{u}_i' = u_i' / \overline{u}_*$, $\tilde{T}_i = \overline{T}_i / \overline{T}_*$, $\tilde{T}_i' = T_i' / \overline{T}_*$, $\tilde{x} = x / \ell_*$ and $\tilde{t} = t \psi_*$ ($\psi_* = t_*^{-1}$), ℓ_*, $\overline{u}_*$, $\overline{T}_*$ and t_* are characteristic scales of the length, velocity, temperature and time.

The first terms on the left and right-hand sides of Eq. (11.27) are on the order of St and Pe^{-1}, respectively, whereas the second and third terms on left-hand side of Eq. (11.27) have the order of one. When $\mathrm{Pe} \ll 1$ and $\mathrm{St} \gg 1$ it is possible to omit the terms accounting for convective heat transfer due to oscillations and present Eq. (11.17) as follows:

$$\frac{\partial T_i'}{\partial t} = \alpha_i \frac{\partial^2 T_i'}{\partial x^2} + \tilde{q}'_i \,. \tag{11.28}$$

11.3.3 Perturbed Energy Equation for Moderate Peclet Number

When the temperature T_{s} of the interface is constant, and wall heat flux is also constant, temperature oscillations are the result of the meniscus displacement along micro-channel axis. They are expressed as

$$T_i' = x' \frac{\mathrm{d}T_i}{\mathrm{d}x} = x' \frac{\mathrm{d}\overline{T}_i}{\mathrm{d}x} + x' \frac{\mathrm{d}T_i'}{\mathrm{d}x} \,. \tag{11.29}$$

Neglecting the term containing product of oscillations, we obtain

$$T_i' = x' \frac{\mathrm{d}\overline{T}_i}{\mathrm{d}x}, \quad \frac{\mathrm{d}T_i'}{\mathrm{d}x} = x' \frac{\mathrm{d}^2\overline{T}_i}{\mathrm{d}x^2} \,. \tag{11.30}$$

The oscillations of the meniscus position x' can be estimated as follows:

$$x' = \frac{u_*'}{\psi_*} \tag{11.31}$$

where u_*' is the characteristic oscillations velocity (order of liquid oscillation velocity).

Convective heat transfer that is due to oscillations determines the second and the third terms on the left-hand side of Eq. (11.17). Using Eqs. (11.30) and (11.31), we estimate the values of these terms. For this, we consider the ratio of the third term to the second one

$$\left| \frac{\overline{u}_i \dfrac{\partial T_i'}{\partial x}}{u_i' \dfrac{\partial \overline{T}_i}{\partial x}} \right| = \frac{1}{\mathrm{St}_*} \left| \frac{\left(\dfrac{\partial^2 \tilde{T}_i}{\partial \tilde{x}_i^2} \right)}{\left(\dfrac{\partial \tilde{T}_i}{\partial \tilde{x}} \right)} \right| \,. \tag{11.32}$$

The temperature distribution in a heated micro-channel is described by the following correlation (Peles et al. 2001).

$$\tilde{T} = C_1^{(i)} + \vartheta_i (\tilde{x}^* + \mathrm{Pe}_i^{*-1}) + C_2^{(i)} \exp(\mathrm{Pe}_i * \tilde{x}^*) \tag{11.33}$$

where the constants $C_1^{(i)}$ and $C_2^{(i)}$ are expressed as

$$C_1^{(\mathrm{L})} = (1 - C_2^{(\mathrm{L})}) - \vartheta_{\mathrm{L}}/\mathrm{Pe}_{\mathrm{L}}^* \tag{11.34}$$

$$C_2^{(\mathrm{L})} = [(\tilde{T}_{\mathrm{s}} - 1) - \vartheta_{\mathrm{L}}\tilde{x}_{\mathrm{f}}^*]/[\exp(\mathrm{Pe}_{\mathrm{L}}^*\tilde{x}_{\mathrm{f}}^*) - 1] \tag{11.35}$$

$$C_1^{(\mathrm{G})} = \tilde{T}_{\mathrm{s}} - \vartheta_{\mathrm{G}}(\tilde{x}_{\mathrm{f}}^* + \mathrm{Pe}_{\mathrm{G}}^{*-1}) - C_2^{(\mathrm{L})}\exp(\mathrm{Pe}_{\mathrm{L}}^*\tilde{x}_{\mathrm{f}}^*) \tag{11.36}$$

$$C_2^{(\mathrm{G})} = -\vartheta_{\mathrm{G}}/[\mathrm{Pe}_{\mathrm{G}}\exp(\mathrm{Pe}_{\mathrm{G}}^*)] \,. \tag{11.37}$$

Here characteristic length $\ell_* = \ell$ is the length of the capillary tube, $T_* = T_{\mathrm{L.in}}$ is the inlet liquid temperature.

The temperature distribution in a heated micro-channel is not uniform (Fig. 11.2, Peles et al. 2000). The liquid entering the channel absorbs heat from the walls and its temperature increases. As the liquid flows toward the evaporating front it reaches a maximum temperature and then the temperature begins to decrease up to the saturated temperature. Within the vapor domain, the temperature increases monotonically from saturation temperature T_{s} up to outlet temperature $T_{\mathrm{G.0}}$.

The module of ratio of the second-order derivative $\partial^2\tilde{T}_i/\partial\tilde{x}^2$ to the first-order derivative $\partial\tilde{T}_i/\partial\tilde{x}$ is

$$\left| \frac{\left(\dfrac{\partial^2\tilde{T}_i}{\partial\tilde{x}^{*2}}\right)}{\left(\dfrac{\partial\tilde{T}_i}{\partial\tilde{x}^*}\right)} \right| = \left| \frac{C_2^{(i)}\mathrm{Pe}_i^{*2}\exp(\mathrm{Pe}_i^*\tilde{x}^*)}{1 - C_2^{(i)}\mathrm{Pe}_i^*\exp(\mathrm{Pe}_i^*\tilde{x}^*)} \right| . \tag{11.38}$$

The value of the ratio $\chi = (\partial^2\tilde{T}_i/\partial\tilde{x}^{*2})/(\partial\tilde{T}_i/\partial\tilde{x}^*)$ depends on the Peclet number, as well as on meniscus position in stable state $\tilde{x}_{\mathrm{f}}$. The dependence of the meniscus position $\tilde{x}_{\mathrm{f}}^*$ on Pe^* is shown in Fig. 11.3. It is seen that in the range of moderate Peclet number $\tilde{x}_{\mathrm{f}}^* \ll 1$. The values of the right-hand side of Eq. (11.32) determines by ratio χ/St. At moderate values of characteristic frequency $w_* \sim 10^{-2}$ m the Strouchal number has order of 0.1–1. When the term $C_2^{(i)}\mathrm{Pe}_i^{*2}\exp(\mathrm{Pe}_i^*\tilde{x}^*)$ in Eq. (11.38) is more than unit, the parameter χ has order of Pe_i^* i.e. larger than unit. In this case it is possible to omit the second term on the left-hand side of Eq. (11.17) and it takes

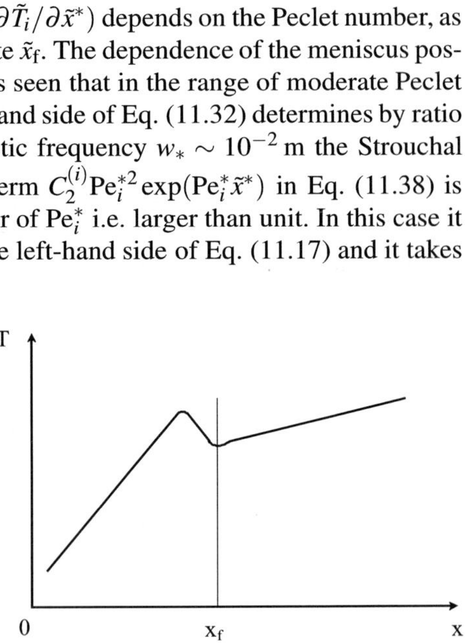

Fig. 11.2 The scheme of temperature distribution along a heated micro-channel. Reprinted from Peles et al. (2000) with permission

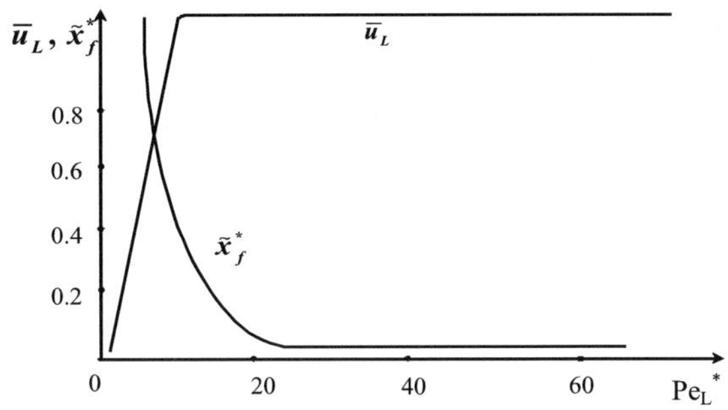

Fig. 11.3 The dependences of $\overline{u}_\mathrm{L}(\mathrm{Pe}^*_\mathrm{L})$and $\tilde{x}^*_\mathrm{f}(\mathrm{Pe}^*_\mathrm{L})$. Reprinted from Hetsroni et al. (2004) with permission

the following form:

$$\frac{\partial T_i'}{\partial t} + \overline{u}_i \frac{\partial T_i'}{\partial x} = \frac{\partial}{\partial x}\left(\alpha_i \frac{\partial T_i'}{\partial x}\right) . \tag{11.39}$$

11.4 Flow with Small Peclet Numbers

11.4.1 The Velocity, Pressure and Temperature Oscillations

The estimations enable us to disregard the minor convective effects, and to consider the problem in the framework of a pure conductive approximation. Neglecting, in Eq. (11.17), the term $u_i' \frac{\partial \overline{T}_i}{\partial x}$ and $\overline{u}_i \frac{\partial T_i'}{\partial x}$ we reduce the problem to

$$\frac{\partial T_i'}{\partial t} = \alpha_i \frac{\partial^2 T_i'}{\partial x^2} + \tilde{q}_i' . \tag{11.40}$$

First we restricted ourselves to considering a particular case of flow in a capillary tube with $q_\mathrm{w} = \mathrm{const.}$ ($q' = 0$). We also neglected the change of the capillary pressure through the changes of the contact angle, due to the motion of the meniscus. Accordingly we assume that $f_\sigma' = 0$ in condition (11.25).

To determine the velocity, pressure and temperature oscillations we use Eqs. (11.15), (11.16) and (11.40). From Eq. (11.15) it follows that

$$u' = u'(t) . \tag{11.41}$$

Thus, the velocity oscillations, in the flow of an incompressible fluid, depend only on time, i.e., the liquid and vapor columns move in the capillary tube, on the whole, similar to a solid body. Bearing this in mind, we present the solution of Eq. (11.15) as follows:

$$u_i' = A_i \exp(\Omega t) \tag{11.42}$$

where A is the amplitude of the velocity oscillations, $\Omega = \xi + i|\psi|$ is the complex frequency, and ξ and ψ are the growth rate and frequency of the velocity oscillations.

The specific friction force F_{i} in laminar flow is expressed as (Yarin et al. 2002)

$$F_{\mathrm{i}} = \frac{32}{d^2}\mu_i u_i x \qquad (11.43)$$

where d is the diameter of the tube, and μ is the viscosity.

In accordance with Eq. (11.43), the oscillations of F_i are

$$F_i' = \frac{32}{d^2}\mu_i u_i' x\,. \qquad (11.44)$$

Taking into account Eqs. (11.42) and (11.44) we can present the pressure oscillations as follows:

$$P_i' = A_i \overline{\rho}_i f_i(x)\exp(\Omega t) + a_i \qquad (11.45)$$

where $f_i(x)$ is some function of x, and the parameter $a_i = a_i(t)$.

From a dimensional consideration, it is necessary to assume that the derivative of the function $f(x)$ is constant: $f_i'(x) = k_i$. Substitution of expressions (11.44) and (11.45) in Eq. (11.16) gives

$$k_i = -\left(\frac{32}{d^2}\nu_i + \Omega\right) \qquad (11.46)$$

where ν is the kinematic viscosity.

The parameter a_i is determined by using the conditions

$$x = 0\,, \quad P_{\mathrm{L}}' = P_{\mathrm{L.in}}' \qquad (11.47)$$

$$x = \ell\,, \quad P_{\mathrm{G}}' = P_{\mathrm{G.0}}'\,. \qquad (11.48)$$

As a result we obtain

$$a_1 = P_{\mathrm{G.0}}' - \overline{\rho}_{\mathrm{G}} A_{\mathrm{G}} k_{\mathrm{G}} \ell \exp(\Omega t) \qquad (11.49)$$

$$a_2 = P_{\mathrm{2.in}}'\,. \qquad (11.50)$$

The oscillations of the phase temperatures can be presented in the following form

$$T_i' = A_i\left(\frac{h_{\mathrm{LG}}}{c_{\mathrm{p}_i}\overline{u}_{\mathrm{L}}}\right)\varphi_i(x)\exp(\Omega t) \qquad (11.51)$$

where h_{LG} is the latent heat of the liquid vaporization, $\overline{u}_{\mathrm{L}} = \overline{u}_{\mathrm{L.in}}$ is the liquid velocity in the stationary flow regime, and $\varphi(x)$ is some function of x that satisfies the condition $\varphi''(x)/\varphi(x) = \mathrm{const}$.

Assuming that $\varphi_i(x) = \exp(n_i x)$, we obtain

$$T_i' = A_i\left(\frac{h_{\mathrm{LG}}}{c_{\mathrm{p}_i}\overline{u}_{\mathrm{L}}}\right)\exp(n_i x* + \Omega t) \qquad (11.52)$$

where $x^* = x - \overline{x}_{\mathrm{f}}$, $n = ik_x$, k_x is the longitudinal component of the wave vector $\mathbf{k}$ ($k_x \neq 0$, $k_y = k_z = 0$).

The substitution of expression (11.52) into Eq. (11.40) gives

$$n_i = \pm \left(\frac{\Omega}{\alpha_i} \right)^{1/2} \tag{11.53}$$

Assuming that the temperature oscillations that are due to the displacement of the interface decrease far from $\overline{x}_{\mathrm{f}}$, the sign in front of Eq. (11.53) is positive for phase L and negative for phase G.

The oscillations of the meniscus position x'_{f} depend only on the time and are expressed as follows:

$$x'_{\mathrm{f}} = C \exp(\Omega t) \tag{11.54}$$

where C is the amplitude.

11.4.2 Dispersion Equation

Using expressions (11.42), (11.45), (11.51) and (11.54) for the velocity, pressure, temperature and meniscus position oscillations, as well as Eqs. (11.46) and (11.53) for k_i and n_i, we arrive at the system of algebraic equations for unknown amplitudes A_{G}, A_{L} and C.

$$\begin{aligned} A_{\mathrm{G}}\tilde{\alpha}_{11} + A_{\mathrm{L}}\tilde{\alpha}_{1.2} + C\tilde{\alpha}_{13} &= 0 \\ A_{\mathrm{G}}\tilde{\alpha}_{21} + A_{\mathrm{L}}\tilde{\alpha}_{22} &= 0 \\ A_{\mathrm{G}}\tilde{\alpha}_{31} + A_{\mathrm{L}}\tilde{\alpha}_{32} + C\tilde{\alpha}_{33} &= 0 \end{aligned} \tag{11.55}$$

where

$$\tilde{\alpha}_{11} = \overline{\rho}_{\mathrm{G}}\,, \quad \tilde{\alpha}_{12} = -\overline{\rho}_{\mathrm{L}}\,, \quad \tilde{\alpha}_{13} = -\Omega(\overline{\rho}_{\mathrm{G}} - \overline{\rho}_{\mathrm{L}})$$

$$\tilde{\alpha}_{21} = (k_{\mathrm{G}}\overline{\rho}_{\mathrm{G}}\overline{x}_{\mathrm{f}} + 2\overline{\rho}_{\mathrm{L}}\overline{u}_{\mathrm{L}})\,, \quad \tilde{\alpha}_{22} = -(k_{\mathrm{L}}\overline{\rho}_{\mathrm{L}}\overline{x}_{\mathrm{f}} + 2\overline{\rho}_{\mathrm{L}}\overline{u}_{\mathrm{L}})$$

$$\tilde{\alpha}_{31} = \left(\overline{u}_{\mathrm{L}} + \rho_{\mathrm{G.L}} \frac{\overline{h}_{\mathrm{G}}}{h_{\mathrm{LG}}} - \alpha_{\mathrm{G}}\rho_{\mathrm{G.L}} n_{\mathrm{G}} \right)\,, \quad \tilde{\alpha}_{32} = \left(\overline{u}_{\mathrm{L}} + \frac{\overline{h}_{\mathrm{L}}}{h_{\mathrm{LG}}} - \alpha_{\mathrm{L}} n_{\mathrm{L}} \right)$$

$$\tilde{\alpha}_{33} = \Omega \overline{u}_{\mathrm{L}} \left(\rho_{\mathrm{G.L}} \frac{\overline{h}_{\mathrm{G}}}{h_{\mathrm{LG}}} - \frac{\overline{h}_{\mathrm{L}}}{h_{\mathrm{LG}}} \right)\,, \quad \rho_{\mathrm{G.L}} = \rho_{\mathrm{G}}/\rho_{\mathrm{L}}\,.$$

Note that the system (11.55) is valid for small deviations of the interface from $\overline{x}_{\mathrm{f}}$ when $n_i x'_{\mathrm{f}} \ll 1$ and $\exp(n_i x'_{\mathrm{f}}) \simeq 1$. Estimations show that the term $C\tilde{\alpha}_{33}$ in the thermal balance equation on the interface is small in comparison with the term $A_{\mathrm{G}}\tilde{\alpha}_{31}$ and $A_{\mathrm{L}}\tilde{\alpha}_{32}$. Moreover, since $\rho_{\mathrm{G.L}}(\overline{h}_{\mathrm{G}}/h_{\mathrm{LG}}) \ll 1$ and $(\overline{h}_{\mathrm{L}}/h_{\mathrm{LG}}) \ll 1$, it is possible to neglect the second term in the expressions for coefficients $\tilde{\alpha}_{31}$ and $\tilde{\alpha}_{32}$ and assume that $\tilde{\alpha}_{31} = (\overline{u}_{\mathrm{L}} - \alpha_{\mathrm{G}}\rho_{\mathrm{G.L}} n_{\mathrm{G}})$, $\tilde{\alpha}_{32} = (\overline{u}_{\mathrm{L}} - \alpha_{\mathrm{L}} n_{\mathrm{L}})$. Then the non-trivial solution of the relations in Eq. (11.55) correspond to the following condition:

$$\begin{vmatrix} \tilde{\alpha}_{11} & \tilde{\alpha}_{12} & \tilde{\alpha}_{13} \\ \tilde{\alpha}_{21} & \tilde{\alpha}_{22} & 0 \\ \tilde{\alpha}_{31} & \tilde{\alpha}_{32} & 0 \end{vmatrix} = 0\,. \tag{11.56}$$

From (11.93) it follows that

$$\tilde{\alpha}_{13}(\tilde{\alpha}_{21}\tilde{\alpha}_{32} - \tilde{\alpha}_{22}\tilde{\alpha}_{31}) = 0\,. \tag{11.57}$$

The case $\tilde{\alpha}_{13} = 0$ corresponds to the condition $\Omega = 0$ (stationary regime), and we obtain the following dispersion equation for $\Omega \neq 0$

$$\tilde{\alpha}_{21}\tilde{\alpha}_{32} - \tilde{\alpha}_{22}\tilde{\alpha}_{31} = 0\,. \tag{11.58}$$

The specific form of the dependence of the complex frequency Ω on parameters of the problem found by Eq. (11.58), is presented as follows:

$$a_*\Omega^{3/2} + b_*\Omega + c_*\Omega^{1/2} + d_* = 0 \tag{11.59}$$

where

$$a_* = \left(1 + \alpha_{\mathrm{G.L}}^{1/2}\right)\alpha_{\mathrm{L}}^{-1/2},$$

$$b_* = -\frac{\overline{u}_{\mathrm{L}}}{\alpha_{\mathrm{L}}}(1 - \rho_{\mathrm{L.G}}),$$

$$c_* = -\left\{-\frac{32}{d^2}\nu_{\mathrm{G}}\left(1 - \nu_{\mathrm{L.G}}\alpha_{\mathrm{G.L}}^{1/2}\right) + 2\rho_{\mathrm{L.G}}\frac{\overline{u}_{\mathrm{L}}}{\overline{x}_{\mathrm{f}}}\left(1 - \rho_{\mathrm{G.L}}\alpha_{\mathrm{L.G}}^{1/2}\right)\right\}\frac{1}{\sqrt{\alpha_{\mathrm{L}}}},$$

$$d_* = -\frac{32}{d^2}\nu_{\mathrm{G}}\frac{\overline{u}_{\mathrm{L}}}{\alpha_{\mathrm{L}}}(1 - \nu_{\mathrm{L.G}}\rho_{\mathrm{L.G}})$$

and the ratio of characteristic parameters, corresponding to liquid and gaseous phases, is expressed as $\alpha_{j.i} = \alpha_j/\alpha_i$, $\rho_{j.i} = \rho_j/\rho_i$, $\nu_{j.i} = \nu_j/\nu_i$.

Introducing the new variable

$$y = \Omega^{1/2} + \frac{b_*}{3a_*} \tag{11.60}$$

we reduce Eq. (11.59) to the form

$$y^3 + 3P_*y + 2q_* = 0 \tag{11.61}$$

where

$$2q_* = \frac{2b_*^3}{27a_*^3} - \frac{b_*c_*}{3a_*^2} + \frac{d_*}{a_*}, \tag{11.62}$$

$$3P_* = \frac{3a_*c_* - b_*^2}{3a_*^2}\,. \tag{11.63}$$

11.4.3 Solution of the Dispersion Equation

Equation (11.61) has three roots: three real, or one real and two complex, depending on the value of determinant $D = q_*^2 + P_*^3$ (Korn and Korn 1968). Since our aim is to determine the complex frequency Ω, we will consider the complex solution of Eq. (11.61) only.

In the case when $q_*^2 + P_*^3 > 0$ and $P_* < 0$, the complex roots of Eq. (11.61) are

$$\Omega_{\mathrm{I}}^{1/2} = \left(r_* \operatorname{Cosh}\frac{\varphi}{3} + i\sqrt{3} r_* \operatorname{Sinh}\frac{\varphi}{3} \right) - \frac{b_*}{3a_*} \tag{11.64}$$

$$\Omega_{\mathrm{II}}^{1/2} = \left(r_* \operatorname{Cosh}\frac{\varphi}{3} - i\sqrt{3} r_* \operatorname{Sinh}\frac{\varphi}{3} \right) - \frac{b_*}{3a_*} \tag{11.65}$$

where $\operatorname{Cosh}\varphi = q_*/r_*^3$, $r_* = \pm\sqrt{|P_*|}$, and the sign of r_* is the same as sign of q_*.

In the case when $P_* > 0$ the complex roots of Eq. (11.61) are

$$\Omega_{\mathrm{I}}^{1/2} = \left(r_* \operatorname{Sinh}\frac{\varphi}{3} + i\sqrt{3} r_* \operatorname{Cosh}\frac{\varphi}{3} \right) - \frac{b_*}{3a_*} \tag{11.66}$$

$$\Omega_{\mathrm{II}}^{1/2} = \left(r_* \operatorname{Sinh}\frac{\varphi}{3} - i\sqrt{3} r_* \operatorname{Cosh}\frac{\varphi}{3} \right) - \frac{b_*}{3a_*} \tag{11.67}$$

where $\operatorname{Sinh}\varphi = q_*/r_*^3$.

Since $\Omega = \xi + i|\psi|$ and

$$\Omega^{1/2} = \left\{ \sqrt{\sqrt{\xi^2 + |\psi|^2} + \xi} + i\sqrt{\sqrt{\xi^2 + |\psi|^2} - \xi} \right\} \tag{11.68}$$

we split Eqs. (11.64), (11.65), (11.66), and (11.67) into the real and imaginary parts. As a result we obtain expressions for the growth rate and frequency of oscillations

$$\xi = \left(r_* \operatorname{Cosh}\frac{\varphi}{3} - \frac{b_*}{3a_*} \right)^2 - 3r_*^2 \operatorname{Sinh}^2\frac{\varphi}{3} \tag{11.69}$$

$$|\varphi| = \left| 4\left(r_* \operatorname{Cosh}\frac{\varphi}{3} - \frac{b_*}{3a_*} \right) \sqrt{3} r_* \operatorname{Sinh}\frac{\varphi}{3} \right| \tag{11.70}$$

for the case $q_*^2 + P_*^3 > 0$, $P_* < 0$, and

$$\xi = \left(r_* \operatorname{Sinh}\frac{\varphi}{3} - \frac{b_*}{3a_*} \right)^2 - 3r_*^2 \operatorname{Cosh}^2\frac{\varphi}{3} \tag{11.71}$$

$$|\varphi| = \left| 4\left(r_* \operatorname{Sinh}\frac{\varphi}{3} - \frac{b_*}{3a_*} \right) \sqrt{3} r_* \operatorname{Cosh}\frac{\varphi}{3} \right| \tag{11.72}$$

for the case $P_* > 0$.

11.4.4 Analysis of the Solution

First we estimate the values of the coefficients a_*, b_*, c_* and d_* for realistic physical values of the characteristic parameters (Table 11.1).

Taking into account the data in Table 11.1 it is possible to simplify significantly the expressions for the coefficients a_* , b_* , c_* and d_*.

$$a_* \simeq \alpha_{\mathrm{G}}^{1/2}\alpha_{\mathrm{L}}^{-1} \tag{11.73}$$

$$b_* \simeq \rho_{\mathrm{L.G}}\frac{\overline{u}_{\mathrm{L}}}{\alpha_{\mathrm{L}}} \tag{11.74}$$

$$c_* \simeq -\left\{-\frac{32}{d^2}\nu_{\mathrm{G}} + 2\rho_{\mathrm{L.G}}\alpha_{\mathrm{L.G}}\frac{\overline{u}_{\mathrm{L}}}{\tilde{x}_{\mathrm{f}}}\right\}\frac{1}{\sqrt{\alpha_{\mathrm{L}}}} \tag{11.75}$$

$$d_* \simeq \frac{32}{d^2}\nu_{\mathrm{L}}\frac{\overline{u}_{\mathrm{L}}}{\alpha_{\mathrm{L}}}\rho_{\mathrm{L.G}}\,. \tag{11.76}$$

For the study of flow stability in a heated capillary tube it is expedient to present the parameters P_* and q_* as a function of the Peclet number defined as $\mathrm{Pe} = (\overline{u}_{\mathrm{L}}d)/\alpha_{\mathrm{L}}$. We notice that the Peclet number in capillary flow, which results from liquid evaporation, is an unknown parameter, and is determined by solving the stationary problem (Yarin et al. 2002). Employing the Peclet number as a generalized parameter of the problem allows one to estimate the effect of physical properties of phases, micro-channel geometry, as well as wall heat flux, on the characteristics of the flow, in particular, its stability.

Using Eqs. (11.73–11.76) and (11.62) and (11.63), we obtain

$$P_* = A_*\mathrm{Pe}_{\mathrm{L}}^2 + B_*\mathrm{Pe}_{\mathrm{L}} + C_* \tag{11.77}$$

$$q_* = A_{**}\mathrm{Pe}_{\mathrm{L}}^3 + B_{**}\mathrm{Pe}_{\mathrm{L}}^2 + C_{**}\mathrm{Pe}_{\mathrm{L}} \tag{11.78}$$

where

$$A_* = -\frac{1}{9}\rho_{\mathrm{L.G}}^2\frac{\alpha_{\mathrm{L}}}{d^2}\alpha_{\mathrm{L.G}}\,,$$

$$B_* = -\frac{2}{3}\rho_{\mathrm{L.G}}\alpha_{\mathrm{L.G}}^{1/2}\frac{\alpha_{\mathrm{L}}}{d^2}\frac{1}{\tilde{x}_{\mathrm{f}}}\,,$$

Table 11.1 Characteristics of phases (saturated state $T = 100\,°\mathrm{C}$) (Vargaftic et al. 1996)

Phase	Parameter						
	ρ (kg/m^3)	μ (kg/ms)	k (J/s m K)	c_{p} (J/kg K)	ν (m^2/s)	α (m^2/s)	P_{r}
Water	958.4	282.5 $\times 10^{-6}$	0.679	4.2 $\times 10^{3}$	0.295 $\times 10^{-6}$	16.8 $\times 10^{-8}$	1.75
Vapor	0.598	12.28 $\times 10^{-6}$	2.5 $\times 10^{-2}$	2.135 $\times 10^{3}$	20.53 $\times 10^{-6}$	19.58 $\times 10^{-6}$	1.05

$$C_* = \frac{1}{3} 32 \frac{\alpha_{\rm L}}{d^2} \alpha_{\rm G.L}^{1/2} \Pr_{\rm G} ,$$

$$A_{**} = \frac{1}{27} \frac{1}{d^3} \rho_{\rm L.G}^3 \alpha_{\rm L.G}^{3/2} \alpha_{\rm L}^{3/2} ,$$

$$B_{**} = -\frac{1}{3} \rho_{\rm L.G}^2 \alpha_{\rm L.G} \alpha_{\rm L}^{3/2} \frac{1}{d^3 \tilde{x}_{\rm f}} ,$$

$$C_{**} = \frac{16}{d^3} \rho_{\rm L.G} \alpha_{\rm L.G}^2 \alpha_{\rm G}^{3/2} \Pr_{\rm L} \left(1 - \frac{1}{3} \nu_{\rm G.L} \alpha_{\rm L.G}^{1/2}\right) ,$$

$$\tilde{x}_{\rm f} = \frac{\overline{x}_{\rm f}}{d} .$$

The form of the solution of the dispersion equation (11.61) depends on the sign of the determinant $D = q_*^2 + P_*^3$, i.e., on the values of the characteristic parameters q_* and P_*. The latter are determined by the physical properties of the liquid and its vapor, as well as the values of the Peclet number. This allows us to use q_* and P_* as some general characteristics of the problem considered here.

The dependence of $P_*(\mathrm{Pe_L})$ and $q_*(\mathrm{Pe_L})$ is shown in Fig. 11.4. The parameter $P_*(\mathrm{Pe_L})$ is a parabola with an axis of symmetry left of the line $\mathrm{Pe_L} = 0$. Since the Peclet number is positive, for any value of the operating parameters, the physical meaning is that only for the right branch of this parabola, which intersects the axis of the abscissa at some critical value of Peclet number, $\mathrm{Pe_L} = \mathrm{Pe_{cr}}$. The vertical line $\mathrm{Pe_L} = \mathrm{Pe_{cr}}$ subdivides the parametrical plane $P_* - \mathrm{Pe_L}$ into two domains, corresponding to positive ($\mathrm{Pe_L} < \mathrm{Pe_{cr}}$) or negative ($\mathrm{Pe_L} > \mathrm{Pe_{cr}}$) values of the parameter P_*. The critical Peclet number is

$$\mathrm{Pe_{cr}} = 3\alpha_{\rm G.L}^{1/2} \left(-\frac{1}{\tilde{x}_{\rm f}} \pm \sqrt{\left(\frac{1}{\tilde{x}_{\rm f}}\right)^2 + \frac{32}{3} \alpha_{\rm G.L}^{1/2} \Pr_{\rm G}} \right) \tag{11.79}$$

Taking into account that $\mathrm{Pe_{cr}} > 0$, we should choose the positive value for the radical in Eq. (11.79). For very small and large $\tilde{x}_{\rm f}$ the following estimates for the

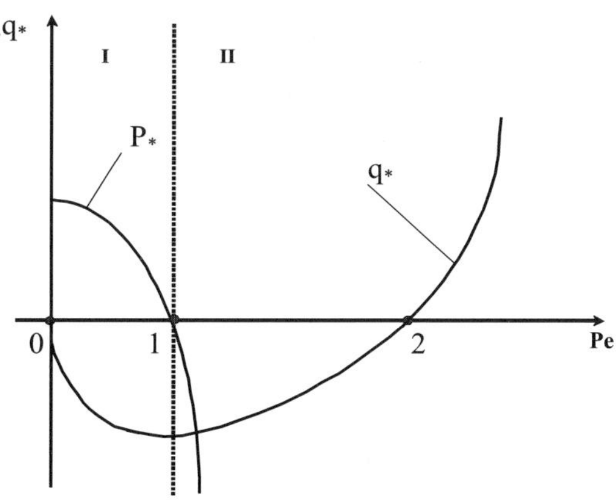

Fig. 11.4 The dependence of P_* ($\mathrm{Pe_L}$) and q_* ($\mathrm{Pe_L}$). The *dotted line* shows $\mathrm{Pe_L} = \mathrm{Pe_{cr}}$: *I* domain $\mathrm{Pe_L} < \mathrm{Pe_{cr}}$, *II* domain $\mathrm{Pe_L} > \mathrm{Pe_{cr}}$, *1* critical point, *2* $q_* = 0$. Reprinted from Hetsroni et al. (2004) with permission

critical Peclet number are valid: $\mathrm{Pe_{cr}} \simeq 16(\alpha_{\mathrm{G.L}}/\rho_{\mathrm{L.G}})\mathrm{Pr_G}\tilde{x}_{\mathrm{f}}$, $\tilde{x}_{\mathrm{f}} \leq 4\times10^{-3}$ and $\mathrm{Pe_{cr}} = (\sqrt{96}/\rho_{\mathrm{G.L}})\,\mathrm{Pr_G^{1/2}}\,\alpha_{\mathrm{G.L}}^{3/4}$, $\tilde{x}_{\mathrm{f}} \geq 4$ ($\tilde{x}_{\mathrm{f}}$ is the dimensionless liquid height in the stable state). In both cases the errors in the calculation $\mathrm{Pe_{cr}}$ do not exceed 5%. The dependence of the critical Peclet number on the dimensionless meniscus position $\tilde{x}_{\mathrm{f}}$ is plotted in Fig. 11.5. An increase of the wall heat flux, which is accompanied by a shift of the interface towards the capillary tube inlet, leads to decreasing $\mathrm{Pe_{cr}}$. At small enough q_{w} (large $\tilde{x}_{\mathrm{f}}$) $\mathrm{Pe_{cr}}$ approaches its asymptotic value $(\mathrm{Pe_{cr}})_{\mathrm{lim}} = \sqrt{96}\alpha_{\mathrm{G.L}}^{3/4}\mathrm{Pr_G^{1/2}}$.

The curve $q_*(\mathrm{Pe_L})$ is a cubic parabola, which passes through the point $O(0,0)$. Since, the Peclet number is positive, the physical meaning has the falling and rising branches of $q_*(\mathrm{Pe_L})$, which are located on the right part of the parameter plane $q_* - \mathrm{Pe_L}$.

Bearing in mind the characteristics of the dependences of $P_*(\mathrm{Pe_L})$ and $q_*(\mathrm{Pe_L})$ we estimate the growth rate of the oscillations in the vicinity of the two characteristic points: $\mathrm{Pe_L} = 0$ and $\mathrm{Pe_L} = \mathrm{Pe_{cr}}$.

1. $\mathrm{Pe_L} = 0$. In the vicinity of this point P_* is close to $C_* > 0$ and q_* is close to zero. Then $\mathrm{Sinh}\,\varphi = q_*/|P_x|^{3/2} \sim 0$, $\varphi \sim 0$, $\mathrm{Sinh}\,\varphi/3 \sim 0$, $\mathrm{Cosh}\,\varphi/3 \sim 1$. Since $b_*/3a_* = 0$ at $\mathrm{Pe_L} = 0$, we obtain

$$\xi = -3P_* = -32\frac{\alpha_{\mathrm{L}}}{d^2}\alpha_{\mathrm{G.L}}^{1/2}\mathrm{Pr_G}\,. \tag{11.80}$$

   Thus at small $\mathrm{Pe_L}$ the growth rate of the oscillations is negative and the capillary flow is stable. The absolute value of ξ sharply increases with a decrease of the capillary tube diameter. It also depends on the thermal diffusivity of the liquid and the vapor, as well as on the value of the Prandtl number.
2. $\mathrm{Pe_L} = \mathrm{Pe_{cr}}$. In the vicinity of this point the parameters P_* and q_* are: $P_* \sim 0$, $q_* \neq 0$. Bearing in mind that the sign of the parameter q_* is the same as that of the parameter r_* we find that ratio $q_*/r_*^3 \gg 1$ and $\varphi \gg 1$ in the vicinity of the point $\mathrm{Pe_L} = \mathrm{Pe_{cr}}$. In accordance with that, at large φ

$$\mathrm{Sinh}\frac{\varphi}{3} = \mathrm{Cosh}\frac{\varphi}{3} = \frac{1}{2^{2/3}}(\mathrm{Sinh}\,\varphi)^{1/3} = \frac{1}{2^{2/3}}\left(\frac{q_*}{r_*^3}\right)^{1/3}. \tag{11.81}$$

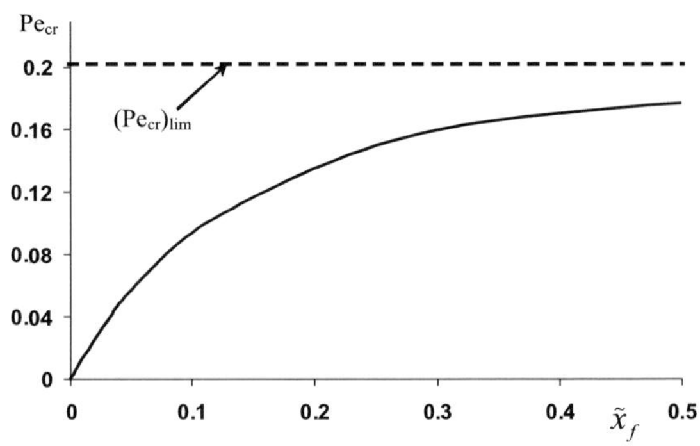

Fig. 11.5 The dependence of $\mathrm{Pe_{cr}}(\tilde{x}_{\mathrm{f}})$. Reprinted from Hetsroni et al. (2004) with permission

Substitution of (11.81) in Eq. (11.71) leads to the following expression for the growth rate of the oscillations:

$$\xi = \frac{q_*^{2/3}}{2^{4/3}}\left[\left(1 - \frac{b_*}{3a_*}\frac{2^{2/3}}{q_*^{1/3}}\right)^2 - 3\right]. \tag{11.82}$$

Since $q_*^{2/3} > 0$, the sign of the growth rate is determined by the difference of the terms in the bracket of Eq. (11.82): (1) $(1-N)^2 > 3$, $\xi > 0$, (2) $(1-N)^2 = 3$, $\xi = 0$, (3) $(1-N)^2 < 3$, $\xi < 0$, where $N = (b_*/3a_*)\left(2^{2/3}/q_*^{1/3}\right)$.

The behavior of the growth rate and the frequency of oscillations of flow parameters in the vicinity of the critical point is illustrated in Fig. 11.6, where the dependencies $\xi(2/q_*)^{2/3} = f(\tilde{x}_\mathrm{f})$ and $(1/2)\left(|\psi|/q_*^{2/3}\right) = \varphi(\tilde{x}_\mathrm{f})$ are plotted. It is seen that there are three ranges of changing meniscus position, which correspond to stable and unstable regimes of the flow. At small enough wall heat fluxes, when $\tilde{x}_\mathrm{f} > \tilde{x}_\mathrm{f}^{(1)}$, the growth rate is negative and the flow in the capillary is stable. An increase of the wall heat flux is accompanied by a displacement of the meniscus towards the inlet ($\tilde{x}_\mathrm{f} \sim 1/q_\mathrm{w}$), and a decrease of the absolute value of ξ. In the vicinity of the point $\tilde{x}_\mathrm{f}^{(1)}$, sharp growth of ξ is observed. The latter leads to a change of the sign of the growth rate and to the transition from stable to unstable regimes. At large heat fluxes when the meniscus reaches the inlet, the growth rate sharply decreases and becomes negative. The flow stabilization at $\tilde{x}_\mathrm{f} < \tilde{x}_\mathrm{f}^{(2)}$ is due to intense heat transfer to the cooling inlet, when the meniscus position and rate of evaporation weakly depend on q_w (Yarin et al. 2002).

It will be noted that applying the present approximation for the analysis of the stability of capillary flow at high heat fluxes corresponding to the domain $0 < \tilde{x}_\mathrm{f} < \tilde{x}_\mathrm{f}^{(L)}$ is purely symbolic, since the general assumption that $\mathrm{Pe_L} \ll 1$ is not valid at

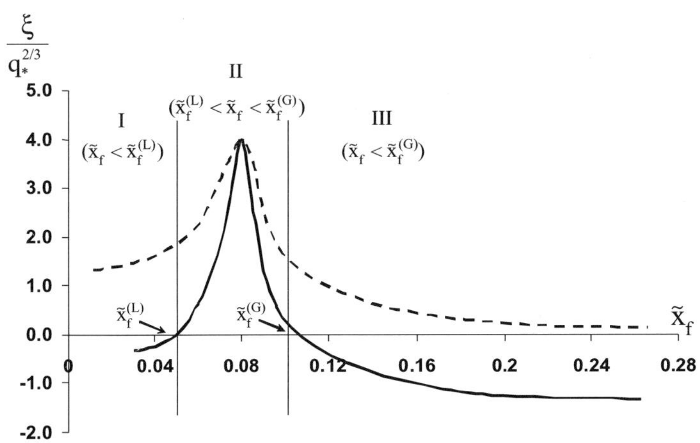

Fig. 11.6 The dependence of the increment (*solid line*) and frequencies (*dotted line*) of oscillations on $\tilde{x}_\mathrm{f}$ in the vicinity of the critical point. Reprinted from Hetsroni et al. (2004) with permission

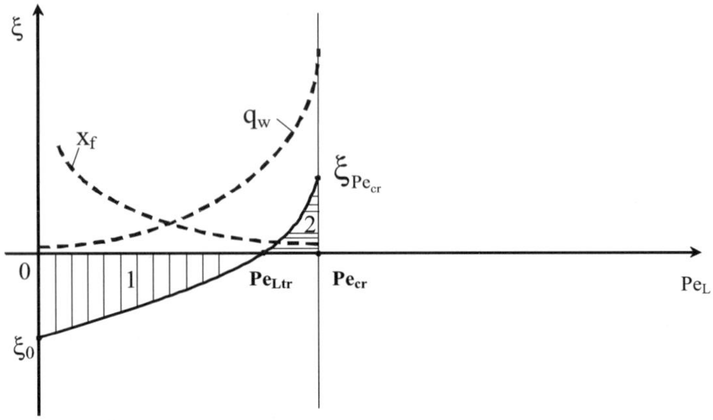

Fig. 11.7 The dependence $\xi(\mathrm{Pe_L})$: *1* domain of stationary steady regimes of flow, *2* domain of unsteady states. $\mathrm{Pe_L} = \mathrm{Pe_{Ltr}}$ point of transition from the stable to unstable flow regime. Reprinted from Hetsroni et al. (2004) with permission

large q_{w}. Thus, in the case considered here, only stable stationary $(\tilde{x}_{\mathrm{f}} > \tilde{x}_{\mathrm{f}}^{(\mathrm{G})})$ or unsteady $(\tilde{x}_{\mathrm{f}} < \tilde{x}_{\mathrm{f}}^{(\mathrm{G})})$ flow occur in capillary tube. The above is also related to the frequency of oscillations. At physically realistic $\tilde{x}_{\mathrm{f}}(\tilde{x}_{\mathrm{f}} > 1)$ only the low-frequency oscillations occur (as estimations show the order of these oscillations does not exceed 10 GHz). The dependence of the growth rate on the Peclet number (moderate q_{w}) is shown in Fig. 11.7. It is seen that at small $\mathrm{Pe_L}$ (small enough q_{w}) the flow is stable. An increase of the wall heat flux leads to an increase of the rate of evaporation, growth of the Peclet number, development of flow instability and transition (at $\mathrm{Pe_L} = \mathrm{Pe_{L,tr}}$) from stable to unstable flow.

11.5 Effect of Capillary Pressure and Heat Flux Oscillations

In this section the influence of the pressure in the capillary and the heat flux fluctuations on the stability of laminar flow in a heated capillary tube is analyzed. All the estimations performed in the framework of the general approach and developed in the previous section are kept also in the present cases. Below we will assume that the single cause for capillary pressure oscillations is fluctuations of the contact angle due to motion of the meniscus, whereas heat flux oscillations are the result of fluid temperature fluctuations only.

11.5.1 Capillary Pressure Oscillations

The present analysis is based on the assumption that the interfacial temperature T_{s} is constant and the capillary pressure is determined by the following expression

$$f_\sigma = \frac{2\sigma}{r}\cos\theta_d \tag{11.83}$$

where θ_d is the dynamic contact angle.

Assuming that the dynamic contact angle θ_d is a sum of its basic value corresponding to stationary flow θ_{st} and small perturbation θ' we arrive at the following relation for the fluctuation of capillary pressure

$$f'_\sigma = \frac{2\sigma}{r}(\cos\theta_{st}\cos\theta' - \sin\theta_{st}\sin\theta') \,. \tag{11.84}$$

For a system in which the contact angle is close to 90° (for example, the water–steel system: $70° < \theta_{st} < 90°$ (Grigoriev and Zorin 1982)) it is possible to assume that $\cos\theta_{st} \sim 0$, $\sin\theta_{st} \sim 1$ and $\sin\theta' \sim \theta'$. Then Eq. (11.84) takes the following form:

$$f'_\sigma = -\frac{2\sigma}{r}\theta' \,. \tag{11.85}$$

There is a number of theoretical and experimental relations determining the dependence of the dynamic contact angle on flow velocity (Dussan 1979; Ngan and Dussan 1982; Cox 1986; Blake 1994; Kistler 1993). Hoffman (1975) expressed the dynamic contact angle as a function solely of dimensionless parameters: capillary number Ca

$$\theta_d = f(\mathrm{Ca}) \tag{11.86}$$

where $\mathrm{Ca} = \mu u/\sigma$.

We estimate the effect of the velocity fluctuations on the capillary pressure, using the Hoffman–Voinov–Tanner law which is valid at $\theta_d \le 135°$ and $\mathrm{Ca} \le 0\,(0.1)$

$$\theta_d^3 = C_\mathrm{T} - \mathrm{Ca} \,. \tag{11.87}$$

where $C_\mathrm{T} \cong 93$, θ_d is in radians.

From Eq. (11.86), we obtain

$$\theta' = -\frac{1}{3}\frac{\mu_\mathrm{L} u'_\mathrm{L}}{\sigma}\left(C_\mathrm{T} - \frac{\mu_\mathrm{L} u_\mathrm{L}}{\sigma}\right)^{-2/3} \,. \tag{11.88}$$

Taking into account that $u_\mathrm{L} = \bar{u}_\mathrm{L} + u'_\mathrm{L}$ and $u'_\mathrm{L} \ll \bar{u}_\mathrm{L}$, we arrive at the following relation for capillary pressure oscillations:

$$f'_\sigma = \frac{2}{3}\frac{\mu_\mathrm{L} u'_\mathrm{L}}{r}\left(C_\mathrm{T} - \frac{\mu_\mathrm{L} \bar{u}_\mathrm{L}}{\sigma}\right)^{-2/3} \,. \tag{11.89}$$

From (11.42), (11.45) and (11.87) we transform Eq. (11.25) to the following form:

$$A_\mathrm{G}\tilde{\alpha}_{21} + A_\mathrm{L}\alpha^*_{22} = 0 \tag{11.90}$$

where $\tilde{\alpha}^*_{22} = -(k_\mathrm{L}\bar{\rho}_\mathrm{L}\bar{x}_\mathrm{f} + 2\bar{\rho}_\mathrm{L}\bar{u}_\mathrm{L} + \varepsilon)$, $\varepsilon = (2\mu_\mathrm{L})/(3r)(C_\mathrm{T} - (\mu_\mathrm{L}\bar{u}_\mathrm{L})/\sigma)^{-2/3}$, $\tilde{\alpha}_{21}$ is the same as $\tilde{\alpha}_{21}$ in Eq. (11.55).

Then the dispersion equation for the problem considered here takes the following form:

$$\tilde{\alpha}_{21}\tilde{\alpha}_{32} - \tilde{\alpha}_{22}^{*}\tilde{\alpha}_{31} = 0 \tag{11.91}$$

where coefficients $\tilde{\alpha}_{31}, \tilde{\alpha}_{32}$ are the same as in Eq. (11.55).

The equation can be presented as follows:

$$a_* \Omega^{3/2} + b_* \Omega + \tilde{c}_* \Omega^{1/2} + \tilde{d}_* = 0 \tag{11.92}$$

where the coefficients a_* and b_* are the same as in Eq. (11.59) and the coefficients $\tilde{c}_*$ and $\tilde{d}_*$ are

$$\tilde{c}_* = -\left\{-\frac{32}{d^2}\nu_{\mathrm{G}}\left(1 - \nu_{\mathrm{L.G}}\alpha_{\mathrm{G.L}}^{1/2}\right) + 2\rho_{\mathrm{L.G}}\frac{\overline{u}_{\mathrm{L}}}{\overline{x}_{\mathrm{f}}}\left(1 - \rho_{\mathrm{G.L}}\alpha_{\mathrm{L.G}}^{1/2} - \varepsilon\frac{\rho_{\mathrm{G.L}}\alpha_{\mathrm{L.G}}^{1/2}}{2\overline{\rho}_{\mathrm{L}}\overline{u}_{\mathrm{L}}}\right)\right\}\frac{1}{\sqrt{\alpha_{\mathrm{L}}}} \tag{11.93}$$

$$\tilde{d}_* = -\left\{-\frac{32}{d^2}\nu_{\mathrm{G}}\frac{\overline{u}_{\mathrm{L}}}{\alpha_{\mathrm{L}}}(1 - \nu_{\mathrm{L.G}}) - \rho_{\mathrm{L.G}}\frac{\overline{u}_{\mathrm{L}}^2}{\alpha_{\mathrm{L}}}\frac{\varepsilon}{\overline{\rho}_{\mathrm{L}}\overline{u}_{\mathrm{L}}}\frac{1}{\overline{x}_{\mathrm{f}}}\cdot\right\} \tag{11.94}$$

Approximate expressions for the parameters $\tilde{c}_*$ and $\tilde{d}_*$ corresponding to realistic values of operating parameters are

$$\tilde{c}_* \simeq -\left\{-\frac{32}{d^2}\nu_{\mathrm{G}} + 2\rho_{\mathrm{L.G}}\frac{\overline{u}_{\mathrm{L}}}{\overline{x}_{\mathrm{f}}}\left(1 - \varepsilon\frac{\rho_{\mathrm{G.L}}\alpha_{\mathrm{G.L}}^{1/2}}{2\overline{\rho}_{\mathrm{L}}\overline{u}_{\mathrm{L}}}\right)\right\}\frac{1}{\sqrt{\alpha_{\mathrm{L}}}} \tag{11.95}$$

$$\tilde{d}_* \simeq \frac{32}{d^2}\nu_{\mathrm{L}}\frac{\overline{u}_{\mathrm{L}}}{\alpha_{\mathrm{L}}}\rho_{\mathrm{L.G}} - \rho_{\mathrm{L.G}}\frac{\overline{u}_{\mathrm{L}}^2}{\alpha_{\mathrm{L}}}\frac{\varepsilon}{\overline{\rho}_{\mathrm{L}}\overline{u}_{\mathrm{L}}}\frac{1}{\overline{x}_{\mathrm{f}}}\,. \tag{11.96}$$

Using Eqs. (11.62) and (11.63) as well as expressions (11.73), (11.74), (11.95) and (11.96), it is possible to transform the dependencies $P_*(\mathrm{Pe_L})$ and $q_*(\mathrm{Pe_L})$ to the canonical form similar to Eqs. (11.77) and (11.78) with coefficients A_*, B_*, C_* and A_{**}, B_{**}, C_{**}.

$$A_* = -\frac{1}{9}\rho_{\mathrm{L.G}}^2\alpha_{\mathrm{L.G}}\,,\quad B_* = -\frac{2}{3}\rho_{\mathrm{L.G}}\alpha_{\mathrm{L.G}}^2\frac{1}{\overline{x}_{\mathrm{f}}}\left(1 - \varepsilon\frac{\rho_{\mathrm{G.L}}\alpha_{\mathrm{G.L}}^{1/2}}{\overline{\rho}_{\mathrm{L}}\overline{u}_{\mathrm{L}}}\right),$$

$$C_* = \frac{32}{3}\frac{\alpha_{\mathrm{L}}}{d^2}\alpha_{\mathrm{G.L}}^{1/2}\rho_{\mathrm{L.G}}$$

$$A_{**} = \frac{1}{27}\frac{1}{d^3}\rho_{\mathrm{L.G}}^3\alpha_{\mathrm{L.G}}^{3/2}\alpha_{\mathrm{L}}^{3/2}\,,\quad B_{**} = 3\rho_{\mathrm{G.L}}\frac{1}{d^3\overline{x}_{\mathrm{f}}}\alpha_{\mathrm{L.G}}\alpha_{\mathrm{L}}^{3/2}\left(\rho_{\mathrm{L.G}} - \frac{5}{2}\alpha_{\mathrm{G.L}}^{1/2}\frac{\varepsilon}{\overline{\rho}_{\mathrm{L}}\overline{u}_{\mathrm{L}}}\right)$$

$$C_{**} = \frac{32}{d^3}\rho_{\mathrm{L.G}}\mathrm{Pr_L}\alpha_{\mathrm{L.G}}^{1/2}\alpha_{\mathrm{L}}^{3/2}\left(1 - \frac{1}{3}\mathrm{Pr_{G.L}}\alpha_{\mathrm{L.G}}^{1/2}\right)\,.$$

In the domain of a very small Peclet number the growth rate of flow oscillations is negative at any values of flow parameters. In the vicinity of the critical point ($\mathrm{Pe_L} = \mathrm{Pe_{cr}}$, $P_* \simeq 0$) the sign ξ is determined by Eq. (11.82). An increase in ε (other parameters are fixed) leads to an increase of the critical value of the Peclet number and expansion of domain of stable flows.

11.5.2 Heat Flux Oscillations

There are two causes for oscillations of the heat flux, with $T_w = \text{const.}$: (1) fluctuations of the heat transfer coefficient due to velocity fluctuations, and (2) fluctuations of the fluid temperature. At small enough Reynolds numbers the heat transfer coefficient is constant (Bejan 1993), whereas at moderate Re (Re $\sim 10^2$) it is a weak function of velocity (Peng and Peterson 1995; Incropera 1999; Sobhan and Garimella 2001). Bearing this in mind, it is possible to neglect the influence of velocity fluctuations on the heat transfer coefficient and assume that heat flux fluctuations are expressed as follows:

$$q'_{w_i} = -h_i T'_i \tag{11.97}$$

where h_i is heat transfer coefficient for stationary flow of the ith phase.

Using Eq. (11.40), as well as Eqs. (11.52) and (11.97) we obtain

$$n_i = \sqrt{\frac{\Omega + \Omega_{0_i}}{\alpha_i}} \tag{11.98}$$

where $\Omega_{0_i} = \frac{4h_i}{\rho_i c_{p_i} d}$.

Using expressions (11.46) and (11.98) we transform Eq. (11.59). Bearing in mind that $h_i = k_i \frac{\text{Nu}}{d}$ and $\frac{\Omega_{0_i}}{\alpha_i} = 4\frac{\text{Nu}}{d^2}$ we arrive at the equation

$$-\left(\frac{32}{d^2}\nu_G + \Omega\right)\sqrt{\frac{\Omega}{\alpha_L} + 4\frac{\text{Nu}}{d^2}} \cdot N_1 + \sqrt{\frac{\Omega}{\alpha_L} + 4\frac{\text{Nu}}{d^2}} \cdot N_2 - \left(\frac{32}{d^2}\nu_G + \Omega\right) \cdot N_3 + N_4 =$$
$$+\left(\frac{32}{d^2}\nu_L + \Omega\right)\sqrt{\frac{\Omega}{\alpha_L} + 4\frac{\text{Nu}}{d^2}} \cdot M_1 + \sqrt{\frac{\Omega}{\alpha_G} + 4\frac{\text{Nu}}{d^2}} \cdot M_2 - \left(\frac{32}{d^2}\nu_L + \Omega\right) \cdot M_3 + M_4 \tag{11.99}$$

where Nu is the Nusselt number

$$N_1 = \overline{\rho}_G \overline{x}_f \alpha_L\,, \quad N_2 = 2\overline{\rho}_L \overline{u}_L \alpha_L\,, \quad N_3 = -\overline{\rho}_G \overline{x}_f \overline{u}_L\,, \quad N_4 = -2\overline{\rho}_L \overline{u}_L^2\,,$$
$$M_1 = \overline{\rho}_L \overline{x}_f \alpha_G \rho_{G.L}\,, \quad M_2 = 2\overline{\rho}_L \overline{u}_L \alpha_G \rho_{G.L}\,, \quad M_3 = -\overline{\rho}_L \overline{x}_f \overline{u}_L\,, \quad M_4 = -2\overline{\rho}_L \overline{u}_L^2\,.$$

Equation (11.99) shows that the effect of heat flux oscillations is not significant in micro-channels with large diameter when the term $4\text{Nu}/d^2$ is small enough.

Presenting the complex frequency as

$$\Omega = \xi + i|\psi| \tag{11.100}$$

we arrive at two equations that determine the increment and frequency of oscillations

$$\begin{aligned}&-f_G F_L N_1 + |\psi|\phi_L N_1 + F_L N_2 - f_G N_3 + N_4 \\ &-f_L F_G M_1 + |\psi|\phi_G M_1 - F_G M_2 + f_L M_3 - M_4 = 0\end{aligned} \tag{11.101}$$

$$\begin{aligned}&-|\psi| F_L N_1 - f_G \phi_L N_1 + \phi_L N_2 - |\psi| N_3 + N_4 \\ &-|\psi| F_G M_1 - f_L \phi_G M_1 - \phi_G M_2 - |\psi| M_3 - M_4 = 0\end{aligned} \tag{11.102}$$

where

$$F_i = \frac{1}{\sqrt{2}}\sqrt{\sqrt{a_i^2+b_i^2}+a_i}\,,$$

$$\phi_i = \frac{1}{\sqrt{2}}\sqrt{\sqrt{a_i^2+b_i^2}-a_i}\,,$$

$$f_i = \frac{32}{d^2}\nu_i + \xi\,,$$

$$a_i = \frac{\xi}{\alpha_i} + 4\frac{\mathrm{Nu}}{d^2}\,,$$

$$b_i = \frac{|\psi|}{\alpha_i}\,.$$

Consider the particular case as corresponding to low frequency. Assuming $b \sim 0$, $\phi \sim 0$ and $F_i \sim a_i^{1/2}$, we arrive at the following equation for increment of oscillations at $\psi \to 0$

$$-f_G a_L^{1/2} N_1 + a_L^{1/2} N_2 - f_G N_3 + N_4 - f_L a_L^{1/2} M_1 - a_L^{1/2} M_2 + f_L M_3 - M_4 = 0\,. \tag{11.103}$$

Transforming Eq. (11.103) we obtain

$$\begin{aligned}\xi = &-\frac{\nu_L}{d^2 \mathrm{Nu}^{1/2}(1+\alpha_{G.L})}\left\{32\mathrm{Nu}^{1/2}(\nu_{G.L}+\alpha_{G.L})\right. \\ &\left.-\mathrm{Pe}_L\left[\frac{\mathrm{Nu}^{1/2}\alpha_{L.G}(\rho_{L.G}-2)}{\tilde{x}_f \mathrm{Pr}_L} + 32\,\underset{L}{\mathrm{Pr}}(\nu_{G.L}-1)\right]\right\}.\end{aligned} \tag{11.104}$$

At small Pe_L ($\mathrm{Pe}_L \to 0$), the growth rate is negative and the flow is stable whereas at relatively large Pe_L the flow is unstable: $\xi > 0$. Assuming in Eq. (11.104) $\xi = 0$ and taking into account that $\rho_{L.G} \gg 1$, $\nu_{L.G} \gg 1$ we find the value of Peclet number corresponding to the transition from stable to unstable flow

$$\mathrm{Pe}_{L,tr} \cong \frac{\mathrm{Nu}^{1/2}\mathrm{Pr}_L(\nu_{G.L}+\alpha_{G.L})}{\dfrac{\mathrm{Nu}^{1/2}}{\tilde{x}_f}\alpha_{L.G}\rho_{L.G} + \nu_{G.L}\mathrm{Pr}_2}\,. \tag{11.105}$$

It is seen that the Peclet number corresponding to transition from stable to unstable flow decreases with increasing wall heat flux (decreasing $\tilde{x}_{\mathrm{f}}$). The increase of the Nusselt number leads to increasing $\mathrm{Pe}_{\mathrm{L,tr}}$.

11.6 Moderate Peclet Number

The perturbed energy equation for moderate Peclet number has (at $q' = 0$) the following form:

$$\frac{\partial T_i'}{\partial t} + \overline{u}_i \frac{\partial T_i'}{\partial x} = \frac{\partial}{\partial x}\left(\alpha_i \frac{\partial T_i'}{\partial x}\right). \tag{11.106}$$

Assuming, as earlier, that u', P' and T' are determined by Eqs. (11.42), (11.45), (11.51) we find k_i and n_i

$$k_i = -\left(\frac{32}{d^2}\nu_i + \Omega\right) \tag{11.107}$$

$$n_i = \frac{1}{2}\left(\frac{\overline{u}_i}{\alpha_i} \pm \sqrt{\left(\frac{\overline{u}_i}{\alpha_i}\right)^2 + 4\frac{\Omega}{\alpha_i}}\right). \tag{11.108}$$

Substituting expressions (11.107), and (11.108) in Eq. (11.91) leads to the dispersion equation

$$\begin{aligned}
&-\left(\frac{32}{d^2}\nu_{\mathrm{G}} + \Omega\right)\frac{1}{2}\left(\frac{\overline{u}_{\mathrm{L}}}{\alpha_{\mathrm{L}}} + \sqrt{\left(\frac{\overline{u}_{\mathrm{L}}}{\alpha_{\mathrm{L}}}\right)^2 + 4\frac{\Omega}{\alpha_{\mathrm{L}}}}\right)N_1 \\
&+\frac{1}{2}\left(\frac{\overline{u}_{\mathrm{L}}}{\alpha_{\mathrm{L}}} + \sqrt{\left(\frac{\overline{u}_{\mathrm{L}}}{\alpha_{\mathrm{L}}}\right)^2 + 4\frac{\Omega}{\alpha_{\mathrm{L}}}}\right)N_2 - \left(\frac{32}{d^2}\nu_{\mathrm{G}} + \Omega\right)N_3 + N_4 \\
&= -\left(\frac{32}{d^2}\nu_{\mathrm{L}} + \Omega\right)\frac{1}{2}\left(\frac{\overline{u}_{\mathrm{G}}}{\alpha_{\mathrm{G}}} - \sqrt{\left(\frac{\overline{u}_{\mathrm{G}}}{\alpha_{\mathrm{G}}}\right)^2 + 4\frac{\Omega}{\alpha_{\mathrm{G}}}}\right)M_1 \\
&+\frac{1}{2}\left(\frac{\overline{u}_{\mathrm{G}}}{\alpha_{\mathrm{G}}} - \sqrt{\left(\frac{\overline{u}_{\mathrm{G}}}{\alpha_{\mathrm{G}}}\right)^2 + 4\frac{\Omega}{\alpha_{\mathrm{G}}}}\right)M_2 - \left(\frac{32}{d^2}\nu_{\mathrm{L}} + \Omega\right)M_3 + M_4.
\end{aligned} \tag{11.109}$$

Transforming this equation we obtain

$$A^{\circ} + B^{\bullet}\Omega^{\circ} + C^{\bullet}F_1(\Omega^{\circ}) + D^{\bullet}F_2(\Omega^{\circ}) + E^{\bullet}\Omega^{\bullet}[F_{\mathrm{L}}(\Omega^{\circ}) + \rho_{2.1}F_1(\Omega^{\circ})] = 0 \tag{11.110}$$

where

$$A^{\circ} = (1 - \mu_{\mathrm{L.G}})\mathrm{Pe}_{\mathrm{L}} + 2\frac{N_3}{N_1}\alpha(1 - \rho_{\mathrm{L.G}}\nu_{\mathrm{L.G}})$$

$$B^{\bullet} = \mathrm{Pe}_{\mathrm{L}}(1 - \rho_{\mathrm{L.G}})$$

$$C^{\bullet}=-\nu_{\mathrm{G.L}}\rho_{\mathrm{L.G}}\mathrm{Pe_L}\left(1+\frac{d^2}{32\nu_{\mathrm{G}}}\frac{N_2}{N_1}\right)$$

$$D^{\bullet}=-\mathrm{Pe_L}\left(1+\frac{d^2}{32\nu_{\mathrm{G}}}\frac{N_2}{N_1}\right)$$

$$E^{\bullet}=\mathrm{Pe_L}$$

$$F_{\mathrm{G}}(\Omega^{\circ})=\sqrt{1+\beta_1\Omega^{\circ}}$$

$$F_{\mathrm{L}}(\Omega^{\circ})=\sqrt{1+\beta_2\Omega^{\circ}}$$

$$\beta_1=128\frac{\mathrm{Pr_G}}{\mathrm{Pe_L^2}}\alpha_{\mathrm{G.L}}^2\rho_{\mathrm{L.G}};\ \beta_2=128\alpha_{\mathrm{G.L}}\frac{\mathrm{Pr_G}}{\mathrm{Pe_L^2}}\,.$$

Assuming $\Omega^{\circ}=\xi^{\circ}+i|\psi^{\circ}|$, where $\xi^{\circ}=\xi\left(d^2/32\nu_{\mathrm{G}}\right)$, $\psi^{\circ}=\psi\left(d^2/32\nu_{\mathrm{G}}\right)$ we obtain from Eq. (11.110) two equations for dimensionless frequency and increment of oscillations:

$$\begin{aligned}&A^{\circ}+B^{\bullet}\xi^{\circ}+\frac{C^{\bullet}}{\sqrt{2}}\sqrt{\sqrt{a_1^{\circ 2}+b_1^{\circ 2}}+a_1^{\circ}}+\frac{D^{\bullet}}{\sqrt{2}}\sqrt{\sqrt{a_2^{\circ 2}+b_2^{\circ 2}}-a_1^{\circ}}\\&+\frac{E^{\bullet}\xi^{\circ}}{\sqrt{2}}\left\{\sqrt{\sqrt{a_2^{\circ 2}+b_2^{\circ 2}}+a_2^{\circ}}+\rho_{\mathrm{L.G}}\sqrt{\sqrt{a_1^{\circ 2}+b_1^{\circ 2}}-a_1^{\circ}}\right\}\\&-\frac{E^{\bullet}|\psi^{\circ}|}{\sqrt{2}}\left\{\sqrt{\sqrt{a_2^{\circ 2}+b_2^{\circ 2}}-a_2^{\circ}}-\rho_{\mathrm{L.G}}\sqrt{\sqrt{a_1^{\circ 2}+b_1^{\circ 2}}-a_1^{\circ}}\right\}=0\end{aligned}\qquad(11.111)$$

and

$$\begin{aligned}&B^{\bullet}|\psi^{\circ}|+\frac{C^{\bullet}}{\sqrt{2}}\sqrt{\sqrt{a_1^{\circ 2}+b_1^{\circ 2}}-a_1^{\circ}}+\frac{D^{\bullet}}{\sqrt{2}}\sqrt{\sqrt{a_2^{\circ 2}+b_2^{\circ 2}}-a_1^{\circ}}\\&+\frac{E^{\bullet}|\psi^{\circ}|}{\sqrt{2}}\left\{\sqrt{\sqrt{a_2^{\circ 2}+b_2^{\circ 2}}-a_2^{\circ}}+\rho_{\mathrm{L.G}}\sqrt{\sqrt{a_1^{\circ 2}+b_1^{\circ 2}}-a_1^{\circ}}\right\}\\&+\frac{E^{\bullet}|\psi^{\circ}|}{\sqrt{2}}\left\{\sqrt{\sqrt{a_2^{\circ 2}+b_2^{\circ 2}}+a_2^{\circ}}+\rho_{\mathrm{L.G}}\sqrt{\sqrt{a_1^{\circ 2}+b_1^{\circ 2}}+a_1^{\circ}}\right\}=0\end{aligned}\qquad(11.112)$$

where $a_1^{\circ}=1+\beta_1\xi^{\circ}$, $a_2^{\circ}=1+\beta_2\xi^{\circ}$, $b_1^{\circ}=\beta_1|\psi^{\circ}|$, $b_2^{\circ}=\beta_2|\psi^{\circ}|$.

Using Eq. (11.112) we estimate the increment of oscillations for low frequencies ($|\psi^{\circ}|\to 0$). Assuming in Eq. (11.112) $b_1\to 0$ we arrive at the equation

$$A^{\circ}+B^{\bullet}\xi^{\circ}+C^{\bullet}a_1^{\circ 1/2}+B^{\bullet}\xi^{\circ}(a_1^{\circ 1/2}+\rho_{\mathrm{L.G}}a_1^{\circ 1/2})=0\,.\qquad(11.113)$$

To find the solution of Eq. (11.113) we use an approximate expression for the coefficients A°, $B^{\bullet}$, $C^{\bullet}$ and $E^{\bullet}$. Characteristic values of the operating parameters are:

$$A^{\circ} \cong \mu_{\mathrm{L.G}} \mathrm{Pe_L} > 0\,, \quad B^{\bullet} \cong -\rho_{\mathrm{L.G}} \mathrm{Pe_L} < 0\,,$$
$$C^{\bullet} \cong -\nu_{\mathrm{G.L}} \rho_{\mathrm{L.G}} \mathrm{Pe_L} (1 + \rho_{\mathrm{L.G}} \alpha_{\mathrm{L.G}} \frac{\mathrm{Pe_L}}{\mathrm{Pr_G}} \frac{1}{\tilde{x}_{\mathrm{f}}}) < 0\,, \quad E^{\bullet} = \mathrm{Pe_L} > 0\,.$$

Consider three particular cases corresponding to very small and large values of ξ°: (1) $\xi^{\circ} \le 10^{-6}$, (2) $\xi^{\circ} \le 10^{-5}$, (3) $\xi^{\circ} \ge 10^{2}$. In the first case $a_1^{\circ 1/2} \sim 1$, $a_2^{\circ 1/2} \sim 1$ and solution to Eq. (11.113) is

$$\xi^{\circ} = -\frac{A^{\circ} + C^{\bullet}}{B^{\bullet} + E^{\bullet}(1 + \rho_{\mathrm{L.G}})}\,. \tag{11.114}$$

Since $\nu_{\mathrm{G.L}} \rho_{\mathrm{L.G}} > \mu_{\mathrm{L.G}}$ and $(1 + \rho_{\mathrm{L.G}} \alpha_{\mathrm{L.G}} (\mathrm{Pe_L}/\mathrm{Pr_G})(1/\tilde{x}_{\mathrm{f}})) > 1$ the sum $A^{\circ} + C^{\bullet} < 0$. The sum $B^{\bullet} + E^{\bullet}(1 + \rho_{\mathrm{L.G}}) = E^{\bullet} > 0$. Accordingly, the ratio $(A^{\circ} + C^{\bullet})/(B^{\bullet} + E^{\bullet}(1 + \rho_{\mathrm{L.G}}))$ is negative and the growth rate is positive $\xi^{\circ} > 0$. Thus, in this case the flow in heated micro-channel is unstable at any values of the Peclet number.

In the second case, the growth rate is expressed as

$$\xi^{\circ} = \frac{1}{2}\left\{\beta_2^{-1} + \sqrt{\beta_2^{-2} - 8\beta_2^{-1}(A^{\circ} + C^{\bullet})}\right\}\,. \tag{11.115}$$

Since $\beta_2 > 0$ and $A^{\circ} + C^{\bullet} < 0$, the growth rate is positive and the flow is also unstable.

In the third case Eq. (11.113) is transformed to a form similar to Eq. (11.92)

$$a\xi^{\circ 3/2} + b\xi^{\circ} + c\xi^{\circ 1/2} + d = 0 \tag{11.116}$$

with the coefficients a, b, c and d expressed as $a = E^{\circ}(\beta_2^{1/2} + \rho_{\mathrm{L.G}}\beta_1^{1/2})$, $b = B^{\bullet}$, $c = \beta_1^{1/2} C^{\bullet}$, $d = A^{\circ}$. Estimations show that the determinant

$$D = q_*^2 + P_*^3 \tag{11.117}$$

which is defined by correlations (11.62) and (11.63) and coefficients a, b, c and d are negative. That means that Eq. (11.116) has three real roots, which are:

$$\begin{aligned}
\xi_{\mathrm{I}}^{\circ 1/2} &= -2r_* \cos\frac{4}{3} - \frac{b}{3a} \\
\xi_{\mathrm{II}}^{\circ 1/2} &= 2r_* \cos\left(60 - \frac{4}{3}\right) - \frac{b}{3a} \\
\xi_{\mathrm{III}}^{\circ 1/2} &= -2r_* \cos\left(60 + \frac{4}{3}\right) - \frac{b}{3a}
\end{aligned} \tag{11.118}$$

where $\cos\varphi = q/r_*^3$; $r_* = \sqrt{|P_*|}$, sign r_* is the same as sign q_*.

At realistic flow conditions $\cos\varphi \sim 0$ and φ is close to $\pi/2$. Under these conditions, in any case, one of the roots of (11.118) is positive. This shows that capillary flow in a capillary tube is unstable at large ξ°.

Summary

The system of quasi-one-dimensional non-stationary equations derived by transformation of the Navier–Stokes equations can be successfully used for studying the dynamics of two-phase flow in a heated capillary with distinct interface.

The following results have been obtained:

1. The quasi-one-dimensional model allows analyzing the behavior of the vapor–liquid system, which undergoes small perturbations. In the frame of the linear approximation the effect of physical properties of both phases, the wall heat flux and the capillary sizes, on the flow instability is studied, and a scenario of the development of a possible processes at small and moderate Peclet number is considered.
2. The boundaries of the stability, subdividing the domains of stable and unstable flows, are outlined, and the values of geometrical and operating parameters corresponding to the transition from stable to unstable flow are estimated.
3. The performed calculations show that flow instability in a heated capillary tube, develops under conditions of high wall heat fluxes, which are the main factor in determining the flow regimes. The evolution of capillary flow is due to changes of heat flux on the wall that may be presented as follows. At relatively small q_{W}, when the rate of liquid evaporation is small and the height of the rising liquid is close to the adiabatic one, a stable laminar flow takes place. In this case the equilibrium of the two-phase system is determined by the equality of gravity and capillary forces, whereas the influence of the friction forces and heat losses to cooling inlet is negligible. On the contrary, at high wall heat fluxes the friction and capillary forces, as well as losses to the inlet play the dominant role. Under these conditions, a small deviation from equilibrium leads to progressive (exponential) growth of disturbances, i.e., development of flow instability. The latter is displayed in oscillations of the velocity and temperature of both phases, as well as oscillations of the position of the meniscus.
4. It is shown that the stability of the flow, with evaporating meniscus, depends (other conditions being equal) on the wall heat flux. The latter determines the rate of liquid evaporation, equilibrium acting forces, meniscus position, as well as the heat losses to the cooling inlet. The stable stationary flow with fixed meniscus position corresponds to low wall heat fluxes ($\mathrm{Pe} \ll 1$). In contrast, at high wall heat fluxes ($\mathrm{Pe} \gg 1$) an exponential increase of small disturbances takes place. That leads to the transition from stable stationary to unstable flow with oscillating meniscus.

References

Adams TM, Abdel-Khalik SI, Jeter SM, Qureshi ZH (1998) An experimental investigation of single-phase forced convection in micro-channels. Int J Heat Mass Transfer 41:851–857

Bailey DK, Ameel TA, Warrington RO, Savoie TI (1995) Single-phase forced convection heat transfer in microgeometries: a review ASME. IECEC paper ES-396:301–310
Bejan A (1993) Heat transfer. Wiley, New York
Blake TD (1993) Dynamic contact angles and wetting kinetics. In: Berg JC (ed) Wettability. Dekker, New York, pp 251–309
Bowers MB, Mudawar I (1994) High flux boiling in low flow rate, low pressure drop mini-channel and micro-channel heat sinks. Int J Heat Mass Transfer 37:321–332
Cox RG (1986) The dynamics of the spreading of liquids on a solid surface. Part 1: Viscous flows. J Fluid Mech 168:169–194
Dussan EBV (1979) On the spreading of liquids on solid surfaces: static and dynamic contact lines. Ann Rev Fluid Mech 11:371–400
Grigoriev VA, Zorin VM (eds) (1982) Heat mass transfer. Thermal experiment reference book. Energoizdat, Moscow (in Russian)
Hetsroni G, Yarin LP, Pogrebnyak E (2004) Onset of flow instability in a heated capillary tube. Int J Multiphase Flow 30:1421–1449
Hoffman R (1975) A study of the advancing interface. I. Interface shape in liquid gas system. J Colloid Interface Sci 50:228–241
Incropera FP (1999) Liquid cooling of electronic devices by single-phase convection. Wiley, New York
Khrustalev D, Faghri D (1996) Fluid flow effect in evaporation from liquid–vapor meniscus. J Heat Transfer 118:725–730
Kistler SF (1993) Hydrodynamics of wetting. In: Berg JC (ed) Wettability. Dekker, New York, pp 311–429
Korn GA, Korn TM (1968) Mathematical handbook. McGraw-Hill, Boston
Landau LD, Lifshitz EM (1959) Fluid mechanics, 2nd edn. Pergamon, London
Morijama K, Inoue A (1992) The thermodynamic characteristics of two-phase flow in extremely narrow channels (the frictional pressure drop and heat transfer of boiling two-phase flow, analytical model). Heat Transfer Jpn Res 21:838–856
Ngan CD, Dussan EBV (1982) On the nature of the dynamic contact angle: an experimental study. J Fluid Mech 118:27–40
Ory E, Yuan H, Prosperetti A (2000) Growth and collapse of vapor bubble in narrow tube. Phys Fluid 12:1268–1277
Peles YP, Yarin LP, Hetsroni G (1998) Heat transfer of two-phase flow in heated capillary. In: Heat Transfer 1998, Proceedings of the 11th International Heat Transfer Conference, Kyongju, Korea, 23–28 August 1998, vol 2, pp 193–198
Peles YP, Yarin LP, Hetsroni G (2000) Thermodynamic characteristics of two-phase flow in a heated capillary. Int J Multiphase Flow 26:1063–1093
Peles YP, Yarin LP, Hetsroni G (2001) Steady and unsteady flow in a heated micro-channels. Int J Multiphase Flow 28:1589–1616
Peng XF, Hu HY, Wang BX (1998) Boiling nucleation during liquid flow in micro-channels. Int J Heat Mass Transfer 41:191–196.3
Peng XF, Peterson GP (1996) Convective heat transfer and flow friction for water flow in micro-channel structure. Int J Heat Mass Transfer 39:2599–2608
Peng XF, Peterson GP (1995) The effect of thermofluid and geometrical parameters on convection of liquid through rectangular micro-channels. Int J Heat Mass Transfer 38:755–758
Peng XF, Peterson GP, Wang BX (1994) Heat transfer characteristics of water flowing through micro-channels. Exp Heat Transfer 7:249–264
Peng XF, Tien Y, Lee DJ (2001) Bubble nucleation in micro-channels: statistical mechanics approach. Int J Multiphase Flow 44:2953–2964
Peng XF, Wang BX (1993) Forced convection and flow boiling heat transfer for liquid flowing through micro-channels. Int J Heat Mass Transfer 14:3421–3427
Reid RC, Prausnitz JM, Poling BE (1987) The properties of gases and liquids. McGraw-Hill, Boston

Sobhan CB, Garimella SV (2001) A comparative analysis of studies on heat transfer and fluid flow in micro-channels. Microscale Thermophys Eng 5:293–311
Tuckerman D (1984) Heat transfer microstructure for integrated circuits. Dissertation, Stanford University, Stanford
Tuckerman D, Pease RFW (1981) High-performance heat sinking for VLSI. IEEE Electron Device Lett EDL-2:126–129
Vargaftic NB, Vinogradov YK, Yargin VS (1996) Handbook of physical properties of liquids and gases, pure substance and mixtures, 3rd augmented revised edn. Begel House, New York
Wang BX, Peng XF (1994) Experimental investigation of liquid forced convection heat transfer through micro-channels. Int J Heat Mass Transfer 37:73–82
Wiesberg A, Bau HH, Zemel JN (1992) Analysis of micro-channels for integrated cooling. Int J Heat Mass Transfer 35:2465–2472
Wu PY, Little WA (1984) Measurement of the heat transfer characteristics of gas flow a fine channels heat exchangers used for microminiature refrigerators. Cryogenics 24:415–420
Yarin LP, Ekelchik LA, Hetsroni G (2002) Two-phase laminar flow in a heated micro-channels. Int J Multiphase Flow 28:1589–1616
Yuan H, Qguz HN, Prosperreti A (1999) Growth and collapse of a vapor bubble in a small tube. Int J Heat Mass Transfer 42:3643–3657

Nomenclature

A	Amplitude of velocity oscillations
Ca	Capillary number
c_p	Specific heat
d	Diameter of the pipe
f_σ	Capillary pressure
F	Specific friction force
h	Enthalpy
g	Acceleration due to gravity
k	Thermal conductivity
ℓ	Length of pipe
P	Pressure
q	Specific rate of volumetric heat absorption
q_w	Heat flux on the wall
R	Radius of interface curvature
r	Radius of the pipe
T	Temperature
u	Longitudinal component of the velocity
v_f	Velocity of interface
$\tilde{V}$	Velocity relative to the interface
x_f	Height of the liquid level in a heated capillary
$x_{f,ad}$	Height of the liquid level in adiabatic capillary
$\mathrm{Nu} = \frac{\alpha d}{k}$	Nusselt number
$\mathrm{Pe} = \frac{ud}{\alpha}$	Peclet number

$\mathrm{Pr} = \dfrac{\nu}{\alpha}$	Prandtl number
$\mathrm{Re} = \dfrac{ud}{\nu}$	Reynolds number
$\mathrm{St} = \dfrac{\ell\psi}{u}$	Strouhal number

Greek symbols

α	Thermal diffusivity
θ	Static contact angle
θ_{d}	Dynamic contact angle
μ	Viscosity
ν	Kinematic viscosity
ξ	Growth rate of velocity oscillations
ρ	Density
σ	Surface tension
ψ	Frequency of velocity oscillations
Ω	Complex frequency

Superscripts

$()'$	Corresponds to perturbed parameter

Subscripts

ad	Adiabatic
G	Vapor
in	Inlet
f	Interface
L	Liquid
cr	Critical
tr	Transition
w	Wall
0	Outlet

Author Index

C

D

T

Subject Index

D

E

Printing: Krips bv, Meppel, The Netherlands
Binding: Stürtz, Würzburg, Germany